MATLAB-based Computations of Chemical Engineering Principles

MATLAB-based Computations of Chemical Engineering Principles

Kamal I. M. Al-Malah
Irbid, Jordan

WILEY

Published by John Wiley & Sons, Inc., Hoboken, New Jersey.
Published simultaneously in Canada.

Library of Congress Cataloging-in-Publication Data Applied for:

Hardback: 9781394308828

Cover Design: www.lamamalah.com and Wiley
Cover Image: © is1003/Shutterstock

Set in 9.5/12.5pt STIXTwoText by Straive, Pondicherry, India

Contents

Acknowledgments

All praise and gratitude are due to **Allah (God)**, the Almighty, for granting me the strength, knowledge, and perseverance to complete this work. His guidance has been my light throughout this journey.

I extend my deepest appreciation to **my parents**, whose unwavering love, sacrifices, and support have shaped the person I am today. Their encouragement and prayers have been the foundation of my success.

I am profoundly grateful to all my **instructors**, from my first elementary school teacher to my esteemed PhD professors, whose wisdom and dedication have enriched my academic and professional path. Each one has played a crucial role in cultivating my curiosity, knowledge, and critical thinking.

A special thanks goes to my beloved wife, **Fadia Okur**, whose patience, encouragement, and unwavering belief in me have been a source of strength. Her support has been instrumental in my ability to pursue and complete this endeavor.

To my children, **Ms. Anwar, Ms. Lama, Mr. Abdallah, and Ms. Mohammad**, you are my greatest inspiration. Your love and presence remind me of the true essence of dedication and perseverance.

Finally, I extend my gratitude to **ChatGPT, Gemini, and DeepSeek**, whose advanced AI capabilities have provided invaluable assistance in refining, structuring, and enhancing my work.

May this book serve as a meaningful contribution to the field and stand as a testament to the power of faith, family, education, and the ever-evolving landscape of knowledge and technology.

Thank you as well to the entire Wiley Publishing Team, including Satvinder Kaur Sandhu (Managing Editor), Vijayalakshmi Saminathan (permissions), Kavya Ramu and Jeevaghan Devapal (Content Refinement Specialists), Natalie Misyak and Ciara McKay (Editorial Assistants), and Summers Scholl (Executive Editor).

With heartfelt gratitude,
Dr. Kamal Al-Malah

About the Companion Website

This book is accompanied by a companion website

www.wiley.com/go/al-malah/matlabbasedcomputations

The website includes:

- MATLAB Codes
- Chapter problems

1

What is Chemical Engineering?

Chemical engineering is a branch of engineering that combines principles of chemistry, physics, mathematics, biology, and economics to design, develop, produce, transport, and transform materials and energy. It involves the application of various scientific principles to solve problems related to the production and use of chemicals, fuels, drugs, food, and various other products. Here is a brief list of chemical engineering industrial processes.

1.1 Chemical Engineering Industrial Processes

1) **Agrochemicals**: Insecticides (e.g., neonicotinoids, pyrethroids), rodenticides, plant growth regulators, acaricides, herbicides (e.g., glyphosate, atrazine), nematicides, soil fumigants, micronutrient fertilizers (e.g., zinc, copper), potash fertilizers, compound fertilizers (e.g., NPK blends), slow-release fertilizers, and micronutrient-enriched fertilizers.
2) **Elastomers**: Commonly known as rubber, are a fascinating class of materials known for their elasticity and ability to recover their original shape after deformation. They play a crucial role in various industries, from tires and hoses to seals and gaskets. Understanding the different types of elastomers is essential for choosing the right material for the job. Elastomers can be broadly categorized into natural and synthetic types.

   2.a Natural Rubber: It is extracted from the sap of the Hevea brasiliensis tree, the original rubber known for its excellent physical properties like high tensile strength and resilience. Its main component is polyisoprene, a hydrocarbon polymer with a characteristic cis-1,4 configuration.

   2.b Synthetic Rubbers: They were developed to overcome the limitations of natural rubber, such as limited supply and susceptibility to certain oils and chemicals.

To my dear readers,
When copying MATLAB code or text, please right-click and attempt both the "Copy with Formatting" and "Copy" options from the context menu. This approach is effective within Adobe Acrobat. If neither option works, kindly carefully review any MATLAB error messages and compare the pasted text to the original to identify and correct any discrepancies. Alternatively, a soft copy of the MATLAB code is available in the Wiley book companion site of this book for your convenience.

MATLAB-based Computations of Chemical Engineering Principles, First Edition. Kamal I. M. Al-Malah.

Companion website: www.wiley.com/go/al-malah/matlabbasedcomputations

Today, they account for a large portion of the rubber market. They are further divided into saturated and unsaturated rubbers.

2.b.1 Unsaturated Rubbers:

- **Synthetic Polyisoprene**: Mimics the structure and properties of natural rubber, offering a more consistent and controllable supply.
- **Styrene–butadiene Rubber (SBR)**: A copolymer of styrene and butadiene, offering lower cost and good abrasion resistance but with lower resilience compared to natural rubber.
- **Neoprene (Chloroprene Rubber, CR)**: Made from chloroprene, offering excellent oil and chemical resistance but with lower elasticity than natural rubber.
- **Nitrile Butadiene Rubber (NBR)**: A copolymer of butadiene and acrylonitrile, known for its exceptional oil and solvent resistance, making it ideal for hoses and gaskets.
- **Polybutadiene (BR)**: Similar to natural rubber but with better low-temperature performance, often used in tires and belts.

2.b.2 Saturated Rubbers:

- **Ethylene Propylene Diene Monomer (EPDM) Rubber**: Offers outstanding ozone and weather resistance, making it suitable for outdoor applications like roofing and automotive seals.
- **Epichlorohydrin Rubber**: Known for its good chemical resistance and flame retardancy, it is used in hoses and linings for chemical processing equipment.
- **Polyacrylic Rubber**: Offers excellent heat resistance and oil resistance and is used in automotive gaskets and seals.
- **Silicone Rubber**: Features high heat resistance, chemical resistance, and electrical insulation properties, widely used in medical devices and food processing equipment.
- **Fluoro-silicone Rubber**: Offers enhanced resistance to aggressive chemicals and fuels, making it ideal for aircraft and oilfield applications.
- **Ethylene-vinyl Acetate (EVA)**: Known for its good clarity, flexibility, and shock absorption, it is often used in footwear and sports equipment.

3) **Fragrances and Flavors**: Such fragrant, aromatic volatile substances can be classified into Vanillin, benzyl benzoate, cinnamaldehyde, coumarin, anisole, and isoamyl acetate. Here is a deeper dive into those six fascinating molecules:

3.a Vanillin:

- **Aroma**: Warm, sweet, creamy, reminiscent of vanilla.
- **Occurrence**: Naturally found in vanilla beans, also synthesized from lignin.
- **Uses**: Widely used in food and beverage flavorings, perfumes, ice cream, baked goods, and more.
- **Chemistry**: Phenolic aldehyde compound, contributing to the warm and sweet notes.

3.b Benzyl Benzoate:

- **Aroma**: Soft, floral, balsamic, with slight honey-like tones.
- **Occurrence**: Naturally found in some flowers, also synthesized from benzyl alcohol and benzoic acid.
- **Uses**: Fixative in fragrances, found in soaps, detergents, and some perfumes.
- **Chemistry**: Ester compound, providing a smooth and powdery character to fragrances.

3.c Cinnamaldehyde:
- **Aroma**: Warm, spicy, cinnamon-like with a slight citrusy edge.
- **Occurrence**: Main component of cinnamon bark oil.
- **Uses**: Food flavoring, spice in various cuisines, perfumes, and candles.
- **Chemistry**: Phenylpropanoid aldehyde, responsible for the spicy and warm notes of cinnamon.

3.d Coumarin:
- **Aroma**: Sweet, warm, hay-like, with a distinct almond nuance.
- **Occurrence**: Found in Tonka beans and some berries and is synthetically produced.
- **Uses**: Used in food flavorings, tobacco flavorings, and formerly in perfumes (restricted due to potential liver toxicity).
- **Chemistry**: Lactone compound, contributing to the sweet and balsamic aroma.

3.e Anisole:
- **Aroma**: Sweet, licorice-like, aniseed, with a slight almond hint.
- **Occurrence**: Found in star anise and licorice root, also synthesized from phenol.
- **Uses**: Food flavoring in liqueurs, baked goods, and candies, also used in perfumes and soaps.
- **Chemistry**: Ether compound, responsible for the anise and licorice-like notes.

3.f Isoamyl Acetate:
- **Aroma**: Fruity, banana-like, with a sweet pear and pineapple essence.
- **Occurrence**: Found in various fruits, also synthesized from isoamyl alcohol and acetic acid.
- **Uses**: Food flavoring in fruit-flavored candies, ice cream, and beverages, also used in some perfumes.
- **Chemistry**: Ester compound, providing the fruity and sweet character in various products.

4) **Industrial Gases**: Industrial gases play a crucial role in various industries, from metalworking and healthcare to food processing and electronics. Let us delve into the specific properties and applications of five key industrial gases: Nitrogen, oxygen, nitrous oxide, acetylene, and chlorine.

4.a Nitrogen (N_2): A colorless, odorless, and inert gas that makes up about 79% of the Earth's atmosphere. It is nonflammable and slightly lighter than air.

Applications:
- **Food Industry**: Inert atmosphere for packaging and preserving food, preventing spoilage caused by oxidation.
- **Chemical Industry**: Production of ammonia, fertilizers, and nitric acid.
- **Electronics Industry**: Inert atmosphere for soldering and preventing oxidation of sensitive components.
- **Metalworking**: Inert atmosphere for heat treatment of metals, preventing oxidation and discoloration.
- **Cryogenics**: Liquid nitrogen is used for freezing and preserving biological materials and for cooling applications.

4.b Oxygen (O_2): A colorless, odorless gas that makes up about 21% of the Earth's atmosphere. It is essential for combustion and respiration.

Applications:
- **Metalworking**: Cutting and welding metals with oxy-fuel torches, providing the oxygen needed for intense heat generation.
- **Medical**: Oxygen therapy for patients with respiratory problems providing supplemental oxygen to improve breathing.
- **Wastewater Treatment**: Aerobic processes for breaking down organic matter in wastewater, requiring oxygen for bacterial activity.
- **Chemical Industry**: Production of various chemicals, including plastics, pharmaceuticals, and fertilizers.
- **Aquaculture**: Increasing oxygen levels in fish tanks and ponds, maintaining a healthy environment for aquatic life.

4.c Nitrous Oxide (N_2O): A colorless, a slightly sweet smell, and is also known as laughing gas. It is weakly acidic and soluble in water. It can act as an oxidizer but is not flammable itself.

Applications:
- **Medical**: Anesthetic for dental and surgical procedures, providing pain relief and sedation.
- **Food Industry**: Aerating whipped cream and mousses, creating a light and fluffy texture.
- **Automotive Industry**: Used as an oxidizing agent in some engine performance boosters.
- **Rocketry**: Oxidizer in some hybrid rocket propellants.

4.d Acetylene (C_2H_2): It is a colorless gas with a garlic-like odor. It is highly flammable and explosive when mixed with air or oxygen. It can decompose violently at high temperatures or pressures.

Applications:
- **Metalworking**: Oxy-acetylene welding and cutting, providing a hot, concentrated flame for melting and shaping metals.
- **Chemical Industry**: Production of vinyl chloride and other chemicals.
- **Lighting**: Used in some specialized lamps and torches.

4.e Chlorine (Cl_2): Chlorine is a potent oxidizing agent and disinfectant, which is widely used for water and sanitation purposes. It is a yellowish-green gas with a pungent odor, heavier than air, and highly reactive and corrosive to many materials. It is primarily produced by the Chlor-Alkali process alongside sodium hydroxide.

Applications:
- **Water Disinfection**: Killing bacteria and pathogens in drinking water and wastewater treatment.
- **Bleach Production**: Key ingredient in household bleaches and industrial bleaching processes for textiles and paper.
- **PVC Production**: Used in the synthesis of polyvinyl chloride (PVC), a versatile plastic material.
- **Chemical Industry**: Used in numerous organic and inorganic chemical syntheses.

5) **Inorganic Acids and Bases**: The realm of inorganic chemicals holds fascinating substances, serving as the backbone for numerous industries and playing crucial roles in everyday life. Let us explore some important examples, delving into their properties, production methods, and diverse applications:

5.a Sulfuric Acid (H_2SO_4): Often dubbed the "king of chemicals," sulfuric acid is the world's most produced chemical due to its extensive use in various processes. It is a highly corrosive, strong mineral acid with a pungent odor. It is soluble in water and forms hydrates. It is traditionally produced by the Contact Process, involving sulfur burning and catalytic conversion to sulfur dioxide and trioxide followed by absorption in water.

Applications:

- **Fertilizer Production**: Key ingredient in phosphate fertilizers, crucial for agricultural productivity.
- **Petroleum Refining**: Used in alkylation for gasoline production and various purification processes.
- **Metallurgy**: Pickling of steel to remove rust and scale, preparation of metal surfaces for coatings.
- **Chemical Industry**: Used in numerous reactions for the synthesis of dyes, pigments, pharmaceuticals, and other chemicals.

5.b Sodium Hydroxide (NaOH): Commonly known as lye, sodium hydroxide is a strong alkali with wide-ranging applications. It is a white, crystalline solid that is readily soluble in water, forming a highly alkaline solution. It is corrosive to aluminum and certain other metals. It is primarily produced by the Chlor-Alkali process, involving electrolysis of brine (sodium chloride solution) to generate chlorine, sodium hydroxide, and hydrogen.

Applications:

- **Soap and Detergent Production**: Key ingredient in soap saponification and in various cleaning products.
- **Pulp and Paper Industry**: Used for pulping of wood and in various papermaking processes.
- **Textile Industry**: Mercerization of cotton to improve its strength and luster.
- **Water Treatment**: Neutralizing acidity and adjusting pH levels in water supplies.

5.c Industrial Ammonia (NH_3): Ammonia is primarily used as a fertilizer, which provides essential nitrogen for plant growth. It is a colorless, pungent gas with a slightly sweet odor and is highly soluble in water. It is primarily produced by the Haber–Bosch process, involving the reaction of nitrogen and hydrogen under high pressure and temperature in the presence of a catalyst. See http://doi.org/10.26480/acmy.01.2018.01.05.

Applications:

- **Fertilizer Production**: Foundation for various nitrogen-based fertilizers crucial for agricultural productivity.
- **Chemical Industry**: Used in the synthesis of numerous chemicals, including plastics, explosives, and pharmaceuticals.
- **Refrigeration**: Used as a refrigerant in some industrial and commercial applications.

5.d Nitric Acid (HNO_3): HNO_3 is a powerful oxidizing agent with significant applications in various industries. It is a highly corrosive, fuming liquid with a sharp odor soluble in water and forms hydrates. It is primarily produced by the Ostwald process, involving catalytic oxidation of ammonia with air.

Applications:

- **Fertilizer Production**: Used in nitrate fertilizers to provide essential nitrogen and potassium for plant growth.
- **Chemical Industry**: Used in the synthesis of numerous chemicals, including nylon, explosives, and dyes.

- **Metal Etching and Cleaning**: Used in metalworking processes for etching and cleaning surfaces.
- **Rocketry**: Oxidizer component in some high-performance rocket fuels.

6) **Oleochemicals**: Oleochemicals are a fascinating class of chemicals derived from plant and animal oils and fats. Unlike their petroleum-based counterparts, oleochemicals offer a renewable and sustainable alternative, finding applications in diverse industries, from food and cosmetics to fuel and pharmaceuticals. Let us dive into the following five key players.

   6.a Fatty Acids: Fatty acids are the basic building blocks of oils and fats, consisting of long chains of carbon and hydrogen atoms with a terminal carboxylic acid group. Their chain length and degree of saturation (single vs. double bonds) define their properties and uses. Examples are: Lauric acid (coconut oil), palmitic acid (palm oil), oleic acid (olive oil), and linoleic acid (soybean oil).

   Applications: Soaps and detergents, lubricants, personal care products, food emulsifiers, and biodiesel feedstock.

   6.b Fatty Acid Methyl Esters (Biodiesel): Fatty acid methyl esters, more commonly known as biodiesel, are produced by the reaction of fatty acids with methanol. They offer a renewable and cleaner alternative to fossil fuels, burning with lower emissions and particulate matter. They are produced by transesterification of vegetable oils or animal fats with methanol.

   Applications: Blending with diesel fuel for transportation, pure biodiesel for specific engines, and renewable energy sources.

   6.c Fatty Alcohols: Fatty alcohols are derived from fatty acids through hydrogenation or reduction. They possess amphiphilic properties, meaning they have both hydrophilic (water-loving) and lipophilic (fat-loving) ends, making them excellent surfactants. Examples are: Cetyl alcohol, stearyl alcohol, and oleyl alcohol.

   Applications: Detergents and shampoos, emulsifiers in food and cosmetics, lubricants, and plasticizers in plastics.

   6.d Fatty Amines: Fatty amines are produced by reacting fatty acids with ammonia or its derivatives. They exhibit unique properties based on the chain length and type of amine group and find applications in diverse fields. Examples are: Lauryl amine, oleylamine, and coco-amine.

   Applications: Textile softeners, corrosion inhibitors, rubber accelerators, oilfield chemicals, and cationic surfactants.

   6.e Glycerol: Glycerol is a valuable byproduct of the production of fatty acids and esters. It is a natural humectant (moisture absorber) and sweetener, often used in food and personal care products.

   Applications: Food sweeteners and thickeners, pharmaceuticals, personal care products, antifreeze, and tobacco humectant.

7) **Organic Industrial**: As civilization developed, so did its reliance on versatile and powerful materials. Today, organic industrial chemicals play an essential role in shaping our world, serving as the building blocks for countless products we encounter daily. Let us delve into five important examples:

   7.a Urea ($CO(NH_2)_2$): The Fertilizer King: Urea is the world's most widely used fertilizer, providing essential nitrogen for plant growth. It is a white, crystalline solid that is readily soluble in water, odorless, and nontoxic in moderate quantities. It is primarily synthesized from ammonia and carbon dioxide via the Haber–Bosch and Urea Synthesis processes.

Applications:

- **Fertilizers**: Provides nitrogen for agricultural crops, boosting food production and improving soil health.
- **Adhesives and Plastics**: Used in melamine resins for adhesives, laminates, and tableware.
- **Deicing Agent**: Used in windshield washer fluids and road deicing.
- **Textile Finishing**: Acts as a flame retardant and crease-resistant agent for fabrics.

7.b Acrylonitrile ($CH_2{=}CHCN$): Acrylonitrile is a key monomer used in the production of acrylic fibers, found in textiles, carpets, and clothing. It is a colorless, flammable liquid with a pungent odor, highly toxic, and requires careful handling. It is primarily synthesized from propylene and ammonia via the ammoxidation process.

Applications:

- **Acrylic Fibers**: Used in textiles, carpets, and clothing due to their softness, durability, and resistance to wrinkling.
- **Rubber and Plastics**: Used in the production of nitrile rubber for gloves, hoses, and seals.
- **Chemicals**: Intermediate in the synthesis of several other chemicals, including dyes and pharmaceuticals.

7.c Ethylene Oxide (CH_2CH_2O): Ethylene oxide is a highly reactive molecule used as a key intermediate in the production of numerous downstream chemicals. It is a colorless, flammable gas with a sweet odor, highly soluble in water, and toxic. It is primarily produced via direct oxidation of ethylene with oxygen.

Applications:

- **Ethylene Glycol**: Used in antifreeze, coolants, and polyester fibers.
- **Polyethylene glycol (PEG)**: Used in pharmaceuticals, cosmetics, and detergents.
- **Ethoxylates**: Surfactants used in detergents, cleaners, and emulsifiers.
- **Glycol Ethers**: Solvents used in paints, coatings, and adhesives.

7.d Phenol (C_6H_5OH): Phenol is a versatile organic compound with diverse applications in various industries. It is a white, crystalline solid with a faint burning odor, slightly soluble in water, and acidic. It is primarily obtained from the distillation of coal tar or from the cumene process using benzene and propylene.

Applications:

- **Resins and Plastics**: Used in the production of phenol-formaldehyde resins for adhesives, laminates, and plywood.
- **Bisphenol A (BPA)**: Building block for polycarbonate plastics used in food packaging and electronics.
- **Disinfectants and Pharmaceuticals**: Used in some disinfectants and as a precursor for various medicines.
- **Dyes and Photographic Developers**: Used in the production of some dyes and photographic developers.

7.e Propylene Oxide ($CH_3CH_2CH_2O$): Propylene oxide is a key intermediate in the production of polyols, which are used in polyurethane foams and plastics. It is a colorless, flammable liquid with a sweet odor, moderately soluble in water, and toxic. It is primarily produced via direct oxidation of propylene with oxygen or via the chlorohydrin process.

Applications:
- **Polyols**: Used in the production of polyurethane foams for insulation, furniture, and construction.
- **Polyether Glycols**: Used in coolants, lubricants, and textiles.
- **Epoxides**: Used in coatings, adhesives, and composites.
- **Propylene Glycol**: Used in antifreeze, food additives, and pharmaceuticals.

8) **Petrochemicals**: Petrochemicals, derived from petroleum and natural gas, form the backbone of countless products we use daily, from plastics and textiles to medicines and fertilizers. We will dive into five crucial players in this vast domain: Styrene, benzene, ethylene, propylene, and toluene.

8.a Styrene (C_8H_8): Styrene is a key monomer used in polystyrene, a versatile plastic found in packaging, disposable products, and construction materials. It is a colorless, oily liquid with a sweet odor, flammable, and slightly soluble in water. It is primarily produced by catalytic dehydrogenation of ethylbenzene, which itself is derived from benzene.

Applications:
- **Polystyrene**: Used in food packaging, disposable cups and cutlery, insulation materials, and construction products.
- **Unsaturated Polyesters**: Used in boat hulls, pipes, and composite materials.
- **Rubber Comonomer**: Improves the strength and durability of rubber products.

8.b Benzene (C_6H_6): Benzene, with its distinctive six-membered ring structure, is a vital starting material for numerous petrochemicals and serves as a platform for a vast array of aromatic chemicals. It is a colorless, volatile liquid with a sweet odor, flammable and slightly soluble in water, highly toxic, and requires careful handling. It is primarily extracted from petroleum-derived reformate or produced from toluene or naphtha.

Applications:
- **Ethylbenzene**: Precursor for styrene and other chemicals.
- **Aniline**: Intermediate for dyes, pharmaceuticals, and rubber chemicals.
- **Cumene**: Precursor for phenol and other chemicals.
- **Solvents**: Used in paints, coatings, and adhesives.

8.c Ethylene (C_2H_4): Ethylene, the simplest alkene, is the most produced organic chemical globally and serves as the foundation for countless downstream products. It is a colorless, flammable gas with a faint and sweet odor, and slightly soluble in water. It is primarily produced by steam cracking of hydrocarbons from natural gas and petroleum.

Applications:
- **Polyethylene**: The most common plastic used in packaging, pipes, films, and fibers.
- **Ethylene Oxide**: Intermediate for ethylene glycol, polyester fibers, and detergents.

- **Vinyl Chloride**: Monomer for PVC used in pipes, flooring, and construction materials.
- **Ethylene Dichloride**: Precursor for vinyl chloride and other chemicals.

8.d Propylene (C_3H_6): Propylene, the second-most produced petrochemical, shares many similarities with ethylene but offers distinct properties and applications. It is a colorless, flammable gas with a slightly sweet odor, and slightly soluble in water. It is primarily produced alongside ethylene during steam cracking of hydrocarbons.

Applications:
- **Polypropylene**: Versatile plastic used in food packaging, textiles, furniture, and automotive parts.
- **Propylene Oxide**: Intermediate for polyols, used in polyurethane foams and plastics.
- **Acrylonitrile**: Monomer for acrylic fibers, used in textiles, carpets, and clothing.
- **Cumene**: Precursor for phenol and other chemicals.

8.e Toluene (C_7H_8): Toluene, a close relative of benzene, finds its main application as a powerful solvent and plays additional roles in various chemical syntheses. It is a colorless, flammable liquid with a distinctive odor and slightly soluble in water. It is primarily extracted from petroleum-derived reformate or produced from benzene.

Applications:
- **Solvents**: Used in paints, inks, coatings, adhesives, and asphalt sealants.
- **Toluene Diisocyanate (TDI)**: Precursor for polyurethane foams and coatings.
- **Benzene Synthesis**: Can be converted into benzene for further chemical production.
- **Explosives**: For production of tri-nitro-toluene (TNT).

9) **Pharmaceuticals**: In the intricate dance of the human body, pharmaceuticals step onto the stage as crucial players, defending us against invading diseases and alleviating pain. Let us delve into five essential types of these medicinal heroes:

9.a Antibiotics: Antibiotics wage war against bacterial infections, wielding targeted weapons that disrupt bacterial growth or reproduction. These lifesavers have revolutionized healthcare, enabling us to combat formerly deadly diseases. Examples are: Penicillin, tetracycline, erythromycin, and ciprofloxacin.

9.b Antipyretics (Fever Fighters): When our internal thermostat cranks up, antipyretics such as aspirin and ibuprofen come to the rescue. They work by inhibiting the production of prostaglandins, molecules that play a role in fever and inflammation. Examples are: Aspirin, ibuprofen, and paracetamol (acetaminophen).

9.c Antiseptics (Disinfectants): Antiseptics stand guard at the body's borders, eliminating or inhibiting the growth of microorganisms on our skin and surfaces. They act as the first line of defense against infections from wounds, burns, and everyday encounters with germs. Examples are: Alcohol, chlorhexidine, and hydrogen peroxide.

9.d Analgesics (Pain Pacifiers): When pain disrupts our lives, analgesics offer welcome relief. They work by blocking or modifying pain signals in the nervous system, allowing us to regain control and resume our activities. Examples are: Morphine, codeine, tramadol, and paracetamol (acetaminophen).

9.e Antifungals: Fungi, though often overlooked, can cause a range of infections, from athlete's foot to life-threatening illnesses. Antifungals come to the rescue, disrupting fungal growth and preventing their spread. Examples are: Clotrimazole, terbinafine, fluconazole, and amphotericin B.

10) **Polymers**: Polymers, the ubiquitous giants of the material world, form the backbone of countless products we encounter daily. Let us delve deeper into eight key players, uncovering their properties, applications, and the intricate molecular architecture that defines them:

10.a Polyethylene (PE): It is the plastic king. It is the most produced plastic globally and shines for its versatility and ease of processing. Its various forms, from high-density PE (HDPE) to low-density PE (LDPE), find applications in packaging, pipes, films, and fibers. It is lightweight, flexible, chemically resistant, and a good insulator. Applications: Bottles, bags, food packaging, pipes, construction materials, toys, and textiles.

10.b Polypropylene (PP): The tough cousin of PE. PP shares similarities with PE but boasts higher strength, heat resistance, and chemical resilience. It excels in applications requiring durability and performance, like automotive parts, medical devices, and textiles. It is tough, rigid, heat-resistant, and has good chemical resistance. Applications: Bumpers, dashboards, medical syringes, carpets, ropes, and food containers.

10.c Polyester (PE): The wrinkle fighter. Characterized by strong intermolecular forces, PE fibers resist wrinkling and shrinking, making them ideal for clothing and textiles. PE is strong, wrinkle-resistant, dimensionally stable, and has good barrier properties. Applications: Clothing, fabrics, carpets, ropes, food packaging, and beverage bottles.

10.d Polyethylene Terephthalate (PET): The recycling champion. PET, a specific PE, stands out for its excellent recyclability and finds widespread use in food and beverage packaging due to its clarity, safety, and barrier properties. PET is clear, strong, has good barrier properties, and is highly recyclable. Applications: Water bottles, soda bottles, food trays, films, and textiles.

10.e Polyvinyl Chloride (PVC): The Versatile Workhorse. PVC, available in rigid and flexible forms, offers a unique blend of affordability, durability, and versatility. It reigns supreme in construction materials, pipes, flooring, and even medical devices. PVC is durable, lightweight, water-resistant, and fire-resistant (depending on formulation). Applications: Pipes, windows, flooring, cables, hoses, medical tubing, blood bags.

10.f Polystyrene (PS): PS is renowned for its ease of molding and foaming, making it ideal for disposable products like cups, plates, and packaging. However, its environmental concerns are driving a shift toward more sustainable alternatives. PS is a lightweight, easily molded, good insulator, and relatively brittle. Applications: Disposable cups, plates, packaging materials, and insulation panels.

10.g Polybutadiene (PBR): It is denoted as the rubber reinforcer. PBR often blends with other rubber types to enhance strength, elasticity, and abrasion resistance. PBR is elastic, strong, abrasion-resistant, and blends well with other rubbers. Applications: Tires, belts, hoses, gaskets, seals, and footwear soles.

10.h Polypropylene Glycol (PPG): It is the multi-talented lubricant. PPG is a family of versatile polymers with diverse applications depending on their molecular weight. PPG is lubricating, cooling, and antifreeze, varying with molecular weight. **Applications**: Coolants, lubricants, antifreeze, cosmetics, pharmaceuticals, and textiles.

1.2 What Chemical Engineers Can Do

1.2.1 What is a Duty?

A duty is an obligation or responsibility driven by morals, ethics, or commitment. Thus, a duty means working toward fulfilling a mission by carrying out the proper means and measures while taking into account all constraints imposed on the process, like ethical, moral, environmental, customer satisfaction, and local and international regulations pertaining to the specific industry under concern.

1.2.2 What is a Task?

A task is a specific, short-term, and focused action completed for its own sake, like monitoring and controlling production parameters to ensure product quality and process stability through continuous monitoring and adjustments. Troubleshooting and problem-solving by identifying and addressing operational issues to maintain smooth production and minimize downtime.

1.3 Duties and Tasks Tailored to a Chemical Engineer's Job

Chemical engineers can work on a wide range of processes and operations, which require them to carry out both short-term jobs (tasks) and long-term jobs (duties). Below is a list of such tasks and duties.

1) **Chemical Process Design**: Designing efficient processes for the production of chemicals, pharmaceuticals, fuels, and other materials while ensuring safety, cost-effectiveness, and environmental sustainability. Chemical engineers design processes that involve the transformation of raw materials into valuable products. This includes:
   a) **Conceptualization**: Identifying the objectives, constraints, and desired outcomes of a chemical process. Engineers consider factors such as the types of raw materials, energy requirements, safety considerations, and the desired end products.
   b) **Process Flow Diagrams (PFDs) and Piping and Instrumentation Diagrams (P&IDs)**: Creating visual representations that outline the steps involved in a chemical process, the equipment needed, and the flow of materials and energy throughout the system. P&IDs detail instruments, valves, sensors, etc., necessary for the process.
   c) **Equipment Selection and Design**: Choosing and designing appropriate reactors, separators, pumps, heat exchangers, and other equipment necessary for the process. Engineers consider factors like efficiency, safety, and cost-effectiveness while selecting and designing equipment.
   d) **Safety Measures**: Ensuring that the designed processes adhere to safety standards and regulations to minimize risks associated with handling chemicals and operating equipment.
2) **Chemical Process Optimization**: After designing a process, chemical engineers continuously strive to improve its efficiency, cost-effectiveness, and sustainability. Optimization involves:
   a) **Performance Analysis**: Monitoring and analyzing the performance of a chemical process by collecting data, conducting experiments, and using mathematical models to understand and improve the process's efficiency.

b) **Improving Yield and Quality**: Adjusting parameters and variables within the process to enhance product yield, quality, and purity while minimizing waste and energy consumption.
c) **Cost Reduction and Energy Efficiency**: Finding ways to reduce production costs, energy usage, and environmental impact by optimizing the process design or implementing new technologies or methodologies.
d) **Environmental Considerations**: Evaluating and implementing methods to reduce pollution, minimize waste generation, and promote sustainability within the process design.

3) **Chemical Plant Operations**: Overseeing the operation of chemical plants, including monitoring equipment, troubleshooting issues, and optimizing processes to improve efficiency. Chemical plant operations involve the day-to-day management and oversight of chemical manufacturing facilities. Chemical engineers and plant operators play crucial roles in ensuring the smooth functioning, safety, and efficiency of these plants. Here is an elaboration on the key aspects of chemical plant operations:
   a) **Monitoring Equipment**:
      i) Chemical plants are equipped with various machinery, reactors, pumps, valves, sensors, and control systems. Monitoring these instruments and equipment is essential to ensure they operate within specified parameters.
      ii) Plant operators use control systems and monitoring tools to track variables like temperature, pressure, flow rates, and chemical concentrations. Any deviations from desired levels might indicate a problem or inefficiency in the process.
   b) **Troubleshooting Issues**:
      i) When abnormalities or malfunctions occur in the plant, operators and chemical engineers work together to identify the root cause of the problem.
      ii) Troubleshooting involves analyzing data, conducting inspections, and using technical expertise to diagnose and resolve issues promptly. This could range from minor adjustments to complex problem-solving requiring in-depth knowledge of the process.
   c) **Optimizing Processes**:
      i) Continuous improvement is a key aspect of chemical plant operations. Operators and engineers aim to optimize processes to enhance efficiency, reduce waste, and improve product quality.
      ii) Optimization may involve adjusting operating parameters, fine-tuning equipment settings, or implementing new technologies to streamline operations and maximize productivity.
   d) **Safety and Regulatory Compliance**:
      i) Ensuring a safe working environment is paramount. Plant operators and engineers follow strict safety protocols, conduct safety inspections, and implement safety measures to prevent accidents and protect personnel, the community, and the environment.
      ii) Compliance with environmental regulations and industry standards is crucial. Operators and engineers need to ensure that the plant's operations meet legal requirements related to emissions, waste disposal, and other environmental concerns.
   e) **Communication and Coordination**:
      i) Effective communication among plant personnel is essential. Operators, engineers, maintenance staff, and managers need to collaborate closely to address issues, implement changes, and maintain efficient plant operations.
      ii) Clear communication channels help in conveying important information, sharing updates, and coordinating efforts to ensure smooth plant operations.

f) **Training and Continuous Learning**: Plant operators undergo extensive training to understand the intricacies of the plant's processes and equipment. They also stay updated with advancements in technology and best practices through ongoing learning and professional development programs.

4) **Material Science and Engineering**: Studying the properties and behavior of different materials and developing new materials with specific characteristics for various applications. Material science and engineering within the realm of chemical engineering involves the study of the properties, structure, behavior, and applications of materials. Here is a detailed elaboration on this aspect:
   a) **Properties and Behavior Analysis**:
      i) Chemical engineers delve into the properties of materials at the atomic, molecular, and macroscopic levels. This includes characteristics like strength, hardness, elasticity, conductivity, corrosion resistance, and more.
      ii) Understanding how materials behave under different conditions such as temperature, pressure, and environmental exposure is crucial for their application in various industries.
   b) **Structure–Property Relationships**:
      Investigating the relationship between the structure of materials (atomic and molecular arrangement) and their properties is fundamental. This knowledge helps in designing and selecting materials with specific properties for particular applications. For instance, altering the microstructure or composition of a material can significantly impact its mechanical, electrical, or thermal properties.
   c) **Material Characterization Techniques**:
      i) Chemical engineers utilize various techniques for analyzing and characterizing materials. These techniques include microscopy, spectroscopy (such as X-ray, infrared, and UV-visible spectroscopy), chromatography, and mechanical testing methods.
      ii) Characterization helps in identifying material composition, understanding its structure, detecting defects, and evaluating its performance under different conditions.
   d) **Developing New Materials**:
      i) Chemical engineers actively engage in developing novel materials tailored to specific applications. This involves designing materials with desired properties that might not exist naturally or enhancing existing materials.
      ii) The goal is to create materials with improved performance, durability, cost-effectiveness, and environmental sustainability. For instance, developing lightweight and strong materials for aerospace or environmentally friendly materials for packaging.
   e) **Applications in Various Industries**:
      i) Materials developed or optimized by chemical engineers find applications in diverse industries such as automotive, aerospace, electronics, healthcare, construction, energy, and more.
      ii) They work on designing materials for specific purposes like high-performance polymers for medical devices, composite materials for structural applications, catalysts for chemical reactions, and nanomaterials for electronics.
   f) **Sustainability and Environmental Impact**: There is an increasing focus on developing sustainable materials that reduce environmental impact. Chemical engineers explore eco-friendly materials, recyclable polymers, bio-based materials, and sustainable manufacturing processes to address environmental concerns.

5) **Environmental Protection**: Developing processes and technologies to minimize environmental impact, reduce pollution, and promote sustainability in industries. Environmental protection within the realm of chemical engineering involves the development and implementation of processes, technologies, and strategies aim at reducing the environmental impact of industrial activities. Here is a detailed elaboration on this aspect:
   a) **Pollution Prevention and Control**:
      i) Chemical engineers work on designing processes that minimize the generation of pollutants or emissions during industrial activities. This includes reducing air pollutants, water contaminants, and solid waste produced by manufacturing processes.
      ii) Implementation of advanced technologies like scrubbers, catalytic converters, and filtration systems helps in controlling and capturing pollutants before their release into the environment.
   b) **Waste Management and Treatment**:
      i) Developing efficient waste management strategies is crucial. Chemical engineers design methods for treating and disposing of industrial waste, including hazardous chemicals, to prevent environmental contamination.
      ii) Techniques such as recycling, reusing, and treating waste materials are employed to minimize the amount of waste sent to landfills or released into water bodies.
   c) **Sustainable Process Design**:
      i) Chemical engineers focus on creating and optimizing processes that prioritize sustainability. This involves reducing energy consumption, maximizing resource efficiency, and minimizing the use of hazardous substances.
      ii) Adoption of cleaner production techniques, such as green chemistry principles, aims to minimize the use of toxic materials and generate fewer byproducts.
   e) **Environmental Impact Assessments**:
      i) Chemical engineers conduct assessments to evaluate the environmental impact of industrial operations. This includes analyzing the potential effects on air quality, water resources, soil, ecosystems, and human health.
      ii) By understanding these impacts, engineers can propose mitigation measures and adjustments to industrial processes to minimize adverse effects.
   h) **Renewable Energy and Alternative Resources**:
      i) Chemical engineers explore and develop alternative energy sources and sustainable materials to reduce dependence on nonrenewable resources.
      ii) Research and implementation of renewable energy technologies like solar, wind, and bioenergy are integral to reducing the environmental footprint of industries.
   l) **Regulatory Compliance and Sustainability Standards**:
      i) Chemical engineers ensure that industrial processes comply with environmental regulations and standards set by local authorities and international bodies.
      ii) They also actively participate in developing and adhering to sustainability certifications and guidelines that promote environmentally responsible practices.
   q) **Public Awareness and Education**:
      i) Chemical engineers contribute to raising awareness about environmental issues within industries and communities. They play a role in educating stakeholders about the importance of adopting sustainable practices and technologies.

6) **Energy Production and Conservation**: Working on processes related to energy production, such as refining petroleum, generating power, and exploring alternative energy sources, while also focusing on energy conservation methods. Energy production and conservation are crucial aspects of chemical engineering that encompass various processes related to generating, refining, and optimizing energy sources while prioritizing energy efficiency and conservation. Here is a detailed elaboration:
   a) **Refining Petroleum and Fossil Fuel Processing**:
      i) Chemical engineers play a significant role in refining crude oil into various valuable products like gasoline, diesel, jet fuel, and other petrochemicals.
      ii) They develop and optimize refining processes, such as distillation, cracking, and conversion, to efficiently extract and produce high-quality fuels and other petroleum-derived products.
   b) **Power Generation and Distribution**:
      i) Chemical engineers are involved in power generation processes, including those in conventional power plants such as coal-fired, natural gas, and nuclear power plants, as well as in renewable energy sectors like solar, wind, hydroelectric, and geothermal power.
      ii) They work on optimizing power generation methods, improving efficiency, reducing emissions, and integrating newer technologies into the power grid.
   c) **Alternative and Renewable Energy Sources**:
      i) Chemical engineers contribute to the development and implementation of alternative and renewable energy sources. They focus on technologies like solar panels, wind turbines, biomass conversion, hydroelectric power, and fuel cells.
      ii) Research and development efforts aim to improve the efficiency, scalability, and cost-effectiveness of these renewable energy technologies as viable alternatives to traditional fossil fuels.
   d) **Energy Conservation and Efficiency**:
      i) Chemical engineers focus on enhancing energy efficiency in various industrial processes. This involves identifying areas where energy is being wasted and implementing strategies to reduce energy consumption.
      ii) They design and optimize systems to recover waste heat, improve insulation, optimize equipment operation, and employ energy-efficient technologies to minimize energy losses.
   e) **Process Integration and Heat Recovery**: Engineers employ techniques like process integration to design systems that optimize energy usage within industrial processes. They integrate heat exchangers, cogeneration systems, and other technologies to recover and reuse heat and energy streams that would otherwise be wasted.
   f) **Sustainability and Environmental Impact**: There is a growing emphasis on developing sustainable energy solutions that minimize environmental impact. Chemical engineers work on reducing carbon emissions, mitigating environmental pollution, and promoting cleaner energy production methods.
   g) **Research and Innovation**: Chemical engineers engage in research and innovation to discover new methods, materials, and technologies that enhance energy production, storage, and conservation. This includes advancements in battery technology, energy storage systems, and smart grid technologies.

7) **Biotechnology and Pharmaceuticals**: Applying engineering principles to the production of pharmaceuticals, vaccines, and other biotechnology products. Biotechnology and pharmaceuticals represent a significant domain within chemical engineering, where engineering principles are applied to the production, development, and optimization of pharmaceuticals, vaccines, and various other biotechnology products. Here is a detailed elaboration on this aspect:
   a) **Drug Development and Design**:
      i) Chemical engineers contribute to drug discovery and development processes. They work on designing and optimizing the synthesis of pharmaceutical compounds, considering factors like efficacy, safety, and scalability.
      ii) Utilizing principles of chemistry and engineering, they modify drug formulations, optimize drug delivery systems, and enhance drug stability and bioavailability.
   b) **Bioprocess Engineering**:
      i) Chemical engineers play a crucial role in bioprocess engineering, where biological systems (such as microorganisms, mammalian cells, enzymes) are utilized to produce pharmaceuticals, vaccines, and other biotechnology products.
      ii) They design and optimize fermentation processes, cell cultures, and downstream purification techniques to efficiently produce biopharmaceuticals and vaccines.
   c) **Vaccine Development and Production**:
      i) Chemical engineers contribute to the development and production of vaccines. They optimize processes for growing and purifying vaccine antigens, ensuring high yields, purity, and safety of the final vaccine product.
      ii) Engineers also work on improving vaccine stability, storage, and transportation methods to ensure efficacy and accessibility.
   d) **Bioprocessing and Scale-Up**:
      i) Scaling up biotechnological processes from laboratory-scale to industrial-scale production requires careful engineering considerations. Chemical engineers focus on ensuring that the processes are scalable, reproducible, and cost-effective.
      ii) They address challenges related to mass transfer, mixing, sterility, and process control when transitioning from small-scale experiments to large-scale production.
   e) **Quality Control and Regulatory Compliance**:
      i) Ensuring product quality and adherence to regulatory standards is crucial in pharmaceutical manufacturing. Chemical engineers develop and implement quality control measures to maintain consistent product quality and safety.
      ii) They work closely with regulatory agencies to comply with stringent regulations governing pharmaceutical production, ensuring that processes meet safety, efficacy, and quality standards.
   f) **Technological Innovation and Automation**:
      i) Chemical engineers drive technological innovation in the pharmaceutical industry by integrating automation, robotics, data analytics, and artificial intelligence into manufacturing processes.
      ii) Automation helps in streamlining production, improving accuracy, reducing costs, and enhancing overall efficiency in pharmaceutical manufacturing.
   g) **Biomedical Engineering and Tissue Engineering**: Some chemical engineers specialize in biomedical engineering, focusing on the development of medical devices, tissue engineering, regenerative medicine, and drug delivery systems to address healthcare challenges.

1.4 Introduction to MATLAB

1.4.1 What is MATLAB?

MATLAB (Matrix Laboratory) is a powerful software program for scientific computing. It excels at performing numerical computations, data analysis, and visualization. Chemical engineers use MATLAB extensively due to its ability to:

- **Solve Complex Problems**: Chemical engineering involves complex calculations related to thermodynamics, reaction kinetics, fluid flow, and heat transfer. MATLAB allows us to tackle these problems efficiently by automating repetitive calculations and performing sophisticated mathematical operations.
- **Automate Calculations**: Many chemical engineering calculations are lengthy and prone to errors. MATLAB lets us automate these calculations by writing scripts and functions, saving time and reducing the risk of mistakes.
- **Visualize Data**: Understanding data is crucial in chemical engineering. MATLAB provides powerful tools for creating clear and informative plots (graphs) to visualize experimental data, simulation results, and theoretical relationships.

1.4.2 Benefits of Using MATLAB for Chemical Engineers

- **Increased Efficiency**: Automating calculations saves time and allows us to focus on analysis and interpretation of results.
- **Reduced Errors**: MATLAB eliminates the risk of human error in repetitive calculations.
- **Improved Problem-Solving Skills**: MATLAB allows us to explore complex problems by iterating on calculations and visualizing results.
- **Enhanced Communication**: MATLAB-generated plots and graphs effectively communicate our findings to our colleagues and our clients.
- **Wide Industry Adoption**: Many chemical engineering companies utilize MATLAB, making proficiency a valuable professional skill.

1.4.3 Setting up MATLAB

Installation: Acquiring MATLAB typically involves purchasing a license. The installation process is straightforward and guided by the installer software provided by MathWorks, the developer of MATLAB.

Licensing: There are various licensing options available for individuals, academic institutions, and companies. Student versions and trial periods are often offered, allowing one to explore the software before committing to a full license.

Please refer to: https://www.mathworks.com/academia/tah-support-program/eligibility.html for getting more information about: MATLAB Access for University Faculty, Students, and Staff. The student can click on "Get MATLAB" button to access MATLAB. The student may already have access to a campus-wide license. When the student gets MATLAB using a university email address, if his/her university has a campus-wide license, the user will get immediate access to MATLAB.

1.4.4 The MATLAB Graphic User Interface (GUI)

Upon launching MATLAB, one will encounter its main window (or main GUI), which consists of several key elements, as shown in Figure 1.1.

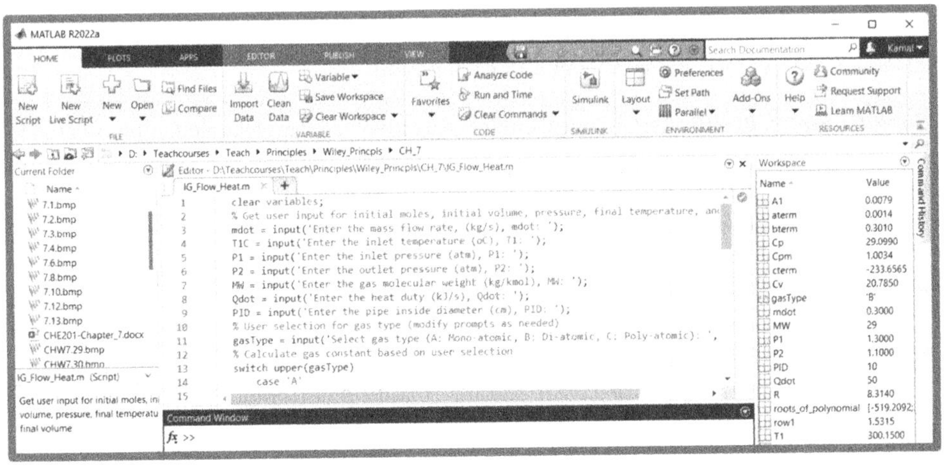

Figure 1.1 The main GUI (window) of the MATLAB platform.

1) **Quick Access Toolbar**: The Quick Access Toolbar in MATLAB is a customizable feature designed to provide users with convenient access to commonly used commands and actions. Located at the top-right corner of the MATLAB window (shown as a thin ribbon or strip in Figure 1.1), this toolbar allows users to streamline their workflow by placing frequently used tools within easy reach. Here is a description of how users can utilize and customize the Quick Access Toolbar:
   a) **Default Commands**: The Quick Access Toolbar comes preloaded with a set of default commands, including "Undo," "Redo," "Save," and "Open". These commands are commonly used in various tasks and can be accessed with a single click.
   b) **Customization**: Users can personalize the Quick Access Toolbar by adding their preferred commands or actions. To customize the toolbar, users can click on the dropdown arrow at the right end of the toolbar, just left to "Search Documentation" Edit box, and select "Customize Toolbar..." sub-menu. This opens "Preferences" dialog box where users can choose from a list of available commands or search for specific ones to add to the toolbar. Alternatively, right-clicking anywhere within the Quick Access Toolbar and selecting "Customize..." the sub-menu will open up the same "Preferences" dialog box.
   c) **Selecting/Deselecting Commands**: In the "Preferences" dialog box, users can browse through categories or use the search bar to find commands they want to add. Once they have located the desired command, they can select it and click the "Apply" button to include it in the Quick Access Toolbar. Users can also rearrange the order of commands by clicking the "Left" or "Right" arrows in the dialog box. For example, the user can select from the left pane "Toolbars" sub-item and further select/deselect "Clear Workspace" radio button followed by clicking on "Apply" button to confirm the selection/deselection step. Finally, clicking on "OK" button will close the open dialog box.
   d) **Accessibility**: The Quick Access Toolbar remains accessible regardless of which tab or toolstrip is currently active in MATLAB, providing users with consistent access to their favorite commands throughout their workflow.

2) **Toolstrip**: From top left to top right, we have the toolstrip (or ribbon) contains: "HOME," "PLOTS," "APPS," "EDITOR," "PUBLISH," and "VIEW" tab. Under each tab, there are additional features lumped in what is called a group box. For example, "Home" tab shows "FILE," "VARIABLE," "CODE," "SIMULINK," "ENVIRONMRNT," and "RESOURCES" group box. Each of these features or sub-features serves as a shortcut link to carry out a specific task.
3) **Command Window**: This is the primary working area where one interacts with MATLAB by typing commands and expressions. Think of it as a command prompt where we can enter instructions for MATLAB to execute. Running arithmetic operations here makes the MATLAB platform like a smart calculator.
4) **Workspace**: This is the behind-the-scenes area where MATLAB stores variables and data we create during our work. We can access and manage the Workspace through separate windows. We can further click on a given Workspace variable to edit and/or view its content.
5) **Help Features**: MATLAB provides extensive documentation and tutorials accessible through menus and toolbars. These resources are invaluable for learning new functions and troubleshooting any issues we may encounter. Those help features are shown in the top right corner of the top tool bar.
6) **Browser (Explorer)**: This left panel serves as a typical file/folder explorer (or browser) being used in Microsoft Windows. By default, it shows the content of the current folder chosen by the user, or default folder set by MATLAB or by the user. The user, however, can change the current folder via using the available icons shown below the toolstrip.

In addition, at any location within those sub-windows (or panels), if the user right clicks the mouse, a context-sensitive menu will pop up where the user can further select from submenus. Finally, clicking on ***fx*** icon (or pressing Shift+F1 key combination) appearing at the top left of "Command Window" will open up a new window where the user can search for a specific function via entering the name or the first segment of the function name.

1.5 Basic MATLAB Operations

Before we go into explaining MATLAB's main features, I would like to draw the user's attention to online learning resources that MATLAB owner, MathWorks company, offers:

1) **Onramp Tutorials**: The student can learn the essentials of MATLAB, Simulink, and more through the Onramp tutorials created by MathWorks.
   These two-to-three-hour tutorials are free and provide a hands-on learning environment with automated assessments and contextual feedback. To get started, the student has to visit: https://matlabacademy.mathworks.com.
2) **MATLAB Online and Simulink Online**: MATLAB Online and Simulink Online allow the user to access the latest version of MATLAB and Simulink through his/her web browser, without any downloads or installations. To learn more about these products, the user should visit: www.mathworks.com/products/matlab-online.html www.mathworks.com/products/simulink-online.html

1.5.1 Numbers, Variables, and Data Types

Numbers: We can work with various numerical data types in MATLAB, including integers (whole numbers), floating-point numbers (decimals), and complex numbers (numbers with a real and imaginary part).

Variables: Variables act like containers that store data. You assign a name (e.g., temperature, pressure) to a variable and then assign a value (e.g., 25, 101,325) to it. Naming conventions involve using letters, numbers, and underscores, but cannot start with a number. The variable name is case-sensitive in terms of lower vs. upper case.

Data Types: MATLAB automatically assigns data types based on the value you enter. Understanding data types is crucial for performing calculations correctly. For example, adding an integer and a floating-point number results in a floating-point number.

1.5.2 Performing Basic Mathematical Operations

MATLAB supports all basic mathematical operations: Addition (+), subtraction (−), multiplication (*), and division (/). We can perform these operations directly on numbers or between variables. MATLAB also offers advanced mathematical functions like exponentiation (^), square root (sqrt), and absolute value (abs).

Figure 1.2 shows sample calculations involving arithmetic operations that can be executed upon defining A, B, and C variables and then arithmetically expressing D, E, F, G, H, I, and J variables in terms of previously defined variables.

NOTE #1

The statement: % Define A, B, and C, written in the first code line of Figure 1.2, is not executed by MATLAB; it is treated as a comment statement. Thus, the "%" symbol tells MATLAB what comes after that is a comment. The **bold face** indicates an entered command.

Example 1.1 Usage of Arithmetic Operations

Figure 1.2 shows some numerical arithmetic operations that can be carried out in Command window.

```
clear variables; % clears Workspace variables.
A=5.0; B=7.0; C=9.0; % Define A, B, and C.
D=A+B
D = 12
E=A-B
E = -2
F=B*C
F = 63
G=F/D
G = 5.2500
H=F^G
H = 2.7960e+09
I=sqrt(H)
I = 5.2877e+04
J=abs(E)
J = 2
```

Figure 1.2 Arithmetic operations can be carried out within "Command Window" at the command prompt ">>".

Workspace

Name	Value	Size	Class	Min	Max
A	5	1x1	double	5	5
B	7	1x1	double	7	7
C	9	1x1	double	9	9
D	12	1x1	double	12	12
E	-2	1x1	double	-2	-2
F	63	1x1	double	63	63
G	5.2500	1x1	double	5.2500	5.2500
H	2.7960e+09	1x1	double	2.7960e+09	2.7960e+09
I	5.2877e+04	1x1	double	5.2877e+04	5.2877e+04
J	2	1x1	double	2	2

Figure 1.3 Workspace variables are created by executing commands listed in Figure 1.2.

Figure 1.3 shows Workspace variables created as a result of executing the commands listed in Figure 1.2. In addition to variable names, it also shows some properties, like size, class type, minimum, and maximum value.

1.6 Built-in Mathematical Functions

MATLAB provides a rich library of built-in mathematical functions for various calculations. These functions cover trigonometry (sin, cos, tan), logarithms (log, log10), exponentials (exp), and many more. You can use these functions directly in your calculations or within expressions. As pointed out earlier, a MATLAB built-in function can be directly searched via ***fx*** icon, shown at the top left of "Command Window," or via search in offline, help documentations. In addition, the user may define his/her own function via MATLAB's scripting language using what is called m-file which must hold the name of the function itself so that it can be recalled by name from within any other written code.

MATLAB has a vast library of built-in mathematical functions that streamline various calculations. Here are some commonly used functions categorized by their purpose:

1) **Trigonometric Functions**:
   - **sin(*x*)**: Calculates the sine of angle *x* (in radians by default).
   - **sind(*x*)**: Calculates the sine of angle *x* (in degrees).
   - **cos(*x*)**: Calculates the cosine of angle *x* (in radians by default).
   - **cosd(*x*)**: Calculates the cosine of angle *x* (in degrees).
   - **tan(*x*)**: Calculates the tangent of angle *x* (in radians by default).
   - **tand(*x*)**: Calculates the tangent of angle *x* (in degrees).
   - **cot(*x*)**: Calculates the cotangent of angle *x* (in radians by default).
   - **cotd(*x*)**: Calculates the cotangent of angle *x* (in degrees).
   - **asin(*x*)**: Calculates the arcsine (inverse sine) of *x* in radians.
   - **acos(*x*)**: Calculates the arccosine (inverse cosine) of *x* in radians.

- **atan(x)**: Calculates the arctangent (inverse tangent) of x in radians.
- **atan2(y, x)**: Calculates the four-quadrant inverse tangent of y/x.

2) **Logarithmic Functions**:
   - **log(x)**: Calculates the natural logarithm (base-e) of x.
   - **log10(x)**: Calculates the base-10 logarithm of x.
   - **exp(x)**: Calculates the exponential of x (e raised to the power of x).
3) **Other Mathematical Functions**:
   - **sqrt(x)**: Calculates the square root of x.
   - **abs(x)**: Calculates the absolute value of x (magnitude).
   - **round(x)**: Rounds x to the nearest integer.
   - **ceil(x)**: Rounds x to the smallest nearest integer greater than or equal to x.
   - **floor(x)**: Rounds x to the largest nearest integer less than or equal to x.
   - **fix(x)**: Rounds x toward zero.
   - **power(x, y)**: Calculates x raised to the power of y (x^y).
   - **factorial(n)**: Calculates the factorial of a non-negative integer n ($n!$).

Example 1.2 Power Function

Calculate $A = 3.5^{4.5}$

```
A1 = 3.5^4.5 %first method
A1 = 280.7412
A2 = power(3.5,4.5) %second method
A2 = 280.7412
```

Example 1.3 Conversion Between Complex and Polar Coordinates

Convert the complex number $(3 + 4i)$ into polar coordinates.

```
clear variables; % clears Workspace variables.
z = 3 + 4i;
r = abs(z)
r = 5
theta1 = atan2(imag(z),real(z)) %first method
theta1 = 0.9273
```

The radius r and the angle theta are the polar coordinate representation of $3 + 4i$. Alternatively, we can use the **angle** function to calculate theta.

```
theta2 = angle(z) %second method
theta2 = 0.9273
```

We can convert r and theta back into the original complex number.

```
z = r*exp(i*theta1)
z = 3.0000 + 4.0000i
```

1.7 Creating and Manipulating Vectors

Vectors are one-dimensional arrays of elements. We can create them by enclosing a list of numbers or variables within square brackets []. Vectors are powerful for storing and manipulating sequences of data. Vectors are fundamental building blocks in MATLAB. They represent one-dimensional arrays of elements, allowing us to store and manipulate sequences of data efficiently.

1.7.1 Creating Row Vectors

There are several ways to create row vectors in MATLAB:

1) **Enclosing a List of Numbers or Variables Within Square Brackets** []: This is the most straightforward way.

```
temperature = [25 30 37.5] % Creates a vector with three temperature values
```

2) **Using the Colon Operator ":"**: We can create a sequence of numbers with a specific increment.

```
time = 0:0.1:1 %Creates a row vector with time value from 0 to 1 with increments
of 0.1
```

3) **Using Built-in Functions**: MATLAB provides functions like **linspace** and **logspace** for creating vectors with evenly spaced values in linear or logarithmic scale, respectively.

```
pressure = linspace(101325, 200000, 5) % Creates a row vector with evenly spaced
5 pressure values between 101325 and 200000
```

1.7.2 Creating Column Vectors

Creating column vectors in MATLAB is a fundamental task, and there are several ways to accomplish it.

1) **Using Square Brackets []**: We can create a column vector by enclosing the elements within square brackets [], separating each element by a semicolon (;) to denote a new row. Since we are creating a column vector, each element will be placed in its own row.

```
column_vector1 = [1; 2; 3; 4; 5] %Creates a column vector of integers
```

2) **Using the Colon Operator (:)**: The colon operator allows one to create a column vector with a sequence of values. We specify the starting value, increment (optional), and ending value within square brackets [].

```
column_vector2 = (1:1:5)' %Creates a column vector of integers from 1 to 5 using
colon and transpose operator
```

3) **Using the Transpose Operator (′) on a Row Vector**: We can also create a row vector first and then transpose it to convert it into a column vector using the transpose operator (′).

```
row_vector = [1 2 3 4 5] % Creates first a row vector
column_vector3 = row_vector' % Then converts row into column vector
```

4) **Using the Linspace Function and Transpose Operator**: The **linspace** function generates a linearly spaced vector of specified length. Using **linspace** function, we can create a row vector and be converted into a column vector via transpose operator.

```
column_vector4 = linspace(1, 10, 5)' % Creates a column vector of 5 elements
between 1 and 10
```

5) **Using the Zeros, Ones, or Rand Functions**: We can create column vectors of zeros, ones, or random numbers using the **zeros**, **ones**, or **rand** functions, respectively.

```
column_vector_zeros = zeros(5, 1) %Creates a column vector of zeros with 5
elements
column_vector_ones = ones(5, 1) %Creates a column vector of ones with 5
elements
column_vector_rand = rand(5, 1) %Creates a column vector of random numbers
with 5 elements
```

1.7.3 Accessing Elements

Individual elements within a vector can be accessed using their index within square brackets. Indexing starts from 1 (not 0), unlike some other programing languages.

```
temperature = [25 30 37.5];
bodyTemp = temperature(3) % Assigns the value at index 3 (37.5) to the variable
bodyTemp
```

In this regard, indexing can also be used to extract sub-sections of vectors. For example, we can specify a range of indices to extract multiple consecutive elements.

```
myVector = [5 12 18 3 9];
subVector = myVector(2:4) % Extracts elements starting from index 2 (12) to
index 4 (3) [12 18 3]
```

1.7.4 Vectorized Operations

One of the significant advantages of vectors is the ability to perform operations on them element-wise. This means the same operation is applied to each corresponding element between two vectors of the same size.

1) **Arithmetic Operations**: Addition, subtraction, multiplication, and division can be performed directly on vectors of equal size.

```
vector1 = [1 4 7];
vector2 = [2 5 8];
sumVector = vector1 + vector2 % Creates a new vector with element-wise
addition (3, 9, 15)
divVector=vector1./vector2 % Element-wise division. Notice the dot "." sign
before "/"
multipVector=vector1.*vector2 % Element-wise multiplication. Notice the ".".
```

2) **Scalar Operations**: A scalar value can be applied to each element of a vector.

```
celsiusTemps = [10 15 20];
fahrenheitTemps = celsiusTemps * 1.8 + 32;
disp('Temperatures in Celsius:');
disp(celsiusTemps);
disp('Temperatures in Fahrenheit:');
disp(fahrenheitTemps);
```

1.7.5 Common Vector Operations

MATLAB provides functions for various vector manipulations:

- **length(vector)**: Returns the number of elements in the vector.
- **max(vector)**: Returns the largest element in the vector.
- **min(vector)**: Returns the smallest element in the vector.
- **sort(vector)**: Sorts the elements of the vector in ascending order.
- **sort(vector,'descend')**: Sorts the elements of the vector in descending order.
- **find(condition)**: Returns the indices of elements that meet a specific condition.

1.8 Creating and Manipulating Matrices

Matrices are two-dimensional arrays arranged in rows and columns. We create them by using square brackets [] with commas separating rows and semicolons separating columns. Matrices are essential for representing data in tables, systems of equations, and other multidimensional applications.

1.8.1 Indexing in Matrices: Accessing Elements with Precision

Indexing is a fundamental concept in MATLAB for accessing and manipulating specific elements within vectors and matrices. It allows us to pinpoint and work with individual values stored in these data structures.

1) **Indexing Basics**:
Indexing uses square brackets [] to specify the position of the element we want to access. However, unlike many programming languages where indexing starts from 0, MATLAB indexing starts from 1. Unlike vectors, matrices require two indices separated by a comma within the square brackets. The first index specifies the row number, and the second index specifies the column number.

```
dataMatrix = [23 7 14; 5 1 10; 12 18 6]; % Each row represents three elements
topLeftElement = dataMatrix(1, 1) % Assigns the value at row 1, column 1 (23) to the variable topLeftElement
bottomRightElement = dataMatrix(3, 3) % Assigns the value at row 3, column 3 (6) to the variable bottomRightElement
```

2) **Extracting Subsets of Elements**:
Indexing can also be used to extract sub-sections of matrices. For example, we can specify a range of indices to extract multiple consecutive elements. We can create a new matrix by combining elements from different rows and columns using separate index lists.

```
dataMatrix = [23 7 14; 5 1 10; 12 18 6];
selectedElements = [dataMatrix(2, 1) dataMatrix(1, 3) dataMatrix(3, 2)] % Creates a new row vector with elements from row 2, col 1 (5); row 1, col 3 (14); and row 3, col 2 (18)
```

We can use colons ":" as shortcuts to represent all elements in a row or column.

```
dataMatrix = [23 7 14; 5 1 10; 12 18 6];
firstRow = dataMatrix(1, :) % Extracts all elements from the first row [23 7 14]
secondColumn = dataMatrix(:, 2) % Extracts all elements from the second column [7 1 18]
```

NOTE #2
Out-of-bounds indexing (referencing a nonexistent row or column) will result in an error.

1.8.2 Matrix Operations

MATLAB allows for addition, subtraction, multiplication, and other operations on matrices with compatible dimensions.

1) **Addition and Subtraction:**
Here is a MATLAB code to demonstrate matrix addition or subtraction, where adding or subtracting matrices with the same dimensions is straightforward:

```
clear variables; % clears Workspace variables.
A = [1 2; 3 4];
B = [5 6; 7 8];
```

```
C = A + B; % Element-wise addition
D = A - B; % Element-wise subtraction
disp(['C matrix elements are:', mat2str(C)]);
disp(['D matrix elements are:', mat2str(D)]);
```

This will display the matrices C and D, containing the sum and difference of corresponding elements in A and B.

2) **Scalar Multiplication**:

We can multiply a matrix by a scalar (a single number):

```
clear variables; % clears Workspace variables.
A = [1 2; 3 4];
scalar = 3;
E = scalar * A; % Multiply each element of A by scalar
disp(['E matrix elements are:', mat2str(E)]);
```

This will create a new matrix E where each element in A is multiplied by the scalar 3.

3) **Matrix Multiplication**:

Matrix multiplication is different from element-wise multiplication. To multiply two matrices, A (of size $m \times n$) and B (of size $n \times q$), their inner dimensions (number of columns in A and number of rows in B) must be equal. The resulting product matrix will have dimensions $m \times q$.

```
clear variables; % clears Workspace variables.
A = [1 2 3; 4 5 6]; % Matrix A is 2×3
B = [1 2; 3 4; 5 6];  % Matrix B is 3×2
C= A * B; % A and B are of compatible dimensions.
disp(['C matrix elements are:', mat2str(C)]);
D= B * A; % A and B are of compatible dimensions.
disp(['D matrix elements are:', mat2str(D)]);
```

NOTE #3

In general, $(C = A*B) \neq (D = B*A)$. In fact, in the previous demonstration C will be 2×2 matrix and D will be 3×3 matrix.

4) **Element-wise Multiplication/Division**:

To perform multiplication/division on each corresponding element of two matrices, use the element-wise multiplication/division operator:

```
clear variables; % clears Workspace variables.
A = [1 2; 3 4];
B = [5 6; 7 8];
C = A .* B; % Element-wise multiplication
```

```
D = A ./ B; % Element-wise division
disp(['C matrix elements are:', mat2str(C)]);
disp(['D matrix elements are:', mat2str(D)]);
```

5) **Other Operations**:

MATLAB offers various other matrix operations, including:

- **Transpose**: **A.′** Swaps rows and columns.
- **Inverse**: **inv(A)** Assuming A is square and nonsingular, gives the inverse of A.
- **Determinant**: **det(A)** Calculates the determinant of a square matrix.

1.9 Introduction to Scripts and Functions

Scripts and functions are essential tools for organizing one's work and making code reusable. Scripts are a collection of MATLAB commands stored in a file (∗.m file extension). Scripts allow the user to automate repetitive tasks and improve readability. On the other hand, functions are reusable blocks of code that take input arguments and return an output. Functions promote modularity and make the user's code more efficient. In addition to MATLAB built-in library functions, user-defined functions can be invoked from within another script or function ∗.m file.

Figure 1.4 shows the code of a function called **circleArea** where it accepts a numerical input as the value of the radius and it outputs the area of the circle. The ∗.m file must hold the same name as the function name itself. It can be saved anywhere within MATLAB-defined paths. It can then be recalled from the MATLAB platform while having the radius value as an input argument.

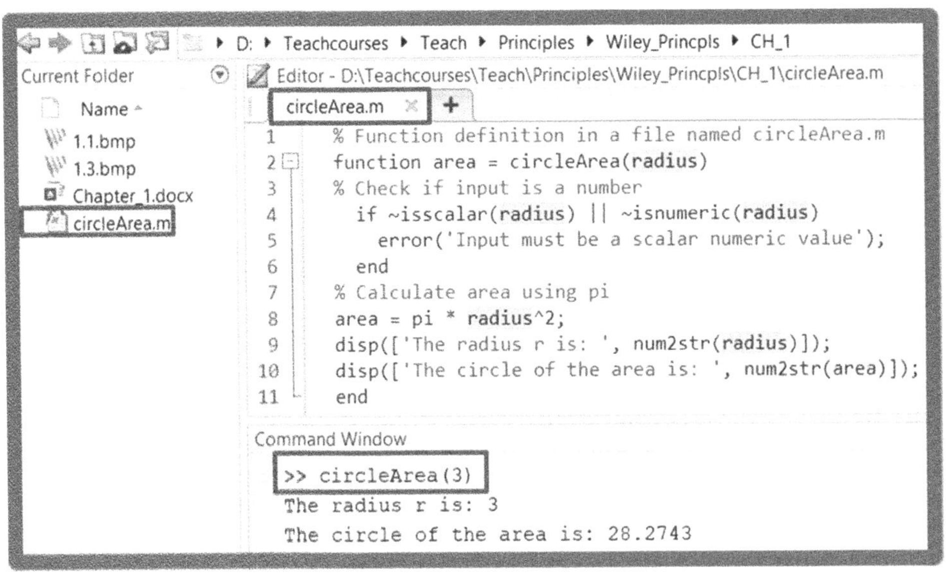

Figure 1.4 A user-defined function for circle area calculation, which can be reused.

The example in Figure 1.4 serves as a basic example. As the user delves deeper, he/she will explore more advanced features like control flow statements (if-else, loops) and error handling, making the code more robust and versatile.

1.10 Data Analysis and Visualization

MATLAB offers several ways to import data from various sources, including text files and spreadsheets. Here are some common methods:

1) **Import Tool (Interactive)**:

   This is a user-friendly graphical interface for importing data. We access it from the "**HOME**" tab | "**VARIABLE**" section | Import Data. Alternatively, use **uiimport** ('mydata.txt') in the command prompt of MATLAB.
   - Select the data file (text or spreadsheet) and preview the data.
   - The Import Tool can handle different delimiters (tabs, commas, spaces), header rows, and data types.
   - It allows the user to preview and configure how the data should be imported before creating a table or variables in MATLAB Workspace.
2) **Text File Import Functions**:

   MATLAB provides functions for reading text files depending on their format. These functions provide flexibility for handling different text file structures. Here are some codes to use in this regard:

```
% First code for reading space separated, not tabbed, data with a header.
clear variables; % clears Workspace variables
filename = 'mydata.txt'; % the data are space separated with data header.
delimiterIn = ' ';
headerlinesIn = 1;
A = importdata(filename,delimiterIn,headerlinesIn);

% Second code for comma-separated data with header
filename2 = 'mydata2.txt';% the data are comma-separated with data header.
delimiterIn = ',';
headerlinesIn = 1;
A2 = importdata(filename2,delimiterIn,headerlinesIn);

% Third code for copying data from the system clipboard. Clipboard must have data
with a header for each column.
A3 = importdata('-pastespecial'); %Loads data from the system clipboard rather
than from a file. Be sure your last copy activity is data, say, of x and y.
```

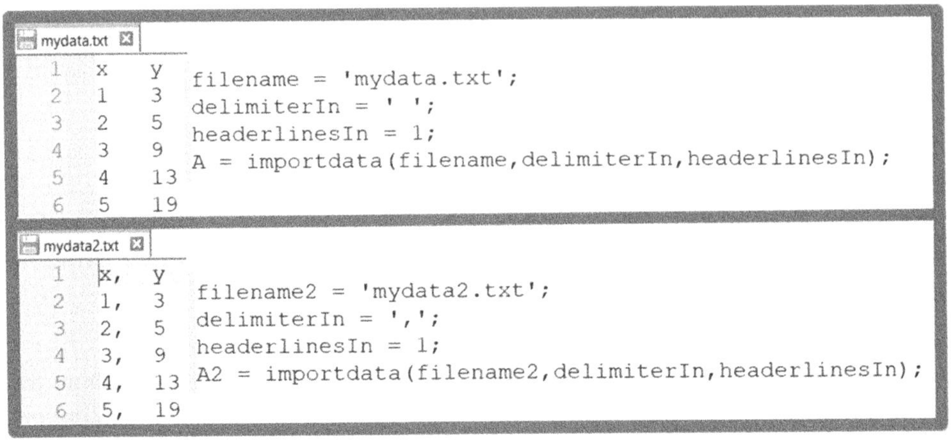

Figure 1.5 Data formats that can be read by **importdata** function. (*top*) space-separated data and (*bottom*) comma-separated data.

Figure 1.5 shows samples of data to be read by **importdata** function, which can be space-separated data (*top*) or comma separated data (*bottom*).

3) Spreadsheet Import Functions:
   MATLAB supports importing data from spreadsheets like Excel files. These functions allow the user to import data from specific sheets or ranges within the spreadsheet.

```
T2 = readtable('Temp.xlsx','Format','auto');
```

Figure 1.6 shows the original Excel sheet where it has the data to be imported (left) and the table **T2** created in the MATLAB workspace (right).

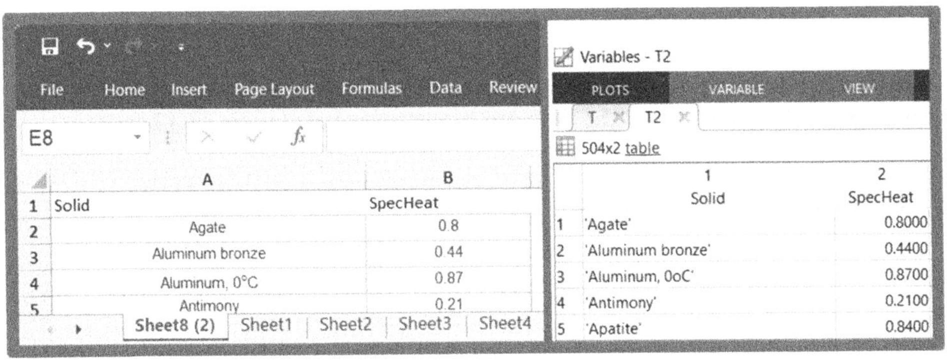

Figure 1.6 The Excel sheet containing data (*left*) and the corresponding created table is shown in the MATLAB workspace (*right*).

NOTE #4

By default, MATLAB reads the first available Excel data sheet, that is, **Sheet8(2)**, in Figure 1.6. If a sheet other than the first is sought, then the name of the sheet has to be spelled out, as in the following command:

T3 = readtable('Temp.xlsx','Sheet','Sheet2','Format','auto');

This will read the content of the Excel sheet named **"Sheet2,"** the third sheet in Figure 1.6. To reference the second row and first column data (Figure 1.6 right), that is, **'Aluminum bronze,'** we can use:

T2.Solid(2)
ans = 1×1 cell array
  {'Aluminum bronze'}

Alternatively,

T2(2,1)
ans = table
  Solid
  {'Aluminum bronze'}

1.11 Data Manipulation Techniques

MATLAB provides a rich set of tools for manipulating data in tables and matrices, including sorting, filtering, and averaging.

1) **Sorting**:
   To rearrange data rows based on specific criteria, we use the **sortrows** function:

   sortedData = sortrows(dataTable, sortVariable);

   - **dataTable**: The table or matrix we want to sort.
   - **sortVariable**: The variable (column) to use for sorting (can be a string or numeric).
   - We can sort in ascending or descending order using additional arguments to **sortrows**.

   Figure 1.7 shows MATLAB code for carrying out a row-wise data sorting based on the age in ascending (default) fashion.

```
clear variables; % clears Workspace variables.
% Original data
Name = {'Alice'; 'Bob'; 'Charlie'; 'David'};
Age = [25, 30, 22, 38]';
Score = [85, 70, 90, 65]';
% Create a table
T = table(Name, Age, Score, 'VariableNames', {'Name', 'Age', 'Score'});
disp('The Unsorted Table');
disp(T);
% Sort by age (ascending order)
sortedData = sortrows(T, 'Age');
% Display sorted table
disp('The Sorted Table by Age');
disp(sortedData);
```

Figure 1.7 MATLAB code for sorting data row-wise based on "Age" as a criterion.

The results are shown below:

```
The Unsorted Table
      Name          Age       Score
    ________       ___       _____

    {'Alice'  }     25         85
    {'Bob'    }     30         70
    {'Charlie'}     22         90
    {'David'  }     38         65
The Sorted Table by Age
      Name          Age       Score
    ________       ___       _____

    {'Charlie'}     22         90
    {'Alice'  }     25         85
    {'Bob'    }     30         70
    {'David'  }     38         65
```

2) **Filtering**:
   The goal is to get a subset of data based on certain conditions. Logical indexing with comparison operators is used to fulfill a certain constraint or condition. Figure 1.8 shows that out of the original data, only a subset is selected such that the age >30 and at the same time the score <65.

```
clear variables; % clears Workspace variables.
% Sample data
Age = [25; 35; 40; 28; 50; 60; 70; 80];
Score = [60; 75; 65; 55; 70; 70; 65; 55];
dataTable = table(Age, Score);
% Define the threshold values
ageThreshold = 30;
scoreThreshold = 65;
% Filter the data based on conditions
filteredData = dataTable(dataTable.Age > ageThreshold & dataTable.Score
< scoreThreshold, :);
% Display the filtered data
disp(filteredData);
```

Figure 1.8 Filtering out the original data based on age >30 and score<65.

The results are shown below:

```
    Age    Score
    ___    _____

    80      55
```

3) **Average, Minimum, and Maximum**:

We can calculate various measures of central tendency (mean, median, mode), minimum, and maximum, using MATLAB built-in functions:

- **mean(data)**: Calculates the average (arithmetic mean) of numeric data.
- **median(data)**: Finds the middle value when data is sorted.
- **mode(data)**: Identifies the most frequent value(s).
- We can specify dimensions (rows or columns) for averaging using additional arguments.
- We can find the minimum and maximum of data, using **min** and **max** functions, being applied row-wise, column-wise, or matrix-wise.

If we apply the previous built-in function to "**Age**" vector, we have:

>> Age_mode=mode(Age)	**Age_mode = 25**
>> Age_mean=mean(Age)	**Age_mean = 48.5000**
>> Age_median=median(Age)	**Age_median = 45**
>> Age_min=min(Age)	**Age_min= 25**
>> Age_max=max(Age)	**Age_max= 80**

1.12 Creating Basic Plots (Scatter Plot and Line Plot)

MATLAB provides powerful plotting capabilities for creating various types of plots, including scatter plots, line plots, bar charts, and others. For this book, we need to show how to create a scatter and line plot.

1) **Scatter Plot**:

Scatter plots are used to visualize the relationship between two continuous variables. Each data point is represented as a dot on the plot. Figure 1.9A shows the code to make a typical scatter plot while making some cosmetic changes in terms of font size, the line width of the circle that represents the datapoint, and legend location.

```
clear variables; % clears Workspace variables.
% Sample data
x = randn(100,1); % Random x-values
y = 2*x + randn(100,1); % Random y-values with some noise
% Create a scatter plot
scatter(x, y, 'SizeData', 50, 'LineWidth', 2);
title('Scatter Plot', 'FontSize', 16); % Set title font size to 16
xlabel('X-axis Label', 'FontSize', 14); % Set xlabel font size to 14
ylabel('Y-axis Label', 'FontSize', 14); % Set ylabel font size to 14
legend('Scattered Data', 'Location', 'southeast', 'FontSize', 14);
```

Figure 1.9A Creation of a scatter plot while making some cosmetic changes.

Alternatively, one can also make use of the get current axis (**gca**) property by getting a handle to the current graph and modify it accordingly. This is shown in Figure 1.9B.

```
clear variables; % clears Workspace variables.
% Sample data
x = randn(100,1); % Random x-values
y = 2*x + randn(100,1); % Random y-values with some noise
% Create a scatter plot
scatter(x, y, 'SizeData', 50, 'LineWidth', 2);
title('Scatter Plot');
xlabel('X-axis Label');
ylabel('Y-axis Label');
legend('Scattered Data', 'Location', 'southeast');
ax = gca; % Get handle to the current axes
% Set new font size for all axes text (including labels)
ax.FontSize = 16;
```

Figure 1.9B An alternative means of making cosmetic changes using **gca** property.

Figure 1.10 shows the generated scatter plot upon executing commands in Figure 1.9B.

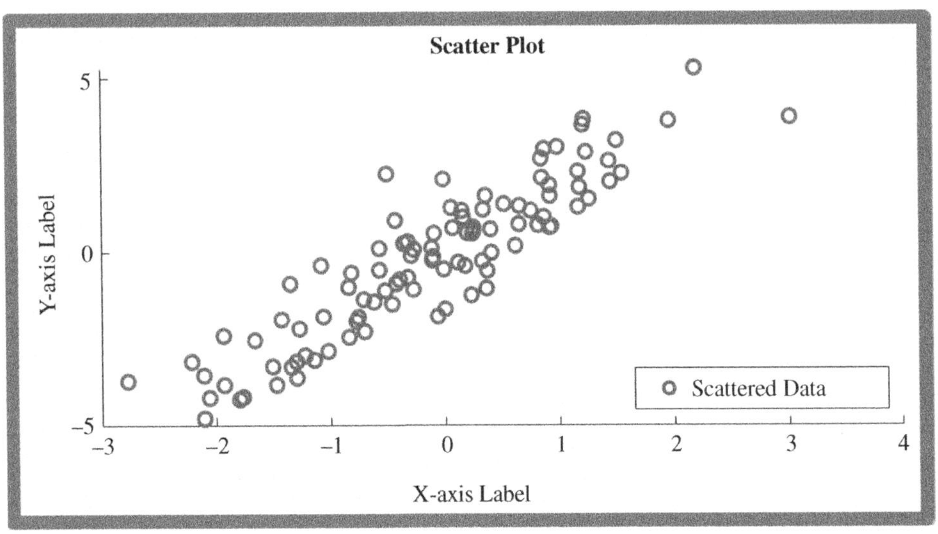

Figure 1.10 Generation of a scatter plot with cosmetic changes.

2) **Line Plot**:

Line plots are used to visualize data points connected by straight lines. They are commonly used to show trends or changes over time. Figure 1.11 shows the code for generating a line plot for sin(*x*) over a range of radian *x* while making some cosmetic changes to presenting the plot.

```
clear variables;
% Sample data
x = 0:0.1:10; % x-values from 0 to 10
y = sin(x); % Corresponding y-values
% Create a line plot
plot(x,y, 'LineWidth', 3);
title('Sin(x) Plot ');
xlabel('X-axis Label ');
ylabel('Y-axis Label ');
legend('Sin(x) ', 'Location', 'southeast');
ax = gca; % Get handle to the current axes
% Set new font size for all axes text (including labels)
ax.FontSize = 16;
```

Figure 1.11 Code for generating a line plot for sin(*x*) over a range of *x*.

Figure 1.12 shows the generated line plot of sin(*x*) upon executing the code in Figure 1.11.

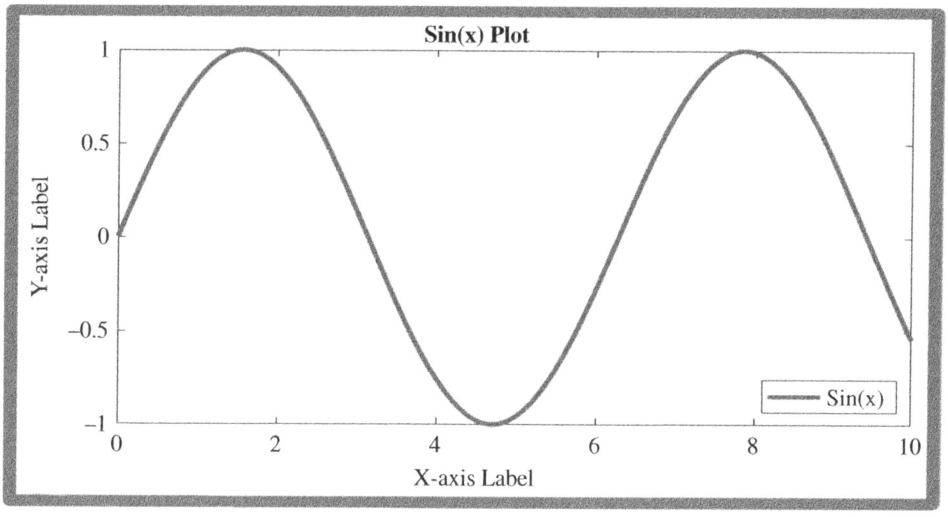

Figure 1.12 A line plot of sin(*x*) for a range of radian *x*.

Notice that two figures can be plotted in one plot; like a scatter plot for experimental data and a regressed-line (or curve) for the data themselves in one plot. In this regard, see Figure 2.19 for MATLAB code and Figure 2.20 for the two-in-one plot.

So far, I have covered MATLAB basics that will entitle the user to proceed with MATLAB codes, which are shown later in subsequent chapters. More MATLAB code implementations are to come in later chapters regarding utilization of MATLAB computational power in solving many problems in chemical engineering calculations.

End of Chapter Problems

CHW 1.1 Which of the following statements about MATLAB's built-in trigonometric functions is INCORRECT?

A sin(x) and cos(x) calculate the sine and cosine of an angle in degrees by default
B tan(x) calculates the tangent of an angle, and cot(x) is its inverse function (cotangent)
C asin(x), acos(x), and atan(x) are used to find the inverse trigonometric functions (arcsin, arccos, and arctan) and return values in radians
D atan2(y, x) calculates the arctangent considering all four quadrants based on the signs of y and x

CHW 1.2 Consider the following MATLAB code

```
clear variables;
angles = pi/2:0.1:pi;
sineValues = sin(angles);
cosineValues = cos(angles);
```

What is the expected range of values for the variables sineValues and cosineValues after running this code?

A 0 to −1 for both sineValues and cosineValues
B −1 to −1 for both sineValues and cosineValues
C Only sineValues will range from −1 to 1, while cosineValues will range from 0 to −1.
D Only cosineValues will range from −1 to 1, while sineValues will range from 0 to −1

CHW 1.3 Which of the following statements about the behavior of MATLAB's trigonometric functions is true?

A All trigonometric functions (sin, cos, tan, etc.) require angles to be provided in radians
B Trigonometric functions can directly operate on complex numbers as inputs
C atan2(y, x) will always return a value between 0 and pi/2 (0° and 90°)
D The cot(x) function is not available in MATLAB

CHW 1.4 Which of the following statements about MATLAB's logarithmic functions is true?

A log(x) and log10(x) both work for negative values of x
B exp(x) may returns a negative value, if the input x is negative
C log10(x) is equivalent to log(x)/log(10)
D All logarithmic functions in MATLAB require their arguments to be greater than 1

CHW 1.5 The factorial(n) function calculates the factorial of a non-negative integer n. What is the result of the following MATLAB code: result = factorial(101);

- **A** The result will be an error, as factorials are not defined for numbers greater than 100
- **B** The result will be a very large positive number
- **C** The result will be a decimal value between 0 and 1
- **D** The result will depend on the specific data type of n in the workspace

CHW 1.6 What is the output of the following MATLAB code snippet?

```
value = -3.14;
absoluteValue = abs(value);
```

- **A** The output, absoluteValue, will be a negative number (−3.14)
- **B** The output will be an error, as abs only works with positive numbers
- **C** The output will be −3
- **D** The output, absoluteValue, will be the positive version of the input value (3.14)

CHW 1.7 What is the result of the following MATLAB expression:

```
result = power(2, 3)
```

- **A** The result will be 8
- **B** The result will be 6
- **C** The result will be 9
- **D** The result will be 5

CHW 1.8 Which of the following expressions will give a value of 5?

- **A** fix4_49=fix(4.49)
- **B** ceil4_49=ceil(4.49)
- **C** round4_49=round(4.49)
- **D** floor4_49=floor(4.49)

CHW 1.9 Creating a Temperature Vector

You need to create a row vector in MATLAB to store the daily high temperatures for a week. The temperatures range from 22°C on Monday to 28°C on Sunday

- **A** Using square brackets, create a row vector named dailyHighs to represent these temperatures.
- **B** How would you modify the code using the colon operator if you only knew the starting temperature (22°C) and the increment was 2°C?

CHW 1.10 Creating a Time Vector for Sampling

You design a data acquisition system that needs to sample data at a rate of 100 Hz for a duration of two seconds. You need a row vector named timeStamps to store the time points at which the samples are taken. Using the linspace function, create the timeStamps vector with 201 elements (one for each sample point) starting from 0 seconds and ending at 2.00 seconds.

CHW 1.11 Random Noise Generation

You simulate a noisy signal and need a row vector named noise with 100 random numbers between −1 and 1.

A Using the rand function, create the noise vector with 100 elements.

B How would you modify the code to generate 100 random numbers between 0 and 5 instead?

CHW 1.12 Consider the following vectors:

vector1 = [2 5 3]; vector2 = [4 1 7];

What is the result of the following operation: differenceVector = vector1 − vector2;

A differenceVector = [2 −4 4]

B differenceVector = [2 4 −4]

C differenceVector = [−2 4 −4]

D differenceVector = [−2 −4 4]

CHW 1.13 The following code snippet converts temperatures from °F to °C:

fahrenheitTemps = [32 50 77];

Which of the following statements is true about finding the corresponding T in °C?

A celsiusTemps = fahrenheitTemps – 32.0/1.8;

B celsiusTemps = fahrenheitTemps + 32.0/1.8;

C celsiusTemps = (fahrenheitTemps – 32.0)/1.8;

D celsiusTemps = (fahrenheitTemps + 32.0)/1.8;

CHW 1.14 Which of the following functions is used to find the element in a vector with the highest absolute value?

A length(vector)

B max(vector)

C min(vector)

D sort(vector)

CHW 1.15 Logical Indexing:

Consider a vector data = [1 −2 5 3 0]. How can one select the elements from this vector that are greater than 0?

A PositiveData = (data)>0

B PositiveData = positive(data)

C PositiveData = data(data > 0)

D PositiveData = sort(data)

CHW 1.16 Given the following matrix:

A = [4 8 12; 16 20 24; 28 32 36]

1) Using indexing, find the element in the second row and third column. Call it Element.

2) Using indexing, extract the second column of the matrix A. Call it Column2.

3) Using indexing, create a new matrix B by selecting elements from the first two rows and columns of matrix A.

CHW 1.17 Matrix Addition and Subtraction:

Consider the following matrices:

matrix1 = [1 4; 2 5]; matrix2 = [3 2; 1 0];

What is the result of the following operation: differenceMatrix = matrix1–matrix2;

A differenceMatrix = [−2 2; 1 5];
B differenceMatrix = [2 −2; −1 −5];
C differenceMatrix = [1 5; −2 2];
D differenceMatrix = [−1 −5; 2 −2];

CHW 1.18 Scalar Multiplication

The following code snippet scales a matrix by a value:

```
matrix = [2 -1; 4 3]; scaledMatrix = 2 * matrix;
```

What is the value of the element at row 2, column 1 of the scaledMatrix?

A 2
B 4
C 6
D 8

CHW 1.19 Matrix Multiplication:

Which of the following statements is true about matrix multiplication in MATLAB?

A Matrix multiplication is always commutative ($A*B = B*A$)
B The resulting product matrix always has the same dimensions as the input matrices
C To multiply two matrices, their corresponding inner dimensions (number of rows and columns) must be equal
D Matrix multiplication is element-wise multiplication between corresponding elements

CHW 1.20 4. Matrix Multiplication Order:

Consider the following matrices:

```
matrixA = [1 2]; matrixB = [1; 2];
```

What is the result of matrixA * matrixB compared to matrixB * matrixA?

A Both multiplications will result in the same matrix dimension
B matrixA * matrixB will result in an error, while matrixB * matrixA will succeed
C matrixB * matrixA will result in an error, while matrixA * matrixB will succeed
D Both multiplications will result in different matrix dimensions with non-zero elements

CHW 1.21 Triangular Area Calculation

Create an m-file with the function triangleArea defined in a file named triangleArea.m. This function accepts two numerical inputs, representing the base and height of a triangle, and outputs the area of the triangle. Test the code calling from the MATLAB platform by providing values for the base and height as input arguments.

CHW 1.22 Cylindrical Volume Calculation

Create an m-file with the function cylinderVolume defined in a file named cylinderVolume.m. This function accepts two numerical inputs, representing the radius and height of the cylinder, and outputs the volume of the cylinder.

Test the code calling from the MATLAB platform by providing values for the radius and height as input arguments.

CHW 1.23 Cubic Volume Calculation

Create an m-file with the function volumeCube defined in a file named volumeCube.m. This function accepts a single numerical input representing the length of the side of a cube and outputs the volume of the cube.

Test the code calling from the MATLAB platform by providing values for the length of the side of the cube as an input argument.

CHW 1.24 Data Sorting

Consider the following three matrices, representing the name of the person, his height (cm), and weight (kg_f).

```
Name = {'Alice'; 'Bob'; 'Charlie'; 'David'; 'Emil'; 'Ford'};
Height = [160, 175, 168, 180, 170, 178]';
Weight = [55, 70, 65, 90, 79, 88]';
```

Sort the given data by the variable: 'Height'. Display the original and sorted data.

CHW 1.25 Scatter Plot

Given the following *x* and *y* data, make a scatter plot. Also, make some cosmetic changes in terms of font size, the line width of the circle that represents the datapoint, and legend location.

```
x = linspace(0, 10, 100);  % Creating evenly spaced x-values
y = sin(x) + randn(size(x));  % Creating y-values with noise
```

CHW 1.26 Line Plot

Given the following *x* and *y* data, create a line plot. Also, make some cosmetic changes in terms of font size, line width, and legend location.

```
x = linspace(-5, 5, 100);  % Creating evenly spaced x-values
y = x.^2;  % Creating y-values
```

2

Measurement, Analysis, and Presentation of Engineering Data

2.1 Dimensions and Units

Quantities that are quantified or counted possess both a numerical value and a unit. This principle is highly relevant in engineering computations and is integral to everyday shopping activities, necessitating the inclusion of both the numerical value and its corresponding unit for every quantity expressed in an equation. Examples are:

5 m, 1/3 seconds, 1.7 kg apple, 11 gold rings, 17.5 mm, 5 bottles of orange juice, and 3 Kb.

A dimension refers to a measurable attribute such as length, time, mass, or temperature. It can also result from multiplying or dividing other dimensions, such as length over time (velocity), length squared (area), length cubed (volume), or mass divided by length cubed (density). Measurable units, distinct from countable units, represent specific values of dimensions established by convention, tradition, or regulation. Examples include grams for mass, seconds for time, and centimeters or feet for length.

Units can be treated like algebraic variables when quantities are added, subtracted, multiplied, or divided. *The numerical values of two quantities may be added or subtracted only if the units are the same.*

$$7\ \text{cm} - 3\ \text{cm} = 4\ \text{cm} \qquad (7x - 3x = 4x) \tag{2.1}$$

but

$$7\ \text{m} - 3\ \text{cm} = ? \qquad (7x - 3y = ?) \tag{2.2}$$

On the other hand, *numerical values and their corresponding units may always be combined by multiplication or division.* Here are some examples:

To my dear readers,
When copying MATLAB code or text, please right-click and attempt both the "Copy with Formatting" and "Copy" options from the context menu. This approach is effective within Adobe Acrobat. If neither option works, kindly carefully review any MATLAB error messages and compare the pasted text to the original to identify and correct any discrepancies. Alternatively, a soft copy of the MATLAB code is available in the Wiley book companion site of this book for your convenience.

MATLAB-based Computations of Chemical Engineering Principles, First Edition. Kamal I. M. Al-Malah.

Companion website: www.wiley.com/go/al-malah/matlabbasedcomputations

Example 2.1
A man walks a distance of 12 m for six seconds. What is his average speed?

$$\text{Average speed} = \frac{\text{Distance travelled}}{\text{Elapsing time}} = \frac{12\ \text{m}}{6\ \text{s}} = 2\ \frac{\text{m}}{\text{s}} \tag{2.3}$$

So, velocity has dimensions of length over time.

Example 2.2
A horizontal force with a magnitude of 12 N is applied to an object. If the object horizontally moves 2 m, what will be the work done on the object by this external force?

$$\text{Work} = \overrightarrow{\text{force}}.\overrightarrow{\text{distance}} = 12\ \text{N} \times 2\ \text{m} = 24\ \text{N.m} = 30\ \text{J} \tag{2.4}$$

So, work has dimensions of newton times meter, which amounts to Joule.

Example 2.3
A student travels with his car from Hail to Braidah with an average speed of 120 km/h, how long will it take him to reach Braidah, given that the distance is 265 km?

$$\text{Speed} = \frac{\text{Distance}}{\text{Time}}, \quad \text{Time} = \frac{\text{Distance}}{\text{Speed}}, \quad \text{Distance} = \text{Speed} \times \text{Time} \tag{2.5}$$

$$\text{Time} = \frac{\text{Distance}}{\text{Speed}} = \frac{265\ \text{km}}{120\ \frac{\text{km}}{\text{h}}} = 2.21\ \text{h} = 2\ \text{h and } 12.6\ \text{min.}$$

Example 2.4
A man has a mass of 70 kg_m, what is his weight in *Newton*? What is his weight in kg_f?

$$\text{Weight} = \text{Mass} \times \left(\frac{g}{g_c}\right) = 70\ \text{kg}_m \times \left(\frac{9.8\ \text{m/s}^2}{1\ \text{kg}_m \cdot \text{m/s}^2/1\text{N}}\right) = 686\ \text{N} \tag{2.6}$$

$$\text{Weight} = \text{Mass} \times \left(\frac{g}{g_c}\right) = 70\ \text{kg}_m \times \left(\frac{9.8\ \text{m/s}^2}{9.8\ \text{kg}_m \cdot \text{m/s}^2/\text{kg}_f}\right) = 70\ \text{kg}_f \tag{2.7}$$

Notice that: 1 kg_f = 9.8 N and both represent units of force or weight.

2.2 Units Conversion

A quantity that has been measured can be represented using units that match its specific dimension. For instance, a given velocity could be presented in units such as feet per second (ft/s), miles per hour (miles/h), kilometers per year (km/yr), centimeters per second (cm/s), or any other combination of a length unit and a time unit. The numerical value assigned to the velocity naturally varies depending on the units used for measurement. The equivalence between two expressions of the same quantity may be defined in terms of a ratio:

$$\left(\frac{1\ \text{cm}}{10\ \text{mm}}\right), \left(\frac{1\ \text{m}}{100\ \text{cm}}\right), \left(\frac{100\ \text{cm}}{1\ \text{m}}\right), \left(\frac{1\ \text{km}}{1000\ \text{m}}\right), \left(\frac{1\ \text{m}}{1000\ \text{mm}}\right), \left(\frac{1\ \text{ft}}{30.48\ \text{cm}}\right), \left(\frac{1\ \text{ft}}{0.3048\ \text{m}}\right) \tag{2.8}$$

Ratios in the form of Eq. (2.8) are known as conversion factors.
Here are some examples.

Example 2.5
A man has a mass of 70 kg_m, what is his mass expressed in lb_m?

DATA: $\mathbf{lb_m = 0.4536\, kg_m}$. This means that the conversion factor is $\left(\frac{1\ \text{lbm}}{0.4536\ \text{kgm}}\right)$ or $\left(\frac{0.4536\ \text{kgm}}{1\ \text{lbm}}\right)$

$$\text{Mass of the man} = 70\ \text{kg}_\text{m} \times \left(\frac{1\ \text{lb}_\text{m}}{0.4536\ \text{kg}_\text{m}}\right) = 154.32\ \text{lb}_\text{m} \tag{2.9}$$

A man has a mass of 160 lb_m, what is his mass in kg_m?

$$\text{Mass of the man} = 160\ \text{lb}_\text{m} \times \left(\frac{0.4536\ \text{kg}_\text{m}}{1\ \text{lb}_\text{m}}\right) = 72.6\ \text{kg}_\text{m} \tag{2.10}$$

Example 2.6
A man has 1000 Saudi Arab Riyals (SAR), how much does he have in Jordanian Dinars (JD)?

DATA: 1 SAR = 0.186 JD. This means that the conversion factor is $\left(\frac{1\ \text{SAR}}{0.186\ \text{JD}}\right)$ or $\left(\frac{0.186\ \text{JD}}{1\ \text{SAR}}\right)$

$$\text{The man has : } 1000\ \text{SAR} \times \left(\frac{0.186\ \text{JD}}{1\ \text{SAR}}\right) = 186\ \text{JD} \tag{2.11}$$

If the man has 500 JD, what will be his wealth in SAR?

$$\text{His wealth} = 500\ \text{JD} \times \left(\frac{1\ \text{SAR}}{0.186\ \text{JD}}\right) = 2688.2\ \text{SAR}$$

Exercise 2.1
If a man is 170 cm, what will be his height in feet (ft)? Data: 1 ft = 30.48 cm

Exercise 2.2
If a plane travels at an altitude (elevation) of 33,000 ft above the ground, what will be its altitude, expressed in km?

Exercise 2.3
If a train travels at a speed of 100 km/h, what will be its speed, expressed in m/s?

2.3 Systems of Units

A system of units has the following components:

1) *Base units* for mass, length, time, temperature, electrical current, and light intensity.

2) *Multiple units,* considered as multiples or fractions of base units, encompass measurements like minutes, hours, milliseconds, kilograms, milligrams, kilometers, and centimeters, all defined in relation to the base unit of time, mass, or length. These derived units are established for ease of use rather than out of necessity. For instance, it is more practical to express a duration of 1 year rather than 31,536,000 seconds.
3) *Derived units,* obtained by one of two means:
   a) By multiplying and dividing base or multiple units (cm^2, ft/min, $kg{\cdot}m/s^2$, etc.). Derived units of this type are referred to as compound units.
   b) As defined, equivalents of compound units (e.g., 1 dyne $\equiv$ (1 g cm/s^2), 1 $lb_f \equiv 32.174$ $lb_m{\cdot}ft/s^2$).

The "Systeme Internationale d'Unites," abbreviated as **SI**, has been widely adopted within the scientific and engineering realms. This course will not cover two fundamental SI base units: The ampere, which measures electrical current, and the candela, which quantifies luminous intensity. Later in the course, we will delve into the discussion of the third base unit, the Kelvin, used to measure temperature. The remaining base units covered in this course include the meter (m) for length, the kilogram (kg) for mass, and the second (s) for time.

Prefixes are used in **SI** to indicate powers of 10. The most common of these prefixes and their abbreviations are giga (**G**) for 10^9, mega (**M**) for 10^6 (1 megawatt = 1 MW = 10^6 watts), *kilo* (k) for 10^3, *centi* (**c**) for 10^{-2}, *milli* (**m**) for 10^{-3}, *micro* (**μ**) for 10^{-6}, and *nano* (**n**) for 10^{-9}.

The **cgs** system is almost identical to **SI**, the principal difference being that *grams* (**g**) and *centimeters* (**cm**) are used instead of kilograms and meters as the base units of mass and length, respectively.

The base units of the American Engineering system are the foot (ft) for length, the pound-mass (lb_m) for mass, and the second (s) for time. This system has two principal difficulties. The first is the occurrence of conversion factors (such as 1 ft/12 in), which, unlike those in the metric systems, are not multiples of 10; the second, which has to do with the unit of force, is discussed in the next section. Appendix A contains several tables that help the student carry out interconversion between the Metric (SI) and Imperial (American Engineering) systems. You may refer to them upon attempting to solve end-of-chapter problems for this or for subsequent chapters.

The principal units of the **SI, cgs, and Imperial (American Engineering)** system are shown in Table 2.1.

Table 2.1 SI, cgs, and Imperial (American Engineering) system of units.

Quantity	SI Unit	cgs Unit	American Engineering
Length	Meter (m)	Centimeter (cm)	Foot (ft), Inch (in)
Mass	Kilogram (kg)	Gram (g)	Pound (lb), Ounce (oz)
Moles	Mole (mol)	Mole (mol)	lb-mole
Time	Second (s)	Second (s)	Second (s)
Temperature	Kelvin (K)	Kelvin (K)	Rankine (°R)
Electric current	Ampere (A)	Ampere (A)	Ampere (A)
Light intensity	Candela (cd)	Candela (cd)	Candela (cd)

Table 2.2 shows the derived units in SI, cgs, and Imperial systems, which are either compound or defined units expressed in terms of the base units.

Table 2.2 The derived units are in SI, cgs, and the Imperial system.

Quantity	SI Unit	cgs Unit	Imperial Unit
Volume	Cubic meter (m^3), L $1\ m^3 = 1000$ L	Cubic centimeter (cm^3)	US gallon (gal), Cubic inch (in^3)
Force	Newton (N)	Dyne (dyn)	Pound-force (lbf)
Pressure	Pascal (Pa)	Barye (Barad, Barrie) (Ba) = 0.1 Pa $1\ \text{Ba} = \dfrac{\text{g}}{\text{cm.s}^2}$	Pound-force per square inch (psi)
Energy (work)	Joule (J)	Erg (erg)	Foot-pound (ft-lb), British thermal unit (Btu)
Power	Watt (W = J/s)	Erg per second (erg/s)	Foot-pound per second (ft-lb/s), Horsepower (hp)

2.4 MATLAB for SI/AE Conversion

Figure 2.1 shows MATLAB code (SI_AE_Units.m) for carrying out unit interconversion between SI and AE systems for the most important physical quantities involved in chemical engineering applications.

```
clear variables;
% Define conversion factors within a structure (using explicit fields)
unitConversionFactors = struct(...
  'energy', struct('conversionFactor', 1 / 0.73756, 'SIlabel', 'Joule','AElabel','lbf-ft'), ...
  'force',  struct('conversionFactor', 4.44822, 'SIlabel', 'Newton','AElabel','lbf'), ...
  'length', struct('conversionFactor', 1/3.28084, 'SIlabel', 'Meter','AElabel','ft'), ...
  'mass', struct('conversionFactor', 1/2.20462, 'SIlabel', 'Kilogram','AElabel','lbm'), ...
  'power', struct('conversionFactor', 1/(1.34102e-3), 'SIlabel', 'Watt','AElabel','hp'), ...
  'pressure', struct('conversionFactor', 1 / 0.00014503768, 'SIlabel',
'Pascal','AElabel','psia'), ...
  'temperature', struct('conversionFactor', 1/1.8, 'SIlabel', 'Kelvin','AElabel','Rankine'), ...
  'volume', struct('conversionFactor', 1/0.035314, 'SIlabel', 'Liter','AElabel','ft^3') ...
);
% Get user input for unit type and system
disp('Select a unit type by entering the corresponding number:');
unitTypes = fieldnames(unitConversionFactors);
for i = 1:numel(unitTypes)
  disp([num2str(i) '. ' unitTypes{i}]);
end
```

Figure 2.1 MATLAB code (SI_AE_Units.m) to convert from and to SI and AE system.

```
unitTypeInput = input('Enter your choice: ');

% Validate unit type input
while ~ismember(unitTypeInput, 1:numel(unitTypes))
  unitTypeInput = input(['Invalid choice. Please enter a number between 1 and '
num2str(numel(unitTypes)) ': ']);
end

disp('Select a unit system by entering the corresponding number:');
disp('1. SI System');
disp('2. American Engineering System');
unitSystemInput = input('Enter your choice: ');
% Validate unit system input
while ~ismember(unitSystemInput, [1, 2])
  unitSystemInput = input('Invalid choice. Please enter either 1 or 2: ');
end
% Determine unit prompt based on user selection
if unitSystemInput == 1
  unitPrompt = 'Enter the value in SI units (';
  unitLabelInfo = unitConversionFactors.(unitTypes{unitTypeInput});
unitLabel = unitLabelInfo.SIlabel;  % Access the 'label' field using dot notation
% Complete unit prompt with the selected unit type and label
unitPrompt = [unitPrompt unitLabel ')'];
disp(unitPrompt);
elseif unitSystemInput == 2
  unitPrompt = 'Enter the value in AE units (';
  unitLabelInfo = unitConversionFactors.(unitTypes{unitTypeInput});
unitLabel = unitLabelInfo.AElabel;  % Access the 'label' field using dot notation
% Complete unit prompt with the selected unit type and label
unitPrompt = [unitPrompt unitLabel ')'];
disp(unitPrompt);
end
% Access unit label based on selection (always use the first element)

% Prompt for value to convert (ensure numerical input)
valueToConvert = input('Enter a numerical value to convert: ');
while ~isscalar(valueToConvert) || ~isnumeric(valueToConvert)
  valueToConvert = input('Invalid input. Enter a numerical value to convert: ');
end
% Access conversion factor directly using dot notation
conversionFactor = unitConversionFactors.(unitTypes{unitTypeInput}).conversionFactor;
% Perform conversion: Multiplication for AE to SI, Division for SI to AE
if unitSystemInput == 2
  convertedValue = valueToConvert * conversionFactor;  % Multiply for AE to SI
else
  convertedValue = valueToConvert / conversionFactor;
end
if unitSystemInput ==  1
myunit=unitLabelInfo.AElabel;
else
  myunit=unitLabelInfo.SIlabel;
end
disp(['Converted value: ', num2str(convertedValue), ' ',myunit]);
```

Figure 2.1 (Continued)

Sample calculations are shown below:

```
>> SI_AE_Units
Select a unit type by entering the corresponding number:
1. energy
2. force
3. length
4. mass
5. power
6. pressure
7. temperature
8. volume
Enter your choice: 5
Select a unit system by entering the corresponding number:
1. SI System
2. American Engineering System
Enter your choice: 1
Enter the value in SI units (Watt)
Enter a numerical value to convert: 1000
Converted value: 1.341 hp
```

2.5 Force and Weight

According to *Newton's second law of motion*, force is proportional to the product of mass and acceleration (length/time^2). *Natural force units* are, therefore, kg·m/s^2 (SI), g·cm/s^2 (cgs), and lb_m·ft/s^2 (American Engineering). To avoid having to carry around these complex units in all calculations involving forces, *derived force units* have been defined in each system. In the metric systems, the derived force units (the Newton in SI, the Dyne in the cgs system) are defined to equal the natural units:

$$1\,\mathrm{N} \equiv \left(\frac{\mathrm{kg_m} \cdot \mathrm{m}}{\mathrm{s}^2}\right) \tag{2.12}$$

$$1\,\mathrm{kg_f} \equiv 9.8\left(\frac{\mathrm{kg_m} \cdot \mathrm{m}}{\mathrm{s}^2}\right) = 9.8\,\mathrm{N} \tag{2.13}$$

$$1\,\mathrm{Dyne} \equiv \left(\frac{\mathrm{g_m} \cdot \mathrm{cm}}{\mathrm{s}^2}\right) \tag{2.14}$$

$$1\,\mathrm{lb_f} \equiv 32.17\left(\frac{\mathrm{lb_m} \cdot \mathrm{ft}}{\mathrm{s}^2}\right) \tag{2.15}$$

Equations (2.12) through (2.15) define conversion factors between natural and derived force units.

For example, the force needed to accelerate a mass of 5.00 kg at a rate of 9.00 m/s^2 is:

$$F = \mathrm{m} \cdot \mathrm{a} = 5\,\mathrm{kg_m} \times 9\frac{\mathrm{m}}{\mathrm{s}^2} \times \left(\frac{1\mathrm{N}}{\left(\frac{\mathrm{kg_m} \cdot \mathrm{m}}{\mathrm{s}^2}\right)}\right) = 45\,\mathrm{N}$$

$$g = 9.8066\frac{\mathrm{m}}{\mathrm{s}^2}$$
$$g = 980.66\frac{\mathrm{cm}}{\mathrm{s}^2} \tag{2.16}$$
$$g = 32.174\frac{\mathrm{ft}}{\mathrm{s}^2}$$

Exercise 2.4
Calculate the **force** needed, expressed in $\mathrm{lb_f}$, to accelerate a mass of 10 $\mathrm{lb_m}$ at a rate of 30 ft/s^2.

Exercise 2.5
Calculate the **force** needed, expressed in dyne, to accelerate a mass of 10 $\mathrm{g_m}$ at a rate of 30 cm/s^2.

Exercise 2.6
Calculate the **force** needed, expressed in N, to accelerate a mass of 10 $\mathrm{kg_m}$ at a rate of 30 m/s^2.

Exercise 2.7
Calculate the **force** needed, expressed in $\mathrm{kg_f}$, to accelerate a mass of 10 $\mathrm{kg_m}$ at a rate of 30 m/s^2.

Exercise 2.8

1) What is a force of 3 kg·m/s^2 equivalent to in newtons? What is a force of 3 lbm·ft/s^2 equivalent to in lbf?
2) If the acceleration of gravity at a point is g = 9.8 m/s^2 and an object is resting on the ground at this point, is this object accelerating at a rate of 9.8 m/s^2?
3) Suppose an object weighs 9.8 N at sea level. What is its mass? Would its mass be greater, less, or the same on the moon? How about its weight?
4) Suppose an object weighs 2 $\mathrm{lb_f}$ at sea level. What is its mass? Would its mass be greater, less, or the same at the center of the earth? How about its weight? Hint: Due to the Earth's uneven mass distribution, the gravitational force at the center is slightly weaker than on the surface. Estimates suggest a decrease of about 0.4%.

Example 2.7
Water has a density of 62.4 lbm/ft^3. How much does 2.0 ft^3 of water weigh (1) at sea level, (2) at the surface of the moon where the gravitational acceleration is 6.0 ft/s^2, and (3) at a point far away from the earth where there is no magnetic force of attraction?

The mass of water: $M = \rho \cdot V = 62.4\frac{\mathrm{lbm}}{\mathrm{ft}^3} \cdot 2.0\,\mathrm{ft}^3 = 124.8\,\mathrm{lbm}$

The Weight of Water:

1) **@sea level**: $W = M \cdot g = 124.8\,\text{lbm} \cdot g = 124.8\,\text{lbm} \cdot 32.174 \frac{\text{ft}}{\text{s}^2} \times \left(\frac{\text{lbf}}{32.17\left(\frac{\text{lb}_\text{m} \cdot \text{ft}}{\text{s}^2}\right)} \right) = 128.8\,\text{lbf}$

2) **@moon**: $W_{\text{moon}} = M \cdot g_{\text{moon}} = 124.8\ \text{Ibm} \cdot g_{\text{moon}} = 124.8\ \text{Ibm} \cdot 6.0 \frac{\text{ft}}{\text{s}^2} \times \left(\frac{\text{Ibf}}{32.17\left(\frac{\text{Ib}_\text{m} \cdot \text{ft}}{\text{s}^2}\right)} \right)$

$= 23.3\ \text{lbf}$

3) **@no-magentic attraction point**: $W = M \cdot g = 124.8\ \text{lbm} \cdot g = 124.8\ \text{lbm} \cdot 0.0 \frac{\text{ft}}{\text{s}^2} \times$
$\left(\frac{\text{lbf}}{32.17\left(\frac{\text{lb}_\text{m} \cdot \text{ft}}{\text{s}^2}\right)} \right) = 0.0\ \text{lbf}$

Exercise 2.9

Calculate the **density** $\left(\frac{\text{lbm}}{\text{ft}^3}\right)$ of a *wooden material* given that a cube of 1000 cm^3 has a mass of 800 g.

2.6 Accuracy and Precision: A Closer Look

Understanding the difference between accuracy and precision is crucial in various fields, from science and engineering to everyday life. Let us delve deeper into each term and explore their relationships through different measurement scenarios:

2.6.1 Accuracy

Accuracy reflects how close a measurement is to the true value, while precision measures its consistency or repeatability. While precise measurements (repeating the same value) are necessary for good accuracy, a systematic error (consistent bias) can still lead to an inaccurate result, even if the measurements are precise. Think of accuracy as shooting the arrow on a target. A highly accurate measurement lands closer to the center than an inaccurate one.

Often, accuracy is expressed as either absolute error or percent relative error.

$$\text{Absolute Error}\,(AE) = X_i - X_{\text{True}} \tag{2.17}$$

$$\text{Percent Relative Error (PRE\%)} = \frac{|X_i - X_{\text{True}}|}{X_{\text{True}}} \times 100\% \tag{2.18}$$

2.6.2 Precision

Precision signifies how consistent or reproducible repeated measurements are, regardless of their true value. Imagine firing multiple arrows at the target. Even if they miss the bullseye (center of the target), precise arrows cluster tightly together, while imprecise ones are scattered widely. Thus, precision describes the reproducibility of measurements (closeness of results to others) obtained in exactly the same way and under the same experimental conditions.

Three terms are widely used to describe precision: Standard deviation (S), variance, and coefficient of variation. All are functions of deviation from the mean. See the upcoming section on such sample statistics.

NOTE #1
Being precise is like always being punctual on time, but being accurate is actually arriving at the right place. You can be consistently early (precise) but still miss the meeting (inaccurate) by going to the wrong place.

2.6.3 Measurements Scenarios of Accuracy and Precision

Now, Let us Analyze Different Measurements Scenarios:

1) **Low Accuracy and Low Precision**:
   The worst case! Measurements are both far from the true value (inaccurate) and scattered all over the place (imprecise). This could indicate faulty equipment, poor technique, or significant external influences. Imagine throwing rocks at a target blindfolded – they'll be both inaccurate and imprecise. Figure 2.2 top left demonstrated this case.
2) **Low Accuracy and High Precision**:
   Measurements are consistent and repeatable (precise), but they consistently deviate from the true value (inaccurate). This could be due to systematic errors like a miscalibrated measuring tool. Imagine a ruler consistently shorter than the actual inch due to a manufacturing error. Figure 2.2 top right demonstrated this case.
3) **High Accuracy and Low Precision**:
   Measurements are consistently close to the true value (accurate), but they vary slightly with each repetition (imprecise). This could be due to random errors or limitations of the measuring instrument. Figure 2.2 bottom left demonstrated this case.
4) **High Accuracy and High Precision**:
   The ideal case! Measurements consistently land near the true value (accurate) and cluster tightly together (precise). Imagine a well-calibrated scale providing consistently correct weight readings. Figure 2.2 bottom right demonstrated this case.

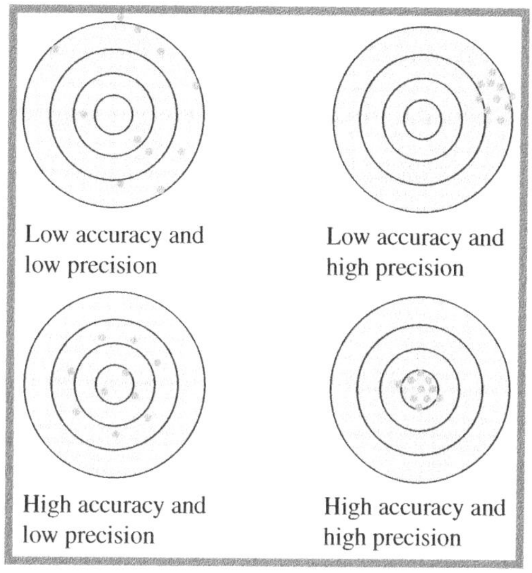

Figure 2.2 Demonstration of the four possible scenarios as far as both accuracy and precision are concerned.

2.7 Mystery of Errors: Types and Their Impact

Definition: Error is the difference between a measured value and the true value. It signifies the deviation from a perfect accuracy. When taking measurements, encountering errors is inevitable. However, understanding the different types and their effects can drastically improve our data analysis and interpretation later. Here is a breakdown of the commonly encountered error categories:

2.7.1 Random (Indeterminate) Errors

The unpredictable fluctuations arising from chance occurrences during measurement (e.g., slight instrument noise). They affect precision, impacting consistency between repeated measurements, even if the average value might be closer to the true value. Indeterminate errors lack a known cause; they are stochastic and yield both high and low results with equal likelihood. They cannot be eradicated or rectified but can be addressed through statistical methods. It might be feasible to identify specific uncontrollable variables near an instrument's performance limit, such as electronic circuit noise, drift, temperature variations, and so forth.

Imagine we have a highly sensitive analytical balance (0.1 mg sensitivity) and want to weigh a small object. We perform multiple weighing measurements, expecting identical results. However, due to inherent randomness in the system (air currents, vibrations, etc.), we might observe slight fluctuations in the final digit of the measurement (e.g., 10.23, 10.24, 10.22 mg). These differences represent stochastic errors.

2.7.1.1 Characteristics of Random Errors

- **Random in Nature**: They occur unpredictably and don't follow any specific pattern.
- **Present even with Good Instruments**: High-precision equipment doesn't eliminate them, but minimizing other types of errors (calibration issues, human error) helps isolate them.

- **Tend to Average out with Enough Measurements**: While individual measurements might vary, the average of many measurements usually gets closer to the true value.

Understanding Random Errors is Crucial Because:

- They limit the absolute precision achievable in a measurement, regardless of the instrument's sensitivity.
- They affect how we analyze and interpret data, requiring statistical methods to account for their randomness.

2.7.2 Systematic Errors (Determinate Errors)

These consistent biases stem from faulty calibration, incorrect methods, or environmental factors. They impact accuracy, causing measurements to consistently deviate from the true value, regardless of repetition. In other words, they are mistakes that can be linked to a specific cause (such as issues with a method), and they have some factors, like: All errors share uniform size, magnitude, and direction. They are reproducible and predictable. Examples are inaccuracies like an improperly calibrated instrument (balance), side reactions in a titration, or fluctuations in temperature that fall into this category. Determinate errors are categorized as methodical, operative, and instrumental based on their origin.

2.7.2.1 Systematic Errors Based on Source (Origin)

- **Instrumental Errors**: These arise from limitations or inaccuracies in the measuring instrument itself. Calibrating equipment regularly can help minimize these errors.
- **Operational (Personal) Errors**: These result from human mistakes or inconsistencies in performing the measurement. Careful technique, training, and attention to detail can reduce their impact.
- **Methodical Errors**: These stem from flaws in the chosen measurement method or procedure. Using standardized methods, running controls, and consulting experts can help overcome these challenging errors.

Systematic Errors can be constant or proportional.

2.7.2.2 Constant Systematic Errors

Constant errors remain unaffected by the size of the analyzed sample. Imagine measuring the weight of people on a scale that consistently adds 2 kg due to faulty calibration. This is a constant error. Graphing the weights would show a horizontal shift compared to the true values. Imagine a straight line representing the actual weights, and all measured weights appear 2 units higher, creating a parallel line shifted upward.

2.7.2.3 Proportional Systematic Errors

Consider measuring the concentration of a solution using a spectrometer with a proportional error of 10%. So, a true concentration of 10 mg/L would be read as 11 and 20 mg/L as 22 mg/L. Graphing these values would show a change in the slope of the line. Imagine the true line representing concentration vs. reading having a certain slope. The proportional error would "stretch" the points upwards, creating a new line with a steeper slope.

2.7.3 Gross Errors

These stand out as outliers, significantly different from other data points. They often result from carelessness, equipment malfunction, or miscalculations. Identifying and excluding them is crucial for reliable data analysis. Here are some examples of gross errors and how they stand out in different scenarios.

Case 2.1 Measuring the Height of People in a Classroom

- Most students' heights fall within a reasonable range (e.g., 150–180 cm).
- A gross error could be recording someone's height as 300 cm (much taller than anyone else). This could be due to misreading the measuring tape, accidentally recording the wrong number, or entering the data incorrectly.

Case 2.2 Monitoring the Temperature of a Chemical Reaction

- The temperature typically increases gradually over time.
- A gross error could be a sudden spike in temperature followed by a return to normal levels. This might indicate equipment malfunction, like a faulty sensor or a cable coming loose.

Case 2.3 Counting the Number of Apples in a Basket

- You count 10 apples, then 8, then 9, and so on.
- A gross error could be counting 20 apples in one basket, significantly higher than the others. This might be due to accidentally double-counting or a change in basket size from the size of other baskets.

Impact of Gross Errors:

- These outliers can significantly skew the overall results of our analysis if not identified and removed.
- Including them can lead to misinterpretations and unreliable conclusions.

Identifying and Addressing Gross Errors:

- Look for data points that are far away from the majority of other values.
- Check for inconsistencies or unrealistic values.
- Investigate the potential causes of the error (e.g., equipment malfunction, data entry mistakes).
- If confident about the error, exclude it for further analysis.

NOTE #2

- Errors are unavoidable, but understanding their types and sources empowers us to interpret data more meaningfully.
- Random errors tend to average out with many measurements, while systematic errors require correction before they affect interpretation.
- Addressing instrumental errors involves calibration, while operational errors demand improved technique and awareness. Methodical errors often require consulting established practices and experts.
- Identifying and removing gross errors ensures data integrity and avoids misleading conclusions.
- Statistics enables us to embrace conclusions with a high probability of accuracy and dismiss those with a low probability.
- Statistical methods are applicable solely to random errors; any determinate errors should be eliminated by the analyst before conducting precise measurements.

2.8 Significant Figures: Rules of Multiplication, Addition, and Rounding Off Numbers

2.8.1 Significant Figures

Process calculations often involve encountering both exceedingly large and exceedingly small numbers. A practical method for representing these numbers is through scientific notation, wherein a number is depicted as the product of a different number (typically ranging from 0.1 to 10) and a power of 10.

Example 2.8

$$120{,}000 = 1.2 \times 10^5 \left(\text{or } 0.12 \times 10^6\right)$$

$$0.000039 = 3.9 \times 10^{-5} \left(\text{or } 0.39 \times 10^{-4}\right)$$

Significant figures are the digits in a number that tell us how accurately it was measured or calculated. These include:

- All nonzero digits
- Zeros between nonzero digits
- Trailing zeros after a decimal point (if the measurement or reporting resolution warrants it)

Additional Points:

- Leading zeros are not significant. These zeros indicate the magnitude of the number, not the level of accuracy.

- Zeros used to place the decimal point are significant.
- Scientific notation doesn't change the number of significant figures.

Example 2.9
Without a Decimal Point:

- **123**: It has three significant figures (all nonzero digits)
- **500**: It has one significant figure (the first nonzero digit)
- **1000**: It has one significant figure (the first nonzero-digit, trailing zeros don't count unless they indicate a specific level of accuracy)

Example 2.10
With a Decimal Point:

- **1.234**: It has four significant figures (all nonzero digits)
- **0.005**: It has one significant figure (nonzero digits)
- **2.500**: It has four significant figures (all nonzero digits and trailing zeros indicate specific accuracy)

Example 2.11
Using scientific notation

$2500 = 2.5 \times 10^3$	It has two significant figures
$2500. = 2.500 \times 10^3$	It has four significant figures
$2500.0 = 2.5000 \times 10^3$	It has five significant figures
$25{,}050 = 2.505 \times 10^4$	It has four significant figures
$0.0355 = 3.55 \times 10^{-2}$	It has three significant figures
$0.03550 = 3.550 \times 10^{-2}$	It has four significant figures.

Significant figures are like a ruler's tick marks, each representing a level of measurement accuracy. The more tick marks, the more precisely we can measure something.

Example 2.12
Measuring a board with a ruler marked only in centimeters tells us its length to the nearest centimeter. This might be written as "10 cm" (one significant figure).

Using a ruler calibrated in millimeters allows measuring to the nearest millimeter. We could write "10.5 cm" (three significant figures).

Employing a ruler with 0.1 mm markings enables measurements to the nearest tenth of a millimeter. The length might be "10.55 cm" (four significant figures).

Remember: Each significant figure tells us how accurately we know a value. More figures imply higher precision, just like a ruler with more markings allows for more precise measurements.

Example 2.13
If we report a mass as 8.3 g (two significant figures), we indicate that the mass lies somewhere between 8.25 g and 8.35 g, whereas if we give the value as 8.30 g (three significant figures) we indicate that the mass lies between 8.295 g and 8.305 g.

2.8.2 Rule for Multiplication and Division of Quantities with Different Significant Figures

> If multiple quantities are multiplied and/or divided together, the resulting number's significant figures should match the lowest count of significant figures among any of the numbers being multiplied or divided.

Example 2.14 Result of Multiplication Having the Lowest Count of Significant Figures

$$(4.59)(5.289) = 24.27651 \Rightarrow 24.3$$

Example 2.15 Result of Division Having the Lowest Count of Significant Figures
Calculate the speed of a bike traveling 8.300 km (four significant figures) in 0.500 hours (three significant figures).

- The number of significant figures in the final answer (speed) will be limited by the measurement with fewer significant figures (0.500 hours)
- Therefore, the speed can only be reported with three significant figures
- The result could be written as (8.300 km/0.500 h) = 16.6 km/h

Remember: Combining measurements with different levels of precision is like mixing ingredients with different levels of accuracy. The final result's precision can't be better than the least precise component.

2.8.3 Rule for Addition and Subtraction of Quantities with Different Significant Figures

> Adding or subtracting measurements is like lining them up by their decimal points. The sum or difference can only be as precise as the least precise measurement. We need to align the rightmost significant digits to ensure the final answer reflects the accuracy of each measurement.

Example 2.16 Result of Addition Having the Lowest Count of Significant Figures
Adding 2.34 m (three significant figures) and 1.0 m (two significant figures):

- We align the numbers by their decimal points:

```
 2.34 m
+1.0 m
---------
 3.34 m
```

- The last significant figure in the sum (3) is in the same position (tenths place) as the last significant figure of the least precise number (1.0 m).
- Therefore, the sum can only be reported with two significant figures: 3.3 m.

Example 2.17 Result of Subtraction Having the Lowest Count of Significant Figures

Subtracting 5.678 g (four significant figures) from 10.0 g (three significant figures):

- We align the numbers by their decimal points:

  10.0 g
  −5.679 g

  4.321 g

- The last significant figure in the difference (2) is in the same position (hundredths place) as the last significant figure of the least precise number (10.0 g).
- Therefore, the difference can only be reported with three significant figures: 4.32 g.

Remember: When adding or subtracting measurements, aligning them by their decimal points ensures the final result accurately reflects the precision of each individual measurement.

2.8.4 Rounding Off Numbers Ending in 5

A rule of thumb for rounding off numbers in which the digit to be dropped is 5 is always to make the last digit of the rounded-off number even:

$$1.65 \Rightarrow 1.6 \qquad 1.15 \Rightarrow 1.2$$

2.9 Dimensional Homogeneity and Dimensionless Quantities

Every valid equation must be dimensionally homogeneous and consistent; that is, all additive terms on both sides of the equation must have the same dimensions and units.

Consider the equation:

$$u\left(\frac{\text{m}}{\text{s}}\right) = uo\left(\frac{\text{m}}{\text{s}}\right) \pm g\left(\frac{\text{m}}{\text{s}^2}\right)t(\text{s}) \tag{2.19}$$

$$\text{Distance(m)} = \text{Speed}\left(\frac{\text{m}}{\text{s}}\right) \times \text{Time(s)} \tag{2.20}$$

Equations like (2.19) and (2.20) are both dimensionally **homogeneous and consistent** in their units. Each additive term has the unit of m/s in Eq. (2.19). The left-hand side of Eq. (2.20) has the same unit as that of its right-hand side.

A dimensionally homogeneous equation is one in which the dimensions (units) of each term on both sides of the equation are consistent and balanced. This means that the units on the left side of the equation match the units on the right side of the equation, ensuring that each term has the same dimensions.

If an equation is dimensionally homogeneous, it means that it satisfies the requirement for dimensional consistency, but this doesn't guarantee its validity in describing physical phenomena. A dimensionally homogeneous equation might still be incorrect in its representation of the underlying physics or relationships between variables. In other words, while dimensional homogeneity is a necessary condition for validity, it is not a sufficient condition. The equation must also accurately reflect the behavior or principles governing the system being described.

Conversely, if an equation is valid in describing a physical phenomenon accurately, it is expected to be dimensionally homogeneous. However, there can be cases where an equation is valid but might not initially appear dimensionally homogeneous due to simplifications, assumptions, or specific mathematical manipulations that maintain the physical integrity despite seeming inconsistencies in units. Yet, such situations are exceptions and generally require further scrutiny or explanation. Overall, a valid equation should ideally be both dimensionally homogeneous and accurately represent the physical reality it describes.

Exercise 2.10

1) *If $y(m/s^2) = a \times z(m^3)$, what are the units of a?*
2) *What is a dimensionless group? What multiplicative combination of d(m), $g(m/s^2)$, and t(s) would constitute a dimensionless group?*
3) *If $Z(lbf) = c \sin(R)$, what are the units of c and R?*

NOTE #3
Refer to Appendix A for interconversion factors between the Metric (SI) and Imperial (American Engineering) system of units.

2.10 Chemical Process Data: Measurement, Analysis, and Presentation

Chemical process data, including measurement, analysis, and presentation involve a comprehensive approach to collecting, interpreting, and communicating data within the realm of chemical engineering. This process is critical for understanding, optimizing, and ensuring the efficiency, safety, and quality of chemical processes. Let us delve into each aspect.

1) **Chemical Process Data: Measurement**:
   - **Identification of Key Variables**: Determine the critical parameters and variables relevant to the chemical process under investigation. This may include physical properties (temperature, pressure, flow rates), chemical compositions, reaction rates, and other process-specific factors.
   - **Instrumentation and Sensors**: Select appropriate instruments and sensors to measure the identified variables accurately. Calibration and regular maintenance of these instruments are crucial to ensure the reliability of the collected data.

- **Sampling Techniques**: Employ suitable sampling techniques, whether continuous or grab sampling, to capture representative data points. Proper sampling is essential for generating reliable and meaningful datasets.

2) **Chemical Process Data: Analysis**:
   - **Descriptive Statistics**: Begin with descriptive statistics to summarize and characterize the main features of the collected data. This includes measures such as mean, median, standard deviation (S), and distribution analysis.
   - **Error Analysis**: Evaluate and quantify measurement errors or uncertainties to enhance the accuracy of the data. Understanding and accounting for errors are crucial for reliable data interpretation.
   - **Thermodynamic Analysis**: Apply thermodynamic principles to analyze heat and material balances, ensuring that the energy and material flows within the system align with the laws of thermodynamics.
   - **Reaction Kinetics**: Investigate the kinetics of chemical reactions by analyzing rate laws, reaction mechanisms, and the factors influencing reaction rates. This is vital for optimizing reaction conditions and understanding reaction pathways.
   - **Computational Fluid Dynamics (CFD)**: Utilize CFD simulations to analyze fluid flow, heat transfer, and mass transfer within chemical processes. CFD provides insights into the behavior of fluids and gases within a system.
3) **Chemical Process Data: Presentation**:
   - **Tabular Representation**: Organize data into tables, providing a structured format for easy reference. This includes arranging experimental results, measurements, and key parameters in a clear and concise manner.
   - **Graphical Representation**: Utilize graphical tools such as charts, graphs, and diagrams to visually represent complex relationships within the data. This includes P–V diagrams, T–S diagrams, and other relevant visualizations.
   - **Process Flow Diagrams (PFDs)**: Develop PFDs to illustrate the overall flow of materials and energy within the chemical process. PFDs are invaluable for conveying the process design and understanding the interconnectedness of various units.
   - **Data Visualization Tools**: Leverage specialized software tools like MATLAB, Excel, or chemical engineering simulation software for advanced data visualization. These tools enhance the interpretability of complex datasets.

2.10.1 Random Variations in a Process Variable Measurement

Measurements can be qualitative or quantitative.

- **Qualitative**: These measurements use descriptive words to express qualities, like "heavy" for an object or "hot" for a beverage. Imagine choosing fruit at the market – you might judge ripeness by "softness" or color.
- **Quantitative**: These involve numbers and involve assigning a specific value to a measurement. We might use a kitchen scale to determine the "weight" of fruit in grams, providing a more precise value.

Illustrations:

- **Buying Products**: When choosing fabric softener, we might use qualitative assessments like "strong scent" or "gentle on clothes." However, for precise dosage, we should rely on the quantitative measurement provided by the milliliter markings on the bottle.

- **Sports Activities**: Monitoring one's workout might involve qualitative observations like "feeling tired" but also quantitative measurements like "running distance" tracked by a fitness app.
- **Cooking**: Following a recipe, one might add "a pinch" of salt qualitatively, but for consistent results, measuring with a teaspoon provides a quantitative value.

Let us handle the inherent (random) variation in a process variable measurement. Suppose that X represents the reaction conversion, extent of reaction, reaction coordinate, or what is called the degree of achievement for a given reaction, say $A \rightarrow B$. It describes this chemical reaction where the percentage of reactant A is converted into product B. X is measured in repeated experiments. While in theory, X should be constant under fixed conditions of pressure, temperature, and feed compositions, even seemingly identical runs yield different X values due to inherent randomness. Let us exemplify the extent of the reaction using the following cookie-baking example.

Example 2.18 Cookie Baking Time

Imagine: We are baking cookies. The recipe says to use 1 cup of flour for 10 minutes at 350°F, but each batch seems to turn out slightly different. Our goal is to understand the "true" baking time that gives the best results.

The Problem: There is no way to perfectly control everything (ingredient amounts, oven temperature, and mixing time). Even slight variations can change the outcome.

NOTE #4

Diversity is the native language of Nature. In particular, it means here variations in the outcome of an experiment. The outcome is usually a property or quality to be examined or assessed by sensing, measuring, observing, weighing, estimating, and later qualitatively, or quantitatively making a final judgement, like giving/assigning a number, factor, or index to the property/quality under scrutiny.

The Data: We bake 10 batches, measuring the time it takes for each to reach the desired doneness. The word "doneness" is the condition of being cooked to the desired degree. While food safety experts stress the internal temperature that ensures the destruction of pathogens in food, "doneness" reflects subjective qualities such as the appearance, texture, and optimum flavor of food.

The outcome of the time measurement of reaching cookie doneness is something like:

Batch 1: 10 minutes 15 seconds
Batch 2: 9 minutes 50 seconds
...
Batch 10: 10 minutes 30 seconds

The Question Is: What is the "true" baking time? Which one among the 10 batches to accept as representative for the baking time?

The "true" baking time is the baking time one should get if everything were perfect (impossible!). Hence, we cannot tell the true time, but we can estimate it by averaging the measured times. Add the times for all batches and divide by 10. This gives us the average baking time, which is likely closer to the "true" time we are looking for than any of the 10 individual batches. In our example, it might be 10 minutes and 17 seconds.

Visualizing the Results: Imagine plotting the baking times for each batch on a graph. The dots will likely scatter around their average time, reflecting the randomness.

More Measurements, Better Estimate: The more batches we bake (more measurements), the closer the average time gets to the true time (assuming no systematic errors). Systematic errors are explained in Section 2.7. But even with many measurements, it is just an estimate, not the absolute truth.

The Key Takeaway: Due to randomness, even under controlled conditions, measurements can vary. We use statistical tools like the sample mean to estimate the underlying true value, but it is always an approximation, and there is always some uncertainty.

So, back to the chemical reaction context, X represents the % conversion, just like baking time represents doneness. In fact, the baking time reflects the biochemical reactions or changes (called organoleptic food properties sensed by the human monster) within the food matrix, which are heavily dependent on temperature and time coordinates. The "true" X is like the "true" baking time – impossible to measure perfectly. The sample mean of measured X values estimates the true X, but there will always be some scatter (uncertainty) due to inherent randomness.

2.10.2 MATLAB for Sample Statistics: Mean ($\bar{X}$), Standard Deviation (S), and Range (R)

The sample mean (or average) and sample S are two fundamental statistics used to summarize and analyze data. They provide valuable insights into the central tendency (middle) and spread (variability) of a dataset, but it is crucial to consider their limitations and how they relate to measurement accuracy and precision.

Sample Mean:

- **Represents the "Typical" Value in the Data**: It is calculated by adding up all the values and dividing by the number of values (N). It tells us where most of the data points are concentrated.
- **Sensitive to Outliers:** A single extreme value can significantly skew the average, making it less representative of the majority of data.
- **Provides a Central Reference Point:** Helps compare different datasets or track changes in a dataset over time.

Sample S:

- **Measures How Spread out the Data is:** It calculates how much, on average, each data point deviates from the average. A larger S indicates more variability, while a smaller one suggests the data is clustered closer to the mean.
- **Not as Sensitive to Outliers as the mean:** While outliers can still affect it, the S is less influenced by a single extreme value compared to the average.
- **Reflects Data Variability:** Useful for understanding the range of values and identifying potential outliers.

Using $\overline{X}$ and S Together:

- **Understanding the Distribution:** Together, the mean and S paint a picture of the data's distribution. Imagine a bell curve: the mean sits at the peak, and the S defines the width of the curve, indicating how much data falls within a certain range around the mean.
- **Identifying Patterns and Trends:** By analyzing changes in the mean and S over time or across different groups, we can uncover patterns and trends in our data.

Accuracy and Precision:

- **Estimates, not Absolutes:** It is important to remember that the mean and S are calculated based on a sample, not the entire population. They are estimates of the population mean and S, which are usually unknown.
- **Sample Size Matters:** The accuracy of these estimates depends on the sample size. Larger, more representative samples tend to provide more accurate estimates.
- **Measurement Limitations:** The accuracy of the mean and S is also limited by the accuracy and precision of the original measurements. If the measurements themselves are not very precise, the calculated statistics will also be less reliable.

Example 2.19 Sample $\overline{X}$, S, and R

Suppose a chemical reaction is carried out, starting with pure A in the reactor at 400°C constant reactor temperature. After three minutes, a sample is withdrawn from the reactor and analyzed to determine X, the percentage of A converted out of A fed to the reactor. Table 2.3 shows 10 different runs presumably carried out under the same circumstances.

Table 2.3 Reaction conversion, X, measured 10 times.

Run #	1	2	3	4	5	6	7	8	9	10
X (%)	70.2	76.2	72.5	70.3	74.1	71.3	72.5	71.1	71.6	73.4

What is the true value of X?

The most common estimate of the true value is the sample mean (or arithmetic mean). We collect N measured values of X ($X_1, X_2, \ldots, X_N$) and then calculate

$$\text{The average, or the mean } \left(\overline{X}\right) = \frac{1}{N}\sum_{i=1}^{N} X_i = \frac{(X_1 + X_2 + \ldots + X_n)}{N} \tag{2.21}$$

$$\left(\overline{X}\right) = \frac{70.2 + 76.2 + 72.5 + 70.3 + 74.1 + 71.3 + 72.5 + 71.1 + 71.6 + 73.4}{10} = 72.32$$

Using MATLAB Command Window:

```
>>Xavg = (70.2+76.2+72.5+70.3+74.1+71.3+72.5+71.1+71.6+73.4)/10
Xavg = 72.3200
```

Also, at the command prompt, we can define X as a row vector:

```
>>X= [70.2  76.2  72.5  70.3  74.1  71.3  72.5  71.1  71.6  73.4];
```

Then, using MATLAB, we can carry out different arithmetic operations:

```
>> Xavg2=sum(X)/10
Xavg2 = 72.3200
```

Or, directly make use of the MATLAB built-in **mean** function:

```
>> Xavg3=mean(X)
Xavg3 = 72.3200
```

On the other hand, let us define the sample variance and S:

$$\text{Variance (Var)} = S^2 = \frac{1}{N-1}\sum_{i=1}^{N}\left(X_i-\overline{X}\right)^2 = \frac{1}{N-1}\left[\left(X_1-\overline{X}\right)^2 + \ldots + \left(X_N-\overline{X}\right)^2\right] \tag{2.22}$$

$$\begin{aligned}\text{Standard Deviation } (S) &= \sqrt{\frac{1}{N-1}\sum_{i=1}^{N}\left(X_i-\overline{X}\right)^2} \\ &= \sqrt{\frac{1}{N-1}\left[\left(X_1-\overline{X}\right)^2 + \ldots + \left(X_N-\overline{X}\right)^2\right]}\end{aligned} \tag{2.23}$$

The range of a given dataset is defined as:

$$\text{Range}, R = X_{\text{max}} - X_{\text{min}} \tag{2.24}$$

Range is another index used to reflect the degree of scatter of the given X data points.

NOTE #5
$\overline{X}$ holds the same dimension or unit of X. Variance holds the square dimension or square unit of X. The S holds the same dimension or unit of X. The range holds the same dimension or unit of X.

Let us use MATLAB built-in features pertaining to variance, S, and range. Figure 2.3 shows MATLAB code for calculating Var, S, and R of the given sample: X.

```
>>X= [70.2 76.2 72.5 70.3 74.1 71.3 72.5 71.1 71.6 73.4];
>> Var=var(X)
Var = 3.4529
>> S=std(X)
S = 1.8582
>>Range1=range(X)
Range1 = 6
>>Range2= max(X)-min(X)
Range2 = 6
```

Figure 2.3 MATLAB code for calculating Var, S, and R of the given sample, X.

Notice that $S = \sqrt{\text{Var}} = \sqrt{3.4529} = 1.8582$. Range1 and Range2 are equal to $X_{max} - X_{min} = 76.2 - 70.2 = 6$.

Figure 2.4 shows MATLAB code to calculate the mean, the S, the ±95% confidence interval around the mean, and the plot of the scattered data points, augmented by the mean line and their 95% confidence interval lower and upper borderline.

```
clear variables;
% Sample data
X = [70.2 76.2 72.5 70.3 74.1 71.3 72.5 71.1 71.6 73.4];
% Calculate mean and standard deviation
mean = mean(X);
std = std(X);
% Calculate 95% confidence interval using 2*std (literally, it should be 1.96*std).
confidence_interval = 2 * std;
% Create the plot
figure;
% Plot data points
% Create the plot and legend in separate steps
plot(X, 'o', linewidth =3);
hold on;  % Keep the plot open for adding more lines
% Plot mean line
x_min = min(X);
x_max = max(X);
% 10 trials or length of X
plot([1 10], [mean mean], 'black', linestyle='-', linewidth =3);
plot([1 10], [mean + confidence_interval, mean + confidence_interval], 'b--', 'LineWidth',
3);
plot([1 10], [mean - confidence_interval, mean - confidence_interval], 'r--', 'LineWidth', 3);
% Create the legend
legend({'Data points', 'Mean', 'Upper bound', 'Lower bound'});
% Customize the plot
xlabel('RUN #', 'FontSize', 16, 'FontWeight', 'bold');
ylabel('X %, Conversion %', 'FontSize', 16, 'FontWeight', 'bold');
title('Data points with mean and 95% confidence interval', fontsize=15);
ax = gca;
% Set new font size for all axes text (including labels)
ax.FontSize = 16;
grid on;
```

Figure 2.4 MATLAB code for estimating statistical parameters of a set of data points, as well as plotting them while showing the mean and the 95% confidence interval lower and upper line.

Figure 2.5 shows the generated plot upon executing Figure 2.4 MATLAB code:

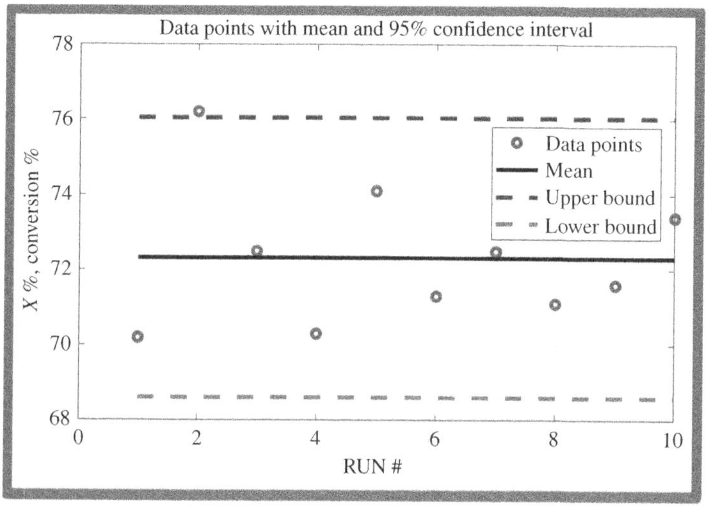

Figure 2.5 A plot of scattered data, their mean, and 95% confidence interval.

General Notes on Data Scattering and Statistical Confidence Interval:

1) The data points are randomly scattered around the mean.
2) The choice of a confidence level for an interval determines the likelihood that the resulting confidence interval will contain (or bracket) the true parameter value. Commonly selected confidence interval areas, denoted as A_{CI}, include 0.90, 0.95, and 0.99. These sub-areas (or levels) correspond to specific percentages of the normal density curve. For instance, a 95% confidence interval encompasses 95% of the normal curve, indicating that the probability of observing a value outside this range is less than 0.05. The symmetry of the normal curve requires that half of the area is under the left tail and the other half is under the right tail. In a confidence interval with level A_{CI}, the area in each tail is $(1-A_{\text{CI}}/2)$, resulting in 0.025 for each tail in a 95% confidence interval, as shown in Figure 2.6.

   The value denoted as Z^*, representing the point on the standard normal density curve where the probability of observing a value greater than Z^* is equal to p, which is known as

Figure 2.6 The confidence interval is drawn around the mean. The red (or gray in a black and white scale) area is a fraction is a fraction of unity, which is frequently set to 0.95, bracketing the 95% confidence interval borders.

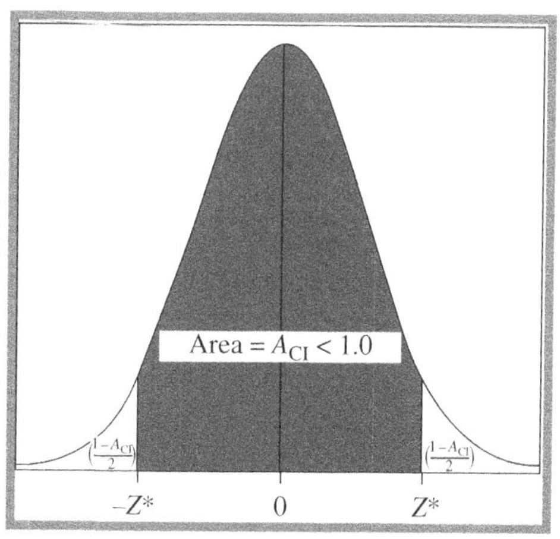

the upper p critical value of the standard normal distribution. For instance, if $p = 0.025$, the value of Z^* such that $P(Z > Z^*) = 0.025$ (or $P(Z < Z^*) = 0.975$) is 1.96. For a confidence interval with level A_{CI}, the value of p is $(1 - A_{CI}/2)$. Therefore, a 95% confidence interval for the standard normal distribution is (−1.96, 1.96) since 95% of the area under the curve falls within this interval (i.e., the red area under the normal distribution curve).

3) The second data point (i.e., Run #2) lies outside the 95% confidence interval upper limit. Given that all ten runs were carried out under similar conditions, the point is worth questioning its authenticity or validity, which means it can be excluded and keep the other nine runs and carry out statistical data analysis.
4) For a typical random variable, roughly two-thirds of all measured values fall within one S; about 95% fall within two S, and about 99% fall within three S above and below the sample mean.
5) If a value of a measured variable is reported with error limits: Such as $X = 71.4 \pm 0.7$. This means a single measured value of X is likely to fall between 70.7 and 72.1. The significance of the given error limits (± 0.7) is not clear unless more information is provided. The interval between 71.4 and 72.1 may represent the range of the X dataset ($X_{max} - X_{min}$) or might represent $1S$, $2S$, or $3S$ of the X dataset of measurements.

Example 2.20 The Volumetric Flow Rate Measurement with Confidence Interval
The volumetric flow rate of a process fluid, $\dot{V}(\mathrm{mL}/\mathrm{s})$, is measured eight times, with the following results, as shown in Table 2.4:

Table 2.4 The volumetric flow rate is measured eight times.

Run #	1	2	3	4	5	6	7	8
$\dot{V}\left(\frac{\mathrm{mL}}{\mathrm{s}}\right)$	230	246	225	239	237	233	236	242

a) Calculate the sample mean ($\overline{\dot{V}}$), range (R), sample variance (Var), and sample standard deviation (S).
b) There is a high probability of 95% that a measured value of $\dot{V}$ will fall within two S of the mean. Report the value of $\dot{V}$, in the form $\dot{V} = a \pm b$, choosing the values of a and b to define the mean and its associated 95% confidence interval limits.

Solution:

a) Using the MATLAB platform, we can carry out all required statistics. Here are the code and results, as shown in Figure 2.7.

```
clear variables;
Vdot = [230 246 225 239 237 233 236 242];
VdotMean=mean(Vdot)
VdotRange=range(Vdot)
VdotVar=var(Vdot)
Vdotstd=std(Vdot)
```

Figure 2.7 MATLAB code for carrying out different sample statistics calculations.

MATLAB results are:

```
VdotMean = 236
VdotRange = 21
VdotVar = 44.5714
Vdotstd = 6.6762
```

b) $\dot{V} = a \mp b = \dot{V} \mp 1.96 * S = 236 \mp 1.96 * (6.6762) = 236 \mp 13.1$. This means that for a 95% confidence interval $\left[222.9 < \overline{\dot{V}} < 249.1\right]$. In this example, all eight data points fall within this range.

Example 2.21 Quality Control Using Sample Statics

In total, 400 batches of chips (electronic, not potato chips) are produced each week. Each batch is subjected to precise electrical testing: Voltage testing to ensure the chip operates correctly at specified voltage levels and current testing to measure the chip's current consumption under different conditions. If a batch does not pass such tests, it will be rejected. X is rejected batches/week. Suppose that the quality assurance (QA) department test results for a 14-week base period are as shown in Table 2.5.

Table 2.5 A number of bad batches out of 400 produced chips batches.

Batch #	1	2	3	4	5	6	7	8	9	10	11	12	13	14
***X*, Bad Batches**	14	22	15	16	18	16	17	20	17	16	18	19	15	21

The company policy is to regard the process operation as normal as long as the number of bad batches produced in a week is not higher than two S above the mean value for the base period (i.e., as long as $X \leq \overline{X} + 2S$). If X exceeds this value, the process will be shut down for corrective maintenance (a long and costly procedure). Significant deviations from the mean may be observed as a natural part of the usual variability in the process. However, these deviations are so uncommon that if they do occur, it is more likely to be attributed to an abnormal issue in the process, which then requires an interruption of the process and carrying out the required surgical operations (i.e., maintenance), before things get even worse and require drastic maintenance or even replacement of total or part of the process pieces of equipment.

How many bad batches in a week would it take to shut down the process?

Solution:

Here are the MATLAB code as shown in Figure 2.8:

```
clear variables;
X= [14 22 15 16 18 16 17 20 17 16 18 19 15 21];
XMean=mean(X)
S =std(X)
```

Figure 2.8 Calculation of X dataset statistic parameters: X_{mean} and S.

MATLAB results are:

XMean = 17.4286
S = 2.3766

$$X \leq \overline{X} + 2S = 17.4 + 2 \times 2.4 = 22.2 \cong 22$$

If we have 22 or more bad batches produced in a week, the process must be shut down for curative maintenance. The QA/control engineer can decide whether 2*S is adequate or not. In other words, he/she may go up to 3*S or down to 1*S. Under each scenario, there is a trade-off between the frequency of requiring off-line maintenance vs. passing the QA test. Less frequent offline maintenance will jeopardize the whole process to face drastic, complete maintenance or over-haul maintenance. Like one's car or one's health, a frequent check-up with either the car mechanic or body biomechanics will help avoid severe mechanical/biological failure by early detection of sources of troubles, which otherwise if ignored might cost the whole capital of a person's asset.

2.10.3 Relative *S* (Coefficient of Variation)

The relative standard deviation (RSD) is useful for comparing the uncertainty between different measurements of varying absolute magnitude. The RSD is calculated from the standard deviation, S, and $\overline{X}$ and is commonly expressed as a percentage (%).

$$\text{RSD} = \frac{S}{\overline{X}} \times 100\% \tag{2.25}$$

RSD ≤ 5% signifies good measurement precision.

If a comparison is made between two sets of sample data for the same measured physical quantity, then the set with the smaller magnitude of RSD will be more precise than its counterpart set.

Example 2.22 RSD Evaluation of Measurements Sample Precision
Consider the following set of X data point measurements, as shown in Table 2.6:

Table 2.6 X value was measured eight times.

Trial #	1	2	3	4	5	6	7	8
X, Value	0.00550	0.00561	0.00551	0.00545	0.00546	0.00556	0.00553	0.00562

Is the measurement set precise?

Figure 2.9 shows MATLAB code to evaluate RSD

```
clear variables;
X = [0.00550 0.00561 0.00551 0.00545 0.00546 0.00556 0.00553 0.00562];
Xmean=mean(X);
S_X=std(X);
RSD=(S_X/Xmean)*100
```

Figure 2.9 MATLAB code for evaluating RSD of X sample.

MATLAB result: RSD = 1.1437

Is the measurement set precise? Yes; RSD = 1.14% < 5%

2.10.4 MATLAB Two-point Linear Interpolation

The equation of the line through (x_1, y_1) and (x_2, y_2) on a plot of y vs. x is:

$$\frac{y - y_1}{y_2 - y_1} = \frac{x - x_1}{x_2 - x_1} \quad (2.26)$$

Equation (2.26) is to estimate y for any x between x_1 and x_2 (interpolation). It may also be used to estimate y for any x outside of this range (extrapolation).

Example 2.23 Two-point Linear Interpolation for $y = f(x)$ or $x = f(y)$
Values of a variable (y) are measured at several times (t), as shown in Table 2.7:

Table 2.7 A set of (Y, t) data points.

y	1	4	8	12	16	20
t	1	2	3	4	5	6

Using the two-point linear interpolation, what is

a) $y(t = 1.5)$
b) $t(y = 19)$

Solution:

We will bracket the given value of t.

a) $t = 1.5$ lies between $t_1 = 1$ and $t_2 = 2$ with $y_1 = 1$ and $y_2 = 4$

Apply Eq. (2.25), we have:

$$\frac{y - y_1}{y_2 - y_1} = \frac{t - t_1}{t_2 - t_1}$$

$$\frac{y - 1}{4 - 1} = \frac{1.5 - 1}{2 - 1} = \frac{0.5}{1} = 0.5$$

$y - 1 = 0.5 \times (4 - 1) = 1.5 \quad \rightarrow y = 1 + 1.5 = 2.5 \quad \rightarrow y(t = 1.5) = 2.5$

b) Here, we will bracket the given value of y

$y = 19$ lies between $y_1 = 16$ and $y_2 = 20$ with $t_1 = 5$ and $t_2 = 6$
Apply Eq. (2.25), we have:

$$\frac{y - y_1}{y_2 - y_1} = \frac{t - t_1}{t_2 - t_1}$$

$$\frac{19-16}{20-16} = \frac{t-5}{6-5} \rightarrow \frac{3}{4} = 0.75 = \frac{t-5}{1}$$

$$t-5 = 0.75 \times (1) = 0.75 \quad \rightarrow t = 5 + 0.75 = 5.75 \quad \rightarrow t(y = 19) = 5.75$$

MATLAB code is shown below for both cases on how to use MATLAB built-in **interp1** function for either finding y for a given x (Figure 2.10) or vice versa, finding x for a given y (Figure 2.11).

```
% Define your sample points and corresponding values
x = [1, 2, 3, 4, 5, 6];
y = [1, 4, 8, 12, 16, 20];
% Specify the query points where you want to interpolate
xq = [1.5];% q stands for query.
% Perform linear interpolation to find yq for the given x=xq.
yq = interp1(x, y, xq) % yq is the interpolated y value
```

Figure 2.10 Two-point linear interpolation built-in **interp1** function for finding y given x.

The result for yq, the interpolated y value, is shown below:

yq = 2.5000

```
% Define your sample points and corresponding values
x = [1, 2, 3, 4, 5, 6];
y = [1, 4, 8, 12, 16, 20];
% Specify the query points where you want to interpolate
yq = [19];
% Perform linear interpolation to find xq for the given y=yq.
xq = interp1(y, x, yq) % xq is the interpolated x value
```

Figure 2.11 Two-point linear interpolation built-in **interp1** function for finding x given y.

The result for xq, the interpolated x value, is shown below:

xq = 5.7500

Notice that both results are the same as those manually calculated.

2.10.5 MATLAB Linear Regression

The following text is quoted from Chapter 4 of my book, Al-Malah [1]:

Linear regression is a parametric regression technique in which the response is modeled using a known formula given in terms of the predictor variables. It is one of the simplest but most powerful regression techniques. Suppose we want to fit a linear regression model to the fuel economy data using the following formula.

$$\text{Fuel Economy} = a + b \times \text{Weight} + c \times \text{Horse Power} + d \times \text{Axle Ratio}$$

In this model, the *fuel economy* is considered as a response (i.e., Y variable), whereas the *weight*, *horse power*, and *axle ratio* are the predictor variables (X variables). The fitting process finds the coefficients a, b, c, and d that minimize the prediction error.

Notice that a linear model simply means that the model regression coefficients: a, b, c, and d are each raised to the power one (or to the power zero for being absent in the model), and it has nothing to deal with the predictor variable X.

Examples of linear regression models are:

$$Y = a + b \times X + c \times X^2 \tag{2.27a}$$

$$Y = a + b \times X + c \times \ln(X) \tag{2.27b}$$

$$Y = a + b \times X + c \times e^{X^2} \tag{2.27c}$$

$$Y = a + \frac{b}{X} + c \times X^{0.5} \tag{2.27d}$$

Examples of nonlinear regression models are:

$$Y = a + b \times \ln(cX) \tag{2.28a}$$

$$Y = a + b \times \mathrm{e}^{cX^2} \tag{2.28b}$$

$$Y = a + \frac{b}{X + c} \tag{2.28c}$$

$$Y = a + \frac{b}{X} + X^c \tag{2.28d}$$

In this chapter, since this textbook is introductory, we will explain in detail the linear regression model as far as calculating the model parameters, or coefficients, their confidence interval, and plotting the data points, their residuals, and the 95% confidence interval for the model prediction.

2.10.5.1 Method of Least Squares

The method of least squares is a powerful statistical technique used to find the line of best fit for a set of data points. Whenever we have a linear relationship between Y and X represented by the equation:

$$Y = a + b * X \tag{2.29}$$

The least squares method helps us determine the optimal values of a (intercept) and b (slope) that best represent the overall trend in the data.

1) **Visualizing the Problem**:
   If we have several data points scattered around a straight line. The method of least squares aims to draw a line through these points in a way that minimizes the total distance between the line and each data point.
2) **Minimizing the Error**:
   Instead of measuring distances directly, the method considers the squared distances between each data point and the estimated line. This emphasizes larger deviations and gives more weight to points farther away from the line.
3) **Mathematical Formulation**:
   For each data point (X_i, Y_i), the squared distance to the estimated line with $Y = a + bX$ is:

$$\mathrm{Sum}(a, b) = \sum_{i=1}^{N} \left[Y_{\mathrm{exp}} - Y_{\mathrm{fit}}\right]^2 = \sum_{i=1}^{N} [Y_i - Y_{\mathrm{fit}}]^2 = [Y_i - (a + bX_i)]^2 \tag{2.30}$$

The method aims to minimize the sum of these squared distances for all data points. That simply means to select the values of both a and b such that the sum exhibits the minimum value. At this condition, we say the a and b are the best values for fitting a linear model vs. scattered (X_i,Y_i) data points.

4) **Finding the Optimum a and b**:
   This minimization problem has a mathematical solution that defines the optimal values of a and b to minimize the sum of squared errors. These solutions involve complex math, but software and calculators can readily solve them.

$$b = \frac{\sum_{i=1}^{N} (X_i - \overline{X})(Y_i - \overline{Y})}{\sum_{i=1}^{N} (X_i - \overline{X})^2} \tag{2.31a}$$

An alternative formula for b is:

$$b = \frac{N \sum_{i=1}^{N} (X_i Y_i) - \left(\sum_{i=1}^{N} Y_i\right)\left(\sum_{i=1}^{N} X_i\right)}{N \sum_{i=1}^{N} X_i^2 - \left(\sum_{i=1}^{N} X_i\right)^2} \tag{2.31b}$$

$$a = \overline{Y} - b\overline{X} \tag{2.32a}$$

An alternative formula for a is:

$$a = \frac{\sum_{i=1}^{N} Y_i}{N} - b\frac{\sum_{i=1}^{N} X_i}{N} \tag{2.32b}$$

This straight line equation (Eq. 2.29), with the best fit a and b values, represents the closest fit to the overall trend in the data, considering the inherent variability in individual points.

Benefits of the Least Squares Method:

- **Objectivity**: Provides a standardized and statistically sound approach to fitting a line to data.
- **Robustness**: Accounts for random errors and deviations in the data.

- **Generalizable**: Applicable to all forms of linear relationships listed, for example, in equations (2.27a–d).

Precaution:
The method assumes a linear relationship between variables. If the relationship is nonlinear, other fitting methods might be needed. Outliers can significantly impact the line of best fit. Identifying and addressing outliers is crucial.

Example 2.24 Linear Fitting Using the Least Squares Method
Consider the following set of eight (X_i,Y_i) data points, as shown in Table 2.8:

Table 2.8 A set of eight (X_i,Y_i) data points.

X_i	1	2	3	4	5	6	7	8
Y_i	5	7	8	12	13	14	17	19

Find the best value of a and b such that $Y = a + b * X$ gives the best fit with minimum error or deviation between fitted and measured Y_i.

Solution:

Let us construct Table 2.9 by adding a third and fourth column to the existing X_i and Y_i columns. The third column is made by squaring each entry (i.e., row) in the first column. The fourth column is made by multiplying the two entries found in the first and second columns. Below each column, there is the vertical sum of the eight records (or entries) for each column.

Finally, $\overline{X}$ is found by dividing the sum of the first column by a number of data points (i.e., 8). So does the case for $\overline{Y}$ where now we divide the sum of the second column by 8.

Table 2.9 Creation of a table made of X_i and Y_i columns plus two new columns and showing the sum of each column at the end.

X_i	Y_i	X_i^2	X_iY_i
1	5	1	5
2	7	4	14
3	8	9	24
4	12	16	48
5	13	25	65
6	14	36	84
7	17	49	119
8	19	64	152
$\sum_{i=1}^{8} X_i = 36$	$\sum_{i=1}^{8} Y_i = 95$	$\sum_{i=1}^{8} X^2{}_i = 204$	$\sum_{i=1}^{8} (X_iY_i) = 511$
$\overline{X} = 4.5$	$\overline{Y} = 11.875$		

Now, we have calculated all terms found in equations (2.31a, b) and (2.32a, b). Let us substitute by referring to the previous table. Let us use Eq. (2.31b):

$$b = \frac{N\sum_{i=1}^{N}(X_iY_i) - \left(\sum_{i=1}^{N}Y_i\right)\left(\sum_{i=1}^{N}X_i\right)}{N\sum_{i=1}^{N}X_i^2 - \left(\sum_{i=1}^{N}X_i\right)^2} = \frac{8(511)-(95)(36)}{8(204)-(36)^2} = \frac{668}{336} = 1.9881$$

Let us use Eq. (2.32a):

$$a = \overline{Y} - b\overline{X} = 11.875 - (1.9881)(4.5) = 2.9285$$

The best-fitted model is:

$$Y = a + b * X = 2.9285 + 1.9881 * X$$

This example is enough to demonstrate the manual calculation of the least squares method fitting parameters: a and b. We will use the MATLAB platform to calculate the linear regression parameters.

MATLAB offers several built-in functions that utilize the least squares method depending on the specific type of model we want to fit and the analysis we are performing. Here are some common options:

1) **Linear Regression**:
   fitlm: This is the most versatile function for fitting linear models in MATLAB. It provides access to various statistics and diagnostics related to the fit.
   polyfit: This function fits a polynomial model of a specified degree to the given data.
   regress: This classic function can also perform linear regression but offers fewer features compared to **fitlm**.
2) **Nonlinear Regression**:
   fitnlm: This function allows us to fit nonlinear models defined by custom equations or built-in models like exponential, logistic, or power functions.
   nlinfit: This function is less user-friendly but offers more control over the optimization process for complex nonlinear models.
3) **Other Least Squares Applications**:
   lsqr: This function solves linear systems of equations using the least squares method.
   robustfit: This function performs outlier-resistant linear regression, which is helpful when the given data might contain outliers.

NOTE #6
Please, refer to MATLAB off-line, built-in help documentation for a given function. Just key in the name of the function, and MATLAB will refresh the user with relevant documents containing the search keyword or phrase. As this course is introductory and not purely a mathematical course, we will focus our attention on **fitlm** function and its applications in engineering applications suitable for a sophomore chemical engineering student. Of course, the **fitlm** function has a lot of features, and if properly exploited, it will give all the information needed with respect to linear regression problems.

Figure 2.12 shows MATLAB code for utilizing MATLAB **fitlm** built-in function. We will use Example 2.24 data for the sake of comparison.

```
clear variables;
% Create X data (independent variable)
X = [1 2 3 4 5 6 7 8];
% Create Y data (dependent variable)
Y = [5 7 8 12 13 14 17 19];
% Fit the model using fitlm
mdl = fitlm(X, Y);
%This line of code fits a linear model of the form Y = beta0 + beta1 * X.
% Get the estimated coefficients.
beta0 = mdl.Coefficients.Estimate(1); % Intercept
beta1 = mdl.Coefficients.Estimate(2); % Slope
% Display the equation
Word1="Equation: Y =  ";
Word2= num2str(beta0);
Word3= num2str(beta1);
Title=Word1 + Word2 + " + " + Word3 + "X " ;
disp(Title);
Model_Coeffs=mdl.Coefficients
AdjR2=mdl.Rsquared
% Access other statistics like R-squared, p-values, etc.
% See fitlm documentation for details
% Plot the data and the fitted line
plot(X, Y, 'o', linewidth =3);
hold on;
plot(X, mdl.Fitted, '-', linewidth =3);
legend('Data', 'Fitted Line');
xlabel('X', 'FontSize', 16, 'FontWeight', 'bold');
ylabel('Y', 'FontSize', 16, 'FontWeight', 'bold');
title(Title, fontsize=16);
ax = gca;
% Set new font size for all axes text (including labels)
ax.FontSize = 16;
grid on;
figure;
plot(Y, mdl.Residuals{:,1}, 'x', linewidth =3)
hold on;
xval=min(Y):0.1:max(Y);
yval=0.0;
plot(xval,yval, '+', linewidth =3)
xlabel('Y-Datapoint', 'FontSize', 16, 'FontWeight', 'bold');
ylabel(' Y_{i} -Y_{fit}', 'FontSize', 16, 'FontWeight', 'bold');
title('Y-Datapoint Residuals, Y_{i} -Y_{fit}', fontsize=16);
hold off; ax = gca;
% Set new font size for all axes text (including labels)
ax.FontSize = 16;
```

Figure 2.12 Code for linear fitting, parameters statics, R^2, plot of the fitted model, and plot of residuals.

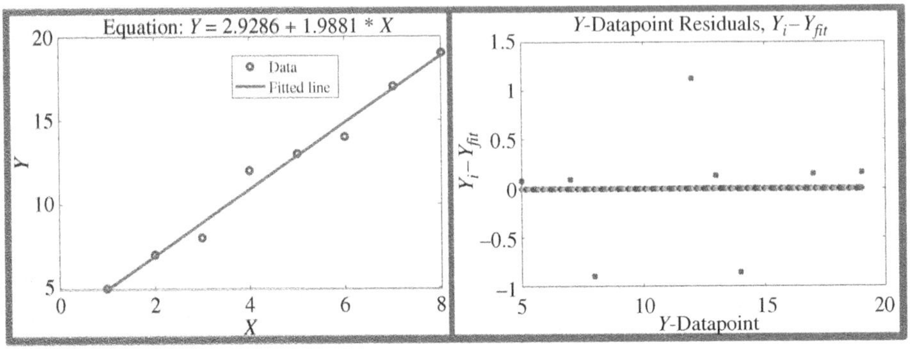

Figure 2.13 (Left) A plot of fitted line vs. experimental data. (Right) A plot of Y-data point residuals.

MATLAB numerical results are shown below, with two plots in Figure 2.13. The point estimate values of a and b exactly match with those manually obtained earlier.

Equation: Y = 2.9286 + 1.9881 * X
Model_Coeffs =

	Estimate	SE	tStat	pValue
(Intercept)	2.9286	0.53881	5.4352	0.0016096
x1	1.9881	0.1067	18.632	1.5422e-06

AdjR2 =
struct with fields:
Ordinary: 0.9830
Adjusted: 0.9802

The results of model coefficients show what is called the point estimate of parameters a and b as well as the standard error and p-value associated with each parameter. The p-values reported (0.0016096 for intercept and 1.5422e-06 for slope) are much lower than the typical significance level of 0.05. This means we can reject the null hypothesis that the respective coefficients are zero, with a high confidence. However, remember that p-values only tell us how likely it is to observe such results if there is no effect (the null hypothesis is true). They don't directly tell us the magnitude or importance of the coefficient itself. The low p-values confirm that both the intercept and slope are statistically significant at the 0.05 level.

In other words, it is important to remember that statistical significance doesn't necessarily imply practical significance. Analyze the confidence intervals and consider the magnitude of the coefficients in the context of the research question (or case study) to draw meaningful conclusions.

In addition, tStat for both model parameters can be given by dividing the point estimate by the corresponding SE value. For example, tStat for the intercept = 2.9286/ 0.53881 = 5.4352.

For a 95% confidence interval bracketing both the intercept and slope, here is the range:

$$\textbf{Intercept}: [(2.9286 - 1.96 \times 0.53881)\ (2.9286 + 1.96 \times 0.53881)] = [1.8725 \quad 3.9847]$$

$$\textbf{Slope}: [(1.9881 - 1.96 \times 0.1067)\ (1.9881 + 1.96 \times 0.1067)] = [1.7790 \quad 2.1972]$$

In addition to the default input arguments for X and Y data vectors, the second input to **fitlm**, we can use one of the predefined models. The most common predefined models are shown in Table 2.10.

Table 2.10 Predefined models to be used with **fitlm** function.

Model Name	Meaning
"linear"	Intercept and linear terms for each predictor.
"interaction"	Intercept, linear terms, and all products of pairs of distinct predictors (no squared terms).
"quadratic"	Intercept, linear terms, interactions, and squared terms.

We can also use a more specific model by providing a formula in Wilkinson–Rogers notation: responseVar ~ terms, as shown in Table 2.11.

Table 2.11 Wilkinson–Rogers notation for specific models.

Operator	Meaning	Example
+	Include this term.	"$y \sim x_1 + x_2$" includes the intercept term, x_1, and x_2: $y = c_0 + c_1\,x_1 + c_2\,x_2$
−	Exclude this term.	"$y \sim x_1 + x_2 - 1$" excludes the intercept term: $y = c_1\,x_1 + c_2\,x_2$
*	Include product and all lower-order terms.	"$y \sim x_1 * x_2$" includes the intercept term, x_1, x_2, and $x_1 * x_2$: $y = c_0 + c_1\,x_1 + c_2\,x_2 + c_3\,x_1\,x_2$
:	Includes the product term only.	"$y \sim x_1{:}x_2$" includes only $x_1 * x_2$: $y = c_1\,x_1\,x_2$
^	Include power and all lower-order terms.	"$y \sim (x_1\hat{}2) + (x_2\hat{}2)$" includes the intercept, x_1, x_2, $x_1\hat{}2$, and $x_2\hat{}2$: $y = c_0 + c_1\,x_1 + c_2\,x_2 + c_3\,x_1^2 + c_4\,x_2^2$

NOTE #7

Forms listed in Tables 2.10 and 2.11 are valid for linear regression of multidimensional X. In other words, X is resolved into x_1, x_2, x_3, etc. However, some forms can be implemented in a one-dimensional X.

Moreover, there is an option for switching on/off the robust feature of the linear regression algorithm. The advantage of using the "**RobustOpts**" option in **fitlm** function in MATLAB is its ability to make the linear regression model more resistant to outliers in data. Here is how it works:

Ordinary Least Squares (OLS): The default fitting method in **fitlm** is OLS. This method minimizes the sum of squared residuals, meaning it heavily weighs points that are far away from the fitted line. This can be problematic if data contains outliers, as these few points can significantly influence the fitted line and lead to a misleading model. On the contrary, when enabling the "**RobustOpts**" option, robust fitting methods down-weigh the influence of outliers. It achieves this by assigning lower weights to data points with large residuals, reducing their impact on the model. This leads to a more accurate representation of the main trend in your data when outliers are present.

Example 2.25 First-order Reaction Kinetic Data
Consider the liquid-phase reaction: $A \rightarrow$ Products where the concentration of A is monitored over the time course of 30 minutes from the start of the run, as shown in Table 2.12. The same data set is saved in an Excel sheet named Conc_Time.xlsx and also as a MATLAB file named CA_t.mat.

Table 2.12 The concentration of, C_A, as a function of time.

Time (min)	C_A (M)	Time (min)	C_A (M)	Time (min)	C_A (M)
0	100	9.5	38.6741	19	14.9569
0.5	95.1229	10	36.7879	19.5	14.2274
1	90.4837	10.5	34.9938	20	13.5335
1.5	86.0708	11	33.2871	20.5	12.8735
2	81.8731	11.5	31.6637	21	12.2456
2.5	77.8801	12	30.1194	21.5	11.6484
3	74.0818	12.5	28.6505	22	11.0803
3.5	70.4688	13	27.2532	22.5	10.5399
4	67.032	13.5	25.924	23	10.0259
4.5	63.7628	14	24.6597	23.5	9.5369
5	60.6531	14.5	23.457	24	9.0718
5.5	57.695	15	22.313	24.5	8.6294
6	54.8812	15.5	21.2248	25	8.2085
6.5	52.2046	16	20.1897	25.5	7.8082
7	49.6585	16.5	19.205	26	7.4274
7.5	47.2367	17	18.2684	26.5	7.0651
8	44.9329	17.5	17.3774	27	6.7206
8.5	42.7415	18	16.5299	27.5	6
9	40.657	18.5	15.7237	30	4.9787

In general, for a constant volume batch process, we can express the rate of change of A moles as a function of C_A:

$$\frac{dC_A}{dt} = -kC_A^n \tag{2.33}$$

n is called the reaction order with respect to A and if A is the only reactant, it will be denoted as the overall reaction order. n can be 0, 1, 2, or a noninteger, rational value.

First-order Reaction:
If $n = 1$, then we can integrate dt in Eq. (2.33) from 0 to any t, while C_A from C_{Ao} to any C_A, to get:

$$\frac{dC_A}{C_A} = -kdt \quad \rightarrow \int_{C_{Ao}}^{C_A} \frac{dC_A}{C_A} = -k\int_0^t dt \rightarrow C_A = C_{Ao}e^{-kt} \tag{2.34}$$

Taking the natural log term for both sides:

$$\ln(C_A) = \ln(C_{Ao}) - kt \tag{2.35}$$

Equation (2.35) represents a linear relationship between $\ln(C_A)$ and t, similar to saying:

$$Y = a + bX$$

$$\text{With } a = \text{Intercept} = \ln(C_{Ao}) \text{ and } b = \text{Slope} = -k \tag{2.36}$$

Figure 2.14 shows MATLAB code for creating a plot of C_A vs. time data shown in Table 2.12. A plot of C_A vs. time will give an exponential decay of C_A, starting from C_{Ao} to final C_A, as shown in Figure 2.15, which is generated by executing Figure 2.14 code.

```
load('CA_t.mat')
t=CA_t.t;
CA=CA_t.CA;
plot(t,CA, linewidth=3)
xlabel('time, min', 'FontSize', 16, 'FontWeight', 'bold');
ylabel(' C_{A}', 'FontSize', 16, 'FontWeight', 'bold');
title('C_{A}vs. Time', fontsize=16);
```

Figure 2.14 MATLAB code for loading C_A vs. time data and making a plot.

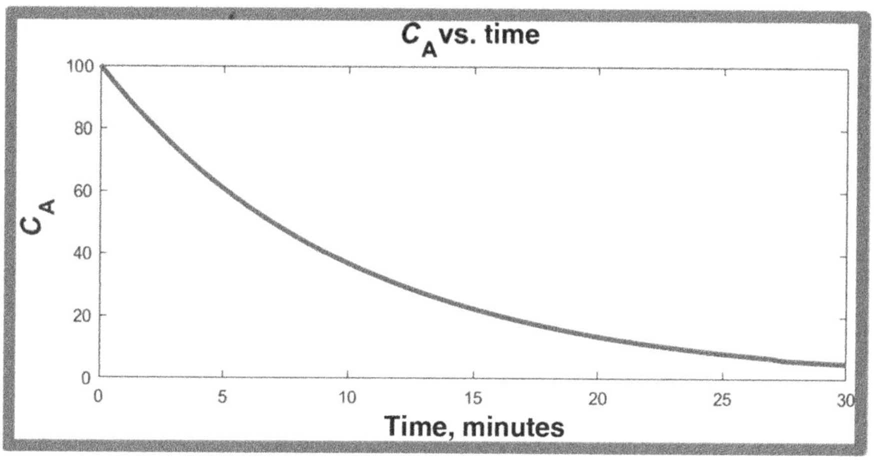

Figure 2.15 C_A vs. t plot showing an exponential decay over time.

NOTE #8
Keep in mind that Eq. (2.34) is nonlinear equation with respect to parameter k as far as regression is concerned. On the contrary, Eq. (2.35) is a linearized version of Eq. (2.34). Linearization makes regression step simpler; yet, in return, a penalty is paid for the reliability of the regressed parameters via the linearized route, compared with that of the nonlinear regression route.

Let us show how to use the **fitlm** function to carry out liner regression. There are two approaches to do that, and they are shown in Figures (2.16) and (2.17).

In Figure 2.16, we carry out a transformation by taking the log value of the C_A column, first, and then carry out linear regression using the transformed Y, called logC_A. In Figure 2.16, mdl statement appears as: mdl = fitlm(t, logC_A). C_{Ao} is obtained by taking the exponent of the intercept value simply because a = Intercept = ln(C_{Ao}). Hence, C_{A0} = exp(lnC_{A0}).

```
% Read data from Excel file
clear variables;
try
  % Attempt to read the Excel file first
  CA_t = readtable('Conc_Time.xlsx', 'Format','auto');catch ME
  % If reading the Excel file fails, try loading the MAT file
  disp('Failed to read Excel file; it will continue with the saved MATLAB file.');
  try
    load('CA_t', 'CA_t');
    disp('Loaded CA time data from MAT file.');
  catch ME2
    % If loading the MAT file also fails, display an error message
    disp('Failed to read MAT file.');
  end
end
t=CA_t.t;
CA=CA_t.CA;
logCA = log(CA);
mdl = fitlm(t, logCA);
disp(mdl);
k = -mdl.Coefficients.Estimate('x1');  % Negative slope represents -k
lnCA0 = mdl.Coefficients.Estimate('(Intercept)'); % Intercept represents ln(CA0)
CA0 = exp(lnCA0);  % Calculate initial concentration (CA0)
fprintf('Rate constant (k): %.4f\n', k);
fprintf('Initial concentration (CA0): %.4f\n', CA0);
```

Figure 2.16 MATLAB code for utilizing **fitlm** function where transformation of Y into log(Y) is done prior to calling **fitlm** function.

MATLAB results are shown below:

```
Linear regression model:
   y ~ 1 + x1 (This means y=a+bx1; 1 means intercept is not zero)
Estimated Coefficients:
                   Estimate        SE           tStat        pValue
                   ________    __________    _______    ___________

   (Intercept)      4.6071      0.0021547     2138.2     5.4623e-137
   x1             -0.10022     0.00013219    -758.15     3.178e-112

Number of observations: 57, Error degrees of freedom: 55
Root Mean Squared Error: 0.00827
R-squared: 1, Adjusted R-Squared: 1
F-statistic vs. constant model: 5.75e+05, p-value = 3.18e-112
Rate constant (k): 0.1002
Initial concentration (CA0): 100.1949
```

The second approach is to use $\log(C_A)$ within the arguments of fitlm function itself. In this way, there is no need to carry out the transformation step prior to calling **fitlm** function. Figure 2.17 shows MATLAB code for carrying out the linear regression and presenting the results. Notice that mdl statement appears as: mdl2 = fitlm(t, $\log(C_A)$). Again, C_{Ao} is obtained by taking the exponent of the intercept value simply because a = Intercept = $\ln(C_{Ao})$.

Figure 2.18 shows that both approaches exactly yield the same numerical results.

Figure 2.19 shows MATLAB code for showing a single plot for both the original, scattered C_A vs. t data and the regressed curve for $C_A = C_{Ao}e^{-kt} = 100.1949 \times e^{-0.1002 \times t}$. Notice here that we must retain the Workspace variables, created via running Figure 2.16 code first, in order to smoothly run Figure 2.19 code.

Figure 2.20 is the generated plot upon executing the code in Figure 2.19.

```
% Read data from Excel file
try
  % Attempt to read the Excel file first
  CA_t = readtable('Conc_Time.xlsx', 'Format','auto');catch ME
  % If reading the Excel file fails, try loading the MAT file
  disp('Failed to read Excel file; it will continue with the saved MATLAB file.');
  try
    load('CA_t', 'CA_t');
    disp('Loaded CA time data from MAT file.');
  catch ME2
    % If loading the MAT file also fails, display an error message
    disp('Failed to read MAT file.');
  end
end
t=CA_t.t;
CA=CA_t.CA;
mdl2 = fitlm(t, log(CA));
```

Figure 2.17 MATLAB code for utilizing **fitlm** function where log(Y) is directly used within the arguments of **fitlm** function.

```
disp(mdl2);
k2 = -mdl2.Coefficients.Estimate('x1');  % Negative slope represents -k2
lnCA02 = mdl2.Coefficients.Estimate('(Intercept)'); % Intercept represents ln(CA02)
CA02 = exp(lnCA02);  % Calculate initial concentration (CA02)
fprintf('Rate constant (k2): %.4f\n', k2);
fprintf('Initial concentration (CA02): %.4f\n', CA02);
```

Figure 2.17 (Continued)

Workspace

Name	Value	Size	Class	Min	Max
CA	57x1 double	57x1	double	4.9787	100
CA0	100.1949	1x1	double	100.1949	100.1949
CA02	100.1949	1x1	double	100.1949	100.1949
CA_t	57x2 table	57x2	table		
k	0.1002	1x1	double	0.1002	0.1002
k2	0.1002	1x1	double	0.1002	0.1002
lnCA0	4.6071	1x1	double	4.6071	4.6071
lnCA02	4.6071	1x1	double	4.6071	4.6071
logCA	57x1 double	57x1	double	1.6052	4.6052
mdl	1x1 LinearModel	1x1	LinearModel		
mdl2	1x1 LinearModel	1x1	LinearModel		
t	57x1 double	57x1	double	0	30

Figure 2.18 MATLAB Workspace variables showing the complete match between the two approaches listed in figures (2.16) and (2.17).

```
% Plot scattered data
figure;
scatter(t, CA, 'o', 'MarkerEdgeColor', 'b', 'MarkerFaceColor', 'b');
hold on;
% Plot the fitted regression curve
t_fit = linspace(min(t), max(t), 100);
CA_fit = CA0 * exp(-k * t_fit);
plot(t_fit, CA_fit, 'r', 'LineWidth', 3);
% Add labels and title
xlabel('Time, minutes', 'FontSize', 16, 'FontWeight', 'bold');
ylabel('Concentration (C_{A})', 'FontSize', 16, 'FontWeight', 'bold');
title('Scattered Data and Fitted Regression Curve', fontsize=16);
% Add legend
legend('Scattered Data', 'Fitted Regression Curve', fontsize=14);
% Display the plot
ax = gca;
% Set new font size for all axes text (including labels)
ax.FontSize = 16;
hold off;
```

Figure 2.19 MATLAB code for showing a plot of both the original, scattered C_A vs. t data and the regressed curve for $C_A = C_{Ao}e^{-kt} = 100.1949 \times e^{-0.1002 \times t}$.

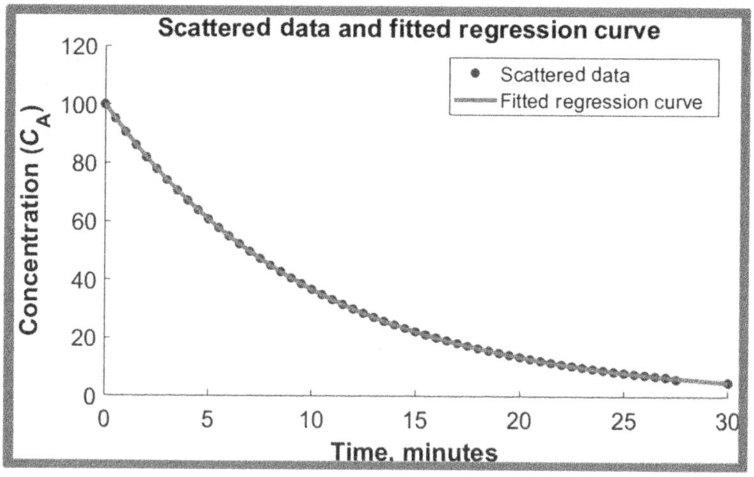

Figure 2.20 A plot of the original, scattered C_A vs. t data and the regressed curve for $C_A = C_{Ao}e^{-kt} = 100.1949 \times e^{-0.1002 \times t}$.

Back to Eq. (2.33), let us integrate for $n \neq 1$.

$$\frac{dC_A}{dt} = -kC_A^n \quad \rightarrow \quad \frac{dC_A}{C_A^n} = -kdt \quad \rightarrow \quad \int_{C_{Ao}}^{C_A} \frac{dC_A}{C_A^n} = \int_0^t -kdt$$

$$\left[\frac{C_A^{1-n}}{1-n} - \frac{C_{Ao}^{1-n}}{1-n}\right] = -kt \quad \rightarrow \frac{C_A^{1-n}}{1-n} = \frac{C_{Ao}^{1-n}}{1-n} - kt \qquad n \neq 1 \tag{2.37}$$

$$C_A^{1-n} = C_{Ao}^{1-n} - (1-n)kt \qquad n \neq 1 \tag{2.38}$$

Generally, for a reaction of nth-order, a graph of C_A^{1-n} vs. t must be a straight line with a slope of $(-(1-n)k)$ and an intercept of C_{Ao}^{1-n}. Unfortunately, n has to be given or be assumed ahead in order to make a plot with a straight-line relationship. If n is not known, then a trial-and-error procedure must be carried out until a straight-line relationship is obtained for a given n.

Let us calculate t when C_A goes to zero.

$$0 = C_{Ao}^{1-n} - (1-n)kt \rightarrow \qquad C_{Ao}^{1-n} = (1-n)kt \qquad \rightarrow t|_{C_A=0} = \frac{C_{Ao}^{1-n}}{(1-n)k} \tag{2.39}$$

A Reaction with Order $n > 1$: The reaction will never attain completion in finite (positive) time; otherwise, t will be negative.

A Reaction with Order $n < 1$: The reactant concentration will fall to zero at a certain finite time and then will become negative after that.

A Straight-line Plot for Special Cases of Eq. (2.37):

For $n = 0$ (zero – order reaction): $C_A = C_{Ao} - kt$ (Plot of C_A vs.t) (2.40)

For $n = 2$ (2nd – order reaction): $\frac{-1}{C_A} = \frac{-1}{C_{Ao}} - kt \rightarrow \frac{1}{C_A} = \frac{1}{C_{Ao}} + kt$ $\left(\text{Plot of } \frac{1}{C_A} \text{ vs.} t\right)$ (2.41)

For $n = 3$ (3rd – order reaction): $\frac{-2}{C_A^2} = \frac{-2}{C_{Ao}^2} - kt \rightarrow \frac{2}{C_A^2} = \frac{2}{C_{Ao}^2} + kt$ $\left(\text{Plot of } \frac{2}{C_A^2} \text{ vs.} t\right)$ (2.42)

From a linear regression point of view, this simply means that:
For $n = 0$, no need for any variable transformation; straight forward, $Y = C_A$ vs. $X = t$.
For case $n = 2$, transform $Y = \frac{1}{C_A}$ vs. $X = t$.
For case $n = 3$, transform $Y = \frac{2}{C_A^2}$ vs. $X = t$.

Example 2.26 Linear Transformation Prior to Using fitlm Function
Show how to carry out the proper linear transformation and find a and b using MATLAB **fitlm** function.

```
x =[1   2   3   4   5   6   7   8   9   10];
y=[ 0.0960  0.1691 0.2266 0.1691 0.0960  0.0558 0.0352  0.0239 0.0171 0.0128];
```

Use $\frac{1}{y} = a(x-3)^2 + b$
After finding a and b, make one plot for both the scattered data and the regressed line (or curve).

Solution:

Figure 2.21 shows MATLAB code for carrying out a linear regression using the **fitlm** function:

```
clear variables;
x =[1      2      3      4      5      6      7      8      9      10];
y=[ 0.0960 0.1691 0.2266 0.1691 0.0960 0.0558 0.0352 0.0239 0.0171 0.0128];
Y=1./y;
X=(x-3.0).^2.0;
%Y=aX+b; Both X and Y are transformed.
mdl = fitlm(X, Y);
disp(mdl);
a = mdl.Coefficients.Estimate('x1');  % a term represents slope of X
b = mdl.Coefficients.Estimate('(Intercept)');  % b term represents Intercept
fprintf('Slope a: %.4f\n', a);
fprintf('Intercept b: %.4f\n', b);
```

Figure 2.21 MATLAB code for y vs. x data transformation, followed by linear regression using **fitlm** function.

MATLAB results are shown below:

```
Linear regression model:
  y ~ 1 + x1
Estimated Coefficients:
                 Estimate      SE          tStat     pValue
                 ________    _________    ______    __________
  (Intercept)     4.3911     0.024314      180.6    9.8888e-16
  x1              1.503      0.0011224    1339.2    1.0828e-22
Number of observations: 10, Error degrees of freedom: 8
Root Mean Squared Error: 0.0571
R-squared: 1, Adjusted R-Squared: 1
F-statistic vs. constant model: 1.79e+06, p-value = 1.08e-22
Slope a: 1.5030
Intercept b: 4.3911
```

The regressed model becomes:

$$Y = aX + b \rightarrow \frac{1}{y} = a(x-3)^2 + b = 1.503 * (x-3)^2 + 4.3911$$

Figure 2.22 shows MATLAB code for making a plot of the scattered and regressed data. Notice here that we must retain the Workspace variables created by running Figure 2.21 code first, in order to smoothly run Figure 2.22 code; i.e., show the y–x scatter plot.

Figure 2.23 shows the generated combined plot of scattered and regressed data points.

```
% Plot scattered data; run after the previous code, please.
figure;
scatter(x, y, 'o', 'MarkerEdgeColor', 'b', 'MarkerFaceColor', 'b');
hold on;
% Plot the fitted regression curve
x_fit = linspace(min(x), max(x), 100);
y_fit = 1./(a * (x_fit-3.0).^2.0 + b);
plot(x_fit, y_fit, 'r', 'LineWidth', 3);
% Add labels and title
xlabel(' x-data points', 'FontSize', 16, 'FontWeight', 'bold');
ylabel('y-data/y_{fit}', 'FontSize', 16, 'FontWeight', 'bold');
title('Scattered Data and Fitted Regression Curve', fontsize=16);
% Add legend
legend('Scattered Data', 'Fitted Regression Curve', fontsize=14);
% Display the plot
ax = gca;
% Set new font size for all axes text (including labels)
ax.FontSize = 16;
hold off;
```

Figure 2.22 MATLAB code for making a plot of both scattered and regressed y vs. x data.

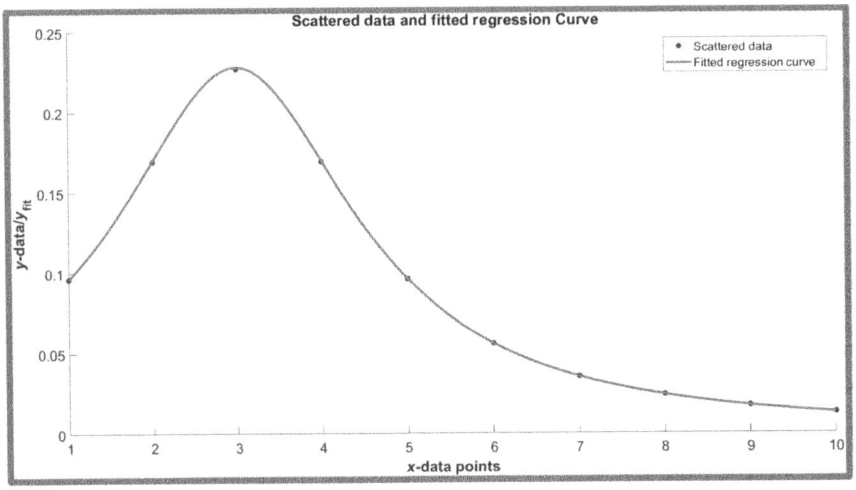

Figure 2.23 The scattered and regressed curve for y vs. x data. $\frac{1}{y} = a(x-3)^2 + b = 1.503 * (x-3)^2 + 4.3911$.

End of Chapter Problems

CHW 2.1 Mass vs. Weight

A man has a mass of 100.0 kg_m.

A What is his mass in lb_m?

B What is his weight in Newton?

C What is his weight in kg_f?

D What is his weight in lb_f?

CHW 2.2 Unit Conversion

Refer to Using dimensional conversion, convert

A One day to seconds

B 45.3 ft/s to miles/h

CHW 2.3 Unit Conversion

Using the table of conversion factors, convert

A 80 miles/h to km/h

B 6.95×10^3 kJ/min to kW

C 960 kg/m^3 to lbm/ft^3

CHW 2.4 Unit Conversion

A Convert 3.5 hours to seconds.

B Convert 75.6 m/s to kilometers/hour.

C Convert 120 miles/h to meters/second.

D Convert 500 kg to pounds.

E Convert 150 L to cubic feet.

CHW 2.5 Which of the following statements is the MOST accurate description of the relationship between accuracy and precision?

A They are always directly proportional to each other
B They are completely independent of each other
C Precise measurements guarantee accurate measurements
D Accurate measurements require but do not guarantee precise measurements

CHW 2.6 Which of the following factors MOST LIKELY contributes to low precision in measurements?

A A well-calibrated instrument
B Consistent measuring technique
C Random fluctuations in the environment
D Systematic bias in the instrument

CHW 2.7 The analogy in the chapter text compares accuracy to:

A Throwing a dart at a target blindfolded
B Using a ruler that is consistently too short
C Always being punctual, regardless of the destination
D Arriving at the correct place on time

CHW 2.8 What is the definition of error in the context of measurements?

A The difference between the highest and lowest values
B The deviation from a perfect accuracy
C The average value of repeated measurements
D The sum of random and systematic errors

CHW 2.9 Which type of error is characterized by unpredictable fluctuations arising from chance occurrences during measurement?

A Systematic Errors
B Determinate Errors
C Random (Indeterminate) Errors
D Gross Errors

CHW 2.10 What characteristic is shared by all systematic errors?

A They tend to average out with enough measurements.
B They are random in nature.
C They cannot be addressed through statistical methods.
D They have a specific cause and are reproducible.

CHW 2.11 Which type of systematic error can be minimized by calibrating equipment regularly?

A Instrumental Errors
B Operational Errors
C Methodical Errors
D Constant Systematic Errors

CHW 2.12 What is the impact of gross errors on data analysis?

A They improve the overall results
B They can significantly skew the overall results
C They have no impact
D They contribute to the precision of the analysis

CHW 2.13 How can gross errors be addressed for reliable data analysis?

A Ignore them, as they don't impact the analysis
B Include them to increase data variability
C Identify and exclude them from further analysis
D Average them with other data points

CHW 2.14 Which of the following numbers has the most significant figures?

A 0.005
B 123.4
C 10,000
D 1.20×10^3

CHW 2.15 How many significant figures does the number 2.500 have?

A 1
B 2
C 3
D 4

CHW 2.16 Multiplication

You multiply two numbers, one with three significant figures and another with four significant figures. How many significant figures should the result have?

A 3
B 4
C 5
D 7

CHW 2.17 Division

You divide a number with two significant figures by another number with four significant figures. What is the maximum number of significant figures the result can have?

A 4
B 3
C 2
D 1

CHW 2.18 Addition

You add two numbers, one with two decimal places and another with three decimal places. How many decimal places should the result have?

A 2
B 3
C 4
D 5

CHW 2.19 Subtraction

Consider the following subtraction: 1.234–0.234. What will be the final answer, taking into account the significant figures in both numbers?

A 1
B 1.0
C 1.00
D 1.000

CHW 2.20 Rounding off Numbers Ending in 5

Which of the following numbers is the correct rounded version of 1.75 according to the "rounding off to even" rule?

A 1.7
B 1.8
C 1.70
D 1.75

CHW 2.21 Cost Efficiency in Fuel Consumption #1:

You are trying to decide which of two automobiles to purchase. The first is American made, costs $15,700, and has a rated gasoline mileage of 30.0 mile/gal. The second car is of Japanese manufacture, costs $22,700, and has a rated mileage of 22 km/L. If the cost of gasoline is $3.30/gal and if the cars actually deliver their rated mileage, estimate how many miles you would have to drive for the lower fuel consumption of the second car to compensate for the higher cost of this car?

CHW 2.22 Cost Efficiency in Fuel Consumption #2:

You're considering purchasing two electric cars. Car *A* costs $32,000 and has an estimated range of 300 miles on a full charge. Car *B* costs $42,500 and has an estimated range of 400 miles on a full charge. Assuming the efficiency for both cars remains constant, determine how many miles you would need to drive Car *B* to match the cost efficiency of Car A.

CHW 2.23 Consumption of Annual Production of Crude Oil by Airplanes:

A supersonic aircraft consumes 25.0 m^3 of kerosene per hour of flight and flies an average of 15 hours per day. It takes roughly seven tons of crude oil to produce 1 ton of kerosene.

The density of kerosene is 0.965 g/cm^3. How many planes would it take to consume the entire annual world production of 6.05×10^9 metric tons of crude oil?

CHW 2.24 Weight of a Wastewater Treatment Pond:

A waste treatment pond is 50 m long and 15 m wide, and has an average depth of 2 m. The density of the waste is 85.3 lb_m/ft^3. Calculate the weight of the pond contents in lb_f.

CHW 2.25 Weight of a Swimming Pool:

A swimming pool has dimensions of 20 m in length, 10 m in width, and a constant depth of 2 m. Calculate the weight of water in the pool if the density of water is 1000 kg/m^3.

CHW 2.26 Environmental Impact:

An industrial plant emits liquid pollutants at a rate of 25 kg/min. If the pollutants have a density of 1.1 g/cm^3, estimate the volume of pollutants emitted in cubic meters in a day (i.e., volumetric flow rate of pollutants).

CHW 2.27 Nitrogen Volume Estimation:

Six hundred lb_m of nitrogen is to be charged into a small metal cylinder at 25°C, at a pressure such that the gas density is 11.5 kg/m^3. Estimate the required cylinder volume.

Show your work.

CHW 2.28 Fluid Flow: Reynolds Number:

The Reynolds number is a dimensionless group defined for a fluid flowing in a pipe as

$$\mathrm{Re} = \frac{\rho v D}{\mu},$$

where D is pipe diameter, v is fluid velocity, ρ is fluid density, and μ is fluid viscosity. When the value of the Reynolds number is less than about 2100, the flow is *laminar* that is, the fluid flows in smooth streamlines. For Reynolds numbers above 2100, the flow is *turbulent,* characterized by a great deal of agitation.

Liquid methyl ethyl ketone (MEK) flows through a pipe with an inner diameter of 2.067 in at an average velocity of 0.60 ft/s. At the fluid temperature of 20°C the density of liquid MEK is 0.810 g/cm^3 and the viscosity is 0.44 centipoise [1 cP = 1.00×10^{-3} kg/(m·s)]. Determine whether the flow is laminar or turbulent. Show your calculations.

CHW 2.29 Rate of Seed Crystal Nucleation:

A seed crystal of diameter ***D*** (mm) is placed in a solution of dissolved salt, and new crystals are observed to nucleate (form) at a constant rate r (crystals/min). The rate of nucleation varies with the seed crystal diameter as: r(crystals/min) = $180D - 9D^2$ (D in mm)

A What are the units of the constants 180 and 9? (Assume the given equation is valid and therefore dimensionally homogeneous).

B Calculate the crystal nucleation rate in crystals/s corresponding to a crystal diameter of 2 mm.

C Find the optimum D such that the rate of nucleation is maximum.

CHW 2.30 Heat Transfer Coefficient: Nusselt Number

The heat transfer coefficient (h) for a particular material is given by the equation: $h = k/L$, where k is thermal conductivity (in W/(m·K)) and L is the thickness of the material (in meters).

A Determine the heat transfer coefficient for a material with a thermal conductivity of 20 W/(m·K) and a thickness of 0.2 in.

B Formulate a dimensionless Nusselt number out of h, k, and L.

CHW 2.31 Radioactive Decay and Half-life:

The SI unit for the decay rate is becquerel per second (Bq/s) and directly represents the number of disintegrations occurring per second. It's the most widely used unit for scientific purposes. The rate of decay (R) of a radioactive substance is given by the equation: $R = R_0 \times e^{-\lambda t}$, where R_0 is the initial decay rate, λ is the decay constant, and t is time. If $R_0 = 100$ Bq/s and $\lambda = 0.01$/s, calculate the decay rate after 100 seconds. Express the final result in Bq per second. Find the half time ($t_{1/2}$) in seconds at which $R = 0.5R_o$.

CHW 2.32 Electrical circuits and Ohm's law:

For an electrical circuit, Ohm's law is represented as $\Delta V = IR$, where V is voltage (in volts), I is current (in amperes), and R is resistance (in ohms). If the voltage across a circuit is 12 V and the current passing through it is 2 amperes, determine the resistance of the circuit.

CHW 2.33 Projectile Motion and Range:

The range (R) of a projectile launched at an angle θ with initial velocity v_0 is given by the equation: $R = {}^{v_o^2 \sin(2\theta)}/_{g}$, where g is the acceleration due to gravity (in m/s^2). Calculate the range of a projectile launched at an angle of 30° with an initial velocity of 20 m/s in a location with 9.8 m/s^2.

Notice that when dealing with a consistent initial velocity, such as that generated by a cannon, the optimal range occurs when $\theta_0 = 45°$. However, this principle holds true solely under conditions that disregard air resistance. If air resistance is accounted for, the maximum angle shifts to approximately 38°. Interestingly, apart from the 45° initial angle, each other angle has a pair that yields an identical range, summing up to 90°.

CHW 2.34 Ideal Gas Equation:

The ideal gas law equation $PV = nRT$ relates pressure (P), volume (V), the number of moles of gas (n), the ideal gas constant (R), and temperature (T). Calculate the volume of 2 moles of an ideal gas at a temperature of 300 K and a pressure of 3 atm using the ideal gas constant $R = 0.0821$ L·atm/(mol·K).

CHW 2.35 Elasticity and Young's Modulus:

Young's modulus (Y) is defined as the ratio of stress to strain in a material ($Y = {}^{\sigma}/_{\varepsilon}$). Calculate Young's modulus for a material subjected to a stress, σ, of 5000 N/m^2 and experiencing a strain, ε, of 0.002 (dimensionless; $\varepsilon = {}^{L_0 + \Delta L}/_{L_o} = 1 + {}^{\Delta L}/_{L_o}$). Express the result in N/m^2 or Pascals. Express also in terms of MPa.

CHW 2.36 Sample Mean, S, Range, and 95% Confidence Interval

We grow 12 crystals from a solution and measure the length of each crystal in millimeters. Here is the recorded data:

$$9, 3, 5, 6, 12, 7, 8, 11, 9, 3, 8, 10$$

A Calculate the sample mean, S, and range of the length of the crystals.

B Using MATLAB, plot the scattered data points, augmented by the mean line and their 95% confidence interval lower and upper border line.

CHW 2.37 Precision of Sampling Methods

A calibration curve for the detection of an analyte in a sample and the following data were obtained using two different methods:

Concentration (ppm)	Method #1	Method #2
1.0	3.2	0.9
2.0	6.9	1.5
3.0	11.0	2.5
4.0	14.0	4.0
5.0	17.0	5.0
6.0	20.0	6.0
7.0	23.0	7.1
8.0	26.3	8.1
9.0	28.9	9.1
10.0	31.9	10.1

Sensitivity is defined as the ability to measure the analyte concentration accurately and specifically in the presence of other components (or solutes) that are present in the same sample matrix. Which one of the two methods has a better sensitivity? Use RSD as an index for sensitivity.

CHW 2.38 Linear Interpolation

Consider the rotameter calibration data (volumetric flow rate, $\dot{V}$ vs. rotameter reading, R) as follows:

The volumetric flow rate, $\dot{\boldsymbol{V}}$ vs. rotameter reading, R.

$\dot{V}$ ($^{L}/_{\min}$)	10	20	35	50	85	100	118	150
Rotameter Reading, *R*	5	10	20	30	50	60	70	90

A Calculate the flow rate that corresponds to a rotameter reading, $R = 40$, using the two-point linear interpolation. Use manual calculation or MATLAB code.

B Predict the rotameter reading for a volumetric flow rate $\dot{V} = 110$ L/min, using the two-point linear interpolation. Use manual calculation or MATLAB code.

CHW 2.39 Nonlinear Regression Models

Two of the following forms are nonlinear regression models:

A $Y = a + bX + cX^2$

B $Y = ax_1 + bx_2 + cx_3$

C $Y = ax_1x_2 + bx_2x_3 + cx_1x_3$

D $Y = a + b\sin(x_1) + c\sin(x_2)$

E $Y = a + b\sin(cx_1)$
F $Y = a + bx_1 + x_1^c$

CHW 2.40 Linear Transformation

Show how to carry out the proper linear transformation and find a and b using MATLAB fitlm function.

$x = [1 \quad 2 \quad 3 \quad 4 \quad 5 \quad 6 \quad 7 \quad 8 \quad 9 \quad 10];$

$y = [5.1270 \quad 5.9554 \quad 6.5911 \quad 7.1270 \quad 7.5991 \quad 8.0260 \quad 8.4185 \quad 8.7838 \quad 9.1270 \quad 9.4515];$

Use $y = ax^{1/2} + b$

After finding a and b, make in one plot both the scattered data and the regressed line (or curve).

CHW 2.41 Given the following C_A vs. time data for $A \rightarrow B$:

C_A	t, min	C_A	t, min
100.0	0	1.0527	10
9.6658	1	0.9579	11
5.0630	2	0.8787	12
3.4298	3	0.8116	13
2.5933	4	0.7541	14
2.0848	5	0.7041	15
1.7430	6	0.6604	16
1.4975	7	0.6218	17
1.3126	8	0.5874	18
1.1684	9	0.5567	19

The reaction order is to be determined first. Make the proper plot to tell whether the reaction order is one, two, or three. See Eqs. (2.35, 2.40–2.42). After finding the right reaction order, carry out a linear regression using the **fitlm** function and make a plot of the scattered and regressed data.

The data in the table are also given as row vectors to facilitate the MATLAB coding step

```
t=[0,1,2,3,4,5,6,7,8,9,10,11,12,13,14,15,16,17,18,19];
CA=[100.0, 9.6658, 5.0630, 3.4298, 2.5933, 2.0848, 1.7430, 1.4975, 1.3126, 1.1684, ...
1.0527, 0.9579, 0.8787, 0.8116, 0.7541, 0.7041, 0.6604, 0.6218, 0.5874, 0.5567];
```

Reference

1 Al-Malah, K. (2023). *Machine and Deep Learning Using MATLAB Algorithms and Tools for Scientists and Engineers*, Wiley, and Sons, Inc. https://onlinelibrary.wiley.com/doi/book/10.1002/9781394209118 https://www.amazon.com/Machine-Deep-Learning-Using-MATLAB/dp/1394209088/".

3

Precision Measurement: Instruments for Process Variables

A process encompasses any set of actions aimed at achieving a specific goal. Within this textbook, our focus lies on actions inducing physical or chemical alterations in substances or mixtures. The substance entering a process is termed the input or feed, while what exits is termed the output or product. Processes typically involve several stages, each executed within a designated process unit. Each process unit is linked to a distinct set of input and output process streams.

As a chemical engineer, you may be tasked with the design or operation of a process. Designing encompasses creating a process layout and detailing individual process units (like reactors, separation tools, heat exchangers) along with their operating specifications. The operation involves the daily management and execution of the process. The primary goal is to ensure that the process, along with all its equipment, consistently generates products at a predetermined rate and with specified qualities.

Furthermore, the chemical engineer's role might involve preserving the economic viability of the process by pinpointing strategies to minimize expenses, whether through reducing raw material usage or optimizing energy consumption. Occasionally, issues may arise causing the process to underperform, leading you to engage in troubleshooting efforts to identify and resolve the problem. Fluctuating market demands might necessitate a production rate exceeding the capabilities of current equipment under existing operating conditions, calling for debottlenecking measures. Conversely, situations may arise where production rates need to decrease, prompting concerns regarding turndown.

3.1 Mass and Volume

The **density** of a substance is defined as the mass per unit volume of the substance (kg/m^3, g/cm^3, lb$_m$/ft^3, etc.) On the contrary, the **specific volume** of a substance is the volume occupied by a unit mass of the substance; thus, it is the inverse of density. The densities of pure solids and liquids remain largely unaffected by pressure and exhibit minor changes in

To my dear readers,
When copying MATLAB code or text, please right-click and attempt both the "Copy with Formatting" and "Copy" options from the context menu. This approach is effective within Adobe Acrobat. If neither option works, kindly carefully review any MATLAB error messages and compare the pasted text to the original to identify and correct any discrepancies. Alternatively, a soft copy of the MATLAB code is available in the Wiley book companion site of this book for your convenience.

MATLAB-based Computations of Chemical Engineering Principles, First Edition. Kamal I. M. Al-Malah.

Companion website: www.wiley.com/go/al-malah/matlabbasedcomputations

response to temperature. These temperature-induced variations can go in either direction. For instance, the density of liquid water increases from 0.99987 g/cm^3 at 0°C to 1.0000 g/cm^3 at 3.98°C, then decreases to 0.95838 g/cm^3 at 100°C. Techniques for estimating the densities of gases and liquid mixtures are discussed in Chapter 5 of this textbook.

The density of a substance can be used as a conversion factor to relate the mass and the volume of a quantity of the substance.

Example 3.1

What is the mass of 1 L of liquid water at 4°C?

$$\text{Mass} = \text{density} \times \text{Volume} = 1\ \frac{\text{g}}{\text{cm}^3} \times 1\ \text{L} \times \left(\frac{1{,}000\ \text{cm}^3}{1\ \text{L}}\right) = 1{,}000\ \text{g} = 1\ \text{kg} \qquad (3.1)$$

The **specific gravity** of a substance is the ratio of the density ρ of the substance to the density

ρ_{ref} of a reference substance at a specific condition:

$$SG = \frac{\rho}{\rho_{\text{ref}}} \qquad (3.2)$$

where ρ_{ref} can be the density of any reference substance. The density of water is commonly used as a reference for solids and liquids.

$$\left[\rho_{\text{ref}} = \rho_{H_2O(l)}@4°\text{C} = \begin{array}{c} 1.0\ \text{g/cm}^3 \\ 1{,}000\ \text{kg/m}^3 \\ 62.43\ \text{lb}_\text{m}/\text{ft}^3 \end{array}\right] \qquad (3.3)$$

Note that the density of a liquid or solid in g/cm^3 is numerically equal to the specific gravity of that substance. The notation: $SG = 0.8^{20°\text{C}}/_{4°\text{C}}$, indicates that the specific gravity of a substance at 20°C with reference to water at 4°C is 0.8.

If we are given the specific gravity of a substance, multiply it by the reference density in any units to get the density of the substance in the same units. For example, if the specific gravity of a liquid is 1.50, its density will be 1.50×10^3 kg/m^3; 1.50 g/cm^3; or 1.5×62.43 lb$_\text{m}$/ft^3 = 93.64 lb$_\text{m}$/ft^3. Densities of both common liquids and solids are given in Appendix B | Table B.1 and Table B.2, respectively.

3.1.1 MATLAB Code for Searching and Presenting Density of Liquids and Solids

The density data for both liquids and solids, shown in Appendix B | Table B.1 and Table B.2 are combined as a database for searching the density of either liquid or solid and presenting the numerical value in both SI system as kg/m^3 and AE system as lb$_\text{m}$/ft^3.

Appendix B | B.3 "MATLAB Code for Searching Table B.1 and Table B.2 and Reporting Solid/Liquid Density" section shows how to carrying out the above mission, where it reads density data from either an Excel sheet or from a saved MATLAB *.mat file. First, the user has to select whether the material is solid or liquid. Second, the search keyword, which represents the name of the material, must be entered by the user. Third, the database will be searched for all occurrences of the search keyword and are presented before the user. Fourth, the user has to select the material of his/her choice by entering its row number. Fifth and last, the density is reported both in the SI system as kg/m^3 and in the AE system as lb$_\text{m}$/ft^3.

Sample calculations are shown below:

```
>> SLD_LIQ_DENS
Enter s or S for solid material, l or L for liquid material: s
You selected the solid material.
Enter a search keyword: gold
Matching entries:
 - gold (Row: 127)
Enter the row number for the compound: 127
   {'The compound names is: '}    {'gold'}
Solid Density: 19290 kg/m3
Solid Density: 1203.696 lbm/ft^3
>> SLD_LIQ_DENS
Enter s or S for solid material, l or L for liquid material: l
You selected the liquid material.
Enter a search keyword: water
Matching entries:
   - seawater Temperature is: 25 (Row: 212)
   - water - pure Temperature is: 4 (Row: 243)
   - water - sea Temperature is: 77oF (25oC) (Row: 244)
   - water, heavy Temperature is: 11.6 (Row: 245)
Enter the row number for the compound: 212
   {'The compound names is: '}    {'sea water'}
Liquid Density: 1025 kg/m3
Liquid Density: 63.96 lbm/ft^3
```

Exercise 3.1

1) What units does specific gravity measure?
2) A liquid possesses a specific gravity of 0.60. What is its density in g/cm³? What is its specific volume in cm³/g? What is its density in lbm/ft³? What is the mass of 4.0 cm³ of this liquid? What volume is occupied by 20 g?
3) If both substance A and substance B have a specific gravity of 1.45, is it necessary for 4 cm³ of A to have the same mass as 4 cm³ of B?
4) Does the density of liquid mercury increase or decrease with increasing temperature? Justify your answer using a thermometer as an illustration.

Example 3.2
The density of mercury, from Appendix B | Table B.1, is given as 13,595 kg/m³ at room temperature. Calculate the volume in ft³ occupied by 200 kg of mercury.

Solution: We will use Eq. (3.3) to carry out the conversion between SI and the American Engineering system.

$$\rho_{Hg} = 13{,}595\frac{\text{kg}}{\text{m}^3} \times \left(\frac{62.43\dfrac{\text{lb}_\text{m}}{\text{ft}^3}}{1000\dfrac{\text{kg}}{\text{m}^3}}\right) = 848.7\frac{\text{lb}_\text{m}}{\text{ft}^3}$$

$$\text{Volume of 200 kg mercury} = \frac{\text{Mass}}{\text{Density}} = \frac{200\,\text{kg} \times \left(\dfrac{1\,\text{lbm}}{0.4536\,\text{kg}}\right)}{848.7\,\dfrac{\text{lb}_\text{m}}{\text{ft}^3}} = \frac{440.92\,\text{lbm}}{848.7\,\dfrac{\text{lb}_\text{m}}{\text{ft}^3}} = 0.519\,\text{ft}^3 \tag{3.4}$$

As stated earlier, temperature and pressure do not have large influences on the densities of solids and liquids. Nevertheless, the fact that mercury in a thermometer rises or falls with changing temperature shows that the effect of temperature on liquid density is measurable. For example, the Handbook (***Perry's Chemical Engineers' Handbook***) gives the dependence of the volume of mercury on temperature as:

$$V(T) = Vo\left(1 + 0.18182 \times 10^{-3}T + 0.0078 \times 10^{-6}T^2\right), \tag{3.5}$$

where $V(T)$ is the volume of a given mass of mercury at temperature T(°C) and Vo is the volume of the same mass of mercury at 0°C.

Example 3.3

It was found in Example (3.2) that the volume (ft^3) of 200 kg mercury at 25°C is 0.519 ft^3, (1) What volume would the mercury occupy at 100°C? (2) Suppose the mercury is contained in a cylinder having a diameter of 0.30 in. What change in height would be observed as the mercury is heated from 25°C to 100°C?

Solution: From Eq. (3.5)

$$V(T) = Vo\left(1 + 0.18182 \times 10^{-3}(100°\text{C}) + 0.0078 \times 10^{-6}\left(100°\text{C}^2\right)\right) \tag{3.6}$$

So, Vo is unknown. However, applying Eq. (3.5) at $T = 25°\text{C}$, one may get the value at 0°C.

$$V(25°\text{C}) = 0.519 = V_\text{O}\left(1 + 0.18182 \times 10^{-3}(25°\text{C}) + 0.0078 \times 10^{-6}\left(25°\text{C}^2\right)\right) \tag{3.7}$$

$$V(25°\text{C}) = 0.519 = V_\text{O}\left(1 + 0.00455 + 0.000005\right) = V_\text{O}(1.0046) \tag{3.8a}$$

$$V(0°\text{C}) = V_\text{O}\left(1 + 0.18182 \times 10^{-3}(0°\text{C}) + 0.0078 \times 10^{-6}\left(0°\text{C}^2\right)\right) = V_\text{O} \tag{3.8b}$$

$$V(0°\text{C}) = V_\text{O} = \frac{0.519}{1.0046} = 0.517\text{ft}^3 \tag{3.9}$$

Substituting Vo in Eq. (3.6), one reaches at:

$$V(100°\text{C}) = 0.517\left(1 + 0.018182 \;\; + 0.000078\right) = 0.517(1.01826) = 0.5264\,\text{ft}^3 \tag{3.10}$$

$$\Delta V = \Delta(A.h) = A\Delta h \tag{3.11}$$

$$\Delta h = \frac{\Delta V}{A} = \frac{(0.5264 - 0.519)\text{ft}^3}{\dfrac{\pi(0.3/12)^2}{4}\text{ft}^2} = \frac{0.0074\text{ft}^3}{4.91 \times 10^{-4}\text{ft}^2} = 15.07\text{ft} = 15.07\,\text{ft} \times \frac{30.48\,\text{cm}}{1\,\text{ft}} = 459.5\,\text{cm} \tag{3.12}$$

Of course, this huge height change is for a mass of 200 kg mercury. Notice that for laboratory thermometers, a range of mercury content, depending on their size and purpose, can vary from 0.5 to 10 g. Moreover, the typical diameter of the capillary of a laboratory thermometer can vary depending on its specific purpose and desired sensitivity; however, a general range is between 0.1 and 0.5 mm.

So, for placing 1 g of mercury in a pipe of 0.1 mm, the height change, between 25°C and 100°C, will be:

$$\Delta h = \frac{\Delta V/\text{mass}}{A/\text{mass}} = \frac{\dfrac{0.0074\text{ft}^3}{200\text{ kg}}}{\dfrac{\dfrac{\pi\left(\dfrac{0.1}{304.8}\right)^2}{4}\text{ft}^2}{0.001\text{ kg}}} = \frac{3.7\times10^{-5}}{8.454\times10^{-5}}\text{ft} = 0.4376\text{ ft}\times\frac{30.48\text{ cm}}{1\text{ ft}} = 13.34\text{ cm} \tag{3.13}$$

Notice that the volume change per unit mass of substance due to change in temperature, will be the same for a small and large size of a sample of the same substance. This is scaling up/down of a system sample or size. The specific properties, like density or specific volume will be size independent given the same pressure and temperature for both sample sizes. Such properties are called intensive properties from thermodynamics point of view.

3.2 Mass and Volumetric Fluid Flow Rate

The majority of processes involve transferring materials from one location to another, occasionally between different process units or from a production facility to a transportation depot. The speed at which a material moves along a process line determines its flow rate.

The flow rate of a process stream may be expressed as a mass flow rate (mass/time) or as a volumetric flow rate (volume/time). Suppose a fluid (gas or liquid) flows in the cylindrical pipe shown in Figure 3.1, where the red circular area represents a section perpendicular to the direction of flow.

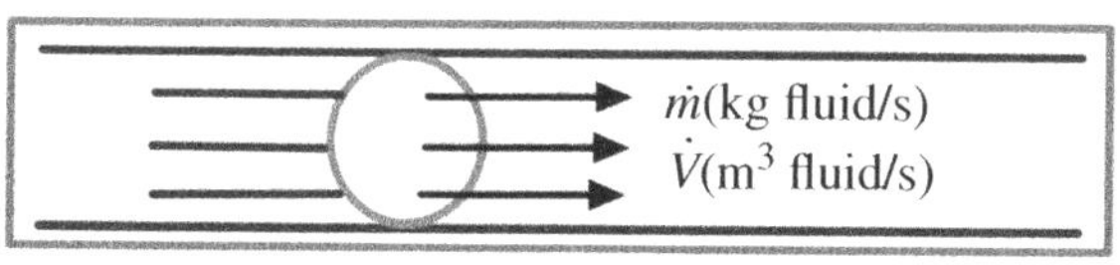

Figure 3.1 The fluid stream passes through a red circular cross-section perpendicular to the flow direction.

If the mass flow rate of the fluid is $\dot{m}$(kg/s), then every second m kilograms of the fluid pass through the cross-section. If the volumetric flow rate of the fluid at the given cross-section is $\dot{V}$(m^3/s), then every second V cubic meters of the fluid pass through the cross-section. However, the mass m and the volume V of a fluid-in this case, the fluid that passes through the cross-section each second are not independent quantities but are related through the fluid density, ρ:

$$\rho = \frac{m}{V} = \frac{\dot{m}}{\dot{V}} \Rightarrow \dot{m} = \rho\cdot\dot{V} = \rho\cdot A.v \tag{3.14}$$

A is the cross-sectional area of the fluid conveyer or carrier (i.e., pipe), and v is the average velocity of the fluid through the conveyer. Thus, the density of a fluid can be used to convert a known volumetric flow rate of a process stream to the mass flow rate of that stream or vice versa.

Exercise 3.2

1) The mass flow rate of n-hexane ($\rho = 0.66\,\text{g/cm}^3$) in a pipe is $\dot{m} = 6.6\,\text{g/s}$. What is the volumetric flow rate, $\dot{V}$, of n-hexane?
2) The volumetric flow rate of $CC1_4$ ($\rho = 1.6\,\text{g/cm}^3$) in a pipe is $\dot{V} = 120.0\,\text{cm}^3/\text{min}$. What is the mass flow rate, $\dot{m}$, of the $CC1_4$?

3.3 Fluid Flow Rate Measurement

A fluid flow measuring device refers to any instrument or apparatus designed to quantify the rate, volume, velocity, or mass of fluid passing through a system, pipeline, or channel. These devices are used across various industries to monitor and control the flow of liquids, gases, or steam in different applications. Flow measuring devices come in diverse types and principles of operation, including differential pressure-based meters like orifice plates, venturi meters, and flow nozzles, as well as positive displacement meters, electromagnetic flow meters, ultrasonic flow meters, turbine flow meters, and others.

The primary function of a fluid flow measuring device is to provide accurate and reliable measurements of the quantity or rate at which a fluid moves within a given system. They play a critical role in industrial processes, aiding in process optimization, quality control, and resource management. These devices help ensure the proper functioning of systems, maintain operational efficiency, and enable precise monitoring and management of fluid flow rates for a wide range of purposes across industries like oil and gas, chemical manufacturing, water treatment, heating, ventilation, and air conditioning (HVAC), and more. A YouTube video about flow measuring devices is available at: https://www.youtube.com/watch?v=f9LFJOno96c.

3.3.1 Rota Meter

A flow meter is a device mounted in a process line that provides a continuous reading of the flow rate in the line. A rotameter is a type of flow-measuring device widely used to gauge the flow rate of liquids or gases within a system. Its fundamental design consists of a tapered tube, usually made of glass or plastic, with a float or rotor inside. As the fluid passes through the tube, the float rises, and the height of the float within the tube indicates the flow rate. The larger the flow rate, the higher the float rises in the tube.

The principle behind this measurement is based on balancing the gravitational force acting on the float with the drag force caused by the moving fluid. The position of the float within the tube provides a visual indication of the flow rate, which can be read directly from a scale marked on the tube, as shown in Figure 3.2.

One of the significant advantages of a rotameter is its simplicity and ease of use. It doesn't require external power sources or complex mechanisms to operate, making it cost-effective

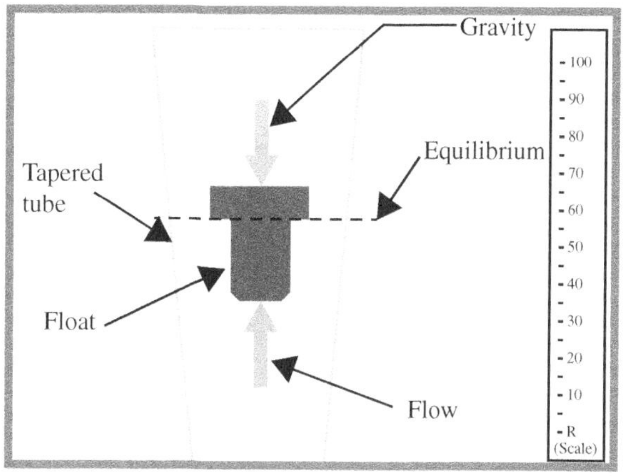

Figure 3.2 Rotameter is a flow-measuring device based on equilibrium (balance) between gravity force (down) and drag force (up).

and straightforward for measuring flow rates in various applications. However, while it is suitable for general flow measurement purposes, rotameters may not be as accurate or precise as some other flow meters, particularly in scenarios where high accuracy is essential or where the flow conditions fluctuate significantly. Despite this limitation, the rotameter remains a popular choice for its simplicity, reliability, and ability to provide a quick visual indication of flow rates in many industrial processes.

3.3.2 Differential Pressure Flow Meters

3.3.2.1 Orifice Meter

An orifice meter is a commonly used flow-measuring device designed to determine the rate of fluid flow within a pipeline. It comprises a plate known as an orifice plate, which is installed in the pipeline. This plate has a precisely sized hole (or orifice) drilled through it. As fluid passes through the orifice, it creates a pressure drop, and this drop in pressure is measured to calculate the flow rate. Typically, pressure taps placed upstream and downstream of the orifice plate are connected to a differential pressure gauge or transmitter. The differential pressure sensed across the orifice plate correlates directly with the flow rate and can be calculated using established equations, such as the Bernoulli equation, to accurately determine the fluid flow, as shown in Figure 3.3. The fluid pressure drops (decreases) from the upstream side of the orifice to the downstream side; the pressure-drop (which may be measured with a number of devices, including a differential manometer) varies with the flow rate; the greater the flow rate, the larger the pressure drop.

Orifice meters are favored for their simplicity, reliability, and relatively low cost compared to some other flow-measuring devices. They are used across various industries to measure the flow of liquids, gases, and steam. However, while orifice meters are widely utilized, it is important to note that they may have limitations in accuracy and can be sensitive to changes in fluid properties, temperature, and pressure. Proper installation, calibration, and consideration of these factors are crucial for obtaining accurate flow measurements using orifice meters.

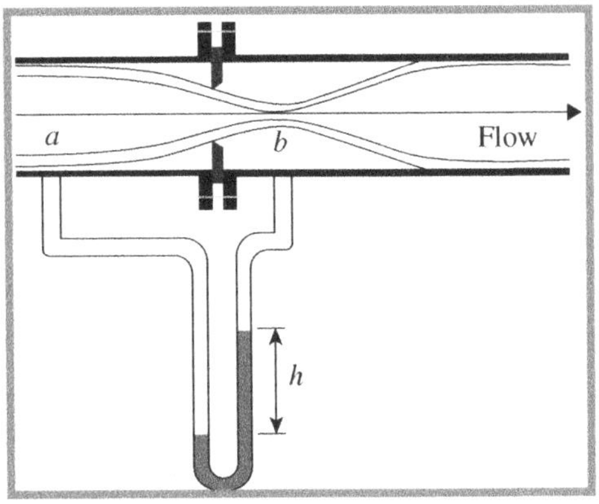

Figure 3.3 Orifice-meter, a flow measuring device based on installing an obstruction that results in pressure drop, reflecting the velocity or flow rate.

3.3.2.2 Venturi Meter

A venturi meter is a flow-measuring device utilized to determine the rate of fluid flow within a pipeline. It consists of a constricted section in the pipeline shaped like a venturi tube, as shown in Figure 3.4. As the fluid passes through this constricted portion, its velocity increases while the pressure decreases. This change in pressure is measured using pressure taps located before and within the venturi tube. By calculating the pressure differential between these two points, the flow rate of the fluid can be accurately determined. The design of the venturi meter allows for more precise flow measurement compared to other devices, as it minimizes energy loss and offers improved accuracy, especially in applications with varying flow rates and fluid viscosities.

Venturi meters are known for their accuracy and reliability in measuring flow rates of liquids, gases, and steam in various industries. They provide advantages such as low-pressure loss, wide operational range, and less susceptibility to erosion or damage compared to other flow meters. However, the installation and maintenance of venturi meters require expertise to ensure accurate measurements, and they can be more expensive to install than simpler flow measurement devices like orifice meters. Despite this, venturi meters remain a popular choice for applications requiring precise and reliable flow measurement.

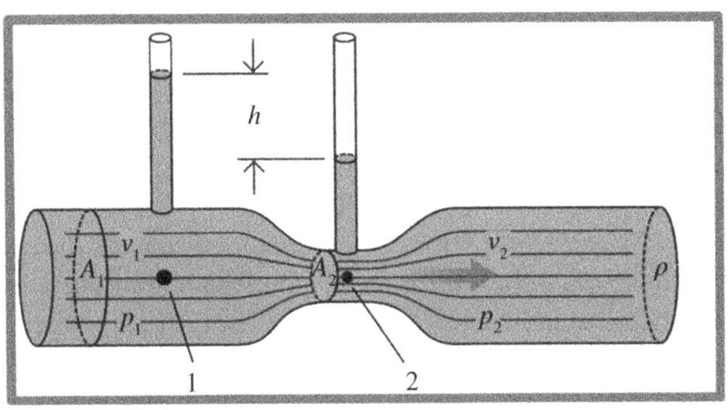

Figure 3.4 Venturi meter, a flow measuring device based on installing a specimen made of converging, throat, and diverging sections, which results in pressure drop, reflecting the velocity or flow rate. *Source:* https://en.wikipedia.org/wiki/File:Venturi5.svg.

3.3.2.3 Flow Nozzle

It is a compromise between an orifice plate and a venturi meter. It is more accurate than an orifice plate, but it is not as expensive or as difficult to install as a venturi meter. A flow nozzle serves as a flow-measuring device utilized to gauge the rate of fluid flow within a pipeline. It comprises a specially designed constricted section, resembling a nozzle, inserted within the pipeline. As the fluid passes through this nozzle-shaped section, its velocity increases while the pressure decreases. The pressure drop across the flow nozzle is measured using pressure taps situated before and within the nozzle. This pressure differential is then used to calculate the flow rate of the fluid. Flow nozzles are valued for their ability to handle high flow rates and provide reliable measurements across various applications, including those involving steam, gases, and liquids.

One of the notable advantages of flow nozzles is their accuracy in flow measurement, especially in high-velocity flows, as they experience lower pressure losses compared to other flow meters. They are often preferred in industrial settings due to their robustness, versatility, and capability to accurately measure flow rates in challenging operating conditions. However, similar to other differential pressure-based flow meters, precise installation, calibration, and consideration of fluid properties is vital to ensuring accurate flow measurements using flow nozzles. Despite potential complexities in installation and maintenance, flow nozzles remain a trusted choice for industries requiring precise and reliable flow measurements. Figure 3.5 shows a sketch for a flow nozzle.

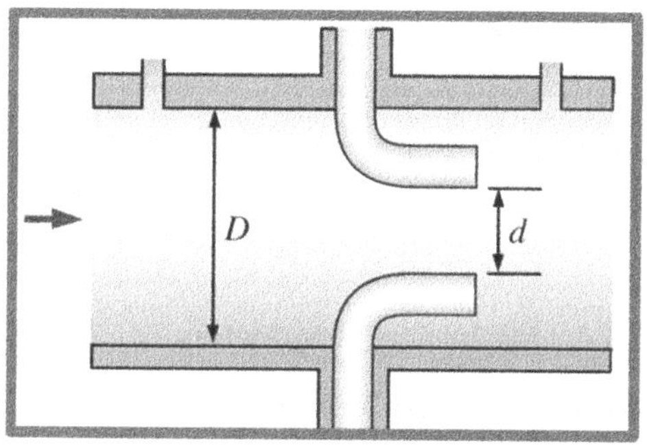

Figure 3.5 Flow nozzle, a flow measuring device based on installing a specimen made of converging and throat section, which results in pressure drop, reflecting the velocity or flow rate. *Source:* https://electricalworkbook.com/flow-nozzle/, with permission of Electrical Workbook.

3.3.3 Positive Displacement Flow Meters

A positive displacement flow meter operates on the principle of measuring the volume of fluid passing through by repeatedly filling and emptying fixed-volume chambers within the meter. These chambers move the fluid in discrete increments, counting the number of these increments to calculate the total volume of fluid passing through the meter. This metering mechanism is highly accurate and reliable, particularly for viscous fluids or those with varying flow rates. Positive displacement flow meters come in various designs, such as *gear meters, oval gear meters, piston meters, and rotary vane meters*, each utilizing chambers or gears to measure the fluid volume passing through them.

One of the key advantages of positive displacement flow meters is their exceptional accuracy, particularly in low-flow applications and when dealing with viscous fluids. They are capable of precise measurements regardless of changes in temperature, pressure, or viscosity of the fluid being measured. However, their design often leads to higher pressure drops compared to some other flow meters. In addition, they may not be suitable for high-flow applications due to their limited capacity and the potential for increased wear and tear over time. Despite these limitations, positive displacement flow meters remain widely used in industries requiring accurate measurement of fluid flow rates, such as in the pharmaceutical, food and beverage, and chemical sectors.

Figure 3.6 shows a positive displacement flow meter known as the oval gear type which comprises a pair of interlocking, oval-shaped gears with closely fitting teeth that rotate together. As these gear shafts turn, a precise volume of liquid is propelled through the meter. By tracking the rotations of the shafts, the rate of liquid flow can be determined. These meters demonstrate high accuracy when the minimal slippage occurring between the housing and the gears is meticulously controlled.

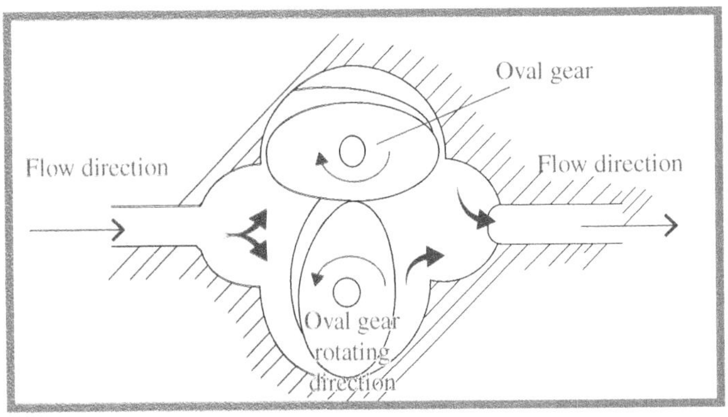

Figure 3.6 An oval gear type positive displacement flow meter.
Source: https://blog.industrialguide.co.in/2021/05/positive-displacement-flow-meter-parts.html, with permission of Industrial Guide.

3.3.4 Mechanical Flow Meters

A mechanical flow meter is a type of flow-measuring device that quantifies the rate of fluid flow within a system using mechanical components and principles. These meters typically involve mechanical parts like gears, turbines, or pistons that are actuated by the flow of the fluid. As the fluid moves through the meter, it causes these internal components to rotate, creating a measurement mechanism based on the rotations or displacements. The frequency or magnitude of these movements directly correlates to the volume or rate of fluid passing through the meter. Various types of mechanical flow meters exist, such as *positive displacement meters, turbine meters, and gear meters*, each employing distinct mechanisms to measure fluid flow rates accurately.

Mechanical flow meters are renowned for their reliability, durability, and accuracy in measuring the flow of both liquids and gases in a wide array of industrial applications. They are suitable for various flow rates and often have minimal sensitivity to changes in the fluid's properties, such as viscosity or temperature. However, these meters might experience limitations in measuring extremely low flows or highly viscous fluids due to potential limitations in their design. Despite these considerations, mechanical flow meters remain popular and widely used due to their robustness, simplicity, and effectiveness in providing accurate measurements of fluid flow rates in diverse industries.

3.3.4.1 Turbine Flow Meter

A turbine flow meter is a type of flow-measuring device that utilizes a turbine-like rotor placed within a pipe or conduit to measure the rate of fluid flow. As the fluid passes through the meter, it impinges on the blades of the rotor, causing it to rotate. The rotational speed of the turbine rotor is directly proportional to the velocity of the fluid passing through the meter. This rotational movement is then converted into a flow rate measurement. The meter often includes sensors that detect the rotations, translating them into volume or flow rate readings. Turbine flow meters are highly efficient for measuring flow rates in both liquids and gases across various industries due to their accuracy, versatility, and ability to handle a wide range of flow rates.

One of the primary advantages of turbine flow meters is their accuracy, particularly in measuring higher flow rates. They offer a relatively wide measurement range and can maintain precision in various flow conditions. However, these meters might be susceptible to damage by abrasive or viscous fluids, potentially affecting their accuracy over time. Proper installation, calibration, and consideration of the fluid properties are essential to ensure accurate measurements with turbine flow meters. Despite some limitations, they remain a popular choice for applications requiring precise measurement of fluid flow rates, including in industries such as petroleum, chemical processing, water management, and more. A YouTube video about a turbine flow meter is available at: https://youtu.be/-RvwXGzzv4c.

3.3.4.2 Nutating Disk Flow Meter

A nutating disk flow meter is a type of positive displacement flow measuring device used to gauge the flow rate of liquids passing through a pipeline. Its design consists of a disk or wobble plate positioned off-center in the meter chamber, causing it to nutate (a periodic variation in the inclination of the axis of a rotating object) or wobble as the fluid flows through. This movement creates a changing chamber volume, displacing a fixed quantity of liquid with each nutation. The number of disk rotations is then measured to determine the volume of liquid passing through the meter, which enables the calculation of the flow rate. Nutating disk flow meters are known for their accuracy, reliability, and suitability for measuring the flow of low-viscosity liquids in various industries such as water treatment, chemical processing, and utilities.

One of the significant advantages of nutating disk flow meters is their ability to maintain accuracy across a broad range of flow rates, especially in applications where consistent and precise measurement of low-flow rates is crucial. These meters also exhibit minimal pressure loss, making them efficient in energy consumption. However, they may be less suitable for highly viscous fluids due to potential challenges in the nutation mechanism's operation. Despite this limitation, nutating disk flow meters are valued for their reliability, versatility, and accurate measurement capabilities in diverse industrial settings. A YouTube video about the nutating disk flow meter is available at: https://www.youtube.com/watch?v=TYX884HI6kw.

3.3.5 Other Types of Flow Meters

Since such types are based on a combination of electric, magnetic, and mechanical components, the description is given below for each sub-type and a relevant YouTube video link is provided for the sake of demonstration and simplifying the basic principle underlying the prototype of each flow meter.

3.3.5.1 Electromagnetic Flow Meter

An electromagnetic flow meter is a type of flow-measuring device used to determine the flow rate of conductive fluids, such as water or other liquids, by utilizing electromagnetic principles. This meter comprises a pipe section with electrodes on the walls and a magnetic field generator typically surrounding the pipe. When the conductive fluid flows through the pipe, the meter applies a magnetic field perpendicular to the flow direction. As the fluid moves, it generates a voltage proportional to its velocity, which is measured by the electrodes. The voltage induced is directly related to the flow rate, allowing the electromagnetic flow meter to accurately measure the velocity of the fluid and compute the volume flow rate. This type of flow meter is particularly effective for measuring the flow of highly conductive liquids and slurries in various industries such as water treatment, wastewater management, and chemical processing.

One of the primary advantages of electromagnetic flow meters is their capability to provide accurate and reliable measurements in challenging flow conditions, including those involving corrosive, abrasive, or viscous fluids. Since these meters don't have moving parts within the flow stream, they are suitable for applications where minimal obstruction and maintenance are essential. However, electromagnetic flow meters may have limitations in measuring nonconductive fluids or those with low conductivity. Despite this restriction, they are favored for their accuracy, durability, and versatility, making them a popular choice for industries requiring precise measurement of fluid flow rates in various environments. A YouTube video of an electromagnetic flow meter is available about an electromagnetic flow meter at: https://www.youtube.com/watch?v=f949gpKdCI4.

3.3.5.2 Coriolis Flow Meter

A Coriolis flow meter is a highly precise and sophisticated flow-measuring device that operates based on the Coriolis effect, which involves the apparent deflection of moving objects caused by the rotation of the Earth. This meter consists of a tube or tubes through which the fluid flows. As the fluid moves through these tubes, the meter induces oscillations or vibrations, causing the tubes to twist or deform. Simultaneously, the meter generates an alternating current or voltage to drive the tubes' vibration. Sensors on the tubes detect the degree of deformation, and the phase shift between the input signal and the detected vibration provides information about the mass flow rate, density, and temperature of the fluid. Coriolis flow meters excel in accurately measuring flow rates of both liquids and gases, and they are widely used across industries such as oil and gas, pharmaceuticals, and food processing due to their exceptional accuracy and ability to handle diverse fluids.

One of the primary advantages of Coriolis flow meters is their unparalleled accuracy in measuring fluid flow rates and properties. Unlike some other flow meters, Coriolis meters measure mass flow directly, which remains unaffected by changes in temperature, pressure, or fluid properties like density or viscosity. This makes them particularly valuable for applications requiring precise measurements and where stringent control over fluid properties is essential. However, Coriolis flow meters can be relatively expensive and may have limitations in measuring extremely high flow rates or fluids with high concentrations of solids, which might cause wear on the meter components. Despite these considerations, their unparalleled accuracy and ability to measure multiple parameters simultaneously make Coriolis flow meters a top choice in industries where precision and reliability are critical. A YouTube video about the Coriolis flow meter is available at: https://www.youtube.com/watch?v=31jYXlnu-hU.

3.3.5.3 Ultrasonic Flow Meter

An ultrasonic flow meter is a nonintrusive flow measuring device used to determine the rate of fluid flow within a pipeline without requiring direct contact with the flowing substance. This meter operates by transmitting ultrasonic waves through the fluid and measures the time it takes for these waves to travel between transducers placed at different points along the pipe. In a transit-time ultrasonic flow meter, one transducer sends ultrasonic pulses, and another transducer receives these pulses, calculating the difference in time taken for the waves to travel with and against the flow direction. The meter then uses this time differential to compute the flow rate. Doppler ultrasonic flow meters, on the other hand, rely on the frequency shift of ultrasonic waves bouncing off the moving particles within the fluid to determine flow velocity. Ultrasonic flow meters are versatile, applicable to various types of liquids and gases, and are commonly used in industries such as water management, wastewater treatment, and oil and gas for their noninvasive nature and capability to measure flow in large-diameter pipelines.

One of the key advantages of ultrasonic flow meters is their nonintrusive nature, allowing measurement without interrupting the flow or requiring insertion into the pipeline. They are suitable for a wide range of fluids, including clean and dirty liquids as well as gases. Moreover, ultrasonic flow meters offer versatility in their installation, working well in pipes of different materials, sizes, and shapes. However, certain factors like the presence of air bubbles, suspended solids, or significant changes in fluid composition or temperature can impact the accuracy of measurements. Despite potential limitations, ultrasonic flow meters remain popular due to their noninvasive design, adaptability, and capability to measure flow rates accurately in diverse industrial applications. A YouTube video about ultrasonic flow meters is available at: https://www.youtube.com/watch?v=Bx2RnrfLkQg.

3.3.5.4 Vortex Flow Meter

Vortex flow meters are instruments used to measure the flow rate of fluids within a pipeline by employing the principle of the von Kármán effect. These meters consist of a bluff body or a shedder bar installed perpendicular to the fluid flow. As the fluid moves past this obstruction, it creates alternating vortices or swirls downstream of the bluff body. Sensors detect these vortices, and the frequency of these vortex-shedding patterns is directly proportional to the fluid velocity. By measuring this frequency, the vortex flow meter calculates the flow rate. These meters are effective for measuring the flow of both liquids and gases in various industries due to their reliability, accuracy, and ability to handle fluctuating flow conditions.

One of the primary advantages of vortex flow meters is their capability to provide accurate measurements even in challenging flow conditions, such as those with varying flow rates or densities. They are suitable for applications requiring precise flow measurements, and their design often allows for minimal pressure loss compared to some other flow meters. However, the accuracy of vortex flow meters can be affected by factors like fluid properties, installation conditions, and the presence of particulates or bubbles in the fluid stream. Despite these considerations, vortex flow meters are widely utilized in industries such as oil and gas, chemical processing, and water management due to their robustness, versatility, and ability to provide accurate flow rate measurements in various operating conditions. A YouTube video about the vortex flow meter is available at: https://www.youtube.com/watch?v=GmTmDM7jHzA.

Example 3.4
A steadily flowing stream of water is funneled into a graduated cylinder for exactly 30.0 seconds, during which time 60.0 mL is collected. What is the volumetric flow rate of the stream? What is the mass flow rate?

Solution:

$$\dot{V} = \frac{60.0\ \text{mL}}{30.0\ \text{s}} = 2.00\ \frac{\text{mL}}{\text{s}} = 2.00\ \frac{\text{cm}^3}{\text{s}} \tag{3.15}$$

$$\dot{m} = \rho \cdot \dot{V} = 1\frac{\text{g}}{\text{cm}^3} \times 2.00\ \frac{\text{cm}^3}{\text{s}} = 2.00\ \frac{\text{g}}{\text{s}} = 2.00\ \frac{\text{g}}{\text{s}} \times \left(\frac{\text{kg}}{10^3\text{g}}\right) = 0.002\frac{\text{kg}}{\text{s}} \tag{3.16}$$

Exercise 3.3
If the mass flow rate of mercury ($\rho = 13.59\ \text{g/cm}^3$) in a pipe is $\dot{m} = 0.1\ \text{g/s}$. What is the volumetric flow rate, $\dot{V}$, of the mercury?

3.4 Chemical Composition

Almost all substances found in nature and within chemical process systems are combinations of different elements or compounds. The characteristics of a mixture/solution are significantly influenced by its specific composition. In this section, we will examine diverse approaches for indicating mixture compositions. Later in the book, we will detail techniques for predicting the physical properties of a mixture/solution based on the properties of its individual components.

3.4.1 Moles and Molecular Weight

The element's atomic weight represents the mass of an atom measured against a scale. For example, ^{12}C means that the carbon isotope is housing six protons and six neutrons within its nucleus and thus is precisely assigned a mass of 12 atomic mass unit (AMU). The periodic table, displayed in Figure 3.7, lists the atomic weights of all elements according to their naturally occurring isotopic proportions. The molecular weight of a compound is derived from the summation of the atomic weights of the constituent atoms forming a molecule of the compound. For instance, atomic oxygen (O) bears an approximate atomic weight of 16, leading to a molecular weight of approximately 32 for molecular oxygen (O_2).

On the other hand, a gram-mole (abbreviated as g-mole or mol in SI units) of a substance represents the quantity of that substance whose mass in grams numerically matches its molecular weight. When referring to an element, technically, it is accurate to use the term gram-atom instead of gram-mole. However, we will use moles for both elements and compounds without distinguishing between them. Other mole units like kg-moles (kmol), lb-moles, or ton-moles follow similar definitions. For instance, considering carbon monoxide (CO) with a molecular weight of 28, 1 mol of CO comprises 28 g, while 1 lb-mole consists of 28 lbm, and 1 ton-mole contains 28 tons, and so forth.

If the molecular weight of a substance is ***M****, then there are* ***M*** kg/kmol. ***M*** *g/mol, and* ***M*** lbm/lb-mole *of this substance*. The molecular weight may thus be used as a conversion factor that

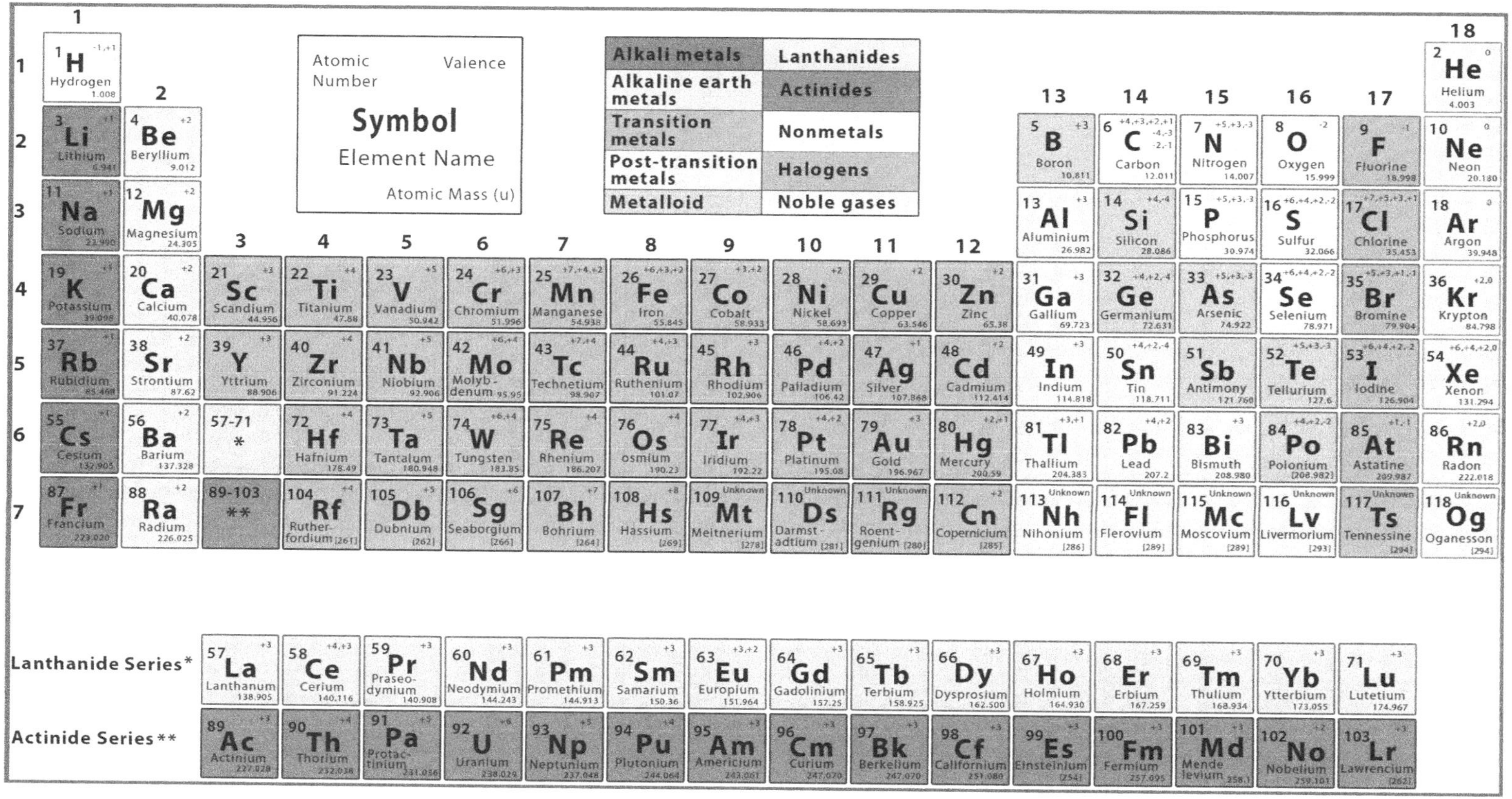

Figure 3.7 The periodic table of elements in their naturally occurring isotopic proportions.

relates the mass and the number of moles of a quantity of the substance. For example, 60.0 kg of ammonia (NH_3: $M = 17.0$) is equivalent to:

$$60.0\,\text{kg}\,NH_3 \times \left(\frac{\text{kmol}\,NH_3}{17.0\,\text{kg}NH_3}\right) = 3.53\,\text{kmol}\,NH_3 \tag{3.17}$$

and 12.0 lb-moles of ammonia is equivalent to:

$$12.0\,\text{lb} - \text{mol}\,NH_3 \times \left(\frac{17.0\,\text{lbm}\,NH_3}{1\,\text{lb} - \text{mol}\,NH_3}\right) = 204\,\text{lbm}\,NH_3 \tag{3.18}$$

The same factors used to convert masses from one unit to another must be used to convert the equivalent molar units: For example, there is 454 g/lb_m; hence, there is 454 g-mol/lb-mole, regardless of the substance involved. *One gram-mole of any species contains approximately* 6.022×10^{23} *(Avogadro's number) molecules of that species.*

Example 3.5
How many of each of the following are contained in 500.0 g of H_2O (***M*** = 18.02)? (a) mol H_2O; (b) lb-moles H_2O; (c) mol O; (d) mol H_2; (e) mol H; (f) g O; (g) g H_2; (h) g H; and (i) molecules of H_2O

Solution

a) Moles of H_2O = $500.0\,\text{g}\,H_2O \times \left(\frac{\text{mol}\,H_2O}{18.02\,\text{g}\,H_2O}\right) = 27.75\,\text{mole}\,H_2O$

b) lb-moles of H_2O = $27.75\,\text{mole}\,H_2O \times \left(\frac{Ib - \text{mol}}{454.0\,\text{mol}}\right) = 0.06111\,\text{lb-mol}$

> Each molecule of H_2O contains one atom of O, one molecule of H_2, or two atoms of H. Therefore, each 6.022×10^{23} molecules of H_2O (1 mol) contain **1** mol **O**, **1** mol **H_2**, or **2** mol **H**. Thus,

c) Moles of O = $27.75\,\text{mole}\,H_2O \times \left(\frac{1\,\text{mol}\,O}{1\,\text{mol}\,H_2O}\right) = 27.75\,\text{mol}\,O$

d) Moles of H_2 = $27.75\,\text{mole}\,H_2O \times \left(\frac{1\,\text{mol}\,H_2}{1\,\text{mol}\,H_2O}\right) = 27.75\,\text{mol}\,H_2$

e) Moles of H = $27.75\,\text{mole}\,H_2O \times \left(\frac{2\,\text{mol}\,H}{1\,\text{mol}\,H_2O}\right) = 55.50\,\text{mol}\,H$

f) Mass of O = $27.75\,\text{mol}\,O \times \left(\frac{16.0}{1\,\text{mol}\,O}\right) = 444\,\text{g}$

g) Mass of H_2 = $27.75\,\text{mol}\,H_2 \times \left(\frac{2.0\,\text{g}}{1\,\text{mol}\,H_2}\right) = 55.5\,\text{g}$

h) Mass of H = $55.5\,\text{mol}\,H \times \left(\frac{1.0\,\text{g}}{1\,\text{mol}\,H}\right) = 55.5\,\text{g}$

i) Molecules of H_2O = $27.75\,\text{mol}\,H_2O \times \left(\frac{6.022 \times 10^{23}\text{molecules}}{1\,\text{mol}\,H_2O}\right) = 1.671 \times 10^{25}\text{molecules}$

3.4.2 Mass and Mole Fractions and Average Molecular Weight

It is less likely that process streams contain a single substance, yet more commonly, they comprise mixture of liquids or gases, or solutions where one or more solutes are dissolved in a liquid solvent. The subsequent vocabulary is often utilized to describe the makeup of a combination of substances encompassing a particular species, denoted as A.

$$\text{Mass fraction: } x_A = \frac{\text{mass of A}}{\text{total mass}} \left[\left(\frac{\text{kg A}}{\text{kg total}} \right), \left(\frac{\text{g A}}{\text{g total}} \right), \left(\frac{\text{lbm A}}{\text{lbm total}} \right) \right] \quad (3.19)$$

$$\text{Mole fraction: } X_A = \frac{\text{moles of A}}{\text{total moles}} \left[\left(\frac{\text{kmol A}}{\text{kmol total}} \right), \left(\frac{\text{mol A}}{\text{mol total}} \right), \left(\frac{\text{lb} - \text{mol A}}{\text{lb} - \text{mol total}} \right) \right] \quad (3.20)$$

The percent by mass of A is $100x_A$, and the mole percent of A is $100X_A$.

Example 3.6 Conversions Using Mass and Mole Fractions

A solution contains 35% **A** by mass ($x_A = 0.35$) and 40 mole% **B** ($X_B = 0.40$).

a) Calculate the mass of **A** in 200 kg of the solution.
Basis: 200 kg solution made of both A and B.
Mass of A in the 200 kg-solution $= x_A \times \text{total mass} = 0.35\left(\text{kg A}/\text{kg total}\right) \times 200\ \text{kg total} = 70\ \text{kg A}$

b) Calculate the mass flow rate of **A** in a stream of solution flowing at a rate of 60 lbm/h.
The mass flow rate of A $= x_A \times \text{total mass flow rate} = 0.35\left(\text{lbm A}/\text{lbm total}\right) \times 60\ \text{lbm total}/h = 21\ \text{lbm A}/h$

Note

If a mass or molar unit such as lbm in 60 lbm/h is not followed by the name of a species, the unit should be understood that it refers to the total mixture or solution rather than to a specific component.

c) Calculate the molar flow rate of **B** in a stream flowing at a rate of 1000 mol/min.
The molar flow rate of **B** $= X_B \times \text{total molar flow rate} = 0.40\left(\text{mol B}/\text{mol total}\right) \times 1000\ \text{mol total}/\text{min} = 400\ \text{mol B}/\text{min}$

d) Calculate the total solution flow rate that corresponds to a molar flow rate of 60 kmol ***B***/s.
The total solution flow rate $= 60\left(\text{kmol B}/\text{s}\right) \times \left(1\ \text{kmol total}/0.40\ \text{kmol B}\right) = 150\ \text{kmol}/\text{s}$

e) Calculate the mass of the solution that contains 700 lbm of **A**.
The total mass of the solution $= 700\ \text{lbm A} \times \left(\text{lbm total}/0.35\ \text{lbm A}\right) \times = 2000\ \text{lbm total}$

Note

The *numerical value of a mass or a mole fraction does not depend on the mass units in the numerator and denominator as long as these units are the same*. If the mass fraction of benzene (C_6H_6) in a mixture is 0.40, then $x_{C_6H_6}$ equals 0.40 (kg C_6H_6/kg total), 0.40 (g C_6H_6/g total), 0.40 (*lbm* C6H6/lbm total), and so on.

Example 3.7 Conversion from a Composition by Mass to a Molar Composition
A mixture of gases has the following composition by mass, as shown in Table 3.1:

Table 3.1 Gas composition on a mass basis.

Component *i*	Percent Mass Composition (%)
O_2	20
CO	5.0
CO_2	25
N_2	50
Total	100

What is the molar composition of this gas mixture?

Solution: *Basis: 100 g of the mixture.*
See Table 3.2 for tabulation of results.

Table 3.2 Conversion of mass composition into molar composition for the gas mixture.

Component *i*	Percent Mass Composition (%)	$x_i = \dfrac{g\,i}{g\,\text{total}}$	Mass of *i* $m_i = x_i \times m_{tot}$	Molecular weight, M_i $\left(\dfrac{g\,i}{\text{mol of } i}\right)$	Moles of *i*, $n_i = \dfrac{m_i}{M_i}$	Mole fraction, $X_i = \dfrac{n_i}{n_{tot}}$
O_2	20	0.20	$0.20 \times 100 = 20$ g	32	0.62	0.20
CO	5.0	0.05	$0.05 \times 100 = 5$ g	28	0.18	0.06
CO_2	25	0.25	$0.25 \times 100 = 25$ g	44	0.57	0.18
N_2	50	0.50	$0.50 \times 100 = 50$ g	28	1.78	0.56
Total	100	1.0	100 g		$3.15 = n_{tot}$	1.00

The **average molecular weight** (or mean molecular weight) of a mixture, $\overline{M}$ (kg/kmol, lbm/lb-mole, etc.), is the ratio of the mass of a sample of the mixture (m_{tot}) to the number of moles of all species (n_{tot}) in the sample.

$$\overline{M} = \frac{m_{tot}}{n_{tot}} = \sum_{i=1}^{N} \frac{m_i}{n_{tot}} = \sum_{i=1}^{N} \frac{n_i M_i}{n_{tot}} = \sum_{i=1}^{N} \left(\frac{n_i}{n_{tot}}\right) M_i = \sum_{i=1}^{N} X_i M_i \tag{3.21a}$$

$$\overline{M} = \sum_{i=1}^{N} X_i M_i = X_1 M_1 + X_2 M_2 + \dots + X_N M_N \tag{3.21b}$$

Alternatively,

$$\frac{1}{\overline{M}} = \frac{n_{tot}}{m_{tot}} = \sum_{i=1}^{N} \frac{n_i}{m_{tot}} = \sum_{i=1}^{N} \frac{\frac{m_i}{M_i}}{m_{tot}} = \sum_{i=1}^{N} \frac{m_i/m_{tot}}{M_i} = \sum_{i=1}^{N} \frac{x_i}{M_i} \tag{3.22a}$$

$$\frac{1}{\overline{M}} = \sum_{i=1}^{N} \frac{x_i}{M_i} = \frac{x_1}{M_1} + \frac{x_2}{M_2} + \dots + \frac{x_N}{M_N} \tag{3.22b}$$

Example 3.8 Calculation of an Average Molecular Weight
Calculate the average molecular weight of air

1) From its approximate molar composition of 79% N_2, 21 % O_2

2) From its approximate composition by mass of 76.7% N_2, 23.3% O_2.

Solution:

$$\overline{M} = \sum_{i=1}^{N} X_i M_i = X_{O2}M_{O2} + X_{N2}M_{N2} = 0.21 \times 32 + 0.79 \times 28 = 28.8\frac{\text{g}}{\text{mol}} = 28.8\frac{\text{lbm}}{\text{lb}-\text{mol}}$$

$$= 28.8\frac{\text{kg}}{\text{kmol}}$$

$$\frac{1}{\overline{M}} = \sum_{i=1}^{N} \frac{x_i}{M_i} = \frac{x_{O2}}{M_{O2}} + \frac{x_{N2}}{M_{N2}} = \frac{0.233}{32} + \frac{0.767}{28} = 0.00728 + 0.02739 = 0.03467$$

$$\overline{M} = \frac{1}{0.03467} = 28.8 = 28.8\frac{\text{g}}{\text{mol}} = 28.8\frac{\text{lbm}}{\text{lb}-\text{mol}} = 28.8\frac{\text{kg}}{\text{kmol}}$$

Note
Air contains small amounts of carbon dioxide, argon, and other gases that have been neglected in this calculation, but whose presence does not significantly affect the calculated value of $\overline{M}$.

Exercise 3.4

1) The molecular weight of atomic hydrogen is approximately 1, and that of atomic bromine is 80. What are (1) the mass fraction and (2) the mole fraction of bromine in pure HBr?
2) If 100 lb_m/min of A ($M_A = 30$) and 200 lb_m/min of B ($M_B = 40$) flow through a pipe, what are the mass fractions and mole fractions of A and B, the mass flow rate of A, the molar flow rate of B, the total mass flow rate, and the total molar flow rate of the mixture?

3.4.3 MATLAB Code for Conversion from Mass % to Molar % Composition

Appendix B | B.4 "MATLAB Code for Conversion from Mass % to Molar % Composition for a Multi-component Mixture" section shows MATLAB code called **Mass2Mole.m** file (Appendix B | Figure B.2) for carrying out conversion from mass % to molar % composition. The algorithm is explained in the indicated section.

Here are MATLAB results using the data in Table 3.2. The results are in harmony with the manual calculations, shown in Table 3.2, of the molar composition of the mixture with known mass composition and molecular mass data.

```
>> Mass2Mole
Enter the number of components (2, 3, 4, or 5): 4
Enter the mass percent of component A: 20
Enter the mass percent of component B: 5
Enter the mass percent of component C: 25
Enter the mass percent of component D: 50
Enter the molecular mass of component A: 32
Enter the molecular mass of component B: 28
```

```
Enter the molecular mass of component C: 44
Enter the molecular mass of component D: 28
Original mass composition:
mA = 20.00
mB = 5.00
mC = 25.00
mD = 50.00
Calculated molar composition:
XA = 19.79%
XB = 5.66%
XC = 17.99%
XD = 56.56%
Molecular weight average (MWavg) = 31.671
```

3.4.4 MATLAB Code for Conversion from Molar % to Mass % Composition

Appendix B | B.5 "MATLAB Code for Conversion from Molar % to Mass % Composition for a Multi-component Mixture" section shows MATLAB code called **Mole2Mass.m** file (Appendix B | Figure B.3) for carrying out conversion from mass % to molar % composition. The algorithm is explained in the indicated section.

Here are MATLAB results using the data in Table 3.2. To check the code consistency, we use the last results shown above as known molar % compositions of the mixture to let **Mole2-Mass.m** file find the mass % composition for the same compounds. The calculated mass compositions are in harmony with the initially given values, shown in Table 3.2, of mass composition of the mixture.

```
>> Mole2Mass
Enter the number of components (2, 3, 4, or 5): 4
Enter the mole percent of component A: 19.8
Enter the mole percent of component B: 5.66
Enter the mole percent of component C: 18.0
Enter the mole percent of component D: 56.56
Enter the molecular mass of component A: 32
Enter the molecular mass of component B: 28
Enter the molecular mass of component C: 44
Enter the molecular mass of component D: 28
Original molar composition:
nA = 19.80
nB = 5.66
nC = 18.00
nD = 56.56
Calculated mass composition:
xA = 20.00%
xB = 5.00%
xC = 25.00%
xD = 49.99%
Molecular weight average (MWavg) = 31.6776
```

3.4.5 Concentration

The mass concentration of a constituent in a mixture or solution is determined by its mass per unit volume of the mixture, measured in units such as g/cm^3, lbm/ft^3, kg/m^3, and so forth. Meanwhile, the molar concentration of a component refers to the quantity of moles of that component within a given volume of the mixture, expressed in units like kmol/m^3, lb-moles/ft^3, and others. Specifically, the **molarity** of a solution represents the molar concentration of the solute stated in gram-moles of solute per liter of solution. For example, a three-molar solution of component A indicates the presence of three moles of A within each liter of the solution.

The substance concentration in a mixture or solution serves as a conversion factor facilitating the relationship between the mass (or moles) of a component within a sample of the mixture concerning the sample volume. Similarly, it aids in establishing a connection between the mass (or molar) flow rate of a constituent in a continuous stream and the overall volumetric flow rate of the stream. For instance, let us consider a solution of NaOH with a concentration of 0.05 mol NaOH per liter (i.e., a 0.05-molar NaOH solution). When dealing with 5 L of this solution, the content of NaOH can be calculated as follows: $5\,\text{L} \times 0.05 \dfrac{\text{mol NaOH}}{\text{L}} = 0.25\,\text{mol NaOH}$.

If a stream of this solution flows at a rate of 4 L/min, the molar flow rate of NaOH is:

$$4 \frac{\text{L}}{\text{min}} \times 0.05 \frac{\text{mol NaOH}}{\text{L}} = 0.20 \frac{\text{mol NaOH}}{\text{min}}$$

Exercise 3.5 A solution with volume **V**(L) contains **n**(mol) of solute A with a molecular weight of $\boldsymbol{M_A}$(g A/mol).

In terms of V, n, and M_A:

1) *What is the molar concentration of A?*
2) *What is the mass concentration of A?*

In terms of C_A (mol A/L) and c_A (g A/L):

3) What volume of the solution contains 30 mol of A?
4) What is the mass flow rate of A in a stream whose volumetric flow rate is 150 L/h?

Example 3.9 Conversion between Mass, Molar, and Volumetric Flow Rates of a Solution

A 0.50-molar aqueous solution of sulfuric acid flows into a process unit at a rate of 1.80 m^3/min.

The specific gravity of the solution is 1.04 at 25°C. Calculate:

1) The mass concentration of H_2SO_4 in kg/m^3,
2) The total mass flow rate in kg/s,
3) The mass flow rate of H_2SO_4 in kg/s,
4) The mass fraction of H_2SO_4.

DATA: Molecular weight of H_2SO_4: $2 + 32 + 4*16 = 2 + 32 + 64 = 98$

Solution:

1) H_2SO_4 Mass concentration $(kg/m^3) = 0.5\frac{mol\ H_2SO_4}{L} \times \left(\frac{98g\ H_2SO_4}{mol\ H_2SO_4}\right) \times \left(\frac{kg\ H_2SO_4}{1000\ g\ H_2SO_4}\right) \times \left(\frac{1000\ L}{1\ m^3}\right) = 49\frac{kg\ H_2SO_4}{m^3}$

2) The total mass flow rate $(kg/m^3) = 1.8\frac{m^3}{min} \times \left(\frac{min}{60\ s}\right) \times \left[\left(\frac{1.04g}{cm^3}\right) \times \left(\frac{kg}{1000\ g}\right) \times \left(\frac{1{,}000{,}000\ cm^3}{1\ m^3}\right)\right] = 31.2\frac{kg}{s}$

3) The mass flow rate of H_2SO_4 in kg/s $= 1.8\frac{m^3}{min} \times \left(\frac{min}{60\ s}\right) \times \left[49\frac{kg\ H_2SO_4}{m^3}\right] = 1.47\frac{kg}{s}$

4) The mass fraction of $H_2SO_4 = \frac{\text{Mass flow rate of } H_2SO_4}{\text{Total mass flow rate}} = \frac{1.47}{31.2} = 0.047\frac{Kg\ H_2SO_4}{kg\ solution}$

3.4.6 Parts per Million and Parts per Billion

The terms parts per million (**ppm**) and parts per billion (**ppb**) serve as measurement units to indicate the concentrations of trace species, which are substances existing in minimal quantities within gas or liquid mixtures. These measurements can refer to mass ratios, commonly employed for liquids, or mole ratios, typically used for gases. They convey the quantity of the species present within a million or billion parts of the mixture, denoting the number of parts (in grams or moles) of the species against a million or billion parts (in grams or moles) of the entire mixture.

If Y_i is the fraction of component i, then by definition

$$ppm_i = Y_i \times 10^6 \tag{3.23}$$

$$ppb_i = Y_i \times 10^9 \tag{3.24}$$

For example, suppose air in the vicinity of a power plant is said to contain 20 ppm SO_2 (air mixture of 20 parts SO_2 and 999,980 parts of air). Assuming that a molar basis has been used (typically for gases), this statement says that every million moles (or molecules) of air mixture contains 20 moles (molecules) of SO_2, or equivalently, that the mole fraction of SO_2 in the air is 20×10^{-6}.

3.4.7 Concentration Assessment

Let us explore some of the common instruments/techniques used in different fields:

Chemistry:

- **Spectrophotometers**: These workhorses analyze the interaction of light with molecules, revealing the concentration of specific substances in a solution. Different types exist, like UV–Vis for organic compounds and atomic absorption for metals.
- **Titrators**: These automate the process of measuring how much of one solution is needed to react completely with another, providing precise concentration values.
- **Electrodes**: Specific electrodes measure the concentration of ions in a solution, like pH meters for acidity or conductivity meters for total dissolved solids.

Biotechnology:

- **ELISA (Enzyme-linked Immunosorbent Assay)**: This powerful technique uses antibodies to detect and quantify specific proteins or other biomolecules in a sample.
- **PCR (Polymerase Chain Reaction)**: This method amplifies tiny amounts of DNA or RNA, allowing for the detection and quantification of even trace amounts of genetic material.
- **Flow Cytometry**: This technology analyzes individual cells one at a time, measuring various parameters like size, fluorescence, and protein expression, which can be used to infer concentrations of different components within the cell. Flow cytometry uses a laser beam to illuminate individual cells suspended in a fluid stream. As each cell passes through the beam, it scatters light and emits fluorescence depending on its characteristics. Detectors capture the scattered light and fluorescence, providing information about the cell's:
  - **Size**: Measured by the amount of light scattered.
  - **Granularity**: Measured by the variations in the scattered light intensity.
  - **Fluorescence**: Measured by the intensity and color of the emitted light.
  - **Concentrations**: Cells can be stained with fluorescently labeled antibodies specific to certain proteins or other molecules. The intensity of the emitted fluorescence from the stained cells indicates the relative concentration of the target molecule within each cell. By analyzing the distribution of fluorescence intensities in a population of cells, researchers can:
    - Identify subpopulations of cells with different levels of the target molecule.
    - Estimate the average concentration of the target molecule in the entire population.
    - Compare the concentration of the target molecule in different cell types or under different experimental conditions.

Environmental Science:

- **Gas Chromatography**: This technique separates and analyzes the components of a gas mixture, allowing for the determination of concentrations of pollutants or other airborne chemicals.
- **High-performance Liquid Chromatography (HPLC)**: This method separates and analyzes the components of a liquid mixture, helping to identify and quantify contaminants in water or soil samples.
- **Mass Spectrometry**: This powerful tool identifies and quantifies individual molecules based on their mass and charge, offering extremely detailed analysis of complex mixtures.

3.5 Pressure

Pressure in a fluid refers to the force exerted per unit area by the molecules of the fluid as they collide with the walls of their container or any surface in contact. It represents the intensity of the distributed forces that the fluid applies perpendicular to the surfaces it encounters. Mathematically, pressure (P) is calculated as the force (F) acting perpendicular to an area (A), depicted as $P = F/A$. In a fluid, pressure plays a fundamental role in determining the fluid's behavior, influencing its flow, compression, and distribution within a confined space, and is crucial in various scientific, engineering, and practical applications across different industries.

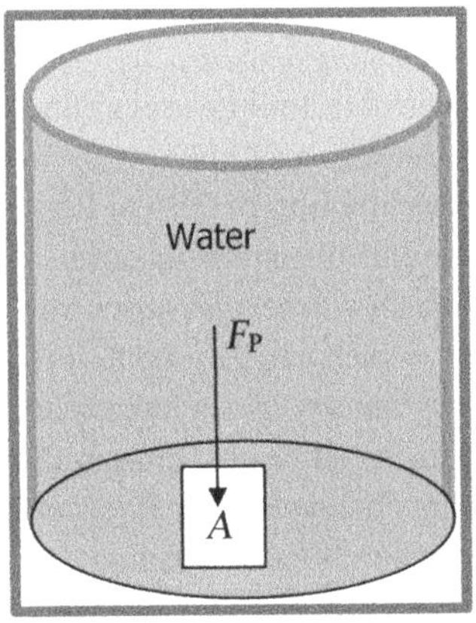

Figure 3.8 The pressure force is defined as the normal inward (compressive) force divided by the unit area upon which it acts.

3.5.1 Fluid Pressure and Hydrostatic Head

A pressure is the ratio of a force to the area on which the force acts. Accordingly, pressure units are force units divided by area units (e.g., N/m^2, $dynes/cm^2$, and $lb_f/in.^2$ or psi). The SI pressure unit, N/m^2, is called a pascal (Pa).

Figure 3.8 shows a vertical tank filled with liquid water and how the pressure force, acting on the floor of the tank, is defined as the normal (perpendicular) inward compressive force divided by the unit area upon which the force acts. Thus, the fluid pressure may be defined as the ratio $P = F_P/A$.

Suppose a vertical column of fluid is h(m) high and has a uniform cross-sectional area $A(m^2)$. Further suppose that the fluid has a density of $\rho(kg/m^3)$, and that a pressure $P_o(N/m^2)$ is exerted on the upper surface of the column (See, Figure 3.9). The pressure P of the fluid at the base of the column, called the hydrostatic pressure of the fluid, is by definition, the force F exerted on the base divided by the base area A. The pressure at the bottom surface is equal to the pressure on the top surface plus the weight of the fluid in the column divided by the acted-upon area. It is not difficult to show that:

$$P = P_o + \frac{\text{Weight of water column}}{\text{Base area}} = P_o + \frac{mg}{A} = P_o + \frac{\rho V g}{A} = P_o + \frac{\rho A h g}{A} = P_o + \rho g h \tag{3.25}$$

In addition to being expressible as a force per unit area, a pressure may be expressed as a head of a particular fluid, that is, as the height of a hypothetical column of this fluid that would exert the given pressure at its base if the pressure at the top were zero. We can thus

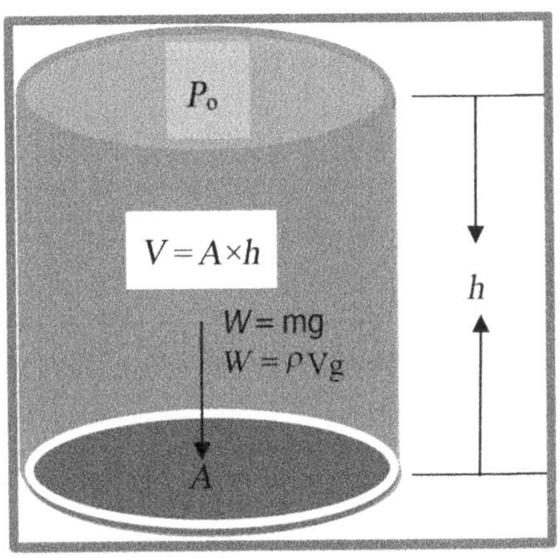

Figure 3.9 The hydrostatic pressure at the bottom surface is equal to the top surface pressure plus the weight of the water column divided by the base area of the cylindrical tank.

speak of a pressure of 14.7 psi or equivalently of a pressure (or head) of 33.9 ft of water (33.9 ft H_2O) or 760 mm of mercury (760 mm Hg). *The equivalence between a pressure P (force/area) and the corresponding head h (height of a fluid) is given by* Eq. (3.25) *with* $P_o = 0$:

$$P\left[\frac{\text{force}}{\text{m}^2}\right] = \rho g\left[\frac{\text{force}}{\text{m}^3}\right] \times h[\text{head of fluid, m}] \tag{3.26}$$

3.5.2 Atmospheric Pressure, Absolute Pressure, and Gauge Pressure

The atmospheric pressure can be likened to the pressure exerted at the bottom of a fluid (air) column situated at a particular measurement location (such as at sea level). Equation (3.25) enables the calculation of atmospheric pressure by considering the scenario where the pressure at the column's top (P_o) is regarded as 0, while the variables ρ and g represent the average values of air density and the gravitational acceleration within the region spanning from the top of the atmosphere to the point of measurement.

The standard pressure of 1 atmosphere, typically noted as 760.0 mm Hg at sea level, serves as a benchmark value for atmospheric pressure. This pressure equivalence in various units can be found in the conversion table provided in Appendix A | Table A.8. The pressures discussed until now relate to absolute pressures, wherein a pressure of zero signifies a complete vacuum. Numerous devices for pressure measurement indicate gauge pressure, signifying the pressure relative to the atmospheric pressure. A gauge pressure reading of zero implies that the fluid's absolute pressure matches the atmospheric pressure. The formula for converting between absolute and gauge pressure is as follows:

$$P_{\text{absolute}} = P_{\text{gauge}} + P_{\text{atmospheric}} \tag{3.27}$$

The abbreviations *psia* and *psig* are commonly used to denote *absolute* and *gauge* pressure in $lb_f/in.^2$. Also, it is common to refer to negative gauge pressures (absolute pressures less than atmospheric) as positive amounts of vacuum: for example, a gauge pressure of −10 cm Hg (means 66.0 cm Hg absolute if atmospheric pressure is 76.0 cm Hg) may also be called 10 cm of vacuum.

3.5.3 Fluid Pressure Measurement

Basic Principles:

- **Pascal's Law**: Pressure applied at any point of a confined incompressible fluid is transmitted equally in all directions throughout the fluid.
- **Hydrostatic Pressure**: Pressure due to the weight of thc fluid above a point and is given by the relationship between pressure, height, and density: $P = \rho g h$.

3.5.4 Fluid Pressure Measuring Devices

3.5.4.1 Manometers

a) **U-tube**: Simple and reliable for low pressure ranges. U-tube manometers are simple and versatile devices used for measuring fluid pressure differences. They consist of a U-shaped glass or transparent tube partially filled with a liquid (such as water, mercury, or oil). The tube is bent into a "U" shape with one end open to the atmosphere or a reference pressure and the other end connected to the point where pressure is being measured.

The principle behind a U-tube manometer is based on the balance of pressure between the two arms of the U-tube. When pressure is applied to one arm, the fluid level in that arm rises, causing a difference in height between the two liquid columns. By measuring this difference in height (known as the "manometer reading") and accounting for the density of the liquid used, one can calculate the pressure difference between the two points. U-tube manometers come in various configurations, such as the simple manometer, differential manometer, and inverted U-tube manometer, each suitable for different pressure measurement scenarios. They are relatively inexpensive, easy to set up, and provide accurate measurements for a wide range of applications in industries including HVAC, laboratory experiments, and fluid systems, where precise pressure measurements are essential. However, they might not be suitable for measuring high pressures or for applications where portability and durability are critical due to the risk of breakage of the glass tube. Three YouTube videos about the principles of U-tube manometers are available at:

1) https://www.youtube.com/watch?v=rzJ80cDjV4Q.
2) https://www.youtube.com/watch?v=gxrkLkJybnA.
3) https://www.youtube.com/watch?v=MpXhXVF9-HM.

We will have some examples below.

b) **Inclined U-tube**: Such manometers are a variation of traditional U-tube manometers designed to measure pressure differences using a tilted or inclined tube rather than a vertical one. They operate on the same fundamental principles as standard U-tube manometers but offer advantages in specific applications where space constraints or sensitivity to small pressure changes are factors. The inclined configuration allows for a longer vertical measurement range within a confined space compared to a standard vertical U-tube manometer. This design is especially useful when precise measurement of small pressure differentials is required, but the available vertical height is limited. By tilting the tube at an angle, the longer length of the inclined arm compensates for the reduced vertical displacement of the liquid column. This design allows for a longer measurement range within a confined or limited space, making it advantageous for applications where precise measurement of small pressure differences is required in constrained environments. These manometers are commonly used in various industries, such as HVAC systems, where precise measurements of small pressure changes in restricted spaces are necessary. Figure 3.10 shows three types of U-tube manometers: (1) Differential pressure

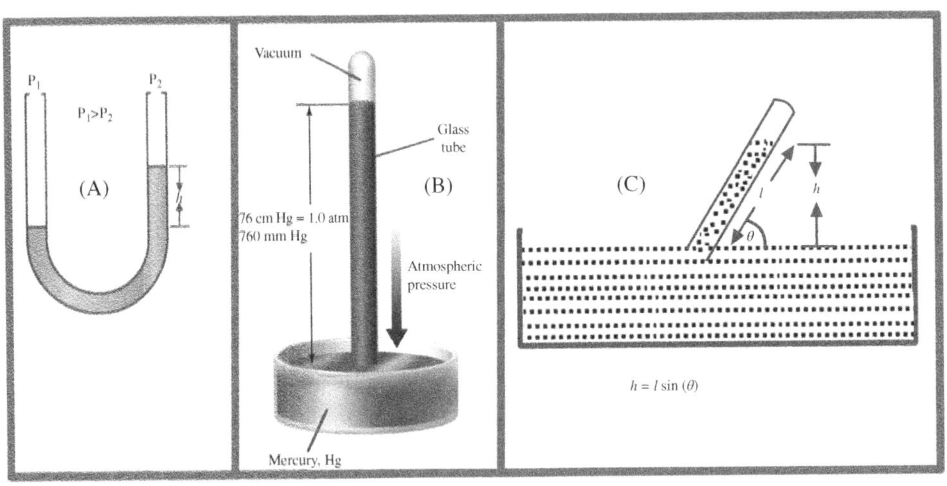

Figure 3.10 Three types of U-manometers: (A) U-tube. (B) Barometer; and (C) Inclined.

(i.e., $P_2 - P_1$); (2) Against barometric pressure where there is one-end that is closed and has zero absolute pressure; and (3) The inclined leg barometer where the it gives a longer distance, compared with the vertical leg barometer, for the same applied pressure value.

c) **Micromanometer**: Micromanometers are precision instruments designed to measure very small pressure differences or low pressures with high accuracy. They are specialized devices used in various fields such as fluid mechanics, aerodynamics, HVAC, and other industries where precise measurement of minute pressure variations is critical. Micromanometers can detect and quantify pressure differences within a specific range, often in units of Pascals (Pa), millimeters of water column (mmWC), or inches of water column (inWC). These devices are essential in applications that require meticulous pressure monitoring, calibration, and analysis, ensuring accuracy in systems and processes that involve fine adjustments or sensitive pressure control. Two YouTube videos about micromanometers are available at:
   1) https://www.youtube.com/watch?v=jYm0qWr8zZU
   2) https://www.youtube.com/watch?v=YgelbVtojGc

d) **McLeod**: Measures very low pressures using mercury vapor. The McLeod gauge, named after its inventor Herbert McLeod (1871–1937), is a high-precision vacuum gauge used to measure very low pressures or high vacuums. It operates based on the principle of compressing a gas within a small, known-volume capillary tube connected to a mercury manometer. By measuring the difference in mercury levels between the evacuated capillary and the reference side of the manometer, the McLeod gauge can determine the pressure of the gas by the known volume of the capillary and the principles of Boyle's law. This type of gauge is highly accurate for measuring pressures in the range of 10^{-6}–10^{-1} torr (1 torr is 1 millimeter of mercury), making it suitable for precise vacuum measurements in various scientific and industrial applications. A YouTube video about the McLeod manometer is available at: https://www.youtube.com/watch?v=yUDOt5B5M_k.

3.5.4.2 Mechanical Gauges

a) **Bourdon Tube**: The most common type that deflects due to pressure. One end of the tube is fixed and sealed, while the other end is free to move. When pressure is applied to the inside of the tube, it tends to straighten due to the elastic properties of the metal. The movement of the free end is proportional to the applied pressure. The movement of the free end is typically linked to a gear or rack-and-pinion mechanism. This mechanism converts the linear movement of the tube into the rotary motion of a pointer or dial. The pointer or dial indicates the pressure on a calibrated scale.

   Types of Bourdon Tubes:
   - **C-shaped**: Simple and economical, suitable for low-to-medium pressure ranges.
   - **Helical**: More sensitive and compact, used for high-pressure ranges.
   - **Spiral**: Provides a wider range of motion for increased sensitivity.

   Advantages:
   - **Simple and Reliable Design**: Easy to manufacture and maintain.
   - **Wide Pressure Range**: Can be used from very low pressures (vacuum) to high pressures.
   - **Durable**: Can withstand harsh environments and pressure fluctuations.
   - **Cost-effective**: Relatively inexpensive compared to other pressure gauges.
   - **Versatile**: Used in diverse applications across various industries.

Limitations:

- **Accuracy**: Not as accurate as some other types of pressure gauges, especially at low pressures.
- **Temperature Sensitivity**: The accuracy can be affected by temperature changes.
- **Limited Range of Motion**: The maximum deflection of the tube can limit the pressure range.

Applications:

- **Industrial Processes**: Monitoring pressure in pipelines, tanks, reactors, etc.
- **HVAC Systems**: Measuring pressure in air ducts and boilers.
- **Automotive Industry**: Gauging tire pressure and engine oil pressure.
- **Medical Equipment**: Measuring blood pressure and respiratory function.
- **Environmental Monitoring**: Tracking air and water pressure.

Two YouTube videos, about bourdon tube, are available at:

1) https://www.youtube.com/watch?v=fmOnrEZ_z6k.
2) https://www.youtube.com/watch?v=uBdTxTp_qLI.

b) **Diaphragms**: These thin, flexible membranes play a crucial role in various instruments, providing a reliable and versatile way to detect and quantify pressure changes. Imagine a thin, flexible membrane clamped between two rigid supports. When pressure is applied to one side of the membrane, it deflects or bulges outwards. The amount of deflection is directly proportional to the applied pressure. This deflection can be measured and converted into a pressure reading. Measuring the physical displacement of the diaphragm with a mechanical linkage or optical sensor. For example, using the strain on the diaphragm to alter the resistance of a strain gauge, generating an electrical signal proportional to pressure.

Types of Diaphragms:

- **Metal**: Durable and robust for high-pressure applications but less sensitive.
- **Polymer**: More flexible and sensitive, ideal for low-to-medium pressure ranges.
- **Composite**: Combining metal and polymer properties for improved performance across wider ranges.

Advantages of Diaphragms:

- **High Sensitivity**: Can detect even minute pressure changes, making them ideal for low-pressure applications.
- **Wide Pressure Range**: Different materials and configurations cater to various pressure needs.
- **Leak-proof**: The sealed design eliminates fluid leakage risks.
- **Corrosion Resistance**: Polymer diaphragms offer good resistance to corrosive fluids.
- **Versatility**: Applicable in diverse industries and applications.

Limitations:

- **Temperature Sensitivity**: Some materials can be affected by temperature changes, impacting accuracy.
- **Fatigue and Wear**: Diaphragms can wear over time, requiring periodic maintenance.
- **Material Limitations**: Not suitable for all fluids due to compatibility concerns.

Applications:

- **Medical**: Measuring blood pressure, respiratory function, and intracranial pressure.
- **Industrial**: Monitoring pressure in pipelines, reactors, and tanks.
- **Aerospace**: Measuring aircraft cabin pressure and engine performance.

- **Environmental**: Tracking atmospheric pressure and water pressure.
- **Food and Beverage**: Monitoring pressure in processing tanks and packaging systems.

Two YouTube videos about diaphragms are available at:

1) https://www.youtube.com/watch?v=5ScB2nEcfOs.
2) https://www.youtube.com/watch?v=JnPSeEW-EeA.

c) **Bellows**: Bellows, are those accordion-like chambers. Their unique ability to expand and contract in response to pressure changes makes them versatile and reliable instruments across various industries. Bellows consist of a series of flexible, gas-tight chambers connected together. When pressure is applied to one side, the chambers expand, pushing against the spring force within. The movement of the bellows is proportional to the applied pressure. The movement of the bellows directly drives a pointer or dial on a calibrated scale. For example, the expansion can be transferred to a lever system to amplify the movement and drive the indicator. Alternatively, the pressure-induced strain on the bellows can be measured by strain gauges, generating an electrical signal for further processing.

Types of Bellows:
- **Single-ply**: Simple and cost-effective, but less resilient and suitable for lower pressure ranges.
- **Multi-ply**: Offer higher strength and pressure resistance, ideal for demanding applications, like process control, environmental monitoring, and medical equipment.
- **Metal**: Durable for high-pressure environments but less flexible and more susceptible to corrosion.
- **Polymer**: More flexible and sensitive but may have limitations in temperature and pressure ranges.

Advantages of Bellows:
- **High Sensitivity**: Can detect small pressure changes over a wide range.
- **Leak-proof**: The sealed design minimizes the risk of fluid leakage.
- **Large Displacement**: Can accommodate significant pressure changes without exceeding their limits.
- **Versatility**: Suitable for various pressure ranges and applications.
- **Low Maintenance**: Simple design with minimal moving parts.

Limitations:
- **Higher Cost**: Compared to some other pressure gauges, bellows can be more expensive.
- **Temperature Sensitivity**: Some materials may be affected by temperature fluctuations, impacting accuracy.
- **Hysteresis**: A small difference in the pressure readings during increasing and decreasing pressure cycles.

Applications:
- **Industrial Processes**: Monitoring pressure in reactors, pipelines, and tanks.
- **Medical Equipment**: Measuring blood pressure and respiratory function.
- **Aerospace**: Gauging engine performance and cabin pressure.
- **Environmental Monitoring**: Tracking air and water pressure.
- **Vacuum Technology**: Measuring low pressures in vacuum systems.

Two YouTube videos about bellows are available at:

1) https://www.youtube.com/watch?v=WJb3kFBayQI.
2) https://www.youtube.com/watch?v=lMbNApD6vWw.

d) **Piston-cylinder**: Piston-cylinder assemblies are known for their high accuracy and precision in pressure measurement, but there is no free lunch, and they do come with some drawbacks related to their bulkiness.

Advantages:

- **Extreme Accuracy and Precision**: Due to minimal friction and leakage potential, piston-cylinder assemblies can achieve incredibly accurate and repeatable pressure measurements, often exceeding other types of gauges by orders of magnitude.
- **Wide Pressure Range**: They can handle a vast range of pressures, from high vacuum to extremely high hydraulic pressures, making them adaptable to diverse applications.
- **Direct Calibration**: The calibration can be easily traced back to fundamental physical principles like mass and length, ensuring high reliability and traceability.
- **Robust Construction**: Their solid build makes them suitable for harsh environments and demanding industrial applications.

Disadvantages:

- **Bulkiness**: Compared to other pressure gauges, piston-cylinder assemblies are physically large and heavy, requiring more space and installation considerations.
- **Cost**: Their complex design and precision manufacturing make them significantly more expensive than many other pressure gauges.
- **Maintenance**: Regular calibration and maintenance are crucial to maintain their accuracy, increasing operational costs.
- **Dynamic Limitations**: They may not be suitable for highly dynamic pressure environments due to inertia and potential lag in response.

Applications:

Despite their drawbacks, piston-cylinder assemblies remain the choice for applications where absolute accuracy and precision are paramount. Some examples include:

- **Calibration Standards**: Providing a reference point for calibrating other pressure gauges.
- **High-pressure Research**: Measuring pressure in reactors, chemical processes, and nuclear power plants.
- **Metrology**: Establishing and maintaining accurate pressure standards for scientific and industrial purposes.
- **Aerospace Testing**: Evaluating engine performance and airframe integrity in simulated flight conditions.

A YouTube video about piston-cylinder pressure gauge is available at: https://www.youtube.com/watch?v=6xLLbM2InE0.

3.5.4.3 Electronic Pressure Transducers

a) **Piezoelectric**: Piezoelectric pressure measurement relies on the piezoelectric effect, a phenomenon exhibited by certain materials, typically crystals or ceramics, that generate an electric charge or voltage when subjected to mechanical stress or pressure. These materials possess a unique property where their atomic structure distorts when pressure is applied, causing a separation of positive and negative charges within the crystal lattice, thereby generating an electric potential.

In piezoelectric pressure sensors, these specially designed crystals or ceramics are incorporated into the sensor's structure. When pressure is applied to the sensor, it mechanically deforms the piezoelectric material, resulting in the generation of an electric charge or voltage proportional to the applied pressure. This voltage can then be measured and correlated to the magnitude of the pressure.

These sensors are commonly used in various applications such as industrial pressure measurement, automotive systems, medical devices, and scientific instruments due to their fast response time, high sensitivity, and ability to accurately measure dynamic pressure changes. Piezoelectric pressure sensors are advantageous for capturing rapid pressure fluctuations and high-frequency pressure events, making them valuable in scenarios where precise and real-time pressure monitoring is critical. Two YouTube videos about piezoelectric pressure sensors are available at:

1) https://www.youtube.com/watch?v=JoRVcys5PB8.
2) https://www.youtube.com/watch?v=6xHJYQreZnM.

b) **Capacitance Manometers**: Capacitive pressure sensors operate on the principle of changes in capacitance due to variations in the distance between two conductive plates. These sensors consist of two closely spaced plates, often made of conductive materials, separated by a dielectric material. When pressure is applied to the sensor, the distance between these plates alters, causing a change in the capacitance of the sensor.

As pressure deforms the sensor, the distance between the plates changes, consequently modifying the capacitance of the system. An increase in pressure compresses the sensor, reducing the separation between the plates and resulting in an increase in capacitance. Conversely, a decrease in pressure allows the plates to move apart, leading to a decrease in capacitance.

The change in capacitance is directly proportional to the applied pressure, allowing for the measurement of pressure variations. By measuring the capacitance alterations, the sensor can determine the magnitude of the pressure being applied. Capacitive pressure sensors are utilized in various applications due to their ability to provide accurate and precise pressure measurements, particularly in environments where high sensitivity and minimal drift over time are essential, such as in industrial, automotive, and medical devices. Two YouTube videos are available about capacitive pressure sensors at:

1) https://www.youtube.com/watch?v=Xsi2udzHYn8.
2) https://www.youtube.com/watch?v=FfPVeUcgT3I.

c) **Strain Gauge**: A strain gauge is a type of sensor used to measure the mechanical strain or deformation in a material caused by an applied force or pressure. In the context of pressure measurement, a strain gauge can be incorporated into a pressure sensor to detect and quantify pressure-induced deformation.

The principle behind a strain gauge involves the use of a wire, usually made of a conductive material like copper or nickel, configured in a specific pattern. When the material experiences mechanical deformation due to pressure, the strain gauge undergoes a change in shape, length, or volume, leading to alterations in its electrical resistance.

As the pressure deforms the material containing the strain gauge, the wire within the gauge also deforms, resulting in a change in its electrical resistance. This change in resistance is directly proportional to the amount of strain or deformation caused by the pressure. By measuring the variations in electrical resistance using Wheatstone bridge circuits or other measurement techniques, the strain gauge can accurately determine the applied pressure.

Strain gauge-based pressure sensors are widely used in different industries due to their high sensitivity, accuracy, and ability to measure small pressure changes. They find applications in pressure transducers, load cells, and various devices where precise pressure measurements are critical, such as in aerospace, automotive, and industrial systems. YouTube videos about strain gauge pressure sensors are available at:

1) https://www.youtube.com/watch?v=BH8hAWRDTkA
2) https://www.youtube.com/watch?v=_z82yWBLbhE
3) https://www.youtube.com/watch?v=CRa9djnZRu0

d) **Vibrating Wire**: Vibrating wire pressure sensors rely on the principle that an applied pressure alters the tension and consequently changes the natural frequency of a vibrating wire or element. These sensors consist of a wire or filament that is clamped at both ends and set into a vibration mode by applying an initial tension.

When pressure is exerted on the sensor element, it causes mechanical deformation, which can result in a change in the tension of the wire. As the tension of the wire changes due to the applied pressure, the natural frequency at which the wire vibrates also changes. Typically, an increase in pressure causes a slight increase in tension, leading to an increase in the wire's natural vibration frequency.

By measuring the change in the natural frequency of the vibrating wire, the sensor can determine the magnitude of the applied pressure. This change in frequency can be detected and translated into pressure readings, providing an accurate measurement of the pressure being exerted on the sensor.

Vibrating wire pressure sensors are known for their high accuracy, stability, and reliability in measuring a wide range of pressures in various environments. They are used in geotechnical monitoring, structural health monitoring, and industrial applications where precise and robust pressure measurements are required. Two YouTube videos about vibrating wire pressure sensors are available at:

1) https://www.youtube.com/watch?v=Uj7s19gj2qQ.
2) https://www.youtube.com/watch?v=cG51Ab0jzzA.

3.5.4.4 Selection Criteria

a) **Pressure Range**: Choose a method suitable for the expected pressure range.
b) **Accuracy and Precision**: Consider the required level of accuracy and repeatability.
c) **Fluid Properties**: Some methods may be incompatible with certain fluids.
d) **Cost and Complexity**: Balance performance with cost and maintenance requirements.

U-tube Manometer Pressure Difference

The formula linking the pressure difference $P_1 - P_2$ with the distance readings in manometer fluid levels is derived from the principle that within a continuous fluid, the pressure remains constant at any two points situated at the same elevation (where the fluid is contiguous). Specifically, the pressure at the level of the lower surface of a manometer fluid remains uniform across both arms of the manometer. By formulating and setting up equations that represent the pressures at points (a) and (b) in Figure 3.11 as equal, we derive the fundamental equation for a manometer:

$$P_1 + \rho_1 g d_1 = P_2 + \rho_2 g d_2 + \rho_M g h \tag{3.28}$$

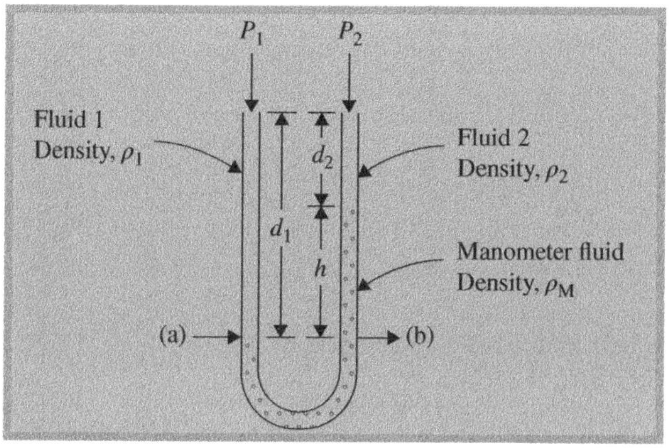

Figure 3.11 U-tube manometer where it shows that the pressure at point (a) is equal to that at point (b) as both lie at the *same level* (i.e., height) and *have one continuous fluid* in between.

In a *differential manometer*, fluids 1 and 2 are the same, and consequently $\rho_1 = \rho_2 = \rho$. Moreover, $d_1 = d_2 + h$. The general manometer equation (Eq. 3.28) then reduces to:

$$P_1 + \rho g d_1 = P_2 + \rho g d_2 + \rho_M g h \Rightarrow P_1 - P_2 = +\rho g d_2 - \rho g d_1 + \rho_M g h$$

$$\Rightarrow P_1 - P_2 = -\rho g h + \rho_M g h \Rightarrow P_1 - P_2 = (\rho_M - \rho) g h \tag{3.29}$$

If the flowing (process) fluid is a gas, then its density (ρ) will be much smaller than that of manometer fluid (ρ_M) and in this case $(\rho_M - \rho) \cong \rho_M$. Equation (3.29) reduces to:

$$P_1 - P_2 = \rho_M g h \tag{3.30}$$

In terms of pressure head, Eq. (3.30) becomes:

$$\frac{P_1 - P_2}{\rho_M g} = h \tag{3.31}$$

Example 3.10 Pressure Measurement with Manometers

A differential manometer is used to measure the drop in pressure between two points in a process line containing water (See Figure 3.12). The specific gravity of the manometer fluid is 13.6. The measured levels in each arm are shown below.

a) Calculate the pressure drop between points 1 and 2 in Pa.

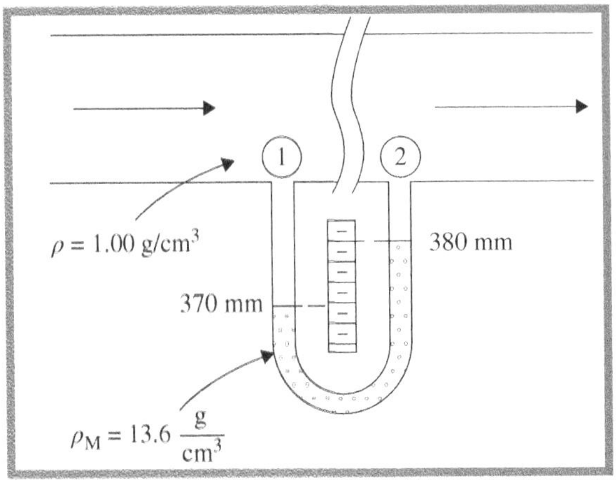

Figure 3.12 A sketch of the manometer for Example 3.10.

$h = (380 - 370)\text{mm} = 10\text{ mm} = 0.01\text{ m}$. From Eq. (3.29)

$$P_1 - P_2 = (\rho_M - \rho)gh = (13.6 - 1.00)\frac{\text{g}}{\text{cm}^3} \times \left(\frac{1000\left(\frac{\text{kg}}{\text{m}^3}\right)}{1.0\left(\frac{\text{g}}{\text{cm}^3}\right)}\right) \times 9.8\frac{\text{m}}{\text{s}^2} \times 0.01\text{ m}$$

$$\times \left(\frac{\text{N}}{\text{kg}\cdot\frac{\text{m}}{\text{s}^2}}\right) = 1235\frac{\text{N}}{\text{m}^2} = 1235\text{ Pa} = 1.235\text{ kPa}$$

Example 3.11

The pressure of gas being pulled through a line by a vacuum pump is measured with an open-end mercury manometer. A reading of −**4** in. is obtained. See Figure 3.13. What is the gas gauge pressure in inches of mercury? What is the absolute pressure if $P_{atm} = 29.9$ in. Hg?

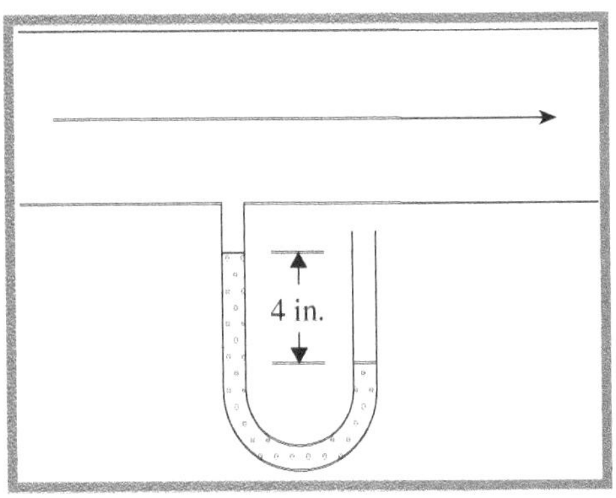

Figure 3.13 A sketch of a manometer for a vacuum pump.

From Eq. (3.30): $P_1 - P_2 = \rho_M gh$

$$P_2 = P_{atm}; \Longrightarrow P_1 - P_{atm} = \rho_M gh$$

$$\frac{P_1 - P_{atm}}{\rho_M g} = h = \frac{P_{gage}}{\rho_M g} = -4\text{ in.Hg}$$

$$P_1 = P_{atm} - 4\text{ in.Hg} = 29.9\text{ in.Hg} - 4\text{ in.Hg} = 25.9\text{ in.Hg}$$

3.6 Temperature

Temperature is a fascinating concept with multiple ways to understand it! Here are some different perspectives:

1) **Physical Definition**: Temperature is a physical quantity that expresses the hotness or coldness of a system. It reflects the average kinetic energy of the particles (atoms or molecules) composing the system.
   - Higher temperature = higher average kinetic energy of particles = more "movement" in the system.
   - Lower temperature = lower average kinetic energy of particles = less "movement" in the system.

2) **Everyday Understanding**: We often sense temperature through touch. A hot object feels warmer because its particles are transferring more energy to our skin, while a cold object feels cooler because it is absorbing energy from our skin.
3) **Scientific Measurement**: Temperature is measured with instruments like thermometers, which rely on various principles like thermal expansion or electrical resistance changes to quantify the average kinetic energy of the particles. Different temperature scales exist, like Celsius, Fahrenheit, Rankine, and Kelvin, each with their own reference points and conversion factors.
4) **Importance in Nature and Technology**: Temperature plays a crucial role in many natural phenomena, like:
   - **Phase Changes**: Water changing from ice to liquid to vapor with increasing temperature.
   - **Chemical Reactions**: Reaction rates generally increase with rising temperature.
   - **Biological Processes**: Enzyme activity and organism functioning depend on temperature ranges.
   - **Weather and Climate**: Global temperature patterns influence weather patterns and climate change.

   Temperature also has Immense Significance in Technology, for Example:
   - **Material Properties**: Many materials change their properties like conductivity or strength with temperature.
   - **Engine Efficiency**: Engine performance depends on optimal temperature control.
   - **Food Preservation**: Keeping food at appropriate temperatures reduces spoilage.
   - **Electronics**: Temperature management is crucial for the efficient and safe operation of electronic devices.

   In essence, temperature is not just a simple measure of hotness or coldness but a fundamental concept influencing everything from the microscopic behavior of particles to the vast dynamics of weather systems and technological advancements. It is a key parameter in understanding and interacting with the world around us.

   Temperature scales are established based on characteristics of a standard substance, like freezing and boiling, occurring at fixed pressures and temperatures. A specific temperature scale is created by arbitrarily assigning numerical measures to two consistently measurable temperatures. For instance, setting 0 as the value for water's freezing point and 100 for its boiling point at 1 atm establishes the Celsius (or centigrade) scale: where the freezing point is labeled as 0°C (T_f) and the boiling point as 100°C (T_b).

Absolute zero (thermodynamically the lowest temperature attainable in nature) on this scale falls at −273.15°C. Imagine the coldest place you can think of. A blizzard on Mount Everest? The depths of outer space? While these are certainly frigid environments, they still hold a substantial amount of heat energy in their atoms and molecules. But there exists a theoretical point where all thermal motion stops – absolute zero. At absolute zero, the thermal kinetic energy of particles is minimized, but they still retain a small amount of kinetic energy due to quantum zero-point motion. This means no vibrations, no rotations, and no movement at the atomic and molecular levels. It is the temperature equivalent of perfect stillness.

Reaching absolute zero in a real-world setting is practically impossible. The closer we get, the harder it becomes to extract even more heat energy from the system. Even the most sophisticated cooling technologies can only approach this elusive point asymptotically. The Celsius scale, widely used in science and everyday life, defines its zero point as the freezing point of water at standard pressure (around 0°C). But this scale also extends into negative territory, representing temperatures colder than ice. And that is where the connection to absolute zero comes in. Through precise experiments and theoretical calculations, scientists have determined that absolute zero lies at −273.15°C on the Celsius scale. This means that at this temperature, the theoretical minimum of thermal motion is achieved, marking the coldest possible state of matter.

At such a condition, the state of matter is called Bose–Einstein condensate (BEC): This term applies specifically to bosons (particles with integer spin), which occupy the same quantum state at very low temperatures, resulting in a highly coherent and superfluid state. For example, some isotopes of Helium form BECs at temperatures near zero Kelvin. A second term, named degenerate state which, refers to the situation where multiple quantum states have the same energy level, allowing the particles to occupy them simultaneously at 0 Kelvin. This leads to unusual properties like zero entropy and perfect electrical conductivity. A third term, named quantum degeneracy is a broader term encompassing the behavior of particles at very low temperatures, where quantum effects become dominant and classical physics approximations break down.

While absolute zero might seem like a purely theoretical concept, it has significant implications in various fields, including:

- **Physics**: Understanding the behavior of matter at extremely low temperatures helps us develop new materials, study superconductivity, and even explore the origins of the universe.
- **Chemistry**: Reactions slow down considerably as they approach absolute zero, offering insights into reaction mechanisms and potentially enabling the development of novel catalysts.
- **Technology**: Cooling technologies striving for ever-lower temperatures rely on knowledge gained from the pursuit of absolute zero, impacting fields like cryogenics and quantum computing.

The Fahrenheit scale designates a value of 32°F for T_f and 212°F for T_b, with absolute zero positioned at −459.67°F. Meanwhile, the Kelvin and Rankine scales establish absolute zero at a value of 0, where each degree matches the size of a Celsius degree (Kelvin scale) or a Fahrenheit degree (Rankine scale).

The following relationships can be used to convert a temperature expressed in one defined scale unit into its equivalent:

$$T(\text{K}) = T(^\circ\text{C}) + 273.15 \tag{3.32}$$

$$T(^\circ\text{R}) = T(^\circ\text{F}) + 459.67 \tag{3.33}$$

$$T(^\circ\text{R}) = 1.8\,T(\text{K}) \tag{3.34}$$

$$T(^\circ\text{F}) = 1.8\,T(^\circ\text{C}) + 32 \tag{3.35}$$

A degree represents both a specific temperature and a temperature difference (ΔT), causing occasional confusion. For instance, take the temperature range between 0°C and 10°C. Within this range, there are 18 Fahrenheit and 18 Rankine degrees, while only 10°C and 10 Kelvin degrees. Consequently, an interval (Δ) of 1°C or Kelvin encompasses 1.8°F or Rankine degrees, which results in the following conversion factors:

$$\left(\frac{1.8\Delta^\circ\text{F}}{1\Delta^\circ\text{C}}\right), \left(\frac{1.8\Delta^\circ\text{R}}{1\Delta\text{K}}\right), \left(\frac{1\Delta^\circ\text{F}}{1\Delta^\circ\text{R}}\right), \left(\frac{1\Delta^\circ\text{C}}{1\Delta\text{K}}\right) \tag{3.36}$$

Example 3.12 Temperature Conversion
Consider the interval from $T_1 = 50^\circ\text{F}$ to $T_2 = 104^\circ\text{F}$.

1) Calculate the equivalent temperatures in °C and the interval between them.
2) Calculate directly the interval in °C between the temperatures.

Solution:

1) From Eq. (3.35), we have:

$$T(^\circ\text{C}) = \frac{T(^\circ\text{F}) - 32}{1.8} \tag{3.37}$$

$T_1(^\circ\text{C}) = \dfrac{50-32}{1.8} = 10.0^\circ\text{C}$ $T_2(^\circ\text{C}) = \dfrac{104-32}{1.8} = 40.0^\circ\text{C}$ $\Delta T(^\circ\text{C}) = T_2(^\circ\text{C}) - T_1(^\circ\text{C}) = 40.0 - 10.0 = 30.0^\circ\text{C}$

2) From Eq. (3.36): $\Delta T(^\circ\text{C}) = (104-50)\Delta^\circ\text{F} \times \left(\dfrac{1\Delta^\circ\text{C}}{1.8\Delta^\circ\text{F}}\right) = \dfrac{54}{1.8} = 30.0^\circ\text{C}$; *same as the previous result.*

3.6.1 Temperature Measuring Devices

Temperature is a fundamental property of matter, and its measurement is crucial in various fields, from science and engineering to everyday life. To cater to diverse needs, a wide range of temperature measuring devices exist, each with its unique strengths and limitations. Here is an outline to help one understand them:

3.6.1.1 Liquid-in-glass Thermometers

The classic thermometers with mercury, alcohol, or other liquids expanding in a graduated scale. Liquid-in-glass thermometers, the familiar glass tubes with an expanding liquid, have served humankind for centuries.

1) **Working Principle**:
   - The bulb at the bottom contains a liquid (mercury, alcohol, etc.) that expands and contracts with temperature changes.

- This expansion pushes the liquid up the thin capillary tube.
- The level of the liquid is read against a calibrated scale marked on the tube, indicating the temperature.

2) **Types of Liquids**:
   - **Mercury**: Historically the most common, but phased out due to its toxicity and environmental concerns. Mercury has a melting point of −38.9°C and a boiling point of 356.7°C, at standard atmosphere. Thus, mercury thermometers typically measure temperatures ranging from around −39°C (which is approximately −38°F) to 356°C (or roughly 672°F). However, these ranges can vary based on the specific design and construction of the thermometer.
   - **Alcohol**: A safer alternative, good for moderate temperatures but freezes at −114°C. Alcohol thermometers commonly measure temperatures within a range of approximately −115°C (−175°F) to 78°C (172°F). The exact range might vary slightly based on the design and the type of alcohol used in the thermometer.
   - **Gallium**: Gallium thermometers are typically used for high-temperature measurements due to gallium's relatively high melting point. They can measure temperatures within a range of approximately 30°C (86°F) to around 1300°C (2372°F), depending on the specific design and the purity of the gallium used in the thermometer. Speckbrock [1] patented a liquid gallium-based alloy that serves as the indicator fluid within a thermometer. This alloy is composed of indium, zinc, and less than 65% gallium by weight. Free from gases and other metal oxides, the alloy possesses a melting point below 0°C at normal atmospheric pressure and below −10°C in a strong vacuum. Its temperature measurement capabilities span from −15 to +1200°C. The alloy typically comprises 50–65% gallium, 10–24.9% indium, and may also include tin, up to 29% zinc, and up to 10% lead by weight.
   - **Toluene and Other Organic Liquids**: Used for specific ranges and applications. Toluene is typically used for thermometers measuring −80°C to 110°C.
3) **Advantages**:
   - **Simple and Easy to Use**: No batteries or complex technology required.
   - **Wide Range of Available Scales**: Celsius, Fahrenheit, Kelvin, etc.
   - **Relatively Inexpensive**: Cost-effective compared to some electronic devices.
   - **Durable and Long-lasting**: Can last for years with proper care.
4) **Disadvantages**:
   - **Breakable**: Glass can crack or shatter if dropped or mishandled.
   - **Mercury Safety Concerns**: Requires precautions for disposal and handling.
   - **Limited Accuracy**: Not as accurate as some electronic thermometers.
   - **Slow Response Time**: Takes time for the liquid to react to temperature changes.
   - **Not Suitable for Extreme Cold and Hot Temperatures**: Mercury freezes at −38°C, alcohol at −114°C.
5) **Applications**:
   - **Home Use**: Measuring room temperature, water for baths, cooking.
   - **Science Education**: Demonstrating basic temperature principles.
   - **Laboratories**: For specific applications where simplicity and low cost are preferred.
   - **Medical Field**: Measuring body temperature (mercury thermometers being phased out).
   - **Other Uses**: Aquariums, compost piles, weather monitoring (non-mercury).

While liquid-in-glass thermometers hold a nostalgic charm, electronic thermometers offer increased accuracy, speed, and versatility. However, liquid-in-glass thermometers remain valuable tools for their simplicity, affordability, and specific applications. Two YouTube videos about liquid-in-glass thermometers are available at:

1) https://www.youtube.com/watch?v=4ZcQ1HQbQwo.
2) https://www.youtube.com/watch?v=eiPWPyU-KgI.

3.6.1.2 Resistive Temperature Devices (RTDs)

RTDs use the change in electrical resistance of metal wires with temperature.

A) **Working Principle**:
At the heart of an RTD lies a thin wire element, typically made of platinum (Pt), though nickel (Ni) or copper (Cu) are also used in some cases. These materials have a predictable relationship between electrical resistance and temperature. As the temperature increases, the atoms in the metal vibrate more, impeding the flow of electrons, which increases the resistance. This change in resistance is measured by applying a small current and calculating the voltage drop across the element. Specialized electronics then convert the resistance to a temperature reading using precise calibration equations, typically based on the known properties of the material.

B) **Advantages of RTDs**:
   a) **High Accuracy and Precision**: Can achieve better than 0.1°C accuracy, crucial for demanding applications.
   b) **Wide Temperature Range**: Platinum RTDs can function from −200°C to 850°C, making them versatile.
   c) **Stable and Repeatable**: Resistance changes predictably with temperature, ensuring reliable measurements.
   d) **Linear Relationship**: The resistance-temperature curve is nearly linear for many metals, simplifying calculations.
   e) **Resistant to Chemical Corrosion**: Certain materials like platinum offer excellent resistance to harsh environments.

C) **Limitations of RTDs**:
   a) **Cost**: More expensive than some other temperature sensors like thermocouples.
   b) **Self-heating**: The applied current can slightly heat the element, affecting accuracy unless it is corrected for.
   c) **Size and Fragility**: Thin wires can be delicate and require careful handling.
   d) **Limited High-temperature Range**: Certain materials like copper lose accuracy at higher temperatures.

D) **Applications of RTDs**:
   a) **Laboratory and Scientific Measurements**: High accuracy is crucial for research and calibration.
   b) **Industrial Process Control**: Accurate temperature monitoring for chemical reactions, furnaces, and pipelines.
   c) **Aerospace and Automotive Industries**: Reliable sensing for engine, fuel, and environmental temperatures.

d) **Medical Equipment**: Monitoring patient temperature during surgeries and treatments.
e) **Food and Beverage Production**: Ensuring precise temperatures for cooking, freezing, and storage.

E) **Types of RTDs**:
a) **Wire-wound**: Thin element wire wound around a ceramic core, commonly used for accurate measurements.
b) **Thin-film**: Element deposited on a thin-film substrate, good for small spaces and fast response times.
c) **Surface Mount**: RTD element packaged for direct integration into electronic circuits.

Three YouTube videos about RTDs are available at:

1) https://www.youtube.com/watch?v=WNs8FBl3c7M.
2) https://www.youtube.com/watch?v=blnnAEmVXp0.
3) https://www.youtube.com/watch?v=yNryBIe5kEg.

3.6.1.3 Thermocouples

Generate voltage due to the temperature difference between two dissimilar metals in contact. Thermocouples, those simple yet sophisticated sensors, harness the magic of the Seebeck effect to transform temperature differences into electrical voltage.

A) **The Seebeck Effect**:
Imagine two dissimilar metals joined at one end. When a temperature difference exists between this junction and the other ends of the metals, a tiny voltage appears across the junction. This is the Seebeck effect, the heart of every thermocouple. The magnitude of the voltage depends on the temperature difference and the specific combination of metals used.

B) **Building a Thermocouple**:
A thermocouple is essentially two dissimilar metal wires joined at one end, called the "hot junction," where the temperature is measured. The other ends, called "reference junctions," are kept at a known temperature. The voltage generated is proportional to the temperature difference between the hot and reference junction.

C) **Popular Thermocouple Types**:
a) **Type B (Platinum-Rhodium)**: Highest accuracy, suitable for high temperatures up to 1800°C. Below roughly 600°C (1112°F), the accuracy of Type B thermocouples starts to diminish compared to other types.
b) **Type E**: Constructed with chromel (alloy of chromium and nickel) and constantan (alloy of copper and nickel), Type E thermocouples operate within a temperature range of −200°C to +900°C. They are known for their high accuracy in moderate temperatures.
c) **Type J**: This thermocouple is suitable for a range of −40°C to +800°C. It is made of iron and constantan and is known for its relatively high sensitivity.
d) **Type K**: One of the most widely used thermocouples, known for its wide temperature range (−200°C to +1200°C) and high accuracy. It is composed of chromel and alumel (alloy of nickel, manganese, aluminum, and silicon).
e) **Type N**: Composed of nicrosil (nickel, chromium, and silicon) and nisil (nickel, silicon, and magnesium), Type N thermocouples cover a range from −270°C to +1300°C.

They offer better resistance to high-temperature oxidation compared to either Type K or Type J thermocouples.

f) **Type R and Type S**: Both Type R and Type S thermocouples use noble metal alloys (platinum and rhodium) and are designed for high-temperature applications, with ranges up to around 1600°C (Type R) and 1600–1768°C (Type S).

g) **Type T**: Comprising copper and constantan, Type T thermocouples cover a range from −270°C to +400°C. They exhibit good performance in low-temperature applications and have excellent repeatability.

D) **Advantages of Thermocouples**:

a) **Simple and Robust**: No moving parts, minimal maintenance, suitable for harsh environments.
b) **Wide Temperature Range**: Different types cater to diverse temperature needs.
c) **Fast Response Time**: Quickly react to temperature changes.
d) **Relatively Inexpensive**: Compared to some high-precision sensors.
e) **Flexible and Versatile**: Can be easily shaped and adapted to various applications.

E) **Limitations of Thermocouples**:

a) **Require Reference Temperature**: Need a stable reference point for accurate measurement.
b) **Nonlinear Voltage Output**: Requires calibration tables or electronics for accurate temperature conversion.
c) **Susceptible to Electrical Noise**: Proper shielding and grounding are crucial.
d) **Limited Low-temperature Accuracy**: Certain types are less accurate at cryogenic temperatures.

F) **Applications of Thermocouples**:

a) **Industrial Process Control**: Monitoring furnaces, reactors, pipelines, and engines.
b) **Aerospace and Automotive Industries**: Measuring gas turbine, exhaust, and engine temperatures.
c) **Power Generation and Distribution**: Monitoring transformers, cables, and turbines.
d) **Food and Beverage Processing**: Ensuring consistent temperatures in cooking, freezing, and storage.
e) **Scientific Research and Experiments**: Measuring temperatures in diverse environments.

Three YouTube videos about thermocouples are available at:

1) https://www.youtube.com/watch?v=mNoI62URtAk.
2) https://www.youtube.com/watch?v=QOmR6IGQrxI.
3) https://www.youtube.com/watch?v=-WnGrJYlTAU.

3.6.1.4 Bimetallic Strips

A) **The Science Behind the Bend**:

Imagine two strips of different metals bonded together, side-by-side. When heated, each metal expands at its own rate. Since one metal expands more than the other, the composite strip bends in the direction of the less expansive metal. This bending motion can then be used to trigger various actions, like opening or closing a circuit.

B) **Choosing the Right Metals**:

The magic of a bimetallic strip lies in the careful selection of metals. Typically, one metal has a high coefficient of thermal expansion (expands a lot with heat), while the other has a low coefficient (expands less). Common pairings include:

a) **Brass and Invar**: Brass expands readily, while invar (an alloy of iron and nickel) is known for its minimal expansion, making it a popular choice.

b) **Steel and Copper**: A readily available and cost-effective combination offering good sensitivity to temperature changes.

c) **Stainless Steel and Nickel**: Used for high-temperature applications where durability and stability are crucial.

C) **Bimetallic Strips in Action**:

Bimetallic strips often take center stage in:

a) **Thermostats**: The bending action controls the flow of electricity, regulating temperature in homes, ovens, and various appliances.

b) **Circuit Breakers**: They detect overheating through bending and trip the circuit to prevent electrical fires.

c) **Safety Valves**: The bending activates mechanisms to release pressure or stop the flow of fluids in potentially dangerous situations.

D) **Advantages of Bimetallic Strips**:

a) **Simple and Reliable**: No complex electronics or moving parts, making them low-maintenance and long-lasting.

b) **Cost-effective**: Relatively inexpensive compared to many electronic sensors.

c) **Fast Response Time**: React quickly to temperature changes due to the direct mechanical action.

d) **Durable and Versatile**: Can withstand moderate temperatures and are adaptable to various applications.

E) **Limitations of Bimetallic Strips**:

- **Limited Accuracy**: Not as precise as some electronic sensors, especially for wide temperature ranges.
- **Hysteresis**: The bending might not be entirely reversible, leading to slight temperature inconsistencies over time.
- **Susceptible to Wear and Tear**: The constant bending can eventually fatigue the material, affecting its performance.

Three YouTube videos about bimetallic strips are available at:

1) https://www.youtube.com/watch?v=82FPQ6z8vcE.
2) https://www.youtube.com/watch?v=RFWj7L05Yxo.
3) https://www.youtube.com/watch?v=ESlv1F4BYOM.

3.6.1.5 Infrared (IR) Thermometers

Infrared thermometers unlock a unique way to measure temperature by sensing the invisible heat waves objects emit.

A) **The Power of Infrared Radiation**:

Every object with a temperature above absolute zero emits infrared radiation, invisible to our eyes but detectable by special sensors. IR thermometers use these sensors, typically semiconductor-based, to capture this radiation and convert it into an electrical signal.

B) **From Radiation to Temperature**:

The electrical signal from the sensor depends on the intensity and wavelength of the infrared radiation it receives. Advanced algorithms then analyze this signal, taking into account the object's emissivity (its ability to emit radiation), and convert it into a temperature reading displayed on the device.

C) **Advantages of IR Thermometers**:

a) **Noncontact Measurement**: No need to touch the object, ideal for delicate surfaces, hazardous environments, or fast readings.

b) **Wide Temperature Range**: Different models cater to diverse ranges, from sub-zero temperatures to over 1000°C.

c) **Fast Response Time**: Quickly capture temperature changes, perfect for dynamic processes.

d) **Easy to Use**: Often point-and-shoot operation, convenient for various users.

e) **Versatile Applications**: From cooking and engine repair to medical diagnoses and industrial monitoring.

D) **Limitations of IR Thermometers**:

a) **Emissivity Dependence**: Accurate readings require consideration of the object's emissivity, which can vary.

b) **Surface Temperature only**: Measures surface temperature, not the internal temperature of an object.

c) **Distance Matters**: Accuracy can decrease with increasing distance from the object.

d) **Environmental Factors**: Ambient temperature and airflow can affect readings.

E) **Applications of IR Thermometers**:

a) **Cooking**: Checking food temperature for safe consumption and optimal cooking.

b) **Industrial Maintenance**: Monitoring machinery temperatures for preventative maintenance.

c) **Medical Field**: Noninvasive temperature measurement for patients.

d) **Firefighting**: Detecting hotspots and assisting with search and rescue.

e) **Building Inspections**: Identifying energy leaks and thermal inefficiencies.

Three YouTube videos about IR thermometers are available at:

1) https://www.youtube.com/watch?v=b4APOYAA918.
2) https://www.youtube.com/watch?v=cIbteXkAFxw.
3) https://www.youtube.com/watch?v=lsYaVHKfKZI.

3.6.1.6 Thermal Imaging Cameras

Thermal imaging cameras are devices that capture and create images based on the infrared radiation emitted by objects and surfaces, allowing visualization of temperature variations.

A) **Principle of Operation**:

Thermal imaging cameras detect and measure the infrared radiation emitted by objects. All objects with a temperature above absolute zero (−273.15°C or 0 Kelvin) emit infrared radiation. The amount of radiation and its wavelength depend on the object's temperature.

These cameras use infrared sensors, specifically microbolometers, to detect infrared wavelengths and convert the absorbed radiation into electrical signals.

B) **Image Creation**:

The camera then processes these electrical signals and generates a thermal image, also called a thermogram or infrared image. It assigns colors or shades to different temperature ranges, creating a visual representation of temperature variations across the observed surface. The colors usually range from blue or purple (colder areas) to red or white (warmer areas), providing a clear depiction of heat distribution.

C) **Applications**:

a) **Building Inspections**: Detecting heat loss, identifying insulation issues, and locating areas of air leakage in buildings for energy efficiency assessments.
b) **Electrical Inspections**: Finding overheating components in electrical systems, identifying faulty connections, or detecting overloaded circuits to prevent potential hazards.
c) **Mechanical and Industrial Maintenance**: Monitoring equipment and machinery for abnormal heat levels, detecting mechanical failures, and conducting predictive maintenance.
d) **Firefighting and Search and Rescue**: Locating heat sources and victims in smoke-filled environments or identifying hotspots during firefighting operations.
e) **Medical Imaging**: In healthcare, thermography is used for diagnosing medical conditions by detecting differences in skin temperature, such as inflammation or blood flow irregularities.
f) **Wildlife Observation**: Studying animal behavior, locating wildlife in darkness, and conducting ecological research.

Advantages:

a) **Noncontact and Nondestructive**: Thermal imaging allows measurements from a distance without physically touching the object.
b) **Speed**: Quickly scanning large areas to identify temperature differences.
c) **Versatility**: Applicable across various industries and fields, aiding in diagnostics, analysis, and problem-solving.

Limitations:

a) **Accuracy**: Calibration and environmental factors can affect the accuracy of temperature measurements.
b) **Cost**: Thermal imaging cameras can be relatively expensive, depending on the specifications and intended use.

Three YouTube videos about thermal imaging cameras are available at:

1) https://www.youtube.com/watch?v=KASYjRW5VgQ.
2) https://www.youtube.com/watch?v=ZSozrAIfigE.
3) https://www.youtube.com/watch?v=EoDFUQsUeN4.

3.6.1.7 Selection Criteria

a) **Accuracy and Precision**: Different devices have varying levels of accuracy and precision, which is important for specific applications.
b) **Temperature Range**: Choose a device suitable for the expected temperature range you need to measure.

c) **Response Time**: How quickly the device reacts to temperature changes, relevant for dynamic processes.
d) **Cost and Ease of Use**: Consider budget and user-friendliness when choosing a device.

End of Chapter Problems

CHW 3.1 What units does specific gravity have?
A Mass
B Volume
C Density
D No units

CHW 3.2 A liquid possesses a specific gravity of 0.60. What is its density in g/cm^3?
A $1.00\,g/cm^3$
B $0.36\,g/cm^3$
C $0.60\,g/cm^3$
D $1.67\,g/cm^3$

CHW 3.3 If both substance A and substance B have a specific gravity of 1.45, is it necessary for $4\,cm^3$ of A to have the same mass as $4\,cm^3$ of B?
A Yes, always
B No, not necessarily
C Depends on the temperature
D Depends on the pressure

CHW 3.4 Does the density of liquid mercury increase or decrease with increasing temperature? Justify your answer using a thermometer as an illustration.
A Increases
B Decreases
C Remains constant
D Fluctuates

CHW 3.5 Calculate the mass of $4.0\,cm^3$ of a liquid with a specific gravity of 0.60.

CHW 3.6 The density of mercury is given as $13{,}595\,kg/m^3$ at room temperature. Calculate the volume in ft^3 occupied by 200 kg of mercury using the provided information.

CHW 3.7 Calculate densities in lb_m/ft^3 of the following substances:
A A liquid with density of $800\,kg/m^3$.
B a solid with a specific gravity of 6.8.

CHW 3.8 What is the principle behind a rotameter's measurement of flow rate?
A Utilizing electrical conductivity
B Measuring fluid pressure differentials
C Balancing gravitational force and fluid drag force
D Analyzing fluid density changes

CHW 3.9 Which of the following statements best describes a limitation of rotameters compared to other flow meters?

A They require complex external power sources
B They are not suitable for measuring liquid flow rates
C They are not cost-effective
D They may lack precision in scenarios where flow conditions fluctuate significantly

CHW 3.10 What is the primary method used by an orifice meter to measure fluid flow?

A Direct volumetric measurement
B Analysis of fluid viscosity changes
C Measuring pressure drop across an orifice plate
D Calculating fluid velocity using magnetic fields

CHW 3.11 What are potential limitations of using orifice meters for flow measurement?

A High cost and complexity
B Inability to measure flow of gases
C Sensitivity to changes in fluid properties, temperature, and pressure
D Difficulty in installation due to complex mechanisms

CHW 3.12 What distinguishes a venturi meter from other flow measuring devices like orifice meters?

A Its reliance on electrical conductivity for flow measurement
B Utilization of a smooth constricted section to measure fluid pressure difference
C Measurement of fluid density changes across the meter
D Calculation of flow rate using ultrasonic waves

CHW 3.13 What advantages do venturi meters offer over other flow meters?

A Lower accuracy and reliability
B Higher susceptibility to erosion or damage
C Wide operational range and reduced energy loss
D Simplified installation and maintenance processes

CHW 3.14 How does a flow nozzle compare to a venturi meter in terms of accuracy and installation complexity?

A Less accurate and more complex to install than a venturi meter
B More accurate and less complex to install than a venturi meter
C Equally accurate and complex to install as a venturi meter
D Not as accurate as but simpler to install compared to a venturi meter

CHW 3.15 What is a significant advantage of positive displacement flow meters over other flow meters?

A High capacity for handling increased flow rates
B Lower accuracy in low-flow applications

C Exceptional accuracy in low-flow applications and with viscous fluids
D Suitable for high-flow applications without wear and tear

CHW 3.16 What is the primary mechanism utilized by a turbine flow meter to measure fluid flow?
A Differential pressure sensing
B Magnetic field manipulation
C Impingement on turbine blades causing rotation
D Ultrasonic wave propagation

CHW 3.17 Which factor is a significant advantage of turbine flow meters in measuring flow rates?
A High susceptibility to damage by abrasive fluids
B Precision in low flow rate conditions
C Limited measurement range
D Accuracy in measuring higher flow rates

CHW 3.18 In a nutating disk flow meter, the key principle employed for measuring flow rate is:
A Pressure drop across the meter chamber
B Measurement of fluid velocity at a specific point
C Displacement of a fixed volume of liquid with each nutation
D Rotation of the disk caused by the fluid's angular momentum

CHW 3.19 A primary advantage of using a nutating disk flow meter compared to other types is:
A High accuracy even at very high flow rates
B Ability to handle highly viscous fluids without performance issues
C Minimal pressure loss, leading to better energy efficiency
D Wide range of applicability for both liquids and gases

CHW 3.20 What is the primary principle behind the operation of an electromagnetic flow meter?
A Pressure differential measurement
B Ultrasonic wave propagation
C Magnetic field induction and voltage measurement
D Impingement on turbine blades causing rotation

CHW 3.21 Which feature is a significant advantage of electromagnetic flow meters in flow measurement?
A Limited accuracy in challenging flow conditions
B Effectiveness in measuring nonconductive fluids
C High accuracy at very low flow rates
D Ability to handle a wide range of fluid viscosities

E Lack of moving parts in the flow stream, minimizing wear and tear
F Suitability for measuring both liquids and gases

CHW 3.22 What principle does a Coriolis flow meter rely on for measuring fluid flow rates?
A Pressure differential measurement
B Magnetic field induction and voltage measurement
C Bernoulli's principle
D Impingement on turbine blades causing rotation
E Archimedes' principle
F Coriolis effect inducing tube deformation

CHW 3.23 Which feature distinguishes Coriolis flow meters from some other flow meters in terms of measurement accuracy?
A Measure volume flow directly
B Measure pressure changes accurately
C Measure fluid properties like density or viscosity
D Measure flow rates unaffected by changes in temperature or pressure
E Low cost and minimal maintenance requirements
F Ability to handle extremely high flow rates or fluids with high concentrations of solids

CHW 3.24 How does a Doppler ultrasonic flow meter determine flow velocity?
A By calculating the time difference for ultrasonic waves to travel with and against the flow direction
B By transmitting ultrasonic waves through the fluid and measuring the time taken between transducers
C By measuring the difference in ultrasonic wave frequency in different pipe materials
D By detecting frequency shifts of ultrasonic waves bouncing off moving particles in the fluid

CHW 3.25 What is the key advantage of ultrasonic flow meters compared to other flow measuring devices?
A Noninvasive measurement capability without interrupting the flow
B Intrusive nature requiring insertion into the pipeline
C Limitation to measuring only clean liquids
D Incompatibility with different pipe materials and shapes

CHW 3.26 How do vortex flow meters measure fluid velocity to determine flow rate?
A By detecting changes in pressure across the meter
B By detecting the frequency of vortex shedding patterns
C By measuring the time taken for ultrasonic waves to travel between transducers
D By inducing a magnetic field and measuring voltage generated

CHW 3.27 What is a primary advantage of vortex flow meters over some other flow meters?

A Greater susceptibility to pressure loss
B Inability to handle varying flow rates or densities
C Reduced accuracy due to fluid properties and installation conditions
D High accuracy at very low flow rates
E Reliability and accuracy even in fluctuating flow conditions
F Suitability for measuring both liquids and gases at high pressures

CHW 3.28 The specific gravity of gasoline is approximately 0.70.

A Determine the mass (kg) of 60.0 L of gasoline.
B The mass flow rate of gasoline exiting a refinery tank is 1200 kg/min. Estimate the volumetric flow rate in liters/s.
C Express the mass flow in (lb_m/min) of part (b).
D Gasoline and kerosene (specific gravity = 0.82) are blended to obtain a mixture with a specific gravity of 0.75. Calculate the mass fraction of the two compounds in the mixture, assuming $V_{blend} = V_{gasoline} + V_{kerosene}$.

$$\frac{1}{\rho_{solution}} = \frac{1}{\rho_{blend}} = V_{blend} = \sum_{i=1}^{N} x_i V_i = \sum_{i=1}^{N} x_i \frac{1}{\rho_i} = \sum_{i=1}^{N} \frac{x_i}{\rho_i}$$

E Based on part (d), calculate the *volume fraction* of the two compounds in the mixture.

CHW 3.29 At 25°C, an aqueous solution containing 40.0wt% H_2SO_4 has a specific gravity of 1.261. A quantity of the 40wt% solution is needed that contains 200.0 kg of H_2SO_4.

a) Calculate the required volume (L) of the solution using the given specific gravity.
b) Estimate the *percentage error* that would have resulted if *pure-component specific gravities* of H_2SO_4 ($SG = 1.8255$) and water had been used for the calculation instead of the given specific gravity of the mixture.

V_{ideal} for a mixture is calculated based on *pure-component data* which assumes that there is no volume change as a result of mixing. In other words, the volume of the mixture is calculated based on the summation of volumes of all materials to be mixed. This assumption, of course, is not always correct; it is valid only for *ideal solutions.*

CHW3.30 A suspension of calcium carbonate ($CaCO_3$) particles in water flows through a pipe. Your assignment is to determine both the flow rate and the composition of this slurry. You proceed to collect the stream in a graduated cylinder for 2.00 minutes; you then weigh the cylinder, evaporate the collected water, and reweigh the cylinder. The following results are obtained:

Mass of empty cylinder: 80.0 g
Mass of cylinder + collected slurry: 1080 g
Volume collected: 800 mL
Mass of cylinder after evaporation: 315 g

Calculate:

a) The volumetric flow rate and mass flow rate of the suspension.
b) The density of the suspension.
c) The mass fraction of $CaCO_3$ in the suspension.

CHW 3.31 A gaseous mixture of CO, CO_2, CH_4, and N_2 is analyzed with a gas chromatograph. The output appears on a strip-chart recorder, as shown here.

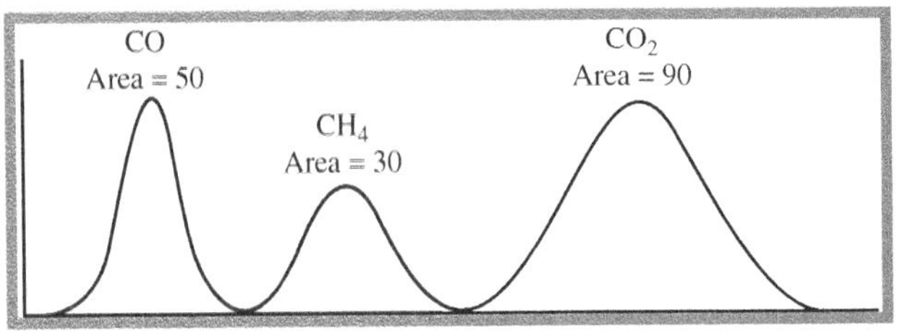

For each of the three species, the area under the peak is approximately proportional to the number of moles of the indicated substance in the sample. From other information, it is known that the *molar ratio of methane (CH_4) to N_2 is 0.250 (i.e., the molar ratio of N_2 to CH_4 is 4).*

A What are the mole fractions of the four species in the output gas mixture? Given that: $y_{N2} = 4y_{CH4}$ $y_{CO2} = \frac{90}{30}y_{CH4}$ $y_{CO} = \frac{50}{30}y_{CH4}$

B What is the average molecular weight of the output gas mixture?

CHW 3.32 A 6.00-wt% aqueous sulfuric acid solution ($\rho = 1.032$ g/mL) flows through a 50-m long pipe with a 5.0 cm diameter at a rate of 90 L/min.

A What is the *molarity* of sulfuric acid in the solution?

B How long (in seconds) would it take to fill a 60-US gallon drum, and how much sulfuric acid (kg) would the drum contain?

C The mean velocity of a fluid in a pipe equals the volumetric flow rate divided by the cross-sectional area normal to the direction of flow. Use this information to estimate how long (in seconds) it takes the solution to flow from the pipe inlet to the outlet.

CHW 3.33 A fluid of unknown density is used in two manometers-one sealed-end, the other across an orifice in a water pipeline. The readings shown here are obtained on a day when barometric pressure is 760 mmHg.

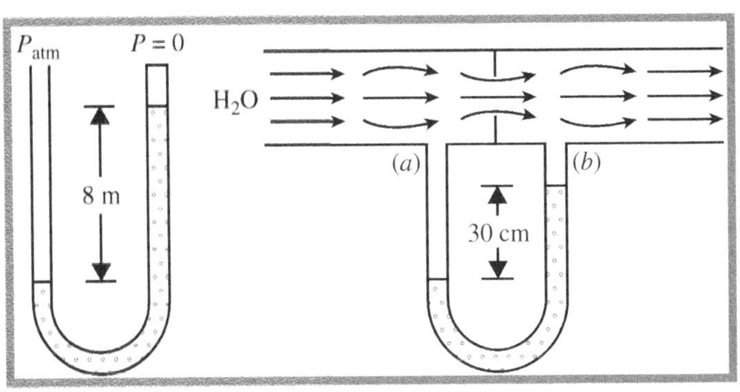

What is the pressure drop (kPa) from point (*a*) to point (*b*)?

Notice that: P_{atm} = 1atm = 14.69psia = 101,325 Pa = 760 mmHg

CHW 3.34 A sample of polluted air contains 0.05 mg of carbon monoxide (CO) per 1.5 kg of air. What is the concentration of CO in the air in ppb (parts per billion)?

CHW 3.35 A closed container filled with an unknown liquid holds a pressure gauge reading 30 kPa at a depth of 2 m. If the gravitational acceleration is 9.81 m/s^2, what is the density of the liquid?

CHW 3.36 Which application might not be suitable for U-tube manometers due to the risk of breakage of the glass tube?

A Measuring low-pressure systems in HVAC
B Laboratory experiments requiring precise pressure measurements
C Industrial fluid systems requiring accurate measurements
D High-pressure measurement scenarios

CHW 3.37 What advantage does the inclined U-tube manometer offer compared to a standard vertical U-tube manometer?

A Greater resistance to breakage
B Capability to measure extremely high pressures.
C Suitable when space constraints or sensitivity to small pressure changes are factors.
D Ease of portability and durability.

CHW 3.38 Which of the following applications is NOT typically suited for a micromanometer?

A Monitoring air pressure within a cleanroom used for sensitive electronics assembly.
B Measuring pressure drop across a filter in a ventilation system.
C Calibrating pressure gauges used in medical equipment.
D Assessing the airspeed of an airplane wing section in a wind tunnel.

CHW 3.39 What pressure range is the McLeod gauge most suitable for measuring?

A 1–100 torr
B 10^{-6}–10^{-1} torr
C 10^{-3}–1 atmosphere
D 100–1000 mm Hg

CHW 3.40 Compared to other pressure gauge types, a key limitation of Bourdon tubes is:

A High cost and complex maintenance requirements
B Lack of suitability for both high- and low-pressure ranges
C Relatively lower accuracy, particularly at low pressures
D Limited versatility due to restricted applications

CHW 3.41 Which of the following types of Bourdon tubes is best suited for measuring extremely high pressure within a hydraulic system?

A C-shaped tube – simple and economical, but with limitations at high pressures

B Helical tube – more sensitive and compact, ideal for high-pressure ranges
C Spiral tube – provides a wider range of motion, but not specifically designed for high pressure
D Any of the above types can be used with appropriate calibration for high-pressure measurement

CHW 3.42 Compared to Bourdon tube gauges, a key advantage of diaphragms is:
A Superior accuracy at high pressure ranges
B Higher sensitivity and ability to detect minute pressure changes at low pressures
C Greater durability and robustness under harsh conditions
D Wider range of applicability for both liquids and gases

CHW 3.43 When selecting a diaphragm for a pressure gauge used in a food processing tank, which material property is most important to consider?
A High sensitivity for precise pressure measurement
B Resistance to fatigue and wear for long-term reliability
C Compatibility with the specific food products and cleaning chemicals
D Durability and robustness to withstand high pressure fluctuations

CHW 3.44 What is a key advantage of using bellows in pressure measurement?
A Resistance to temperature fluctuations
B Low cost compared to other pressure gauges
C Small displacement for subtle pressure changes
D High sensitivity to detect small pressure changes

CHW 3.45 Which type of bellows would be most suitable for high-pressure environments but less flexible and more susceptible to corrosion?
A Metal bellows
B Single-ply bellows
C Multi-ply bellows
D Polymer bellows

CHW 3.46 What is the primary disadvantage of using piston-cylinder assemblies for pressure measurement?
A Limited pressure range
B Low accuracy
C Minimal maintenance requirements
D Bulkiness and size

CHW 3.47 Compared to other pressure gauge types, a key advantage of piezoelectric sensors is:
A Suitability for measuring very low pressures with high accuracy
B Ability to capture rapid pressure fluctuations and dynamic changes
C Durability and resistance to harsh environmental conditions
D Wide range of applicability for both liquids and gases

CHW 3.48 In a medical device monitoring blood pressure, what property of the piezoelectric material within the sensor is most important for accurate readings?

A High resistance to electrical interference
B Strong sensitivity to even small pressure changes
C Long-term stability and resistance to fatigue
D Ability to withstand high pressure and shocks

CHW 3.49 What principle do capacitance manometers operate on?

A Changes in resistance
B Variations in inductance
C Alterations in capacitance
D Adjustments in voltage

CHW 3.50 In a capacitance pressure sensor, what happens to the capacitance as pressure increases?

A It remains unchanged
B It decreases
C It fluctuates randomly
D It increases

CHW 3.51 What property of a strain gauge changes when it undergoes mechanical deformation due to pressure?

A Length
B Temperature
C Width
D Density

CHW 3.52 How does a strain gauge measure pressure change?

A By alterations in temperature
B Through changes in volume
C By modifications in electrical resistance
D Using variations in density

CHW 3.53 How does an increase in pressure affect the natural frequency of a vibrating wire in a pressure sensor?

A Decreases the wire tension
B Increases the wire's natural vibration frequency
C Reduces the wire's vibration amplitude
D Has no impact on the wire's vibration properties

CHW 3.54 What principle do vibrating wire pressure sensors rely on to measure pressure changes?

A Alterations in wire temperature
B Changes in wire density
C Modifications in wire conductivity
D Variations in wire tension

CHW 3.55 Absolute zero is theoretically the point at which particles reach:
A Maximum kinetic energy
B Minimum thermal motion
C Complete ionization
D Supercritical state

CHW 3.56 For some isotopes of Helium, the formation of a Bose-Einstein condensate occurs at temperatures closest to:
A Room temperature (20°C)
B Freezing point of water (0°C)
C Dry ice temperature (−78°C)
D Absolute zero (−273.15°C)

CHW 3.57 Suppose a temperature sensor measures a change of $\Delta T = 36°\text{F}$.
A Convert this temperature change directly to °C using the appropriate conversion factor.
B If the initial temperature was $T_1 = 86°\text{F}$, calculate the final temperature in both °F and °C.

CHW 3.58 Which type of liquid used in liquid-in-glass thermometers offers the widest range of measurable temperatures?
A Mercury
B Alcohol
C Gallium
D Toluene

CHW 3.59 Compared to electronic thermometers, a key disadvantage of liquid-in-glass thermometers is:
A Superior accuracy for scientific applications
B Wider range of available temperature scales
C Faster response time to temperature changes
D Lower cost and affordability

CHW 3.60 Which material is generally considered a safer alternative to mercury for liquid-in-glass thermometers?
A Alcohol
B Gallium
C Toluene
D Zinc

CHW 3.61 Which factor contributes to the increase in electrical resistance in an RTD with rising temperature?
A Decreased atomic vibration
B Enhanced electron flow
C Increased wire thickness
D Increased atomic vibration

CHW 3.62 Compared to liquid-in-glass thermometers, a key advantage of RTDs is:

A Wider range of measurable temperatures
B Higher durability and resistance to breakage
C Significantly superior accuracy and precision
D Lower cost and affordability

CHW 3.63 Which limitation of RTDs poses a challenge for applications requiring fast temperature changes?

A High cost compared to other sensor types
B Self-heating effect potentially impacting accuracy
C Size and fragility of the thin wire element
D Nearly linear relationship between resistance and temperature

CHW 3.64 What is the fundamental principle underlying the generation of voltage in a thermocouple?

A The Seebeck Effect
B Ohm's Law
C Faraday's Law
D Hooke's Law

CHW 3.65 What is a notable limitation of thermocouples that requires compensation for accurate temperature conversion?

A They are susceptible to electrical noise
B They require a stable reference temperature
C They are less accurate at cryogenic temperatures
D They have a nonlinear voltage output

CHW 3.66 For applications requiring accurate temperature measurements across a wide range (−270°C to +1300°C), which type of thermocouple would be the most appropriate choice?

A Type B (Platinum-Rhodium)
B Type E (Chromel-Constantan)
C Type J (Iron-Constantan)
D Type K (Chromel-Alumel)
E Type N (Nicrosil-Nisil)
F Type T (Copper-Constantan)

CHW 3.67 What is the primary reason behind the bending of a bimetallic strip when exposed to heat?

A Both metals expand equally
B One metal expands more than the other
C The strip contracts when heated
D The metals fuse together

CHW 3.68 Which of the following applications commonly utilizes the bending action of bimetallic strips?

A Automotive engines

B Aircraft navigation systems
C Thermostats
D Smartphone circuitry

CHW 3.69 Which of the following factors is NOT significantly important for an IR thermometer to take an accurate temperature reading?
A The intensity of the infrared radiation emitted by the object
B The distance between the IR thermometer and the object
C The temperature of the surrounding environment
D The color of the object being measured

CHW 3.70 What principle do thermal imaging cameras rely on to generate images?
A Detection of ultraviolet radiation
B Measurement of visible light
C Detection and measurement of infrared radiation
D Utilization of X-ray emissions

Reference

1 Speckbrock, G. (2004). Metallic gallium alloy as a thermometer fluid. Patent year. https://patents.google.com/patent/DE10321843A1/en (accessed 6 October 2024).

4

Principles of Material Balance

4.1 Process Classification

Chemical processes are categorized as batch, continuous, or semi-batch and can exist in steady-state or transient states. It is essential to determine the process category before constructing material balances for a system. The density of a substance serves as a conversion factor connecting the mass and volume of the substance quantity.

1) **Batch Process**: In a batch process, the feed is introduced into a vessel at the start, and the contents of the vessel are extracted later. Throughout this duration, no mass crosses the system boundaries from the moment the feed is introduced to the moment the final product is taken out. For instance, this process involves swiftly adding reactants into a tank and extracting the resulting products along with any unreacted materials after the system has reached equilibrium or comes to the end of a prescribed period.
2) **Continuous Process**: In a continuous process, inputs and outputs move continuously throughout the entire process duration. For instance, a mixture of liquids is consistently pumped into a distillation column at a constant flow rate, and product streams are steadily collected from both the top and bottom of the column.
3) **Semi-batch Process**: Any process that doesn't conform to the characteristics of either batch or continuous processes. For example, allowing the contents of a pressurized gas container to be released into the atmosphere. Another example is to gradually mix various liquids in a tank without any withdrawal taking place.

If all the process variables, such as temperatures, pressures, volumes, and flow rates, remain constant over time, except for occasional minor fluctuations around fixed average values, the process is considered to be under a steady-state condition. If any of these variables change over time, it indicates a transient or unsteady-state operation. Batch and semi-batch processes inherently operate in an unsteady-state manner (why?), whereas continuous processes can be either in a steady state or undergo transient changes.

To my dear readers,
When copying MATLAB code or text, please right-click and attempt both the "Copy with Formatting" and "Copy" options from the context menu. This approach is effective within Adobe Acrobat. If neither option works, kindly carefully review any MATLAB error messages and compare the pasted text to the original to identify and correct any discrepancies. Alternatively, a soft copy of the MATLAB code is available in the Wiley book companion site of this book for your convenience.

MATLAB-based Computations of Chemical Engineering Principles, First Edition. Kamal I. M. Al-Malah.

Companion website: www.wiley.com/go/al-malah/matlabbasedcomputations

Batch processing is typically employed to produce relatively small product quantities on specific occasions, while continuous processing is more suitable for high production rates. Continuous processes aim to operate as consistently close to a steady state as possible. Unsteady-state (transient) conditions occur during the process startup and after intentional or unintentional alterations in process operation conditions.

Exercise 4.1

Classify the following processes as batch, semi-batch, or continuous

1) A balloon is filled with air at a steady rate of 5 g/min.
2) A bottle of ice cream is taken from the refrigerator and left on the kitchen table.
3) Alcohol is boiled in an open flask.
4) Initially, ethane and oxygen are continuously introduced into a tubular reactor, maintaining a constant flow rate. These compounds react within the reactor to produce carbon dioxide and steam, while the resulting products and unreacted substances are removed from the opposite end. At the commencement of the process, the reactor contains air. The reactor's temperature remains constant, and the composition, as well as the flow rate of the incoming reactant stream, remain unaffected by time. Determine the process classification (1) at the beginning and (2) after an extended duration has passed.

4.2 The Balance Concept

4.2.1 Introducing the Balance Concept: Mastering Our Measure

Welcome to the fascinating world of the balance concept! This cornerstone principle underpins many scientific and engineering practices, from understanding fundamental laws of physics to optimizing everyday processes. At its core, the balance concept states that within a defined system, **what goes in must equal what comes out**. Understanding this balance is crucial for monitoring key properties like total mass, a specific component, or energy.

Let us journey together as we explore the power of the balance concept. In this chapter, we will lay the foundation, introducing concepts like:

- **Defining the System**: Clearly identifying the boundaries of what we're analyzing, be it a chemical reaction, a nonreactive system, or an industrial operation.
- **Identifying Key Properties**: Pinpointing the specific aspects we want to monitor, such as the total moles of reactants in a reaction, the number of moles (or mole fraction), or mass fraction of a specific element in a mixture, or the energy flow into and out of a machine/vessel.
- **Formulating the Balance Equation**: Expressing the relationship between inputs, outputs, and changes within the system using a simple equation.

In future discussions, we will delve deeper into the practical applications of the balance concept. We will learn how to:

- **Use the Balance Equation to Solve Problems**: Predicting final quantities after a reaction, determining missing information in a system, or optimizing efficiency based on mass or energy constraints.

- **Adapt the Concept to Different Scenarios**: From closed systems where nothing enters or leaves to open systems with continuous exchange, we'll explore how the balance concept adapts to various situations.
- **Apply It to Real-World Examples:** We will see the balance concept in action, monitoring chemical reactions in laboratories, tracking energy consumption in buildings, or ensuring accurate material balances in industrial processes.

Join me on this exciting journey as we unlock the power of the balance concept! Prepare to become a master of measurement, armed with the ability to analyze and understand complex systems through this fundamental principle. Remember, the key to unlocking its potential lies in **defining the system under study, identifying the target properties, and formulating the balance equation.** Are you ready to begin? Since this course serves as an introduction to chemical engineering holy dome, we will focus on balance equations as they pertain to chemical industrial processes.

4.2.2 The General Balance Equation

Suppose water is a component of both the input and output streams of a continuous process unit and in an attempt to determine whether the unit performs as designed, the mass flow rates of water in both streams are measured and found to be different ($\dot{m}_{in} \neq \dot{m}_{out}$), as shown in Figure 4.1.

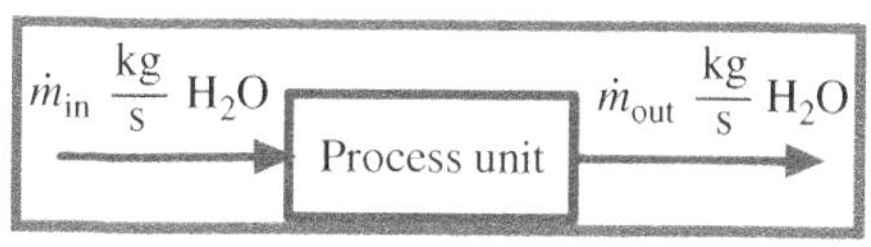

Figure 4.1 A schematic for a general process with different inlet and outlet water flowrate.

There are various potential reasons for the apparent contrast in the recorded flow rates:

1) Water might be used up in a reaction or formed as a product inside the unit.
2) Water could be accumulating within the unit, perhaps being retained within a packing material.
3) Water might be escaping or leaking from the unit.
4) There is a likelihood of inaccuracies in the measurements.

If the measurements are accurate and there are no leaks, the remaining possibilities, generation or consumption in a reaction and accumulation within the processing unit, are the only factors that could explain a discrepancy between the input and output flow rates.

An equation representing the conservation of a quantity (such as overall mass, a specific substance's mass, energy, or momentum) within a system (which could be a single process unit, a group of units, or an entire process) can be formulated as follows:

$$\textbf{Input} + \textbf{Generation} - \textbf{Output} - \textbf{Consumption} = \textbf{Accumulation} \qquad (4.1)$$

Input: Enters through system boundaries.
Generation: Produced within the control system.
Output: Leaves through system boundaries.
Consumption: Consumed within the control system.
Accumulation: Buildup within the control system.

NOTE

A general balance equation can be implemented for any extensive thermodynamic property or countable property. Here are some examples:

Extensive Properties:

- **Mass**: In chemical reactions, the total mass of all reactants before the reaction must equal the total mass of all products after the reaction. This principle forms the basis of the law of conservation of mass.
- **Energy**: In a closed system, the total energy, including internal, potential, and kinetic, remains constant even during transformations. This principle follows the law of conservation of energy.
- **Momentum**: In any collision, the total momentum before the collision equals the total momentum after the collision, even if the direction of the momentum changes.
- **Entropy**: In an isolated system, the total entropy always increases or stays the same, never decreases. This principle forms the basis of the second law of thermodynamics.

Countable Properties:

- **Number of Atoms**: In chemical reactions, the number of atoms of each element must remain constant before and after the reaction, reflecting the law of conservation of atoms.
- **Number of Photons**: In a closed optical system, the total number of photons remains constant, even if their energy or wavelength changes.
- **Number of Particles**: In nuclear reactions, the total number of nucleons (protons and neutrons) involved remains constant, even if they undergo transformations or rearrangements.
- **Number of Electrons**: In a typical chemical reaction, a reaction basically means or involves electron transfer from one species to another. So, the valance electrons (outer shell electrons of an atom) are involved in the process of electron transfer or exchange among different reacting species. The principle of electrical neutrality states that, under normal conditions, the total electric charge within a closed system is zero. In other words, the positive and negative charges balance out. Chemical reactions involve the rearrangement of atoms, but they preserve the total number of protons and electrons, hence maintaining electrical neutrality.
- **Number of Dollars**: In a typical daily life, a man or a woman can keep track of money-related activities, like what is left in one's wallet (accumulation or saving) after a day of spending money on purchasing goods or paying for services (out), money dropped out of his/her pocket or stolen (consumption or loss), and received cash (in).

Example 4.1 General Balance Equation Made by the Shepherd to Count Goat Population in a Pasture

Solution:

The general balance equation can be applied to the number of goats in a typical pasture but with some adaptations due to biological factors. The shepherd, having no mobile phone to be busy with, counts the number of goats early morning and another time late evening to make an assessment.

System: The pasture and the goats within it.

Countable Property: Number of goats (It is also extensive property as it depends on the size of the system).

General balance equation, Eq. (4.1):

$$\text{Accumulation} = \text{Input} - \text{Output} + \text{Generation} - \text{Destruction}$$

Where:

Accumulation: Reflects the change in the number of goats between morning and evening counts (positive if the number increases, negative if it decreases, and zero means no change).

Input: Number of goats entering the pasture during the day.

Output: Number of goats leaving the pasture during the day.

Generation: Number of kid goats born during the day.

Destruction: Number of goats dying during the day due to predation, disease, etc.

Conclusion: The concept of the general balance equation has been early practiced thousands of years ago by humankind since the time of cave men.

Example 4.2 The General Balance Equation

Each year 60,000 people move into a city, 70,000 people move out, 40,000 are born, and 30,000 die.

Write a balance on the population of the city.

Solution:

Recall Eq. (4.1): **Input + Generation – Output – Consumption = Accumulation**

$$60{,}000\frac{\text{Person}}{\text{yr}} + 40{,}000\frac{\text{Person}}{\text{yr}} - 70{,}000\frac{\text{Person}}{\text{yr}} - 30{,}000\frac{\text{Person}}{\text{yr}} = 0.000\frac{\text{Person}}{\text{yr}}$$

Thus, the city's net population *remains constant or steady*. One can also say that the annual rate of population growth is zero.

There are Two Distinct Types of Balances that can be Formulated:

1) **Differential Balances**: These balances depict the ongoing processes within a system at a specific moment in time. Each term within the balance equation signifies a rate (like the rate of input, generation, etc.) and is denoted in units of the quantity being balanced divided by time units (such as people per year, grams of CO_2 per second, and cubic meters per day). This type of balance is typically used for continuous processes (refer to Example 4.2).
2) **Integral Balances**: These balances portray what occurs between two time instances. Each term within the equation represents an amount of the balanced quantity and is measured in corresponding units (for instance, people, grams of CO_2, and cubic meters). Integral balances are commonly employed for batch processes, taking place between time, t_i = zero, once the input quantity is transferred into a batch container and a specified final time, t_f, prior to the product withdrawal. Notice that there is neither mass in nor mass out of the batch container during the batch process. Imagine a housewife cooking a food batch where she sets in the time once the food saucepan is on fire and waits until the food

becomes ready after a prescribed period, dictated by the fire temperature, food amount, and its type/content.

The Following Rules Shall be Used to Simplify the Material Balance Equation:

a) *If the balanced quantity is* the **total mass**, *set both generation = 0 and consumption = 0.* Except in nuclear reactions, mass can neither be created nor destroyed.
b) *If the balanced substance is a nonreactive species (neither a reactant nor a product, like an inert material), set both generation* = 0 *and consumption* = 0.
c) *If a system achieves a steady-state condition, set accumulation* = 0, *regardless of what the balanced quantity is.* By definition, under a steady-state condition, nothing can change with time, including the amount of the balanced quantity.

4.2.3 Balances on Continuous Steady-State Chemical Processes

For continuous processes at a steady state, the accumulation term in the general balance equation, Eq. (4.1) equals zero, and the equation simplifies to:

$$\textbf{Input} + \textbf{Generation} = \textbf{Output} + \textbf{Consumption} \tag{4.2}$$

If the balance is on a nonreactive species or on total mass, the generation and consumption terms equal zero and the equation reduces to *Input* = *Output*.

Example 4.3 Material Balances on a Continuous Distillation Process
One thousand kilograms per hour of a mixture of benzene (B) and toluene (T) containing 60% benzene by mass is separated by distillation into two fractions. The mass flow rate of benzene in the top stream is 550 kg B/h, and that of toluene in the bottom stream is 350 kg T/h. The operation is run at a *steady state (see* Figure 4.2). Write balances on benzene and toluene to calculate the *unknown component flow rates* in the output streams.

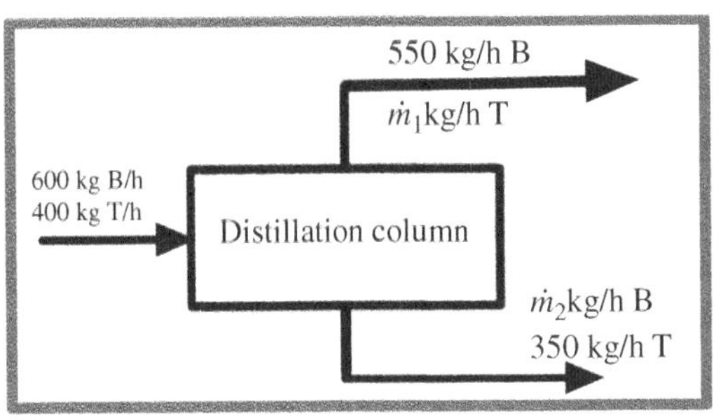

Figure 4.2 A schematic for binary distillation of benzene-toluene mixture.

Solution:

Since the process is at a *steady state* there can be *no buildup* of anything in the system, so the accumulation term equals zero in all material balances. In addition, since *no chemical reaction* occurs, there will be zero generation and/or zero consumption term. For all balances, Eq. (4.2) therefore takes the simple form: *Input* = *Output.*

$$\textit{Benzene Balance}: 600\ \frac{\text{kg B}}{\text{h}} = 550\ \frac{\text{kg B}}{\text{h}} + \dot{m}_2$$

$$\dot{m}_2 = 50 \frac{\text{kg B}}{\text{h}}$$

Toluene Balance: $400 \dfrac{\text{kg T}}{\text{h}} = 350 \dfrac{\text{kg T}}{\text{h}} + \dot{m}_1$

$$\dot{m}_1 = 50 \frac{\text{kg T}}{\text{h}}$$

4.2.4 Integral Balances on Batch Chemical Processes

Ammonia is synthesized from nitrogen and hydrogen within a batch reactor. Initially, at time $t_0 = 0$, there are n_0 moles of NH_3 present in the reactor. At a subsequent time t_f, the reaction reaches equilibrium, and the reactor contents, n_f, moles of ammonia are withdrawn. Throughout the duration between t_0 and t_f, no ammonia enters or exits through the reactor boundaries. Therefore, the general balance equation (Eq. 4.1) becomes simplified to Generation = Accumulation. In addition, the quantity of ammonia that accumulates within the reactor from t_0 to t_f is simply $(n_f - n_0)$, which represents the final amount minus the initial amount.

This same rationale can be applied to any substance involved in a batch process to get:

$$\textbf{Accumulation} = \textbf{Final output} - \textbf{Initial input} \tag{4.3}$$

Applying Eq. (4.1) to a batch process where there is no input and no output:

$$\textbf{Generation} - \textbf{Consumption} = \textbf{Accumulation} \tag{4.4}$$

Equating these two expressions for the accumulation yields:

$$\textbf{Generation} - \textbf{Consumption} = \textbf{Final output} - \textbf{Initial input} \tag{4.5}$$

Rearrangement of terms yields:

$$\textbf{Initial input} + \textbf{Generation} = \textbf{Final output} + \textbf{Consumption} \tag{4.6}$$

Equation (4.6) mirrors Eq. (4.2) previously derived to continuous steady-state processes. However, in this scenario, the terms "input" and "output" signify the initial and final quantities of the substance in balance, differing from the representation of flow rates in continuous feed and product streams. For brevity, the terms "initial" and "final" can be omitted, yet it is crucial to retain the understanding of "input" and "output" in the context of a batch process.

Example 4.4 Balances on a Batch Mixing Process

Two methanol-water mixtures are contained in separate flasks. The first mixture contains 60.0 wt.% methanol, and the second contains 30.0 wt.% methanol. If 300 g of the first mixture is combined with 200 g of the second, what are the mass and composition of the final product? See Figure 4.3 for depicting the process of methanol-water mixing.

Total Mass Balance: Input = Output $300\text{ g} + 200 = m$

$m = 500\text{ g}$

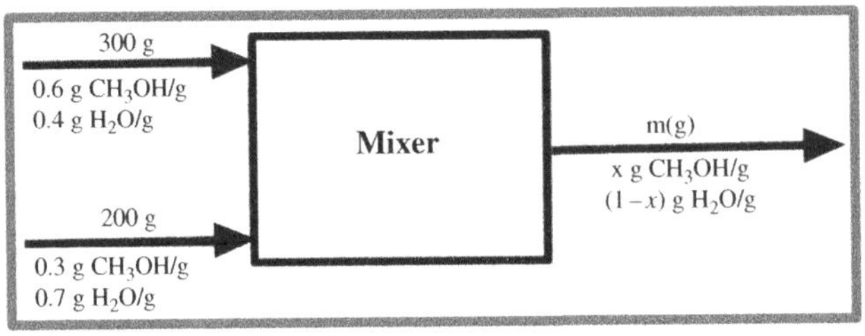

Figure 4.3 A schematic showing the mixing process between two streams with different compositions of methanol solution.

Methanol Balance: Methanol in = Methanol out

$$300\,\text{g}\times\left(\frac{0.6\,\text{g CH}_3\text{OH}}{1\,\text{g}}\right)+200\,\text{g}\times\left(\frac{0.3\,\text{g CH}_3\text{OH}}{1\,\text{g}}\right)=500\,\text{g}\times\left(\frac{x\,\text{g CH}_3\text{OH}}{1\,\text{g}}\right)$$

$$x=\frac{180+60}{500}=0.480$$

Now, everything is specified regarding the product, including its mass fraction of water. Water balance serves as a double-check on the solution.

Water Balance: Water In = Water Out

$$300\,\text{g}\times\left(\frac{0.4\,\text{g H}_2\text{O}}{1\,\text{g}}\right)+200\,\text{g}\times\left(\frac{0.7\,\text{g H}_2\text{O}}{1\,\text{g}}\right)=500\,\text{g}\times\left(\frac{0.52\,\text{H}_2\text{O}}{1\,\text{g}}\right)$$

$$120\,\text{g}+140\,\text{g}=260\,\text{g}=260\,\text{g}\rightarrow \text{L.H.S.}=\text{R.H.S.}$$

4.2.5 Integral Balances on Semi-batch Chemical Processes

Example 4.5 Integral Balance on a Semi-batch Process

Air is bubbled through a drum of liquid heptane (C_7H_{16}) at a rate of 0.200 kmol/min. The gas stream leaving the drum contains 15.0 mole % heptane vapor. *Air may be considered insoluble in liquid heptane.* Use an integral balance to estimate the time required to vaporize 15.0 m^3 of the liquid. See Figure 4.4 for air-driven heptane evaporation process.

NOTE
From a heptane evaporation point of view, the process is considered a semi-batch where heptane is allowed to leave the tank, but at the same time, no backup of heptane exists as a feed. Airflow can be considered as a continuous steady state because air enters the tank and will eventually leave the tank. All in all, it is a semi-batch because the process will stop once the last droplet of liquid heptane vaporizes and the tank becomes empty of liquid heptane.

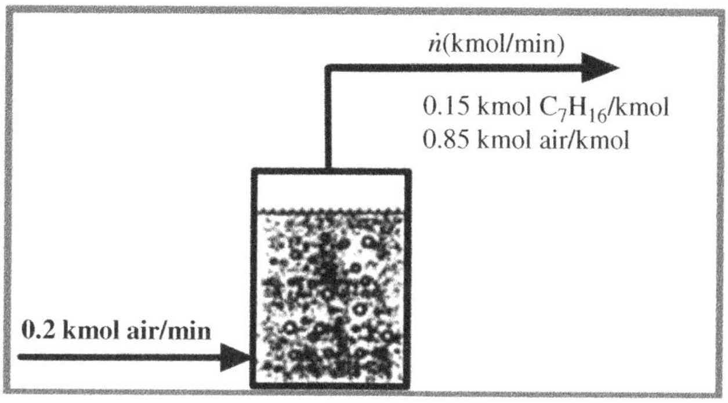

Figure 4.4 A schematic for heptane vaporization via air bubbling system.

Solution: We begin with a differential balance on air. Since we assume that air neither dissolves in the liquid *(accumulation* = 0) nor reacts with heptane in the processing unit (*Generation* = *Consumption* = 0). The balance reduces to *Input* = *Output:*

$$0.200\frac{\text{kmol air}}{\text{min}} = \left(0.85\frac{\text{kmol air}}{\text{kmol}}\right)\times\left(\dot{n}\frac{\text{kmol gas mixture}}{\text{min}}\right) \rightarrow \dot{n} = 0.235\frac{\text{kmol gas mix.}}{\text{min}}$$

The number of moles to be evaporated can be calculated from the volume of heptane to be evaporated.

The specific gravity of liquid heptane (see Appendix B | Table B.1) is given as 0.6795, and the chemical formula is given as C_7H_{16}. $M = 7 * 12 + 16 * 1 = 100$

$$15\,\text{m}^3 \times \left(\frac{679.5\,\text{kg heptane}}{\text{m}^3}\right)\times\left(\frac{\text{kmol heptane}}{100\,\text{kg heptane}}\right) = 101.92\,\text{kmol heptane}$$

This amount will require $(t_f - t_o)$ min to be vaporized.

So, let us write *an integral heptane balance*, proceeding from the initial time, t_o, to the final time, t_f.

Recall Eq. (4.1): **Input + Generation − Output − Consumption = Accumulation**

Keep in mind that there is no generation or consumption of heptane within the vaporization tank. Moreover, there is no input of hexane (i.e., only pure air enters the tank).

So, the general balance equation reduces to:

Output of heptane = Accumulation of heptane

$$-0.15\frac{\text{kmol heptane}}{\text{kmol}}\times\dot{n}\frac{\text{kmol}}{\text{min}}\times(t_f - t_o)\ \text{min} = \Delta n = (n_f - n_o) = (0 - 101.92)\ \text{kmol heptane}$$

$$-0.15\frac{\text{kmol heptane}}{\text{kmol gas mix.}}\times 0.235\frac{\text{kmol gas mix.}}{\text{min}}\times(t_f - 0)\ \text{min} = -101.92\,\text{kmol heptane}$$

Solve for t_f: $t_f = \dfrac{-101.92}{-0.15\times 0.235} = 2891.3\ \text{min} = 48.2\ \text{h}$

Alternatively, we can use a differential equation describing heptane molar balance and then solve it by integration.

$$-0.15\frac{\text{kmol heptane}}{\text{kmol}}\times\dot{n}\frac{\text{kmol}}{\text{min}} = \frac{dn}{dt}$$

Rearrange, we have:

$$\left(-0.15\frac{\text{kmol heptane}}{\text{kmol}}\times\dot{n}\frac{\text{kmol}}{\text{min}}\right)dt = dn$$

At $t = 0$, we have $n_o = 101.92$ kmol, and at $t = t_f$, we have $n_f = 0$. Integrate side by side:

$$\int_0^{t_f}\left(-0.15\frac{\text{kmol heptane}}{\text{kmol}}\times\dot{n}\frac{\text{kmol}}{\text{min}}\right)dt = \int_{no}^{nf} dn = n_f - n_o = 0 - n_o$$

$$\left(-0.15\frac{\text{kmol heptane}}{\text{kmol}}\times\dot{n}\frac{\text{kmol}}{\text{min}}\right)(t_f - 0) = -n_o = -101.92\ \text{kmol heptane}$$

$$t_f = \frac{-101.92\ \text{kmol heptane}}{\left(-0.15\frac{\text{kmol heptane}}{\text{kmol}}\times 0.235\frac{\text{kmol}}{\text{min}}\right)} = 2891.3\ \text{min} = 48.2\ \text{h}$$

4.3 Material Balance Calculations

4.3.1 Flowcharts

Example 4.6 Combustion of Methane (CH_4)

Suppose a gas containing N_2 and O_2 is combined with methane in a batch combustion chamber in which some (but not all) of the O_2 and CH_4 react to form CO_2 and H_2O, and the product is then cooled to condense water. The flowchart of this two-unit process is shown in Figure 4.5.

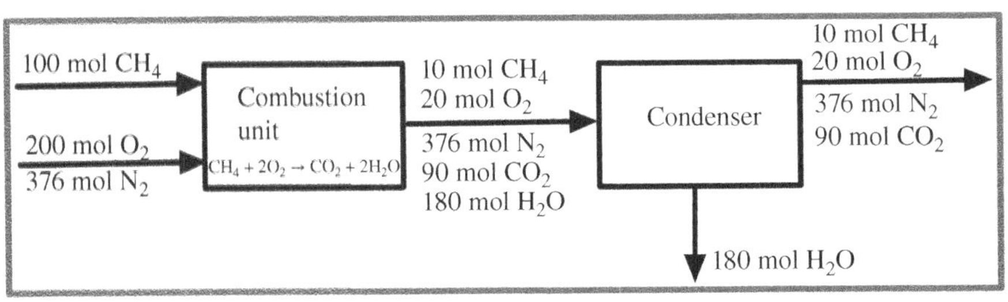

Figure 4.5 Flowchart of a process made of combustion followed by condensation.

The utilization of a process flowchart is instrumental in initiating and maintaining material balance calculations. For its effective use, the flowchart should be comprehensively labeled upon its initial creation, incorporating known process variable values and symbols representing unknown variables for each input and output stream. Subsequently, the flowchart serves as an essential tool throughout problem-solving: As each unknown variable is resolved, its value is updated within the chart, ensuring a continuous visual representation of the solution's progress and remaining tasks. Below are several recommendations for optimizing the labeling of a flowchart to maximize its efficacy in material balance calculations:

1) Write the values and units of all known stream variables at the locations of the streams on the chart. For example, a stream containing 21 mol % O_2 and 79 mol % N_2 at 400°C and 1.5 atm flowing at a rate of 500 mol/h might be labeled as:

500 mol/h
→
0.21 mol O_2/mol

0.79 mol N_2/mol

$T = 400°C$, $P = 1.5$ atm

2) Assign algebraic symbols to unknown stream variables [such as $\dot{m}$ (kg solution/min), x($lb_m N_2/lb_m$), and n (kmol CH_4)] and write these variable names and their associated units on the chart. For example, if the flow rate of the stream described in the first illustration of step 1 is not known, then one may label the stream as:

$\dot{n}$ (mol/h)
→
0.21 mol O_2/mol

0.79 mol N_2/mol

$T = 400°C$, $P = 1.5$ atm

On the other hand, if the flow rate is known and the mole fractions are not, the stream can be labeled as:

500 mol/h
→
y mol O_2/mol

$(1-y)$ mol N_2/mol

$T = 400°C$, $P = 1.5$ atm

Example 4.7 Flowchart of an Air Humidification and Oxygenation Process
To facilitate an experiment concerning the growth rate of specific organisms, an environment characterized by humid air with increased oxygen content is necessary. This setup involves channeling three input streams into an evaporation chamber to generate an output stream with the desired composition.

A: Liquid water, fed at a rate of 30.0 cm^3/min
B: Air (21 mole% O_2, the balance N_2)
C: Pure oxygen, with a molar flow rate of 30% of the molar flow rate of stream **B**

The output gas is analyzed and is found to contain 2.0 mole% water.
Draw and label a flowchart of the process and *calculate* all *unknown stream variables.*

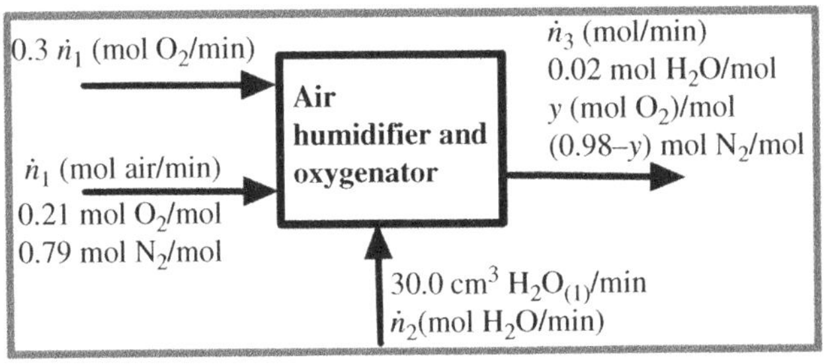

Figure 4.6 Flowchart of an air humidification and oxygenation process.

Notes on Labeling:

1) Since the one known flow rate (30 cm^3 H_2O/min) is given on a per-minute basis, it is most convenient to label all stream flow rates with reference to this basis.
2) Once the variable name ($\dot{n}_1$) is selected for the air flow rate, the given information, about the ratio of the pure O_2 to air stream flow rate, can be used to label the O_2 flow rate = 0.300 $\dot{n}_1$.
3) The mole fractions of the components of any stream must add up to 1. Since the mole fraction of H_2O in the outlet stream is known to be 0.02 and *the mole fraction of O_2 is labeled y*, then that of N_2 must be $1 - (y + 0.02) = (0.98 - y)$ (mol N_2/mol).

The quantity $\dot{n}_2$ can be calculated from the given volumetric flow rate and the density of liquid water:

$$\dot{n}_2 = 30.0\frac{\text{cm}^3\text{H}_2\text{O}}{\text{min}} \times \frac{1\,\text{g H}_2\text{O}}{\text{cm}^3} \times \frac{1\text{mol}}{18.02\,\text{g}} = 1.665\frac{\text{mol H}_2\text{O}}{\text{min}}$$

The three remaining unknowns ($\dot{n}_1, \dot{n}_3,$ and y) can be determined from balances, all of which have the simple form

Input = Output for this nonreactive steady-state process. The balances are easily written by referring to the flowchart shown in Figure 4.6.

$$\text{WATER BALANCE : Water flow rate}_{\text{in}}\left[\frac{\text{mol H}_2\text{O}}{\text{min}}\right] = \text{Water flow rate}_{\text{out}}\left[\frac{\text{mol H}_2\text{O}}{\text{min}}\right] \tag{4.7}$$

$$\dot{n}_2\left[\frac{\text{mol H}_2\text{O}}{\text{min}}\right] = y_{\text{H}_2\text{O}} \times \dot{n}_3 = 0.02\left[\frac{\text{mol H}_2\text{O}}{\text{mol}}\right] \times \dot{n}_3\left[\frac{\text{mol}}{\text{min}}\right]$$

$$\dot{n}_3 = \frac{\dot{n}_2}{y_{\text{H}_2\text{O}}} = \frac{1.665}{0.02} = 83.25\frac{\text{mol}}{\text{min}}$$

$$\text{TOTAL MOLE BALANCE : TOTAL MOLAR FLOW RATE}_{\text{in}}\left[\frac{\text{mol}}{\text{min}}\right] = \text{TOTAL MOLAR FLOW RATE}_{\text{out}}\left[\frac{\text{mol}}{\text{min}}\right] \quad (4.8)$$

$$0.3\dot{n}_1 + \dot{n}_1 + \dot{n}_2 = \dot{n}_3$$

$$0.3\dot{n}_1 + \dot{n}_1 + 1.665 = 83.25 \rightarrow 1.3\dot{n}_1 + 1.665 = 83.25$$

$$\dot{n}_1 = \frac{83.25 - 1.665}{1.3} = 62.76\frac{\text{mol}}{\text{min}}$$

$$\text{N}_2 \text{ BALANCE : N}_2 \text{ flow rate}_{\text{in}}\left[\frac{\text{mol N}_2}{\text{min}}\right] = \text{N}_2 \text{ flow rate}_{\text{out}}\left[\frac{\text{mol N}_2}{\text{min}}\right] \quad (4.9)$$

$$y_{N_2} \times \dot{n}_1 = y_{N_2} \times \dot{n}_3 \rightarrow 0.79 \times 62.76 = y_{N_2} \times 83.25 \rightarrow y_{N_2} = \frac{0.79 \times 62.76}{83.25} = 0.5956$$

$$\text{From the flowchart : } y_{N_2} = 0.5956 = 0.98 - y_{O_2}$$

$$\rightarrow \quad y_{O_2} = 0.98 - 0.5956 = 0.3844$$

$$\sum_{i=1}^{N} y_i = 1 \text{ at the outlet stream} \quad \rightarrow$$

$$y_{N_2} + y_{O_2} + y_{H_2O} = 0.5956 + 0.3844 + 0.02 = 1.000$$

One could also make an O_2 balance instead of an N_2 molar balance but not both.

Notice that we have *three linearly independent balance equations* (Eqs. 4.7–4.9) equivalent to the total number of components which are: H_2O, O_2, and N_2.

In general, if we have N components, we may carry out N *balance equations*, equivalent to the total number of components, N, present in the process or unit. However, if we have *one total material (mass or mole) balance*, then we will have $N - 1$ component balance equations.

This principle is stated in Section 4.3.3.

4.3.2 Balancing a Process

In nonreactive processes, the following guidelines (rules) are applicable:

1) **Rule #1**: The maximum number of independent equations formulated by establishing balances on a nonreactive system corresponds to the number of chemical species present in the input and output streams.
2) **Rule #2**: Begin by formulating balances that encompass the least number of unknown variables.

Example 4.8 Benzene (B) and Toluene (T) Mixer

Suppose 4.0 kg/min of benzene (B) and 2.0 kg/min of toluene (T) are mixed. The process flowchart might be drawn and labeled as in Figure 4.7:

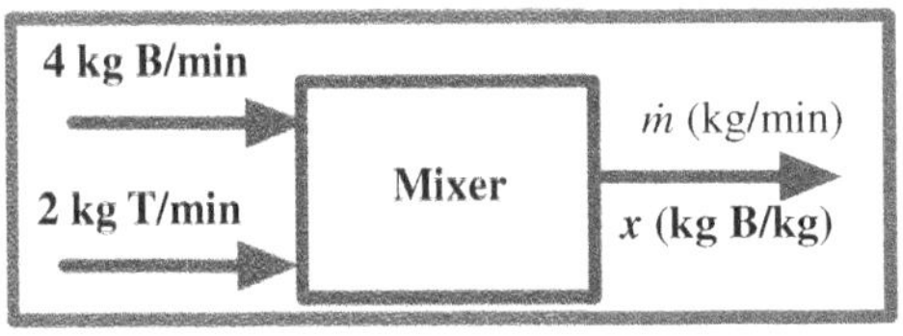

Figure 4.7 Flowchart of the mixing process of benzene and toluene.

Applying **rule #1**, we can write a maximum number of two equations. So, we have to choose two equations out of the three equations that we can write. These *three equations* are: total mass balance, benzene balance, and toluene balance. However, two of them will be linearly independent. If we check the flowchart, we will notice that there are *two unknown quantities*: $\dot{m}$ *and* x_{benzene}. Following **rule #2**, we have to choose the balance equation with the fewest unknown variables.

TOTAL MASS BALANCE:

$$\text{TOTAL MASS FLOW RATE}_{\text{in}}\left[\frac{\text{kg}}{\text{min}}\right] = \text{TOTAL MASS FLOW RATE}_{\text{out}}\left[\frac{\text{kg}}{\text{min}}\right] \tag{4.10}$$

$$4\frac{\text{kg}}{\text{min}} + 2\frac{\text{kg}}{\text{min}} = \dot{m}$$

$$\text{BENZENE BALANCE}: C_6H_6 \text{ flow rate}_{\text{in}}\left[\frac{\text{kg } C_6H_6}{\text{min}}\right] = C_6H_6 \text{ flow rate}_{\text{out}}\left[\frac{\text{kg } C_6H_6}{\text{min}}\right] \tag{4.11}$$

$$4\frac{\text{kg}}{\text{min}} = \dot{m}x_{\text{benzene}}$$

$$\text{TOLUENE BALANCE}: C_7H_8 \text{ flow rate}_{\text{in}}\left[\frac{\text{kg } C_7H_8}{\text{min}}\right] = C_7H_8 \text{ flow rate}_{\text{out}}\left[\frac{\text{kg } C_7H_8}{\text{min}}\right] \tag{4.12}$$

$$2\frac{\text{kg}}{\text{min}} = \dot{m}x_{\text{toluene}} = \dot{m}(1 - x_{\text{benzene}})$$

We can see that Eq. (4.10) directly gives the value of $\dot{m}$. **However, either Eqs. (4.11) or (4.12) does contain the two unknown quantities** $\dot{m}$ ***and*** x_{benzene}. So, both equations need to be solved simultaneously (i.e., two algebraic equations with two unknown quantities to be solved via substitution or elimination method). *If we take Eq. (4.10) as our first choice* *as it contains the least number of unknown terms*, then we can find the outlet mass flow rate as $6\frac{\text{kg}}{\text{min}} = \dot{m}$. Now, we can use this value in either Eqs. (4.11) or (4.12) to find x_{benzene} or x_{toluene}. For example, if we use *benzene balance* then we will get:

$$4\frac{\text{kg}}{\text{min}} = 6x_{\text{benzene}} \qquad \rightarrow x_{\text{benzene}} = \frac{4}{6} = 0.67$$

Of course, $x_{\text{toluene}} = 1 - 0.67 = 0.33$

4.3.3 Degree-of-Freedom Analysis

Before diving into detailed calculations, employing a correctly drawn and labeled flowchart helps ascertain the sufficiency of information for solving a specific problem. This evaluation method is known as a degree-of-freedom analysis. To conduct a degree-of-freedom analysis, sketch and fully label a flowchart, tally the unknown variables indicated on the chart, then count the independent equations linking them. Finally, subtract the latter from the former to determine the number of degrees of freedom in the process, expressed as $n_{\text{df}} = n_{\text{unknowns}} - n_{\text{indep eqns}}$. This assessment leads to three potential scenarios:

1) If $n_{\text{df}} = 0$, there exist n independent equations with n unknowns, indicating that, theoretically, the problem can be solved.
2) If $n_{\text{df}} > 0$, the number of unknowns surpasses the number of independent equations relating to them. This situation requires specifying at least n_{df} additional variable values before determining the remaining ones. It may signify overlooked relationships or an underdefined problem with numerous potential solutions. Engaging in calculations at this point is likely to be unproductive.
3) If $n_{\text{df}} < 0$, there are more independent equations than unknowns. This can result from an incompletely labeled flowchart or an over-specified problem with redundant and potentially inconsistent relations. Once again, attempting to solve it is futile until the equations and unknowns are balanced.

Sources of Equations Relating Unknown Process Stream Variables:

1) **Material Balances**: In a nonreactive process, the maximum number of independent material balances that can be established is limited to n_{ms}, which represents the count of molecular/atomic species (e.g., CH_4, NH_3, O_2) involved in the process. To illustrate, if a distillation column involves benzene and toluene species in the incoming and outgoing streams, it is reasonable to create balances considering benzene, toluene, total mass, atomic carbon, atomic hydrogen, and others. However, at most, two of these balances will be independent. Additional balances beyond this threshold will not be considered independent and, as a result, will not yield any new or additional information beyond the initial ones.
2) **An Energy Balance**: (Chapter 7). If the amount of energy exchanged between the system and its surroundings is specified or if it is treated as one of the unknown process variables, an energy balance will provide a relationship between inlet and outlet material flows and their temperatures.
3) **Process Specifications**: The problem statement may specify how several process variables are interrelated. For example, you may be told that out of the acetone fed to a condenser [flow rate = $\dot{m}_1$ (kg acetone/s)], 45% appears in a condensate stream [flow rate = $\dot{m}_2$ (kg acetone/s)]. A system equation will then be $\dot{m}_2 = 0.45 \times \dot{m}_1$.

4) **Physical Properties and Laws:** If the unknown variables involve both the mass and volume of a stream material, in which case a tabulated specific gravity for liquids and solids or an equation of state for gases (Chapter 5) will provide an equation relating such variables. In other cases, saturation or equilibrium conditions for one or more of the process streams (Chapter 6) can provide the missing relations.
5) **Physical Constraints:** For example, if the mole fractions of the three components of a stream are labeled X_A, X_B, and X_C, then a relation among these variables is $X_A + X_B + X_C = 1$. (If instead of X_C you label the last fraction $1 - X_A - X_B$, then you will have one less variable and one less equation to worry about). Another example is if we have a flash-drum that contains both liquid- and vapor-phase, then the two phase volumes can be inter-related by: $V_{Vessel} = V_{Liquid} + V_{Vapor}$. V_{Vessel} is specified as a design parameter.
6) **Stoichiometric Relations:** If a chemical reaction occurs in a system, the stoichiometric equation of the reaction (e.g., $CH_4 + 2\,O_2 \rightarrow 2H_2O + CO_2$) provides relationships between the moles of the reactants consumed and of the products generated.

4.3.4 General Procedure for Single-Unit Process Material Balance Calculations

Example 4.9 Material Balances on a Distillation Column

A liquid mixture containing 60.0% benzene (B) and 40.0% toluene (T) by mole is fed to a distillation column. A product stream leaving the top of the column (the *overhead product)* contains 97.0 mole% *B,* and a bottom product stream contains 2.0 mole% *B*. The molar flow rate of the feed stream is 100 kmol/h. Determine the molar flow rate and composition of the overhead product stream and the molar flow rate and composition (mole fractions) of the bottom product stream.

1) ***Choose a basis***. Choose the given feed stream flow rate (100 kmol/h) as the basis of calculation.
2) ***Draw and label the flowchart.*** Figure 4.8 shows a flowchart with labels for the binary distillation column.

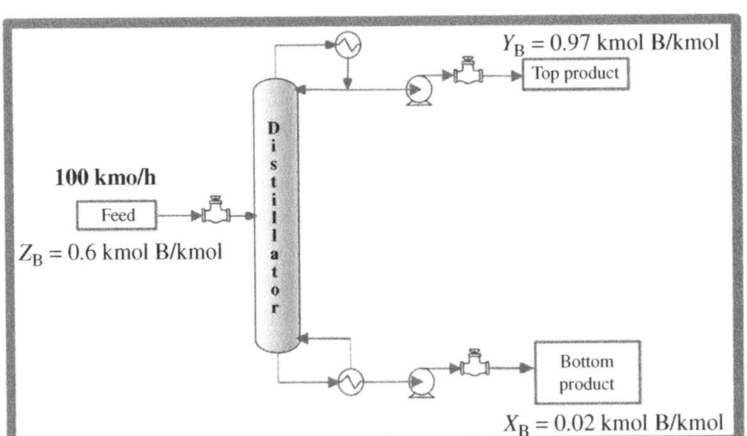

Figure 4.8 Flowchart of binary distillation (molar basis).

3) ***Degree of freedom analysis:***

Two unknown quantities: $\dot{n}_{top}$ *and* $\dot{n}_{bottom}$
Two independent equations: one total molar balance and one-component balance.

$$n_{\text{unknowns}} = 2 \qquad n_{\text{indep eqns}} = 2$$

$$n_{\text{df}} = n_{\text{unknowns}} - n_{\text{indep eqns}} = 2 - 2 = 0 \tag{4.13}$$

The system is specified as $n_{\text{df}} = 0$.

Total Mole Balance:

$$100\frac{\text{kmol}}{\text{h}} = \dot{n}_{\text{top}} + \dot{n}_{\text{bottom}} \tag{4.14}$$

$$\dot{n}_{\text{top}} = 100 - \dot{n}_{\text{bottom}} \tag{4.15}$$

Benzene Balance:

$$100 \times 0.60 = \dot{n}_{\text{top}} \times 0.97 + \dot{n}_{\text{bottom}} \times 0.02 \tag{4.16}$$

So, we have two equations (Eqs. 4.14 and 4.16) with two unknowns ($\dot{n}_{\text{top}}$ *and* $\dot{n}_{\text{bottom}}$). Such equations can be solved by *substitution or elimination method.*

By Substitution Method:

Substitute $\dot{n}_{\text{top}}$ by its equivalence from Eq. (4.15) in Eq. (4.16) and simplify

$$100 \times 0.60 = (100 - \dot{n}_{\text{bottom}}) \times 0.97 + \dot{n}_{\text{bottom}} \times 0.02 \tag{4.17a}$$

$$60 = 97 - 0.97\dot{n}_{\text{bottom}} + 0.02\dot{n}_{\text{bottom}} = 97 - 0.95\dot{n}_{\text{bottom}} \tag{4.17b}$$

$$0.95\dot{n}_{\text{bottom}} = 97 - 60 \tag{4.17c}$$

$$\dot{n}_{\text{bottom}} = \frac{97 - 60}{0.95} = 38.95\frac{\text{kmol}}{\text{h}} \tag{4.17d}$$

Now, use Eq.(4.15) to find $\dot{n}_{\text{top}}$: $\dot{n}_{\text{top}} = 100 - 38.95 = 61.05\frac{\text{kmol}}{\text{h}}$ (4.18)

By Elimination Method:
From Eq. (4.14), we have:
$100\frac{\text{kmol}}{\text{h}} = \dot{n}_{\text{top}} + \dot{n}_{\text{bottom}}$
From Eq. (4.16), we have:
$100 \times 0.60 = \dot{n}_{\text{top}} \times 0.97 + \dot{n}_{\text{bottom}} \times 0.02$

The elimination method depends on **eliminating** one of the variables; let us multiply Eq. (4.14) by 0.97 throughout. Eq. (4.14) becomes:

$$0.97 \times 100 = 0.97 \times \dot{n}_{\text{top}} + 0.97 \times \dot{n}_{\text{bottom}} \tag{4.14a}$$

$$97 = 0.97 \times \dot{n}_{\text{top}} + 0.97 \times \dot{n}_{\text{bottom}} \tag{4.14b}$$

Subtract Eq. (4.16) from Eq. (4.14b) and sum the two equations, side by side:

$$97 = 0.97 \times \dot{n}_{\text{top}} + 0.97 \times \dot{n}_{\text{bottom}}$$

$$-60 = -\dot{n}_{\text{top}} \times 0.97 - \dot{n}_{\text{bottom}} \times 0.02$$

$$97 - 60 = 0.97\dot{n}_{\text{bottom}} - 0.02\dot{n}_{\text{bottom}} = 0.95\dot{n}_{\text{bottom}} \rightarrow \dot{n}_{\text{bottom}} = 38.95\frac{\text{kmol}}{\text{h}} \quad (4.19)$$

Eq. (4.19) is *exactly the same* as Eq. (4.17d). So, we will get the same results.

$$\dot{n}_{\text{top}} = 100 - \dot{n}_{\text{bottom}} = 100 - 38.95 = 61.05\frac{\text{kmol}}{\text{h}}$$

4.3.5 MATLAB for Solution of Algebraic Equations

Let us use the MATLAB platform for solving two simultaneous algebraic equations. Figure 4.9 shows MATLAB code for solving two simultaneous algebraic equations.

```
clear variables;
% Define the system of equations as a matrix of coefficients and the right-hand side vector
%Eq. (4.14): ndottop + ndotbtm = 100
%Eq. (4.16): 0.97 ndottop + 0.02 ndotbtm = 60
A = [1 1; 0.97 0.02];
b = [100; 60];
% Solve the system using matrix inversion
x = A \ b;
% Print the solution for m1 and m3
ndottop = x(1);
ndotbtm=x(2);
disp(['The top stream, ndottop = ', num2str(ndottop), ' kmol/h']);
disp(['The bottom stream, ndotbtm = ', num2str(ndotbtm), ' kmol/h']);
```

Figure 4.9 MATLAB code snippet for solving two simultaneous algebraic equations.

MATLAB results show the calculated molar flow rate of the top and bottom streams, which are exactly the same as those of the manual solution given by Eqs. (4.17d) and (4.18).

The top stream, ndottop = 61.0526 kmol/h
The bottom stream, ndotbtm = 38.9474 kmol/h

Example 4.10 Three-Unit Process

A labeled flowchart of a continuous steady-state three-unit process is shown below (Figure 4.10). Each stream contains two components, ethanol (**E**) and water (**W**), in different proportions. Three streams whose flow rates and/or compositions are not known are labeled as 1, 2, and 3.

> **NOTE**
>
> Since we have a binary mixture made of **E** and **W**, then at any stream, it is implicitly understood that x_W is equal to ($1-x_E$).

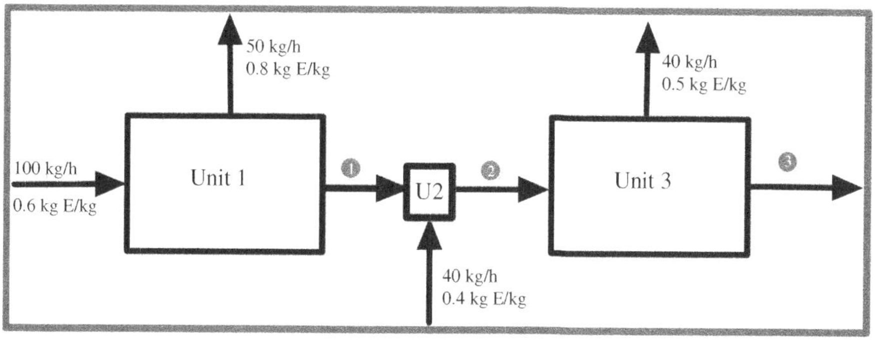

Figure 4.10 Flowchart of three-unit process.

> **NOTE**
>
> For *each number-labeled stream*, there are *two* unknown quantities. Thus, we have, overall, *six* unknown quantities. On the other hand, we can write *two independent equations for each unit: UNIT 1, UNIT 2, and UNIT 3.* Hence, we have, overall, six independent equations:
>
> $n_{\text{unknowns}} = 3 \times 2 = 6 \qquad n_{\text{indep eqns}} = 3 \times 2 = 6$
>
> $n_{\text{df}} = n_{\text{unknowns}} - n_{\text{indep eqns}} = 6 - 6 = 0$
>
> To determine the flow rate and composition of **stream 1**, we carry out material balance around the first unit (UNIT 1).

Total Mass Balance: Mass flow rate$_{\text{in}}$ = Mass flow rate$_{\text{out}}$

$$100\,\frac{\text{kg}}{\text{h}} = 50\,\frac{\text{kg}}{\text{h}} + \dot{m}_1 \tag{4.20}$$

$$\dot{m}_1 = 100 - 50 = 50\,\frac{\text{kg}}{\text{h}} \tag{4.20b}$$

***E*-Balance: (Mass flow rate of *E*)$_{\text{in}}$ = (Mass flow rate of *E*)$_{\text{out}}$**

$$100\,\frac{\text{kg}}{\text{h}} \times 0.6\,\frac{\text{kg}\,E}{\text{kg}} = 50\,\frac{\text{kg}}{\text{h}} \times 0.8\,\frac{\text{kg}\,E}{\text{kg}} + \dot{m}_1 \times x_{E1} \tag{4.21}$$

Substitute $\dot{m}_1$ from Eq. (4.20b) into Eq. (4.21):

$$60 = 40 + 50 \times x_{E_1}$$

$$x_{E_1} = \frac{(60-40)}{50} = 0.40 \qquad (4.21b)$$

$$x_{W_1} = 1 - 0.4 = 0.60$$

Thus, **stream 1** is now specified.

To determine the flow rate and composition of **stream 2**, we carry out material balance around the second unit (U2).

Total Mass Balance: Mass flow rate$_{\text{in}}$ = Mass flow rate$_{\text{out}}$

$$\dot{m}_1 + 40\frac{\text{kg}}{\text{h}} = \dot{m}_2 \qquad (4.22)$$

$$\dot{m}_2 = \dot{m}_1 + 40\frac{\text{kg}}{\text{h}} = 50\frac{\text{kg}}{\text{h}} + 40\frac{\text{kg}}{\text{h}} = 90\frac{\text{kg}}{\text{h}} \qquad (4.22b)$$

E-Balance: (Mass flow rate of *E*)$_{\text{in}}$ = (Mass flow rate of *E*)$_{\text{out}}$

$$\dot{m}_1\frac{\text{kg}}{\text{h}} \times x_{E_1}\frac{\text{kg}\,E}{\text{kg}} + 40\frac{\text{kg}}{\text{h}} \times 0.4\frac{\text{kg}\,E}{\text{kg}} = \dot{m}_2 \times x_{E_2} \qquad (4.23)$$

Substitute $\dot{m}_1$ from Eq. (4.20b), $x_{E_1}$from Eq. (4.21b), and $\dot{m}_2$ from Eq. (4.22b) into Eq. (4.23):

$$50\frac{\text{kg}}{\text{h}} \times 0.40\frac{\text{kg}\,E}{\text{kg}} + 40\frac{\text{kg}}{\text{h}} \times 0.4\frac{\text{kg}\,E}{\text{kg}} = 90 \times x_{E2}$$

$$20 + 16 = 90 \times x_{E_2}$$

$$x_{E_2} = \frac{(20+16)}{90} = 0.40 \qquad (4.23b)$$

$$x_{W_2} = 1 - 0.40 = 0.60$$

Thus, **stream 2** is now specified.

To determine the flow rate and composition of **stream 3**, we carry out material balance around the third unit (Unit 3).

Total Mass Balance: Mass flow rate$_{\text{in}}$ = Mass flow rate$_{\text{out}}$

$$\dot{m}_2 = 40\frac{\text{kg}}{\text{h}} + \dot{m}_3 \qquad (4.24)$$

$$\dot{m}_3 = \dot{m}_2 - 40 = 90 - 40 = 50\frac{\text{kg}}{\text{h}} \tag{4.24b}$$

***E*-Balance: (Mass flow rate of *E*)$_{\text{in}}$ = (Mass flow rate of *E*)$_{\text{out}}$**

$$\dot{m}_2 \times x_{E_2} = 40\frac{\text{kg}}{\text{h}} \times 0.5\frac{\text{kg}\,E}{\text{kg}} + \dot{m}_3 \times x_{E_3} \tag{4.25}$$

Substitute $\dot{m}_2$ from Eq. (4.22b), x_{E_2} from Eq. (4.23b), and $\dot{m}_3$ from Eq. (4.24b) into Eq. (4.25):

$$90\frac{\text{kg}}{\text{h}} \times 0.4\frac{\text{kg}\,E}{\text{kg}} = 40\frac{\text{kg}}{\text{h}} \times 0.5\frac{\text{kg}\,E}{\text{kg}} + 50 \times x_{E_3}$$

$$36 = 20 + 50 \times x_{E_3}$$

$$x_{E_3} = \frac{(36-20)}{50} = 0.32 \tag{4.25b}$$

$$x_{W_3} = 1 - 0.32 = 0.68$$

Thus, **stream 3** is now specified.

Figure 4.11 shows MATLAB code for solving three simultaneous algebraic equations with three unknowns: x_{E1}, x_{E2}, and x_{E3}. Notice that I removed $\dot{m}_1, \dot{m}_2, and\, \dot{m}_3$ from the list of unknowns, as each can be given straight forward by conducting total mass balance around UNIT 1, U2, and UNIT 3, respectively. Also, to simplify the task of using MATLAB code to solve a set of algebraic equations.

```
clear variables;
% Define the system of equations as a matrix of coefficients and the right-hand side vector
%Eq. (4.21): 50*xE1=20
%Eq. (4.23): 50*xE1-90*xE2=-16
%Eq. (4.25): 90*xE2-50*xE3= 20
% System of equations as a matrix of coefficients with zero coefficient for a non-existing xE.
%Eq. (4.21): 50*xE1+0*xE2+0*xE3=20
%Eq. (4.23): 50*xE1-90*xE2+0*xE3=-16
%Eq. (4.25): 0*xE1+90*xE2-50*xE3= 20
A = [50 0 0; 50 -90 0; 0 90 -50];
b = [20;-16;20];
% Solve the system using matrix inversion
x = A \ b;
% Print the solution for m1 and m3
xE1 = x(1);
xE2=x(2);
xE3=x(3);
disp(['Ethanol Mass Fraction in Stream #1, xE1 = ', num2str(xE1)]);
disp(['Ethanol Mass Fraction in Stream #2, xE2 = ', num2str(xE2)]);
disp(['Ethanol Mass Fraction in Stream #3, xE3 = ', num2str(xE3)]);
```

Figure 4.11 MATLAB code snippet for solving three simultaneous algebraic equations.

MATLAB results show the calculated mass fraction of ethanol in three consecutive streams: 1, 2, and 3, which are exactly the same as those of the manual solution given by equations (4.21b), (4.23b), and (4.25b).

```
Ethanol Mass Fraction in Stream #1, xE1 = 0.4
Ethanol Mass Fraction in Stream #2, xE2 = 0.4
Ethanol Mass Fraction in Stream #3, xE3 = 0.32
```

4.4 Chemical Reaction Stoichiometry

Chemistry is a world of transformations, where reactants become products through fascinating processes called chemical reactions. But amidst the seemingly chaotic rearrangements of atoms, there is an underlying orderliness governed by chemical reaction stoichiometry. This essential concept acts as the bookkeeper of reactions, balancing the quantities of atoms and ensuring everything correctly adds up.

What is chemical reaction stoichiometry?

In simpler terms, stoichiometry deals with the quantitative relationships between reactants and products in a chemical reaction. For example, it tells us:

- How much of each reactant do we need to start the reaction for the desired amount of product?
- How much of each product do we expect to get based on the amount of reactants used?
- The ratio of reactants to products involved in the reaction.

The key principle:

Stoichiometry relies on the Law of Conservation of Mass, which states that matter can neither be created nor destroyed in a chemical reaction. In simpler terms, the total number of atoms of each element must remain constant before and after the reaction, even though they rearrange themselves to form new compounds or products.

How does it work?

- **Balancing the Chemical Equation**: This visual representation shows the reactants, products, and their respective coefficients (numbers in front of each formula). Balancing ensures equal numbers of each element type on both sides, reflecting the mass conservation principle.
- **Using the Coefficients**: These numbers act as conversion factors. Once the equation is balanced, the coefficients tell us how many molecules of each reactant react to produce a certain number of product molecules.
- **Calculating Quantities**: We can use the coefficients and mole concept (relating mass to number of particles) to determine the mass of each reactant needed or the mass of the product expected.

Applications of Stoichiometry:

- **Chemical Synthesis**: To optimize the amount of product obtained without wasting precious resources.

- **Environmental Analysis**: To calculate the amount of pollutants produced or the required amount of treatment chemicals.
- **Food Science**: To develop recipes and ensure consistent composition of products.

Beyond the Basics:

- **Limiting Reagents**: In reactions where one reactant runs out before others, it limits the product yield. Stoichiometry helps identify the limiting reagent and calculate the theoretical yield.
- **Percent Yield**: This compares the actual yield to the theoretical yield, reflecting the efficiency of the reaction.
- **Stoichiometry in Complex Reactions**: The core principles apply even to multi-step or equilibrium reactions, requiring careful analysis and application of concepts like reaction mechanisms and reaction rates.

4.4.1 Stoichiometry

Stoichiometry refers to the study of the ratios in which different chemical substances unite with each other. In a chemical reaction, the stoichiometric equation elucidates the specific number of molecules or moles involved as both reactants and products. For instance, consider the following example of a stoichiometric equation:

$$2\,CO + O_2 \rightarrow 2\,CO_2 \tag{4.26}$$

indicates that for every two molecules (g-moles, lb-moles) of CO that react, one molecule (g-mole, kmol, lb-mole) of O_2 reacts to produce two molecules (g-moles, kmol, lb-moles) of CO_2. The numbers that precede the formulas for each species are the stoichiometric coefficients of the reaction components.

For a stoichiometric equation to be considered valid, it must maintain balance, ensuring an equal number of atoms for each atomic species on both sides of the equation, as atoms remain constant and cannot be generated or destroyed during chemical reactions. In Eq. (4.26), we observe a balance with two (2) carbon atoms and four (4) oxygen atoms present on both sides, confirming its validity as a stoichiometric equation. Equation (4.26) can be expressed in the following form:

$$\begin{pmatrix} 2C & \rightarrow & 2C \\ 4O & \rightarrow & 4O \end{pmatrix} \tag{4.27}$$

We can also write the stoichiometric ratios

$$\frac{2\text{ mol }CO_2\text{ generated}}{1\text{ mol }O_2\text{ consumed}} \qquad \frac{1\text{ mol }CO_2\text{ generated}}{1\text{ mol CO consumed}} \qquad \frac{2\text{ mol CO consumed}}{1\text{ mol }O_2\text{ consumed}} \tag{4.28}$$

Exercise 4.2
Consider the reaction

$$C_4H_{10} + \frac{13}{2}O_2 \rightarrow 4\,CO_2 + 5\,H_20$$

1) Is the stoichiometric equation balanced?
2) What is the stoichiometric coefficient of CO_2?
3) What is the stoichiometric ratio of H_2O to O_2? (Include units)

4) How many lb-moles of O_2 react to form 40 lb-moles of CO_2? (Use a dimensional equation)
5) One hundred mol/min of C_4H_{10} is fed into a reactor, and 80% reacts. At what rate is water formed?

4.4.2 Limiting and Excess Reactants, Fractional Conversion, and Extent of Reaction

Two Reactants: *A* and *B* are said to be present in stoichiometric proportion if the ratio (moles *A* present)/(mole *B* present) equals the stoichiometric ratio obtained from the balanced reaction equation. For reactants in the following reaction:

$$2\,CO + O_2 \rightarrow 2\,CO_2 \tag{4.29}$$

to be present in stoichiometric proportion, there must be 2 moles of CO for every mole of O_2 (so that $n_{CO}/n_{O_2} = 2/1$) present in the feed to the reactor.

If reactants are fed to a chemical reactor in stoichiometric proportion and the reaction proceeds to completion, it literally means all of the reactants are completely consumed.

It follows that if we start with 1 mol of O_2 and less than 2 mol of CO (i.e., CO < 2 mol, initially present in the reactor), CO will vanish first, while if more than 2 mol of CO is initially present O_2 will disappear, first.

If a reaction proceeds to completion, the reactant that is consumed first will be called the **limiting** reactant, and the other reactants are termed **excess** reactants. *A reactant will be limiting if it is present in less than its stoichiometric proportion relative to every other reacting species*. If all reactants are present in stoichiometric proportion, then any reactant can be treated as limiting.

Suppose $(n_A)_{feed}$ is the number of moles of an excess reactant, **A**, present in the feed to a reactor and that $(n_A)_{stoich}$ is the **stoichiometric requirement** of **A**, or the amount needed to react completely with the limiting reactant. Then $(n_A)_{feed} - (n_A)_{stoich}$ is the amount by which *A* in the feed exceeds the amount needed to achieve 100% completion of a reaction. The **fractional excess** of the reactant is the ratio of the excess to the stoichiometric requirement:

$$\text{fractional excess of } A = \frac{(n_A)_{feed} - (n_A)_{stoich}}{(n_A)_{stoich}} \tag{4.30}$$

The percentage excess of A is 100 times the fractional excess.

Example 4.11 Fractional Excess of Reactant

Consider the combustion of *n*-hexane:

$$C_6H_{14} + 9\frac{1}{2}\,O_2 \rightarrow 6\,CO_2 + 7\,H20 \tag{4.31}$$

Suppose that 100.0 kmol/h of *n*-hexane and 600.0 kmol/h of oxygen are fed to a reactor. The *stoichiometric ratio of oxygen to n-hexane is 9.5:1* (the ratio of coefficients in the stoichiometric equation), and since the ratio of O_2 to C_6H_{14} in the feed is 6:1 (600:100). Oxygen is fed in a smaller-than-stoichiometric proportion to n-hexane. **Oxygen is therefore the limiting reactant and n-hexane the excess reactant**. If the reaction goes to completion, then the molar flow rate of n-hexane that will be consumed will be given by:

$$(\dot{n}_{hexane})_{stoich} = 600\frac{\text{kmol } O_2 \text{ consumed}}{\text{h}} \times \frac{1 \text{ kmol } n-\text{hexane consumed}}{9.5 \text{ kmol } O_2 \text{ consumed}}$$
$$= 63.16\frac{\text{kmol } n-\text{hexane}}{\text{h}}$$

$$\text{fractional excess of } n-\text{hexane} = \frac{(\dot{n}_A)_{feed} - (\dot{n}_A)_{stoich}}{(\dot{n}_A)_{stoich}} = \frac{100-63.16}{63.16} = 0.583$$

The *percent excess of n-hexane*= fractional excess of n – hexane × 100 % = 58.3%

Notice that the *fractional excess of O_2 will be zero* as it is the *limiting reactant*. We say that there is 58% *excess n-hexane* in the feed.

Chemical reactions do not take place instantaneously and, indeed, often proceed rather slowly. *In such cases, it is not practical to design the reactor for complete conversion of the limiting reactant*; instead, the reactor effluent emerges with some of the limiting reactant still present and is then usually subjected to a separation process to remove the unconverted reactant from the product. The separated reactant is then recycled to the reactor inlet. The **fractional conversion** of a reactant is the ratio:

$$f = \frac{\text{moles reacted}}{\text{moles fed}} \tag{4.32}$$

The un-reacted fraction is accordingly $1-f$. If 100 moles of a reactant are fed and 96 moles react, the fractional conversion is 0.96 (the percentage conversion is 96%) and the fraction unreacted is 0.04. If 30 mol/min of a reactant is fed and the percentage conversion is 80%, then (30) × (0.80) = 24 mol/min has reacted and (30) × (1 – 0.80) = 6 mol/min remains unreacted.

Definition of ν_i: ν_i is a number that represents the stoichiometric coefficient of the ith species in a chemical reaction, making it negative for reactants and positive for products.

Example 4.12 The Stoichiometric Coefficient of Species *i*

Consider the hydrogenation of 1-propyne (methyl acetylene) into n-propane (C_3H_8):

$$CH_3{-}C{\equiv}CH + 2H_2 \rightarrow CH_3CH_2CH_3 \tag{4.33}$$

$$\nu_{CH_3CH_2CH_3} = +1 \qquad \nu_{CH_3-C\equiv CH} = -1 \qquad \nu_{H_2} = -2$$

Example 4.13 The Extent of a Reaction

Consider the same reaction as given by Eq. (4.33)

Suppose that we initially have 3 moles of H_2 ($n^o_{H_2} = 3$ mol), 1 mole of $CH_3{-}C{\equiv}CH$ ($n^o_{CH_3-C\equiv CH} = 1$ mol) and zero moles of $CH_3CH_2CH_3$ ($n^o_{CH_3CH_2CH_3} = 0$ mol).

If half a mole (1/2 mole) of $CH_3{-}C{\equiv}CH$ reacted, then the remaining $CH_3{-}C{\equiv}CH$ would be $1-\frac{1}{2}=\frac{1}{2}$ mol. So, *half a mole of $CH_3{-}C{\equiv}CH$ reacted and another half remained un-reacted.*

The number of moles of H_2 that reacted is given by:

$$\frac{1}{2}\text{mole } (CH_3{-}C{\equiv}CH) \text{ reacted} \times \left(\frac{2 \text{ mol } H_2 \text{ reacted}}{1 \text{ mol } CH_3{-}C{\equiv}CH \text{ reacted}}\right) = 1 \text{ mol } H_2 \text{ reacted} \tag{4.34}$$

So, one mole of H_2 reacted and two moles of H_2 remained un-reacted.
The number of moles of $CH_3CH_2CH_3$ (or C_3H_8) produced will be given by:

$$\frac{1}{2}\text{mole }(CH_3{-}C{\equiv}CH)\text{ reacted} \times \left(\frac{1\text{ mol }CH_3CH_2CH_3\text{ produced}}{1\text{ mol }(CH_3{-}C{\equiv}CH)\text{ reacted}}\right) = \frac{1}{2}\text{ mol }C_3H_8\text{ produced} \tag{4.35}$$

Let us define the change in number of moles of species i as:

$$\Delta n_i = \text{Finally Present (or Remained)} - \text{Initially Present} = n_i - n_i^o \tag{4.36}$$

Table 4.1 summarizes the results of the reaction.

Table 4.1 Number of moles analysis for species involved in a chemical reaction given by Eq. (4.33).

Component	ν_i	Initially, n_i^o	Reacted	Remained, n_i	Δn_i	$\Delta n_i/\nu_i$
$CH_3{-}C{\equiv}CH$	−1	1	½ mol	½ mol	− ½ mol	½ mol
H_2	−2	3	1 mol	2 mol	−1 mol	½ mol
$CH_3CH_2CH_3$	+1	0	½ mol	½ mol	½ mol	½ mol

Notice that the value of the last column is same for all species involved in the reaction and is equal to ½ mol. This term is named the **extent of reaction** and has the units as those of n_i. Thus, the **extent of reaction** is defined as:

$$\xi = \frac{\Delta n_i}{\nu_i} = \frac{\text{Remained} - \text{Initially Present}}{\nu_i} = \frac{n_i - n_i^o}{\nu_i} \tag{4.37}$$

Rearranging Eq. (4.37) gives:

$$n_i = n_i^o + \nu_i \xi \tag{4.38a}$$

For a multi-reaction system, we can generalize Eq. (4.38a) to be:

$$n_i = n_i^o + \sum_{j=1}^{\#\text{of rxnx}} \nu_{ij} \times \xi_j \tag{4.38b}$$

Where;

- $\Sigma_{j=1}^{\#\text{of rxns}}$ represents the summation over all j reactions in the system.
- ν_{ij} is the stoichiometric coefficient of species i in reaction j (positive for products, negative for reactants).
- ξ_j is the extent of reaction j.

Example 4.14 The Extent of Reaction and Reaction Stoichiometry
Acrylonitrile is produced via the following reaction:

$$C_3H_6 + NH_3 + \frac{3}{2}O_2 \rightarrow C_3H_3N + 3\,H_20 \tag{4.39}$$

As shown in Figure 4.12, the feed contains 10.0 mole% propylene, 15.0% ammonia, and 75.0% air. *A fractional conversion of 40.0% of the limiting reactant is achieved.*

Taking 100 mol/min of feed as a basis, determine which reactant is limiting, the percentage by which each of the other reactants is in excess, and the molar amounts of all product gas constituents for a 40% conversion of the limiting reactant.

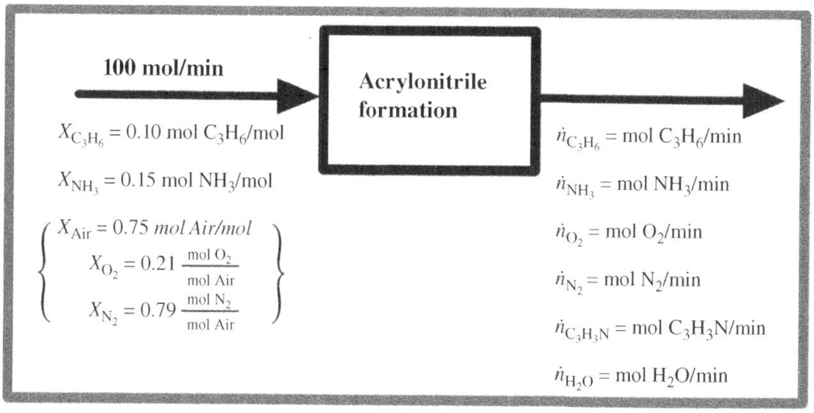

Figure 4.12 Flow chart for a fluidized bed reactor that produces acrylonitrile from ammonia, propylene, and air.

Solution:

Basis: One-minute of operation of 100 mol/min Feed
So, **initially** we have (100 mol/min)*(1min)=100 mol distributed as:

$$10\,\text{mol of } C_3H_6 \left(n^o_{C_3H_6} = 10\,\text{mol}\right)$$
$$15\,\text{mol of } NH_3 \left(n^o_{NH_3} = 15\,\text{mol}\right),$$
$$75\,\text{mol of Air}: \left(n^o_{O_2} = 0.21 \times 75 = 15.75\,\text{mol}\right); \left(n^o_{N_2} = 0.79 \times 75 = 59.25\,\text{mol}\right)$$
$$0.0\,\text{mol of } H_2O \left(n^o_{H_2O} = 0\,\text{mol}\right), \text{and}$$
$$0.0\,\text{mol of } C_3H_3N \left(n^o_{C_3H_3N} = 0\,\text{mol}\right).$$

NOTE

To determine the *limiting reactant*, test for the *smallest ratio* of $\frac{n_i^o}{\|\nu_i\|}$

$$\frac{n^o_{C_3H_6}}{\|\nu_{C_3H_6}\|} = \frac{10}{1} < \frac{n^o_{O_2}}{\|\nu_{O_2}\|} = \frac{15.75}{1.5} = 10.5 < \frac{n^o_{NH_3}}{\|\nu_{NH_3}\|} = \frac{15}{1} = 15$$

Therefore, C_3H_6 is the limiting reactant.
Number of NH_3 moles needed for 10 mol of reacting C_3H_6 is given by:

$$10\text{ mol } C_3H_6 \text{ reacts} \times \frac{1\text{ mol } NH_3 \text{ reacts}}{1\text{ mol } C_3H_6 \text{ reacts}} = 10\text{ mol } NH_3 \text{ reacts}$$

Thus, NH_3 is available in *excess* amount.

$$f\text{fractional excess of } NH_3 = \frac{(n_{NH_3})_{feed} - (n_{NH_3})_{stoich}}{(n_{NH_3})_{stoich}} = \frac{15-10}{10} = \frac{5}{10} = 0.5$$

$$(\%\ \text{excess})_{NH_3} = 100\% \times 0.5 = 50\%$$

Number of O_2 moles needed for 10 mol of reacting C_3H_6 is given by:

$$10\ \text{mol}\ C_3H_6\ \text{reacts} \times \frac{1.5\ \text{mol}\ O_2\ \text{reacts}}{1\ \text{mol}\ C_3H_6\ \text{reacts}} = 15\ \text{mol}\ O_2\ \text{reacts}$$

Thus, O_2 is available in *excess* amount.

$$\text{fractional excess of } O_2 = \frac{(n_{O_2})_{feed} - (n_{O_2})_{stoich}}{(n_{O_2})_{stoich}} = \frac{15.75-15}{15} = \frac{0.75}{15} = 0.05$$

$$(\%\ \text{excess})_{O_2} = 100\% \times 0.05 = 5.0\%$$

Notice that $(\%\text{excess})_{C_3H_6} = 0.0\%$

The **fractional conversion** of the limiting reactant (C_3H_6) is given as 0.4. Using Eq. (4.32):

$$f = \frac{\text{moles reacted}}{\text{moles fed}} = 0.4 \qquad \rightarrow \text{Moles reacted} = 0.4 \times 10\ \text{mol fed} = 4\ \text{mol}$$

$$\text{Moles remained} = \text{Moles fed} - \text{Moles reacted} = 10 - 4 = 6\ \text{mol}$$

Based on the *limiting reactant*, C_3H_6, use Eq. (4.37) to calculate the *extent of reaction*, ξ:

$$\xi = \frac{\Delta n_i}{\nu_i} = \frac{\text{Remained} - \text{Initially Present}}{\nu_i} = \frac{6-10}{-1} = 4$$

Use Eq. (4.38a) to calculate the number of moles for each species:

$$n_i = n_i^o + \nu_i \xi$$

$$n_{C_3H_6} = n^o_{C_3H_6} + \nu_i\xi = 10 + (-1) \times 4 = 6\ \text{mol}$$

$$n_{O_2} = n^o_{O_2} + \nu_i\xi = 15.75 + (-1.5) \times 4 = 9.75\ \text{mol}$$

$$n_{NH_3} = n^o_{NH_3} + \nu_i\xi = 15 + (-1) \times 4 = 11\ \text{mol}$$

$$n_{C_3H_3N} = n^o_{C_3H_3N} + \nu_i\xi = 0 + (+1) \times 4 = 4\ \text{mol}$$

$$n_{H_2O} = n^o_{H_2O} + \nu_i\xi = 0 + (+3) \times 4 = 12\ \text{mol}$$

$$n_{N_2} = n^o_{N_2} + \nu_i\xi = 59.25 + (0) \times 4 = 59.25\ \text{mol}$$

Notice that a fluidized bed reactor does hold several advantages for the production of acrylonitrile via the propylene ammoxidation reaction, explained in Example 4.14:

1) **Enhanced Gas-solid Contact**: As the catalyst particles are suspended in a gas stream within the reactor, a highly turbulent environment is created. This leads to excellent mixing and intimate contact between the gas reactants (propylene, ammonia, and oxygen) and the catalyst surface, significantly improving reaction rates and overall efficiency.
2) **Improved Temperature Control**: The high heat transfer coefficient in a fluidized bed reactor allows for the effective removal of the exothermic heat generated during the

reaction. This helps maintain a well-controlled temperature profile, which is crucial for maximizing acrylonitrile yield and selectivity while minimizing undesired side reactions.

3) **High Catalyst Activity and Stability**: The continuous movement of catalyst particles prevents channeling and hot spots, promoting uniform exposure to reactants and preventing thermal degradation. This can lead to longer catalyst life and sustained productivity.
4) **Scalability**: Fluidized bed reactors can be easily scaled up to large production capacities without significant loss of performance. This makes them particularly attractive for industrial applications.

4.4.3 Chemical Equilibrium and Equilibrium Constant

Certain reactions exhibit an essentially **irreversible** behavior, progressing solely from reactants to products. In these cases, the concentration of the limiting reactant gradually diminishes, reaching near-zero levels, albeit at varying time scales – ranging from seconds in some reactions to years in others. Consequently, the equilibrium composition in such reactions represents the state achieved upon complete exhaustion of the limiting reactant.

Conversely, other reactions, or the previous irreversible reactions but now under altered conditions, demonstrate reversibility. Here, reactants transform into products, while products engage in reverse reactions, reconverting into the initial reactants. For instance, consider the reaction involving the esterification of ethanol and acetic acid to produce ethyl-acetate and water:

$$CH_3COOH + C_2H_5OH \rightleftharpoons CH_3COOC_2H_5 + H_2O \tag{4.40}$$

If we start with ethanol and acetic acid, the forward reaction occurs; then, once ethyl-acetate is present, the reverse reaction begins to take place. As the concentrations of CH_3COOH and C_2H_5OH decrease, the rate of the forward reaction decreases, and as the $CH_3COOC_2H_5$ concentration increases, the rate of the reverse reaction increases. Eventually, a point is reached at which the rates of the forward and reverse reactions are equal. At this point, no further composition change will take place, and the reaction mixture is macroscopically invariant (i.e., exists under **chemical equilibrium** condition).

The position of the equilibrium, meaning how much product is formed before the reaction slows down, depends on several factors like temperature, concentration of reactants and products, and the presence of a catalyst. At room temperature, the equilibrium favors the products, meaning that a big portion of acetic acid and ethanol will be converted to ethyl acetate. By changing the temperature, we can shift the equilibrium to favor the products and increase the yield of ethyl acetate. See Example 4.15.

A full discussion of the relations that can be used to determine equilibrium compositions of reactive mixtures is beyond the scope of this textbook; however, at this point, we have enough knowledge to calculate equilibrium compositions if the relations are given to us.

Let us consider a general *liquid-phase* reaction:

$$aA_{(l)} + bB_{(l)} \rightleftharpoons cC_{(l)} + dD_{(l)} \tag{4.41}$$

The concentration-based equilibrium constant, K_C, can be defined as:

$$K_C \equiv \frac{[C]^c[D]^d}{[A]^a[B]^b} \tag{4.42}$$

Where each bracketed term represents the concentration of the reacting species raised to the power of its stoichiometric coefficient.

On the other hand, for a general *gas-phase* reaction:

$$aA_{(g)} + bB_{(g)} \rightleftharpoons cC_{(g)} + dD_{(g)} \tag{4.43}$$

The concentration-based equilibrium constant, K_C, can be defined as:

$$K_C \equiv \frac{[C]^c[D]^d}{[A]^a[B]^b} \tag{4.44}$$

If we have an ideal gas (low pressure and high temperature) behavior governed by the ideal gas law:

$$PV = nRT \tag{4.45}$$

Then, for a multi-component gas mixture, we can write the component version as:

$$p_iV = n_iRT \tag{4.46}$$

We can define the concentration in the gas-phase as:

$$C_i = \frac{\text{Moles of species } i}{\text{Total volume of the gas}} = \frac{n_i}{V} = \frac{p_i}{RT} \tag{4.47}$$

Substitute the value of C_i in each bracket appearing in Eq. (4.44) and simplify:

$$K_C \equiv \frac{\left[\frac{p_C}{RT}\right]^c\left[\frac{p_D}{RT}\right]^d}{\left[\frac{p_A}{RT}\right]^a\left[\frac{p_B}{RT}\right]^b} \equiv \frac{[p_C]^c[p_D]^d}{[p_A]^a[p_B]^b} \times \left(\frac{1}{RT}\right)^{c+d-a-b} \equiv \frac{[p_C]^c[p_D]^d}{[p_A]^a[p_B]^b} \times \left(\frac{1}{RT}\right)^{\Delta n} \tag{4.48}$$

Where $\Delta n = (c + d - a - b)$ is the change in total number of moles as a result of a chemical reaction. Moreover, the partial pressure-based term is basically called K_P. Equation (4.48) becomes:

$$K_C \equiv K_P \times \left(\frac{1}{RT}\right)^{\Delta n} \tag{4.49}$$

Rearrange to get:

$$K_P \equiv (RT)^{\Delta n} K_C \tag{4.50a}$$

We can further express K_P in terms of the mole-fraction-based equilibrium constant, K_Y:

$$K_P \equiv \frac{[p_C]^c[p_D]^d}{[p_A]^a[p_B]^b} = \frac{[P \times Y_C]^c[P \times Y_D]^d}{[P \times Y_A]^a[P \times Y_B]^b} = (P)^{c+d-a-b} \times \frac{[Y_C]^c[Y_D]^d}{[Y_A]^a[Y_B]^b} = (P)^{\Delta n} \times K_Y \tag{4.50b}$$

Equating Eqs. (4.50a) and (4.50b), we get:

$$(RT)^{\Delta n} K_C = (P)^{\Delta n} \times K_Y \tag{4.51a}$$

Rearrange Eq. (4.51a), we get:

$$K_Y \equiv \left(\frac{RT}{P}\right)^{\Delta n} \times K_C \tag{4.51b}$$

From a thermodynamics point of view, ΔG is expressed as:

$$\Delta G = \Delta G^0 + RT\ln K_C \tag{4.52a}$$

At equilibrium, $\Delta G = 0$ and we have:

$$\Delta G^0 = -\mathrm{RT}\ln K_C \tag{4.52b}$$

Divide both sides by −RT and take the logarithm for both sides, we have:

$$K_C = \mathrm{e}^{-\left(\frac{\Delta G^0}{\mathrm{RT}}\right)} \tag{4.52c}$$

T is the system temperature and ΔG^0 accounts for the standard molar Gibb free energy change for a given reaction at a given reference temperature, T_0. It is either tabulated in literature for some common chemical reactions or can be deduced from the standard molar Gibbs free energy of formation for products and reactants.

$$\Delta G^0 = \sum_{i=1}^{N} \left(\Delta G_f^0\right)_{\text{Products}} - \sum_{i=1}^{N} \left(\Delta G_f^0\right)_{\text{Reactants}} \tag{4.53}$$

The standard molar Gibbs free energy of formation, ΔG_f^0, can be found for some organic and inorganic compounds on the following web pages:

1) https://www.engineeringtoolbox.com/standard-enthalpy-formation-value-Gibbs-free-energy-entropy-heat-capacity-organic-d_1979.html
2) https://www.engineeringtoolbox.com/standard-state-enthalpy-formation-definition-value-Gibbs-free-energy-entropy-molar-heat-capacity-d_1978.html
3) https://en.wikipedia.org/wiki/Standard_Gibbs_free_energy_of_formation
4) https://www.wiredchemist.com/chemistry/data/entropies-inorganic
5) https://www.wiredchemist.com/chemistry/data/entropies-organic
6) https://www.studocu.com/da/document/danmarks-tekniske-universitet/fundamental-chemistry/free-gibbs-energy-enthalpy-and-enthropy/39897479

4.4.4 MATLAB for Finding the Roots of a Polynomial or any *f*(*x*) = 0

Example 4.15 Calculation of K_C and Equilibrium Compositions Utilizing ΔG_o
Consider the reaction:

$$CH_3COOH + C_2H_5OH \rightleftharpoons CH_3COOC_2H_5 + H_2O$$

In notation:

$$AA + EOH \rightleftharpoons EA + H_2O$$

We initially have one mole of AA and one mole of EOH placed in the V (L) solution.

Calculate K_C for this reaction and calculate the equilibrium number of moles of reacting species.

DATA:

$\left(\Delta G_f^0\right)_{EA} = -327.6\frac{\text{kJ}}{\text{mol}}$ (https://www.chemeo.com/cid/57-231-2/Ethyl-Acetate)

$\left(\Delta G_f^0\right)_{H_2O(l)} = -237.14\frac{\text{kJ}}{\text{mol}}$

(https://en.wikipedia.org/wiki/Standard_Gibbs_free_energy_of_formation)

$\left(\Delta G_f^0\right)_{EOH} = -168.4\frac{\text{kJ}}{\text{mol}}$ (https://www.chemeo.com/cid/35-653-8/Ethanol)

$$\left(\Delta G_f^0\right)_{AA} = -376.9\frac{\text{kJ}}{\text{mol}} \text{ (https://www.chemeo.com/cid/51-574-8/Acetic-acid)}$$

Solution:

Refer to Eq. (4.53):

$$\Delta G^0 = \sum_{i=1}^{N}\left(\Delta G_f^0\right)_{\text{Products}} - \sum_{i=1}^{N}\left(\Delta G_f^0\right)_{\text{Reactants}}$$

$$\Delta G^0 = \left(\Delta G_f^0\right)_{H_2O} + \left(\Delta G_f^0\right)_{EA} - \left(\Delta G_f^0\right)_{EOH} - \left(\Delta G_f^0\right)_{AA} =$$

$$\Delta G^0 = -237.14 - 327.6 - (-168.4) - (-376.9) = -19.44\frac{\text{kJ}}{\text{mol}}$$

From Eq. (4.52c)

$$K_C = e^{-\left(\frac{\Delta G^0}{RT}\right)} = e^{-\left(\frac{-19{,}440\frac{\text{J}}{\text{mol}}}{8.314\frac{\text{J}}{\text{mol.K}}(298.15)\text{K}}\right)} = 2.55 \times 10^3$$

Make use of Eq. (4.38a) for the definition of extent of reaction:

$$n_{AA} = n_{AA}^o + \nu_i\xi = 1 - \xi \qquad n_{EOH} = n_{EOH}^o + \nu_i\xi = 1 - \xi$$

$$n_{EA} = n_{EA}^o + \nu_i\xi = 0 + \xi \qquad n_{H_2O} = n_{H_2O}^o + \nu_i\xi = 0 + \xi$$

$$K_C = \frac{[EA][H_2O]}{[AA][EOH]} = \frac{\left[\frac{\xi}{V(L)}\right]\left[\frac{\xi}{V(L)}\right]}{\left[\frac{1-\xi}{V(L)}\right]\left[\frac{1-\xi}{V(L)}\right]} = \left[\frac{1}{V}\right]^{(1+1-1-1)=0}\frac{\xi^2}{(1-\xi)^2} = 2.55 \times 10^3$$

Simplify to end up with a quadrature (i.e. second-order polynomial) with two possible roots.

$$2.55 \times 10^3 \times \left(1 - 2\xi + \xi^2\right) = \xi^2$$

Move the right side to the left and simplify.

$$2550 - 5100\xi + 2549\xi^2 = 0$$

$$c + b\xi + a\xi^2 = 0$$

Figure 4.13 shows MATLAB code for finding the roots of a quadrature (second-order polynomial) with known coefficients.

```
% Define the coefficients of the quadratic equation: 2550 - 5100ξ + 2549ξ²
clear variables;
a = 2549;
b = -5100;
c = 2550;
% Solve for the roots
roots_quadratic = roots([a, b, c]);
% Display the roots
disp(['The first root of the equation is:', num2str(roots_quadratic(1))]);
disp(['The second root of the equation is:', num2str(roots_quadratic(2))]);
```

Figure 4.13 MATLAB code for finding the roots of a quadrature (i.e., second-order polynomial).

Below are the MATLAB results, copied from Command window:

```
The first root of the equation is:1.0202
The second root of the equation is:0.98058
```

Obviously, the second root is the feasible solution as ξ can not be greater than 1; otherwise, we will have negative moles or concentrations for both AA and EOH.

The equilibrium number of moles of each species can be expressed in terms of ξ.

$$n_{AA} = 1 - \xi = 1 - 0.98 = 0.02\,\text{mol} \qquad n_{EOH} = 1 - \xi = 1 - 0.98 = 0.02\,\text{mol}$$
$$n_{EA} = \xi = 0.98\,\text{mol} \qquad n_{H2O} = \xi = 0.98\,\text{mol}$$
$$n_{total} = 0.02 + 0.02 + 0.98 + 0.98 = 2\,\text{moles}$$

Of course, the equilibrium concentration of each species can be calculated by dividing the equilibrium number of moles of each species by the volume of the solution, V, expressed in L. On the other hand, the equilibrium composition (i.e., mole-fraction) can be calculated by dividing the number of moles of each species by the total number of moles, which is 2 moles, as shown above.

Example 4.16 Calculation of K_C and Equilibrium Compositions Utilizing ΔG_o

NOTE

The Haber–Bosch process, the most widely used industrial method for ammonia synthesis, is responsible for over 90% of global production, and operates at high temperature and pressure:

Temperature: 350 – 550°C (662 – 1022°F) and pressure: 150 – 300 bar (2175 – 4350 psi). For the sake of simplicity and using ideal gas law, we reduce the pressure down to 10 atm. By operating the Haber–Bosch process at higher temperatures, even though it may favor the reverse reaction at equilibrium, the rate of the forward reaction is increased. This means that ammonia is produced more quickly in a reasonable timeframe, making the industrial process more efficient. The higher temperature helps overcome the activation energy barrier for the reaction, leading to a faster rate.

Consider the gas-phase ammonia synthesis reaction:

$$N_{2(g)} + 3H_{2(g)} \rightleftharpoons 2NH_{3(g)}$$

Let us assume that initially, we have 3 moles of H_2 and 1 mole of N_2 at $T = 500°C$ and $P = 10$ atm. Find the equilibrium composition.

DATA:

$\left(\Delta G_f^0\right)_{NH_3}$: −16.3 kJ/mol (https://www.wiredchemist.com/chemistry/data/entropies-inorganic).

$\left(\Delta G_f^0\right)_{N_2}$: 0.0 (pure element found in nature)

$\left(\Delta G_f^0\right)_{H_2}$: 0.0 (pure element found in nature)

Solution:

Refer to Eq. (4.53):

$$\Delta G^0 = \sum_{i=1}^{N} \left(\Delta G_f^0\right)_{\text{Products}} - \sum_{i=1}^{N} \left(\Delta G_f^0\right)_{\text{Reactants}}$$

$$\Delta G^0 = 2\left(\Delta G_f^0\right)_{NH_3} - \left(\Delta G_f^0\right)_{N_2} - 3\left(\Delta G_f^0\right)_{H_2} =$$

$$\Delta G^0 = 2(-16.3) - 0 - 3(0) = -32.6\frac{\text{kJ}}{\text{mol}}$$

From Eq. (4.52c)

$$K_C = e^{-\left(\frac{\Delta G^0}{RT}\right)} = e^{-\left(\frac{-32600\frac{J}{mol}}{8.314\frac{J}{mol.K}(773.15)K}\right)} = 159.4$$

Make use of Eq. (4.38a) for the definition of extent of reaction:

$$n_{N_2} = n_{N_2}^o + \nu_i\xi = 1 - \xi \qquad n_{H_2} = n_{H_2}^o + \nu_i\xi = 3 - 3\xi$$

$$n_{NH_3} = n_{NH_3}^o + \nu_i\xi = 0 + 2\xi \qquad n_{\text{total}} = 1 - \xi + 3 - 3\xi + 2\xi = 4 - 2\xi$$

From Eq. (4.51b), we have:

$$K_Y \equiv \left(\frac{RT}{P}\right)^{\Delta n} \times K_C = \left(\frac{0.08206\frac{\text{atm.L}}{\text{mol.K}} \times 773.15\text{ K}}{10\text{ atm}}\right)^{2-3-1} \times K_C = 0.02484\left[\frac{L}{mol}\right]^{-2} \times K_C$$

Notice that $K_C = \dfrac{[NH_3]^2}{[N_2][H_2]^3} = 159.4\left[\dfrac{mol}{L}\right]^{-2}$

Thus, $K_Y = 0.02484\left[\dfrac{L}{mol}\right]^{-2} \times 159.4\left[\dfrac{mol}{L}\right]^{-2} = 3.96[\text{Dimensionless}]$

$$K_Y = \frac{[Y_{NH_3}]^2}{[Y_{N_2}][Y_{H_2}]^3} = \frac{\left[\frac{n_{NH_3}}{n_{\text{total}}}\right]^2}{\left[\frac{n_{N_2}}{n_{\text{total}}}\right]\left[\frac{n_{H_2}}{n_{\text{total}}}\right]^3} = \left(\frac{1}{n_{\text{total}}}\right)^{-2} \times \frac{(2\xi)^2}{(1-\xi)(3-3\xi)^3}$$

$$K_Y = (4-2\xi)^2\frac{(2\xi)^2}{(1-\xi)(3-3\xi)^3} = \frac{(4-2\xi)^2(2\xi)^2}{(1-\xi)(3-3\xi)^3} = 3.96$$

Find ξ that will satisfy the solution. Let us put it in the form of *f(x) = 0*:

$$f(x) = \frac{(4-2x)^2(2x)^2}{(1-x)(3-3x)^3} - 3.96 = 0$$

Figure 4.14 shows MATLAB code for finding the roots of a defined f(x) = 0

% Define the equation: $f(x) = \dfrac{(4-2x)^2(2x)^2}{(1-x)(3-3x)^3} - 3.96 = 0$

```
clear variables;
f = @(x) (((4-2*x)^2)*(2*x)^2) / ((1-x)*(3-3*x)^3) - 3.96;%x=ξ
% Find the root of the equation using `fzero`
x = fzero(f, 0.5); % Initial guess for x
% Display the solution
disp(['The value of x that satisfies the equation is: ', num2str(x)]);
```

Figure 4.14 Finding the root of $f(x) = 0$, using MATLAB code.

MATLAB result is shown below:

The value of *x* that satisfies the equation is: 0.47186

$$\xi = 0.47186$$

$$n_{N_2} = n^o_{N_2} + \nu_i\xi = 1 - 0.47186 = 0.5281 \text{ mol}$$

$$n_{H2} = n^o_{H_2} + \nu_i\xi = 3 - 3\xi = 3 - 3 \times 0.47186 = 1.5844 \text{ mol}$$

$$n_{NH_3} = n^o_{NH_3} + \nu_i\xi = 0 + 2\xi = 2 \times 0.47186 = 0.9437 \text{ mol}$$

$$n_{total} = 1 - \xi + 3 - 3\xi + 2\xi = 4 - 2\xi = 4 - 2 \times 0.47186 = 3.0563 \text{ mol}$$

Of course, the mole fraction can be calculated by dividing the number of moles of species i by the total number of moles.

$$Y_{NH_3} = \frac{n_{NH_3}}{n_{total}} = \frac{n_{NH_3} \text{ mol}}{3.0563 \text{ mol}} = \frac{0.9437 \text{ mol}}{3.0563 \text{ mol}} = 0.3088$$

$$Y_{H_2} = \frac{n_{H_2}}{n_{total}} = \frac{n_{H_2} \text{ mol}}{3.0563 \text{ mol}} = \frac{1.5844 \text{ mol}}{3.0563 \text{ mol}} = 0.5184$$

$$Y_{N_2} = \frac{n_{N_2}}{n_{total}} = \frac{n_{N_2} \text{ mol}}{3.0563 \text{ mol}} = \frac{0.5281 \text{ mol}}{3.0563 \text{ mol}} = 0.1728$$

$$Y_{NH3} + Y_{H2} + Y_{N2} = 1.000$$

Moreover, the equilibrium concentration can be calculated by:

$$C_i\left[\frac{\text{mol}}{\text{L}}\right] = \frac{P_i}{\text{RT}} = \frac{P_{total} \times Y_i}{\text{RT}} = \frac{10 \text{ atm} \times Y_i}{0.08206\frac{\text{atm.L}}{\text{mol.K}} \times 773.15 \text{ K}} = 0.1576 \times Y_i$$

$$C_{NH_3} = 0.1576 \times 0.3088 = 0.0487 \text{ M}$$

$$C_{H_2} = 0.1576 \times 0.5184 = 0.0817 \text{ M}$$

$$C_{N_2} = 0.1576 \times 0.1728 = 0.0272 \text{ M}$$

NOTE

If we start with the same conditions in terms of P, T, and the initial number of moles, but we use

$\frac{1}{2}N_{2(g)} + \frac{3}{2}H_{2(g)} \rightleftharpoons NH_{3(g)}$, instead of $1N_{2(g)} + 3H_{2(g)} \rightleftharpoons 2NH_{3(g)}$, we will end up with a root value that is double the value obtained here; however, the equilibrium conditions will remain invariant. Moreover, $K_C = 159.4$ is double $K_C = 12.6264$ calculated in CHW 4.17 problem. This is left as a homework to try.

Example 4.17 Calculation of Equilibrium Compositions
If the water-gas shift reaction,

$$CO_2(g) + H_2(g) \rightleftharpoons CO(g) + H_2O(g) \tag{4.54}$$

proceeds to **equilibrium** at a given temperature T (K), the mole fractions of the four reactive species will satisfy the relation:

$$K_P(T)_{\text{Equilibrium}} \equiv \frac{p_{H_2O}p_{CO}}{p_{H_2}p_{CO_2}} \tag{4.55}$$

where $K_P(T)$ is the partial pressure-based reaction **equilibrium constant.**
At $T = 1{,}933$ K, $\boldsymbol{K_P} = 4.00$.

Reference:

1) (https://byjus.com/question-answer/the-equilibrium-constant-kp-for-the-reaction-h2g-co2-g-gives-h2o-g-co-g/).
2) (https://www.toppr.com/ask/question/the-equilibrium-constant-kp-for-the-reaction-h2g-co2g-rightleftharpoons-h2og-cog-is/).

Suppose the feed to a reactor contains 1.00 mol of $\mathbf{CO_2}$, 2.00 mol of $\mathbf{H_2}$, and no **CO** or $\mathbf{H_2O}$, and the reaction mixture comes to **equilibrium** at 1933 K. Calculate the equilibrium composition and the fractional conversion of the limiting reactant.

Solution:
To determine the *limiting reactant*, test for the *smallest ratio* of $\frac{n_i^o}{\|\nu_i\|}$

$$\frac{n_{CO_2}^o}{\|\nu_{CO_2}\|} = \frac{1}{1} < \frac{n_{H_2}^o}{\|\nu_{H_2}\|} = \frac{2}{1} = 2$$

Therefore, $\mathbf{CO_2}$ is the limiting reactant.
Let us assume that the extent of reaction is given by: ξ
Using Eq. (4.38a), we can express the number of moles for each species as:

$$n_i = n_i^o + \nu_i\xi$$

$$n_{CO_2} = n_{CO_2}^o + \nu_{CO_2}\xi = 1 + (-1.0) \times \xi = (1 - \xi)\text{mol} \tag{4.56}$$

$$n_{H_2} = n_{H_2}^o + \nu_{H_2}\xi = 2 + (-1.0) \times \xi = (2 - \xi)\text{mol} \tag{4.57}$$

$$n_{CO} = n_{CO}^o + \nu_{CO}\xi = 0 + (1.0) \times \xi = (\xi)\text{mol} \tag{4.58}$$

$$n_{H_2O} = n_{H_2O}^o + \nu_{H_2O}\xi = 0 + (1.0) \times \xi = (\xi)\text{mol} \tag{4.59}$$

$$n_{tot} = \sum_{i=1}^{N} n_i = n_{CO_2} + n_{H_2} + n_{CO} + n_{H_2O} = (1 - \xi) + (2 - \xi) + (\xi) + (\xi) = 3 \tag{4.60}$$

$$Y_{H_2O} = \frac{n_{H_2O}}{n_{tot}} = \frac{\xi}{3} \tag{4.61a}$$

$$Y_{CO} = \frac{n_{CO}}{n_{tot}} = \frac{\xi}{3} \tag{4.61b}$$

$$Y_{H_2} = \frac{n_{H_2}}{n_{tot}} = \frac{(2-\xi)}{3} \tag{4.61c}$$

$$Y_{CO_2} = \frac{n_{CO_2}}{n_{tot}} = \frac{(1-\xi)}{3} \tag{4.61d}$$

Since $\Delta n = (c + d - a - b) = (1 + 1 - 1 - 1) = 0$, *then from* Eq. (4.50b)

$$K_P = 4.0 = (P)^0 \times K_Y \Rightarrow K_Y = K_P = 4.0$$

K_Y is defined as:

$$K_Y = \frac{Y_{H_2O} \times Y_{CO}}{Y_{H_2} \times Y_{CO_2}} \tag{4.62a}$$

Expressing each mole-fraction in terms of ξ, we have:

$$K_Y = \frac{Y_{H_2O} \times Y_{CO}}{Y_{H_2} \times Y_{CO_2}} = \frac{\frac{\xi}{3} \times \frac{\xi}{3}}{\frac{(2-\xi)}{3} \times \frac{(1-\xi)}{3}} = \frac{\xi \times \xi}{(2-\xi) \times (1-\xi)} = 4.0 \tag{4.62b}$$

Simplifying both the numerator and denominator, we get:

$$K_Y = \frac{\xi^2}{2 - 3\xi + \xi^2} = 4.0 \tag{4.62c}$$

It can be further simplified into : $8 - 12\xi + 4\xi^2 = \xi^2 \rightarrow 8 - 12\xi + 3\xi^2 = 0$ (4.62d)

Solve for the two roots:

$$\xi = \frac{-(-12) \mp \sqrt{-12^2 - 4 \times 3 \times 8}}{2 \times 3} = \frac{12 \mp \sqrt{144 - 96}}{6} = \frac{12 \mp \sqrt{48}}{6} = \frac{12 \mp 6.93}{6} \tag{4.63}$$

$$\text{First root} : \xi_1 = \frac{12 + 6.93}{6} = 3.1547 \tag{4.64}$$

$$\text{Second root} : \xi_2 = \frac{12 - 6.93}{6} = 0.8453 \tag{4.65}$$

Based on Eq. (4.56), the maximum extent of reaction, ξ, can be 1.0; otherwise, the concentration of the limiting reactant will be negative for the first root. Hence, $\xi_{\text{Equilibrium}} = \xi_2 = 0.845$.

Figure 4.15 shows MATLAB code for finding the roots of a quadrature (second-order polynomial) with known coefficients.

```
% Define the coefficients of the quadratic equation: 8-12ξ+3ξ²=0.0
clear variables;
a = 3;
b = -12;
c = 8;
% Solve for the roots
roots_quadratic = roots([a, b, c]);
% Display the roots
disp(['The first root of the equation is: ', num2str(roots_quadratic(1))]);
disp(['The second root of the equation is: ', num2str(roots_quadratic(2))]);
```

Figure 4.15 MATLAB code for finding the roots of a quadrature (i.e., second-order polynomial).

Below are the MATLAB results which match with those of the manual solution, shown by Eqs. (4.64, 4.65):

The first root of the equation is: 3.1547
The second root of the equation is: 0.8453

Substituting back in Eqs. (4.56) through (4.59), we get:

$$n_{CO_2} = (1 - \xi) = (1 - 0.845) = 0.155\ \text{mol}$$

$$n_{H_2} = (2 - \xi) = (2 - 0.845) = 1.155\ \text{mol}$$

$$n_{CO} = (\xi) = 0.845\ \text{mol}$$

$$n_{H_2O} = (\xi) = 0.845\ \text{mol}$$

The **fractional conversion** of the limiting reactant (**CO_2**) is given by Eq. (4.32):

$$f = \frac{CO_2\ \text{Moles reacted}}{CO_2\ \text{Moles fed}} = \frac{CO_2\ \text{Moles fed} - CO_2\ \text{Moles remained}}{CO_2\ \text{Moles fed}}$$

$$= \frac{1.0 - 0.155}{1.0} = 0.845$$

Refer to Eqs. (4.61a) through (4.61d), we have:

$$Y_{H_2O} = \frac{n_{H_2O}}{n_{tot}} = \frac{\xi}{3} = \frac{0.845}{3} = 0.2817$$

$$Y_{CO} = \frac{n_{CO}}{n_{tot}} = \frac{\xi}{3} = \frac{0.845}{3} = 0.2817$$

$$Y_{H_2} = \frac{n_{H_2}}{n_{tot}} = \frac{(2 - 0.845)}{3} = 0.385$$

$$Y_{CO_2} = \frac{n_{CO_2}}{n_{tot}} = \frac{(1 - \xi)}{3} = 0.0517$$

Check:
$Y_{tot} = Y_{H_2O} + Y_{CO} + Y_{H_2} + Y_{CO_2} = 0.2817 + 0.2817 + 0.385 + 0.0517 = 1.0001 \cong 1.000$

4.4.5 Multiple Reactions: Yield and Selectivity

Ethylene Production with Multiple Reactions

Consider ethylene production by the dehydrogenation of ethane:

$$C_2H_6 \rightarrow C_2H_4 + H_2 \tag{4.66}$$

Once some hydrogen is produced, it can react with ethane to produce methane:

$$C_2H_6 + H_2 \rightarrow 2\,CH_4 \tag{4.67}$$

Notice that the second reaction goes in parallel with the first reaction as it competes with the first reaction. Of course, initially, the second reaction is weak as no significant hydrogen is present yet in the reaction medium; however, as time goes on, more hydrogen is produced, and now it stands on firm feet to race with the first reaction.

Moreover, ethylene can react with ethane to form propylene and methane:

$$C_2H_4 + C_2H_6 \rightarrow C_3H_6 + CH_4 \tag{4.68}$$

Again, the same argument can be said about the third reaction in terms of progress with time coordinate.

Since the *object of the process is to produce ethylene, only the first of these reactions shall be treated as desirable*; the second one consumes the reactant without yielding the desired product; and the third consumes both the reactant and the desired product.

The engineer who designs the reactor and specifies its operating conditions must consider not only how to maximize the production of the desired product (C_2H_4) but also how to minimize the production of undesired byproducts (CH_4, C_3H_6).

The terms **yield** and **selectivity** are coined to describe the degree to which a desired reaction predominates over competing side reactions.

$$\text{Yield} = \frac{\text{net moles of desired product formed}}{\text{moles that would have been formed without side reactions}} \tag{4.69}$$

the main reaction was complete

$$\text{Selectivity} = \frac{\text{moles of desired product formed}}{\text{moles of undesired products formed}} \tag{4.70}$$

The yield defined by the previous equation is always a fraction; it may also be expressed as a percentage by multiplying by 100%. If **A** is the desired product and **B** is the undesired, one then refers to the *selectivity of A relative to B.* High values of the yield and selectivity signify that the undesired side reactions will be successfully suppressed relative to the desired reaction under the given set of reactor's operating conditions, like: temperature, pressure, feed composition, and presence/absence of an inert or impurity.

Example 4.18 Calculation of Yield and Selectivity

Consider the set of reactions given by Eqs. (4.66–4.68). Since the object of the process is to produce ethylene, only the first of these reactions may be considered desirable; the second one consumes the reactant without yielding the desired product; and the third consumes both the reactant and the desired product.

Now assume that the extent of reaction for the first is ξ_1; the second ξ_2; and the third ξ_3.

To calculate the *yield,* we assume that the fractional conversion, f_1, for the first reaction is 1.00; for the second, f_2, zero; and for the third, f_3, zero. Notice that ethylene (C_2H_4) is mainly produced by the first reaction and consumed by the third reaction; ethylene is not involved in the second reaction. For a multi-reaction system, refer to Eq. (4.38b). The desired product is C_2H_4.

$$\text{Yield} = \frac{(n_{C_2H_4})_{\text{Actual Condition}}}{\left(n_{C_2H_4}\big|_{f_1=1}\right)_{\text{Ideal Condition}}} = \frac{n_{C_2H_4} = n^o_{C_2H_4} + (1)\,\xi_1 + (-1)\,\xi_3}{n_{C_2H_4}\big|_{f_1=1}}$$
$$= \frac{0 + \xi_1 - \xi_3}{n_{C_2H_4}\big|_{f_1=1}} = \frac{\xi_1 - \xi_3}{n^o_{C_2H_6}} \tag{4.71}$$

Notice that the denominator in Eq. (4.71) is equal to the produced number of moles of ethylene (C_2H_4) under the assumption of a complete conversion for the first (main) reaction (i.e., $f_1 = 1.0$) while assuming zero conversion for all other competing reactions (i.e., $f_2 = f_3 = 0.0$). That simply means all initial moles of ethane (C_2H_6) will be completely converted into ethylene (C_2H_4), the desired product.

$$\text{Selectivity} = \frac{(n_{C_2H_4})_{\text{Actual Condition}}}{\text{All moles of udesired products}}$$
$$= \frac{\xi_1 - \xi_3}{(\text{moles of } H_2) + (\text{moles of } CH_4) + (\text{moles of } C_3H_6)} \tag{4.72}$$

Notice that the denominator in Eq. (4.72) basically represents the waste (byproducts) of competing materials formed at the expense of ethylene. Here, we say that not all ethane (C_2H_6) moles are converted into ethylene (C_2H_4), but some of them are converted into H_2, CH_4, and C_3H_6. Note that selectivity can be defined in different ways, depending on what is included in the denominator as undesired products. For a detailed discussion, see my upcoming textbook, MATLAB-Based Chemical Reaction Engineering.

$$\text{Selectivity} = \frac{\xi_1 - \xi_3}{\{(1)\,\xi_1 + (-1)\,\xi_2\} + \{(2)\,\xi_2 + (1)\,\xi_3\} + \{(1)\,\xi_3\}}$$
$$= \frac{\xi_1 - \xi_3}{\{\xi_1 - \xi_2\} + \{2\,\xi_2 + \xi_3\} + \{\xi_3\}} = \frac{\xi_1 - \xi_3}{\xi_1 + \xi_2 + 2\,\xi_3} \tag{4.73}$$

For example, if the initial amount for ethane is 100 moles; ξ_1 is 82 moles; ξ_2 is 10 moles; and ξ_3 is 6 moles. Then

$$\text{Yield} = \frac{82 - 6}{100} = \frac{76}{100} = 0.76$$

$$\text{Selectivity} = \frac{\xi_1 - \xi_3}{\xi_1 + \xi_2 + 2\,\xi_3} = \frac{82 - 6}{82 + 10 + (2 \times 6)} = 0.73$$

4.4.6 Overall and per-Pass Reactor Conversion: Recycle Loop

Recycle reactors find application in scenarios where the reaction exhibits autocatalytic behavior or when maintaining nearly isothermal reactor operation and promoting specific selectivity are crucial. In addition, these reactors are widely employed in biochemical processes. The recycled stream is drawn off at Point 2 and merged with the fresh feed at Point 1. We define the recycle parameter R as the moles recycled per mole of product removed at Point 2, as shown in Figure 4.16.

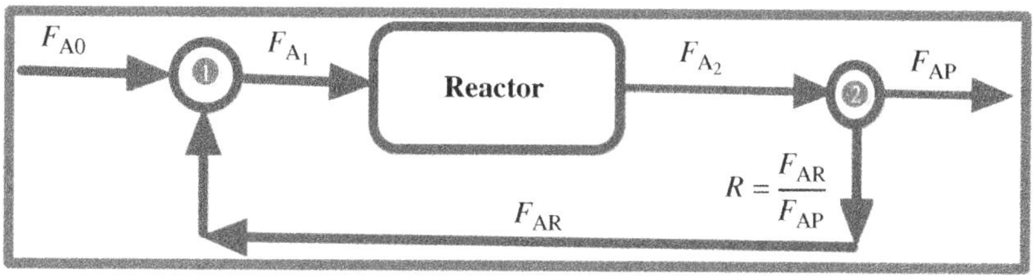

Figure 4.16 A reactor equipped with a recycle stream.

In a typical chemical reactor, the overall conversion is a measure of the extent to which the reactants have been converted into products over the entire reaction process. It is expressed as the ratio of the change in the number of moles of the limiting reactant to the initial number of moles of the limiting reactant. For the limiting reactant, the overall conversion (X_O) is given by the equation:

$$X_O = \frac{-\text{Change in number of moles of the reactant across the process}}{\text{Initial moles of reactant in the fresh feed}}$$

$$X_O = \frac{-(F_{AP} - F_{Ao})}{F_{Ao}} = \frac{F_{Ao} - F_{AP}}{F_{Ao}} = 1 - \frac{F_{AP}}{F_{Ao}} \tag{4.74}$$

When we make stream splits at Point 2, and we recycle unreacted stream to the inlet of the reactor, we are introducing a concept known as recycling. This is done to increase the overall yield and conversion of reactants.

On the other hand, considering the limiting reactant, the conversion per pass (X_P):

$$X_P = \frac{-\text{Change in number of moles of the reactant across the reactor}}{\text{Initial moles of reactant in the combined feed}}$$

$$X_P = \frac{-(F_{A_2} - F_{A_1})}{F_{A_1}} = \frac{F_{A_1} - F_{A_2}}{F_{A_1}} = 1 - \frac{F_{A_2}}{F_{A_1}} \tag{4.75}$$

R is defined as:

$$R = \frac{F_{AR}}{F_{AP}} \tag{4.76}$$

$$\text{At Point 1, we have}: F_{A_1} = F_{Ao} + F_{AR} = F_{Ao} + \text{R}F_{AP} \tag{4.77}$$

$$\text{At Point 2}: \text{we have}: F_{A_2} = F_{AP} + F_{AR} = F_{AP} + \text{R}F_{AP} = (R+1)F_{AP} \tag{4.78}$$

Let us derive the relationship between the overall conversion (X_O), conversion per pass (X_P), and the recycle ratio, R.

Let us start with X_P, we have:

$$X_P = \frac{F_{A_1} - F_{A_2}}{F_{A_1}} = 1 - \frac{F_{A_2}}{F_{A1}} \tag{4.79}$$

Replace F_{A1} by its equivalence from Eq. (4.77) and F_{A2} by its equivalence from Eq. (4.78), we have:

$$X_P = 1 - \frac{F_{A_2}}{F_{A_1}} = 1 - \frac{(R+1)F_{AP}}{F_{Ao} + \text{R}F_{AP}} \tag{4.80}$$

Divide both the numerator and denominator of the last quotient by F_{Ao} and given that $(1 - X_O) = \frac{F_{AP}}{F_{Ao}}$

we have:

$$X_P = 1 - \frac{(R+1)F_{AP}}{F_{Ao} + RF_{AP}} = 1 - \frac{(R+1)\frac{F_{AP}}{F_{Ao}}}{\frac{F_{Ao}}{F_{Ao}} + R\frac{F_{AP}}{F_{Ao}}} = 1 - \frac{(R+1)(1-X_O)}{1+R(1-X_O)} \tag{4.81}$$

Combine the two terms, we have:

$$X_P = \frac{1 + R(1-X_O) - (R+1)(1-X_O)}{1+R(1-X_O)} \tag{4.82}$$

Splitting each term in the numerator while canceling out those of opposite sign, we have:

$$X_P = \frac{1 + R - RX_O - R + RX_O - 1 + X_O}{1+R(1-X_O)} = \frac{X_O}{1+R(1-X_O)} \tag{4.83}$$

Eq. (4.83) gives X_P as $f(X_O, R)$. Let us also express X_o as $f(X_P, R)$.
Simplifying Eq. (4.83), we have:

$$X_O = X_P + RX_P(1-X_O) = X_P + RX_P - RX_PX_O \tag{4.84}$$

Rearrangement of terms, we have:

$$X_O + RX_PX_O = X_P + RX_P \tag{4.85}$$

$$X_O = \frac{X_P + RX_P}{1 + RX_P} = \left\{\frac{(1+R)}{1+RX_P}\right\}X_P \tag{4.86}$$

Since X_P is, in general, less than or equal to 1.0, then the bracketed term $\left\{\frac{(1+R)}{1+RX_P}\right\}$ will always be greater than or equal to 1.0, rendering $X_o \geq X_P$.
If $R = 0$, then $X_o = X_P$.
If $R \to \infty$, then $X_o \to \frac{RX_P}{RX_P} \approx 1$.

Example 4.19 Reactor with a Given *R*
Consider the reaction carried out in a CSTR with a recycle ratio, *R*, as shown in Figure 4.16:

$$aA \rightarrow \text{Products}$$

a) Given $R = 5$, $X_P = 0.6$, Calculate the overall conversion, X_O.
b) Conversely, if $X_O = 0.8$ for the same *R*, what will be X_P?

Solution:

a) Using Eq. (4.86), we have:

$$X_O = \left\{\frac{(1+R)}{1+RX_P}\right\}X_P = \left\{\frac{(1+5)}{1+5*0.6}\right\}0.6 = \left\{\frac{6}{1+3}\right\}0.6 = 0.90$$

b) Using Eq. (4.83), we have:

$$X_P = \frac{X_O}{1 + R(1 - X_O)} = \frac{0.8}{1 + 5(1 - 0.8)} = 0.4$$

Example 4.20 Downstream Separation Followed by Recycle

Consider a reaction with stoichiometry $A \rightarrow B$ that takes place in a process, as shown in Figure 4.17. With a basis of a unit time, the molar flow rates are indicated on each stream.

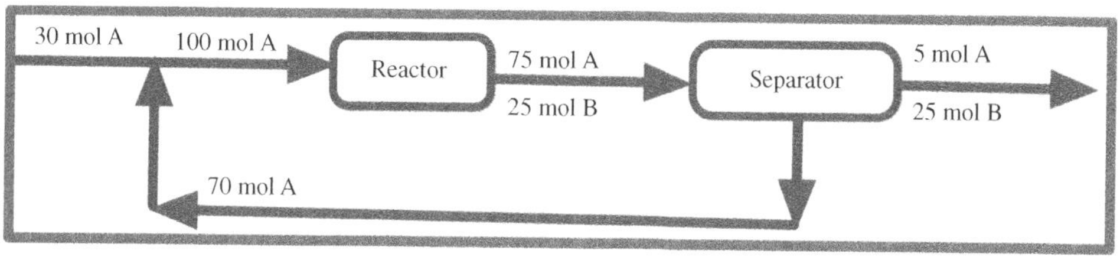

Figure 4.17 A reactor followed by a separator to facilitate recycle of A to the inlet of the reactor.

Here, unlike in Figure 4.16, we carry out the downstream component separation between the reactant A and product B, then we mainly recycle unreacted A back to the inlet of the reactor. The idea in this scheme is to take the unreacted component A, separate it from product B, and then reintroduce the recycle stream into the reactor to undergo further reaction. This is done to increase the overall yield and conversion of reactant.

Based on the indicated stream molar flow rates, calculate the overall conversion and the per-pass conversion in light of the previous definitions, given in Eqs. (4.74) and (4.75).

Solution:

Basis: One unit time of operation

Refer to Eq. (4.74), we have:

$$X_O = \frac{F_{AO} - F_{AP}}{F_{AO}} = 1 - \frac{F_{AP}}{F_{AO}} = 1 - \frac{5}{30} = 0.83$$

Refer to Eq. (4.75), we have:

$$X_P = \frac{F_{A_1} - F_{A_2}}{F_{A_1}} = 1 - \frac{F_{A_2}}{F_{A_1}} = 1 - \frac{75}{100} = 0.25$$

Notice that $X_O = 0.83 > X_P = 0.25$.

Notice the Following Questions and our Reply:

1) The separation unit, along with the recycle pump and piping, incurs significant costs. Is there a rationale for omitting these components and selling the reactor effluent as is? Alternatively, could we retain the separator but forget to recycle the bottom stream?

Feasibility: This might be feasible if there is a market demand for the unseparated mixture containing both *A* and *B*. However, the product's value might be lower compared to selling pure *B* due to the presence of undesired component *A*. On the other hand, retaining the separator but not recycling *A* could be an option if the inert material *C* accumulates in the bottom stream over time, causing issues when recycled. However, abandoning *A* represents a loss of reactant, decreasing process efficiency and increasing costs.

2) If a minute quantity, such as 0.2%, of an inert material C, is present in the fresh feed and all of it remains in the bottom effluent stream that is recycled, what could cause the eventual shutdown of the process? What steps must be taken to ensure its functionality?

Potential Issues Arising from *C* Accumulation: *C* might poison the catalyst in the reactor, reducing its activity and ultimately leading to process shutdown. *C* could also increase reactor pressure if it vaporizes and accumulates in the gas phase, jeopardizing equipment integrity and safety. Accumulation of *C* can interfere with the separation process by affecting its properties (e.g., viscosity, boiling point) and reducing separation efficiency.

Consequently, we regularly purge the bottom stream to remove a fraction of it and prevent *C* from reaching harmful levels. This removes *C* but also leads to some loss of *A*. We have to choose a catalyst less susceptible to poisoning by *C*. We ought to investigate alternative separation techniques (e.g., adsorption) specifically designed for handling *C* if purging is insufficient.

3) Is there a reason not to engineer the reactor to directly generate 5 mol A and 25 mol B from an initial 30 mol A in a single pass or shot, thus obviating the necessity for separation and recycling?

Equilibrium Limitations: Depending on the reaction, reaching a complete conversion (5 mol A–25 mol B) in a single pass or shot might be thermodynamically restricted. The reaction might reach an equilibrium point where conversion stops even with excess A.

Kinetic Limitations: The reaction rate might be too slow to achieve the desired conversion within a single pass at a reasonable reactor size and operating conditions. Increasing the residence time would require a larger reactor, and higher temperatures to further enhance the rate might not be feasible (cost limitations and safety concerns).

Economic Considerations: Depending on the cost of *A* compared to *B*, it might be more economical to operate in multiple passes with *A* recycle, even if additional separation and recycling equipment are needed. This approach can result in better utilization of *A* as a reactant.

End of Chapter Problems

CHW 4.1 Which of the following is NOT a characteristic of a batch process?

- **A** Feed is introduced all at once at the beginning.
- **B** No mass crosses the system boundaries during production.
- **C** Products are extracted together with unreacted materials after reaching equilibrium.
- **D** All process variables remain constant throughout the process.

CHW 4.2 A chemical reaction occurs in a continuously stirred tank reactor with steady input and output flow rates. What is the process classification?

A Batch
B Continuous
C Semi-batch
D Transient

CHW 4.3 Which process type is generally preferred for high production rates and consistent product quality?

A Batch
B Continuous
C Semi-batch
D Transient

CHW 4.4 What distinguishes a steady-state condition from a transient or unsteady-state operation in chemical processes?

A Changes in pressure
B Change in flow rates
C Alterations in process operation conditions
D Constant temperatures and pressures over time

CHW 4.5 According to the General Mass Balance Equation, what happens to the total mass being balanced within a system?

A It always increases over time
B It stays constant only in steady-state systems
C It can be created or destroyed within the system boundaries
D The sum of its inflow and generation must equal its outflow and consumption

CHW 4.6 Which type of balance is typically used for continuous processes?

A Differential balance
B Integral balance
C Both are equally common
D Neither is suitable for continuous processes

CHW 4.7 Which equation represents the conservation of a quantity within a system, considering input, generation, output, consumption, and accumulation?

A Input + Generation − Output − Consumption = Accumulation
B Input − Generation + Output − Consumption = Accumulation
C Input − Generation − Output + Consumption = Accumulation
D Input + Generation + Output − Consumption = Accumulation

CHW 4.8 In the context of steady-state continuous processes, what is the simplified form of the general balance equation (Eq. 4.1)?

A Input = Output
B Input + Generation = Output + Consumption
C Input − Generation = Output − Consumption
D Input + Output = Generation + Consumption

CHW 4.9 What term in the material balances on benzene and toluene for the distillation process is determined as zero due to the process being at a steady state and no chemical reaction occurring?

A Generation only
B Consumption only
C Accumulation
D Output

CHW 4.10 Water enters a 3.00-m^3 tank at a rate of 8.00 kg/s and is withdrawn at a rate of 5.00 kg/s. *The tank is initially half full.*

A Is this process continuous, batch, or semi-batch? Is it a transient or steady state?
B Write a mass balance for the process. Identify the terms of the general balance equation present in your equation and state the reason for omitting any terms.
C How long will it take the tank to have an overflow condition?

CHW 4.11 A liquid-phase chemical reaction $A_{(l)} \xrightarrow{k} B_{(l)}$ is carried out in a well-stirred tank. The concentration of A in the feed is C_{AO} (mol/m^3), and that in the tank and outlet stream is C_A (mol/m^3), as shown in Figure CHW4.11. *Both concentrations remain steady with time.*

The volume of the liquid content in the tank is V (m^3), and the volumetric flow rate of the inlet and outlet streams is $\dot{v}$ (m^3/s). The consumption rate (the rate at which A is consumed by the reaction in the tank) is given by the expression: $kC_AV\left[\frac{\text{mole of } A \text{ consumed}}{\text{s}}\right]$.
where k is called the rate constant, and for the *first-order reaction,* it has units of s^{-1}.

A Is this process continuous, batch, or semi-batch? Is it a transient or steady state?
B What will be the reactant concentration C_A (inside the reactor and at the outlet) equal to if $k = 0$ (no reaction)? What will it approach if $k \to \infty$ (infinitely rapid reaction)?

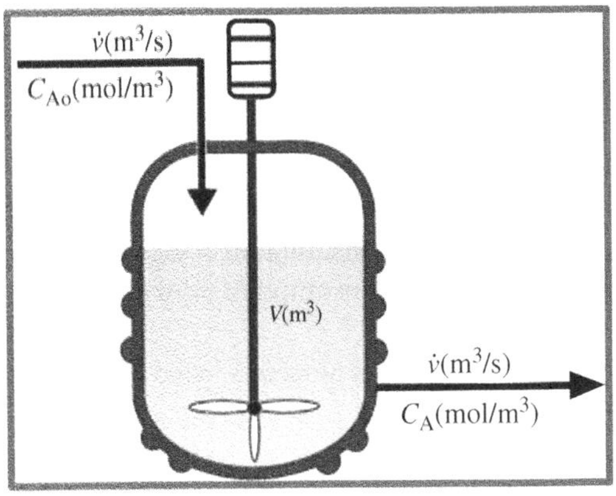

Figure CHW4.11 A steady-state continuous stirred tank reactor (CSTR) with a liquid-phase reaction.

C Write a molar balance on A, stating which terms in the general balance equation: (*accumulation* = *input* + *generation* – *output* – *consumption*) you discarded and why you discarded them. Use the molar balance to derive the following relation between the inlet and outlet reactant concentrations: $C_A = \dfrac{C_{Ao}}{1 + \frac{kV}{\dot{v}}}$.

Verify that this relation predicts the results in part (b).

CHW 4.12 Determining Mass and Composition in a Batch Mixing Process

Two aqueous solutions of methanol are available, as shown in Figure CHW4.12:

- Solution A: 40.0 wt. % methanol, mass $m_A = 450$ g
- Solution B: 70.0 wt. % methanol, mass $m_B = 300$ g

These two solutions are mixed together in a batch process. Determine the mass and composition (wt. % methanol) of the final mixture.

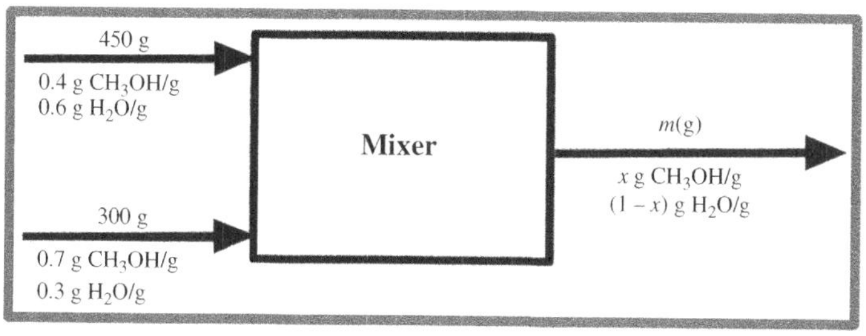

Figure CHW4.12 Mixing of two streams of methanol solutions with different compositions.

CHW 4.13 *Three hundred liters* of a mixture containing 60.0 wt. % methanol (methyl alcohol) and 40% water (mixture specific gravity = 0.874) and a *quantity* of 40.0 wt. % methanol-60% water mixture ($SG = 0.915$) are both blended to produce a mixture containing 50.0 wt. % methanol. The object of this problem is to determine V_2, the required volume of the 40% methanol mixture.

A Draw and label a flowchart of the mixing process and do the degree-of-freedom analysis.

B Use MATLAB code to solve for 2 algebraic equations with two unknowns. Then calculate V_2. You can also use the manual solution by substitution or elimination method.

CHW 4.14 A paint mixture containing 30.0% of a pigment and the balance water sells for \$20.00/kg, and a mixture containing 15.0% pigment sells for \$10.00/kg. If a paint retailer produces a blend containing 20.0% pigment, for how much (\$/kg) should it be sold to yield a 15% profit? A schematic is shown in Figure CHW4.14. Use both the manual solution and MATLAB to solve two algebraic equations.

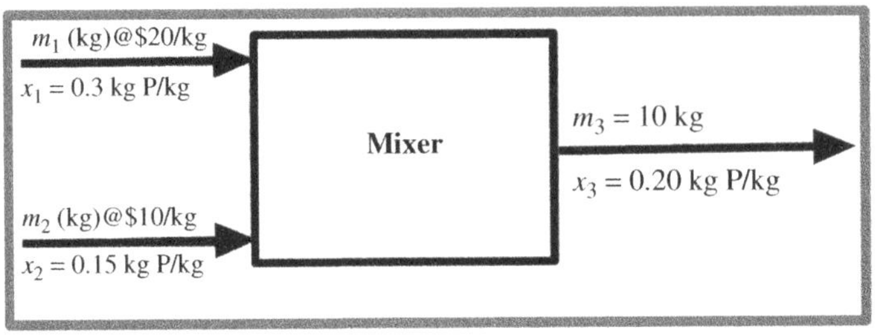

Figure CHW4.14 A flowsheet for paint mixing and preparation for sale.

CHW 4.15 In the Deacon process for the manufacture of chlorine, HCI and O_2 react to form Cl_2 and H_2O. Sufficient air (21 mole% O_2, 79% N_2) is fed to provide *40% excess oxygen,* and the fractional conversion of HCl is 90%. A schematic is shown in Figure CHW4.15.

$$2HCl + \frac{1}{2}O_2 \rightarrow Cl_2 + H_2O$$

A Calculate the mole fractions of the product stream components using the extent of the reaction in the calculation. Use 120 moles of HCl fed to the reactor as the basis.

B A substitute for employing air as the oxygen source involves introducing pure oxygen into the reactor. While utilizing oxygen incurs additional process expenses compared to using air, it also presents the opportunity for substantial savings. Delve into potential costs and savings and identify the factors that would dictate the preferable mode of operation for the process.

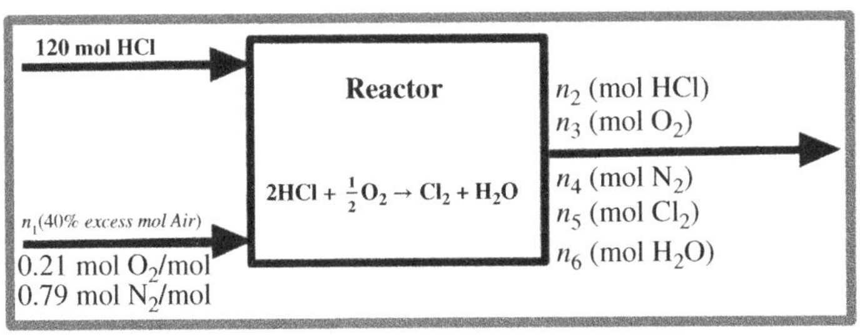

Figure CHW4.15 Deacon process for the manufacture of chlorine, where HCl and O_2 react to form Cl_2 and H_2O.

CHW 4.16 Consider the water-gas shift reaction:

$$CO(g) + H2O(g) \rightleftharpoons CO2(g) + H_2(g)$$

At $T = 1105$ K, the equilibrium constant, $K_C = 1.00$.

Suppose the feed to a reactor contains 1.00 mol of CO, 2.00 mol of H_2O, and *no CO_2 or H_2*, and the reaction mixture comes to *equilibrium* at 1105 K. Calculate the extent of reaction, the fractional conversion of the limiting reactant, an equilibrium composition of the gas mixture.

CHW 4.17 Calculation of K_C and Equilibrium Compositions Utilizing ΔG_o

Consider the gas-phase ammonia synthesis reaction:

$$\frac{1}{2}N_{2(g)} + \frac{3}{2}H_{2(g)} \rightleftharpoons NH_{3(g)}$$

Let us assume that initially, we have 3 moles of H_2 and 1 mole of N_2 at $T = 500°C$ and $P = 10$ atm. Find the equilibrium composition. Show that although the extent of reaction, ξ, has exactly doubled in value, yet the equilibrium number of moles, the gas mole fractions, and the equilibrium concentrations remain the same as those of Example 4.16.

CHW 4.18 Consider the gas-phase formation of ethanol from ethene:

$$C_2H4(g) + H_2O(g) \rightleftharpoons C_2HOH(g)$$

Ethanol is prepared from ethene by hydration in the presence of phosphoric acid acting as a catalyst. Ethene is mixed with heated steam in a fixed molar ratio. They react at 300°C to produce ethanol.

Suppose the feed to a reactor contains 1.00 mol of C_2H_4, 2.00 mol of H_2O, and *no* C_2H_5OH, and the reaction mixture comes to ***equilibrium*** at 573 K and $P = 10$ atm. Calculate the extent of reaction, the fractional conversion of the limiting reactant, and an equilibrium composition of the gas mixture.

$\left(\Delta G_f^0\right)_{C2H5OH(g)}$: −168.3 kJ/mol (https://www.wiredchemist.com/chemistry/data/entropies-organic).

$\left(\Delta G_f^0\right)_{H2O(g)}$: −228.4 kJ/mol (https://www.wiredchemist.com/chemistry/data/entropies-inorganic)

$\left(\Delta G_f^0\right)_{C2H4(g)}$: 68.1 kJ/mol (https://www.wiredchemist.com/chemistry/data/entropies-organic)

CHW 4.19 Methane and oxygen react in the presence of a catalyst to form formaldehyde. In a parallel reaction, methane is oxidized to carbon dioxide and water:

$$CH_4 + O_2 \rightarrow HCHO + H_2O$$

$$CH_4 + 2\,O_2 \rightarrow CO_2 + 2\,H_2O$$

The feed to the reactor contains equimolar amounts of methane and oxygen. Assume a basis of 200 mol feed/s.

A Draw and label a flowchart. Use a degree-of-freedom analysis based on the extents of the reaction to determine how many process-variable values must be specified for the remaining variable values to be calculated.

B Use Eq. (4.38b) to derive expressions for the product stream component flow rates in terms of the two extents of reaction $\xi_1 and\ \xi_2$.

C The *fractional conversion of methane is 0.60,* and the fractional *yield of formaldehyde is 0.85*. Calculate the molar composition of the reactor output stream and the *selectivity* of formaldehyde production relative to carbon dioxide production.

> Consider the following reaction: $C_4H_{10} + \frac{13}{2} O_2 \rightarrow 4 CO_2 + 5 H_2O$
>
> Initially, we have 10 moles of n-butane and 50 moles of oxygen. The extent of reaction, ξ, is measured as 7 moles toward the end of combustion.

Answer the following questions: CHW 4.20 → CHW 4.28

CHW 4.20 What is the stoichiometric coefficient of CO_2 in the given equation?

A 1
B 2
C 4
D 6

CHW 4.21 What is the stoichiometric ratio of H_2O to O_2 in the given reaction? (Include units)

A 5 mol H_2O:13 mol O_2
B 5 mol H_2O:6 mol O_2
C 5 mol H_2O:4 mol O_2
D 10 mol H_2O:13 mol O_2

CHW 4.22 How many lb-moles of O_2 react to form 400 lb-moles of CO_2?

(a) 400 lb-moles
(b) 650 lb-moles
(c) 500 lb-moles
(d) 1300 lb-moles

CHW 4.23 For the given ξ, what is the remaining n_{C4H10}, expressed in mole:

A 10
B 7
C 6.5
D 3
E 1

CHW 4.24 For the given ξ, what is the remaining n_{O2}, expressed in mole:

A 45.5
B 43
C 7
D 4.5
E 0

CHW 4.25 For the given ξ, what is the produced n_{CO2}, expressed in mole:

A 35
B 28
C 7
D 5
E 4

CHW 4.26 For the given ξ, what is the produced n_{H2O}, expressed in mole:

A 35
B 28

C 7
D 5
E 4

CHW 4.27 What is the fractional conversion of C_4H_{10}?
A 1
B 0.7
C 0.5
D 0.3
E 0

CHW 4.28 What is the fractional conversion of O_2?
A 1
B 0.91
C 0.90
D 0.09
E 0

Consider ethylene production by the dehydrogenation of ethane:

$$C_2H_6 \rightarrow C_2H4 + H_2 \quad (1)$$

Once some hydrogen is produced, it can react with ethane to produce methane:

$$C_2H_6 + H_2 \rightarrow 2\,CH_4 \quad (2)$$

Moreover, ethylene can react with ethane to form propylene and methane:

$$C_2H_4 + C_2H_6 \rightarrow C_3H_6 + CH_4 \quad (3)$$

The extent of reaction for the first is $\xi_1 = 82$ mol; the second $\xi_2 = 8$ mol; and the third $\xi_3 = 5$ mol. The initial amount for ethane (C_2H_6) is 100 moles.

Answer the following questions: CHW 4.29 → CHW 4.33

CHW 4.29 The remaining number of moles of ethane (C_2H_6), n_{C2H6}, is:
A 100
B 95
C 82
D 18
E 10
F 5
G 0

CHW 4.30 The remaining number of moles of ethene (C_2H_4), n_{C2H4}, is:
A 100
B 87
C 82
D 77
E 8
F 5
G 0

CHW 4.31 The remaining number of moles of ethene (CH_4), n_{CH4}, is:

- **A** 87
- **B** 79
- **C** 21
- **D** 16
- **E** 13
- **F** 5
- **G** 0

CHW 4.32 Given that the desired product is ethene (C_2H_4), what will be the value of the yield percent (%)?

- **A** 95
- **B** 90
- **C** 87
- **D** 82
- **E** 77
- **F** 23
- **G** 13

CHW 4.33 Given that the desired product is ethene (C_2H_4), what will be the value of the selectivity?

- **A** 1.0
- **B** 0.84
- **C** 0.82
- **D** 0.81
- **E** 0.77
- **F** 0.5
- **G** 0.05

CHW 4.34 In a reaction: $C_2H_6 \rightarrow C_2H_4 + H_2$, if the extent of the reaction is 30 mol, and the initial amount of ethane (C_2H_6) was 100 mol, what are the remaining moles of ethane?

- **A** 70 mol
- **B** 50 mol
- **C** 35 mol
- **D** 75 mol
- **E** 30 mol

CHW 4.35 Given a reaction: $C_2H_6 + H_2 \rightarrow 2\ CH_4$, if the extent of the reaction is 5 mol and 10 mol of H_2 is initially present, what will be the produced moles of methane (CH4)?

- **A** 0 mol
- **B** 5 mol
- **C** 10 mol
- **D** 15 mol
- **E** 20 mol

CHW 4.36 In the reaction: $C_2H_4 + C_2H_6 \rightarrow C_3H_6 + CH_4$, if the extent of the reaction is 7 mol and 20 mol of ethylene (C_2H_4) was initially present, what are the remaining moles of ethylene?

A 13 mol
B 7 mol
C 20 mol
D 25 mol
E 27 mol

CHW 4.37 If the desired product in a reaction is C_2H_4 and the actual yield is 60 mol with an ideal condition yield of 80 mol, what will be the yield percentage?

A 50%
B 60%
C 70%
D 75%
E 80%

CHW 4.38 Reactor with Unknown R

Consider the reaction carried out in a CSTR with a recycle ratio, R, as shown in Figure 4.16:

$$aA \rightarrow \text{Products}$$

A Given $X_P = 0.6$ and $X_O = 0.95$, what will be the recycle ratio R?
B If $X_O = 0.8$ for $R = 10$, what will be X_P?

CHW 4.39 Downstream Separation Followed by Recycle

Consider a reaction with stoichiometry $A \rightarrow B$ that takes place in a process, as shown in Figure CHW4.39. With a basis of a unit time, the molar flow rates are indicated on each stream.

Based on the indicated stream molar flow rates, calculate the overall conversion and the per-pass conversion.

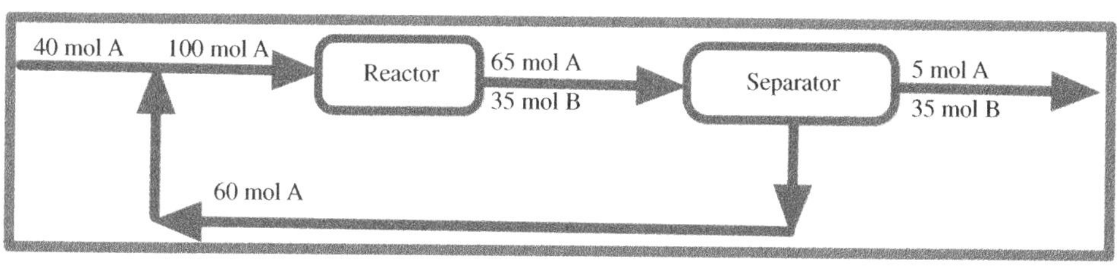

Figure CHW4.39 A reactor followed by a separator to facilitate recycle of A to the inlet of the reactor.

CHW 4.40 Downstream Separation Followed by Recycle

Consider a reaction with stoichiometry $A \rightarrow B$ that takes place in a process, as shown in Figure CHW4.40. With a basis of a unit time, the molar flow rates are indicated on each stream.

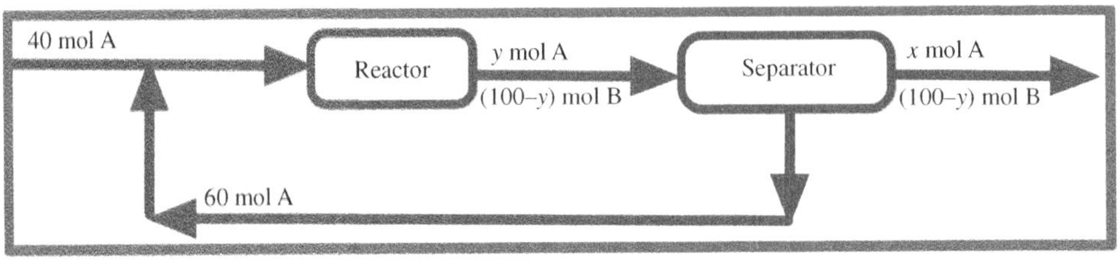

Figure CHW4.40 A reactor followed by a separator to facilitate recycle of *A* to the inlet of the reactor.

The overall conversion is 0.9, and the per-pass conversion is 0.36.

Find the molar flow rate of *A* at the exit of reactor, *y* and at the exit of the process, *x*.

Find the molar flow rate of *B*, as well.

5

Single-Phase Systems

5.1 Solids and Liquids

Solid materials are one of the fundamental states of matter characterized by specific properties and behaviors:

1) **Fixed Shape and Volume**: Solids maintain a definite shape and volume under normal conditions. They resist changes in shape and do not flow readily like liquids or gases.
2) **Strong Intermolecular Forces**: Particles in solids are closely packed and held together by strong intermolecular forces (like ionic, covalent, metallic, or molecular forces), which give solids their structural integrity and rigidity.
3) **Regular Arrangement**: Solids often exhibit ordered and regular arrangements of atoms, ions, or molecules in a crystal lattice structure. This arrangement can be crystalline (highly ordered) or amorphous (less ordered).
4) **Low Compressibility**: Solids are relatively incompressible compared with liquids and gases. When subjected to pressure, the volume change in solids is minimal due to the close arrangement of particles.
5) **High Density**: Solids typically have higher densities compared to liquids and gases due to their tightly packed structure.
6) **Definite Melting Point**: Solids have distinct melting points at which they transition into liquids through the process of melting, where strong intermolecular forces are overcome.
7) **Mechanical Strength**: They possess mechanical strength and can resist deformation and external forces due to their structural integrity.
8) **Vibrational Motion**: While particles in solids are relatively fixed in their positions, they still exhibit vibrational motion about their equilibrium positions due to thermal energy.

To my dear readers,
When copying MATLAB code or text, please right-click and attempt both the "Copy with Formatting" and "Copy" options from the context menu. This approach is effective within Adobe Acrobat. If neither option works, kindly carefully review any MATLAB error messages and compare the pasted text to the original to identify and correct any discrepancies. Alternatively, a soft copy of the MATLAB code is available in the Wiley book companion site of this book for your convenience.

MATLAB-based Computations of Chemical Engineering Principles, First Edition. Kamal I. M. Al-Malah.

Companion website: www.wiley.com/go/al-malah/matlabbasedcomputations

Solids can exist in various forms, such as metals, ceramics, polymers, and composites, each having distinct properties and behaviors based on their atomic/molecular structures and bonding. The specific characteristics and behaviors of solids play a crucial role in various applications across industries, including engineering, materials science, construction, and manufacturing.

On the other hand, liquid materials are one of the fundamental states of matter, possessing specific properties and behaviors that distinguish them from solids and gases:

a) **Fluidity**: Liquids exhibit fluidity and flow, meaning they can adapt to the shape of their container. Unlike solids, fluids cannot resist shear stresses, and fluids continuously deform as long the action of shear stress is sustained.
b) **No Fixed Shape but Fixed Volume**: Unlike solids, liquids do not have a fixed shape. They flow and take the shape of their container. However, they have a constant volume under normal conditions.
c) **Intermolecular Forces**: Liquids have moderate intermolecular forces that allow them to maintain a degree of cohesion, yet these forces are weaker than their solid counterparts. These forces permit molecules to move past one another, enabling flow and fluidity.
d) **Compressibility**: Liquids are generally considered to be incompressible compared to gases but are slightly compressible compared to solids. This property allows them to be compressed to a limited extent under high pressure.
e) **Density**: Liquids have higher densities compared to gases but generally lower densities than solids. This density property leads to liquids sinking in a less dense medium and floating on a denser medium.
f) **Definite Volume**: Liquids have a definite volume, meaning they fill the container they occupy, maintaining a consistent volume regardless of shape.
g) **Surface Tension**: Liquids exhibit a pronounced surface tension due to cohesive forces between their molecules at either the air–liquid or solid–liquid interface, causing the surface to behave like a thin, elastic film.
h) **Viscosity**: Viscosity refers to a liquid's resistance to flow. Some liquids have high viscosity (like honey or motor oil), resisting flow more strongly, while others have low viscosity (like water or alcohol) and flow more easily.
i) **Evaporation and Boiling**: Liquids can evaporate at their surface and boil at specific temperatures, transitioning into gases. Evaporation occurs at the liquid's surface, while boiling happens throughout the liquid at its boiling point.

Understanding the characteristics and behavior of liquids is crucial in various fields such as chemistry, physics, engineering, and industry, particularly in areas involving fluid dynamics, chemical reactions, and material handling.

NOTE #1
Densities of pure liquids and pure solids are tabulated in Appendix B | Table B.1 and B.2, respectively.

5.1.1 Data and Models for Solid Density

Before we dive into the world of density models, let me emphasize that this course is introductory for undergraduate chemical engineering students, and I will keep in mind to simplify and exemplify concepts as much as I can. I will not let the student be stuck with cumbersome mathematical models.

1) **Crystallographic Data and Lattice Parameters**: The density of crystalline solids can be calculated using crystallographic data and lattice parameters obtained from X-ray diffraction or other experimental techniques. These data provide information about the arrangement of atoms or molecules within the crystal lattice, allowing for the calculation of the volume of the unit cell and, thus, the density of the solid.
2) **Theoretical Models**: Some theoretical models based on interatomic or intermolecular potentials aim to predict the properties of solids, including density, based on molecular interactions. These models might include approaches such as density functional theory (DFT) or molecular dynamics simulations. Imagine a bustling party inside a molecule. Electrons, the tiny guests, zip around, constantly interacting and creating a unique atmosphere. DFT is a cool tool that helps us understand this party and predict how it behaves. Here is the gist:
   - **Forget Counting Electrons**: Instead of chasing individual electrons, DFT focuses on the electron density, which is like the overall "crowd or cloud" of electrons at any point in the molecule. It is easier to picture and handle than chasing down specific guests!
   - **Energy is Key**: DFT calculates the total energy of the molecule based on this electron density. Think of it as the party's vibe; the lower the energy, the more stable and happier the molecule.
   - **Minimizing the Party Funk**: DFT finds the electron density that gives the lowest energy, like setting the perfect music for the best party mood. This tells us the most stable arrangement of electrons and, therefore, the molecule's properties.
   - **Like Magic (But It Is Maths)**: DFT uses clever mathematical tricks to handle all the complex interactions between electrons. It is not easy, but it gives us pretty accurate results without getting lost in the crowd.
   - **Accurate Insights**: It provides reliable predictions of various molecular properties, like bond lengths, shapes, and energies.
   - **Versatile Playground**: DFT can be used for different types of molecules, from simple ones like H_2O to complex biomolecules.
3) **Equations of State**: Equations of state like the Murnaghan equation or Birch–Murnaghan equation are used to describe the relationship between volume, pressure, and temperature for solids. These equations, often derived from experimental data, can be used to estimate densities at various conditions based on parameters obtained from fitting experimental data. Here is an exemplified version of the Murnaghan equation of state. Imagine squeezing a squishy ball. The harder you press, the more it resists shrinking, right? The Murnaghan equation of state captures this behavior for solids under high pressure, a crucial tool in fields like earth science and shock physics. It was developed by Francis D. Murnaghan in 1944 (https://en.wikipedia.org/wiki/Murnaghan_equation_of_state), this equation elegantly links a solid's volume to the crushing pressure it experiences. It accounts for a key observation: The tighter you pack a solid, the tougher it gets to squeeze further. Think of a dense brick vs. a soft sponge. The brick, with its tightly packed atoms, requires immense pressure to shrink just a tiny bit. The sponge, however, squishes easily due to its looser structure. But how does the Murnaghan equation work? Based on principles of physics, it relies on two key numbers: The modulus of incompressibility (how resistant a material is to pressure) and its pressure dependence. These are measured at normal pressure and then used to predict the material's behavior at higher pressures. Scientists can determine these values by experimenting:
   - **X-ray Diffraction**: Shining X-rays on a material under pressure reveals how its atoms rearrange, providing clues about its volume change.

- **Shock Tests**: Subjecting a material to powerful shocks mimics high-pressure environments and allows researchers to observe its compression.

  Alternatively, advanced computer simulations called ab initio calculations and molecular dynamics can calculate the energy of a material as its volume changes, providing data for the Murnaghan equation of state. By fitting the equation to experimental or simulation data, scientists can accurately predict how a solid will behave under immense pressures deep within the Earth or during high-speed impacts. This knowledge is crucial for understanding Earth's interior, designing shock-resistant materials, and even simulating planetary collisions.

4) **Empirical Relations**: For some solids, empirical relations may exist between properties like density and temperature/pressure. These relations are often obtained through empirical correlations derived from experimental measurements.

5.1.2 Data and Models for Liquid Density

1) **Empirical Equations**: An empirical equation like the Tait equation (https://www.researchgate.net/publication/37968313_Equation_of_State_of_Water_and_Sea_Water) can describe the variation of liquid density with temperature and pressure. This equation is fitted to experimental data and can be used to predict density under different conditions. Let us exemplify Tait's equation: Imagine we are filling a water tank. As we pour more water in, the pressure at the bottom increases, right? The Tait equation captures this relationship between pressure and the "tightness" of the water, aka its density.

   Think of water molecules as bouncy balls crammed into the tank. The more we push (increase pressure), the closer they pack together, making the liquid denser. But wait, these balls have a bit of springiness (intermolecular forces), so they resist getting squished or squeezed too much. The Tait equation is like a fancy scale that balances this push and pull:

   - **Pressure vs. Volume**: One side considers the pressure pushing down on the balls, causing their volume to shrink. It is like adding weights to the "pressure side" of the scale.
   - **Springiness of the Balls**: The other side accounts for the balls' resistance to compression and intermolecular forces.

     When the Scale Balances: It means the pressure and springiness have reached a sweet spot; the water achieves a specific density at that given pressure. We can then predict the density of the water tank at different pressures. The Tait equation is especially useful for high-pressure conditions, where liquids behave differently than at normal room pressure. It helps chemical engineers in fields like oil and gas extraction or designing high-pressure equipment. In brief, it is just like balancing forces on a scale; pressure pushing down vs. the liquid's "springiness" pushing back.

2) **Modeling Liquid Structure**: Some models focus on the structure of liquids, considering the arrangement and interactions between molecules. This includes approaches such as the radial distribution function, pair distribution functions, or the use of statistical mechanics methods to predict thermodynamic properties, including density.
3) **Mixing Rules**: For mixtures of liquids, there are models like the Margules equation or Redlich–Kister (RK) equation that describe the density of liquid mixtures based on the densities of individual components and their interactions.
4) **PVT Data and Measurements**: Experimental measurements of Pressure–Volume–Temperature (PVT) properties using techniques like densitometry, volumetry, or using PVT tables can provide accurate data for liquid densities under various conditions. Cubic

equations of state like van der Waals (vdW) or RK can be used to solve for roots of the cubic equation, which will give the molar volume of liquid and/or that of vapor. See upcoming section: Cubic Equations of State.

5.1.3 Pressure Effects on Densities of Solids

The effect of pressure on the densities of solids can vary depending on the specific characteristics of the material in question. Here is an elaboration on how pressure affects the densities of solids:

1) **Incompressibility of Solids**: Most solids are relatively incompressible compared to liquids and gases. This means that when pressure is applied to a solid, the change in volume is usually quite negligible, and the density remains practically constant.
2) **Bulk Modulus and Elasticity**: Solids have a property known as the bulk modulus, which quantifies their resistance to changes in volume under pressure. Materials with a high bulk modulus are less affected by changes in pressure, maintaining their structure and density.
3) **Exceptions and Anomalous Behavior**: There are exceptions, especially in certain porous or compressible solids. For example, some materials, like certain types of foams, sponges, or aerogels, have a more open or porous structure, making them more compressible. In these cases, increased pressure can compress the material, reducing its volume and consequently increasing its density.
4) **Phase Transitions**: In some cases, extreme pressure can induce phase transitions in solids. For instance, certain materials under extremely high pressure might undergo structural changes or phase transitions, altering their arrangement of atoms or molecules, which could result in changes in density.
5) **Materials at Extreme Conditions**: Under extreme pressure conditions, such as those found in the Earth's mantle or in laboratory settings using high-pressure equipment, the behavior of solids can deviate significantly. Some materials might exhibit changes in density due to structural modifications caused by pressure.
6) **Applications**: Understanding the response of solids to pressure is crucial in various fields, including materials science, geology, and engineering. It is used to study the behavior of materials under extreme conditions, for designing materials for specific purposes, and for applications like high-pressure technology and the development of new materials.

5.1.4 Pressure Effects on Densities of Liquids

The impact of pressure on the densities of liquids is more pronounced compared to solids, as liquids are generally more compressible than solids. Here is an elaboration on how pressure affects the densities of liquids.

1) **Compressibility of Liquids**: Liquids are much less compressible than gases but more compressible than solids. When pressure is applied to a liquid, its volume can be altered more readily compared to solids, resulting in changes in density.
2) **Effect on Molecular Arrangement**: Pressure affects the molecular arrangement within a liquid. As pressure increases, the intermolecular distances between molecules decrease due to compression, leading to a reduction in the liquid's volume and an increase in its density.

3) **Density Changes**: Typically, an increase in pressure leads to a decrease in the volume of the liquid, causing its density to increase. Conversely, decreasing pressure allows the liquid to expand, resulting in a decrease in density.
4) **Combined Effect of Temperature and Pressure**: Changes in temperature can also influence the effect of pressure on liquid densities. For instance, in certain cases, changes in pressure can influence the phase transition of a liquid (such as from liquid to solid), which can significantly impact its density.
5) **Phase Changes Under Pressure**: Under extremely high pressure, some liquids might undergo phase changes or exhibit different behaviors. For instance, certain substances might become more ordered or adopt denser configurations due to pressure-induced structural changes.
6) **Applications**: Understanding how pressure affects liquid densities is vital in various applications such as chemical engineering, materials science, and geophysics. It is used to model and predict behaviors in industrial processes, study the behavior of liquids in extreme conditions (like deep-sea environments), and understand the behavior of fluids in geological formations.

5.1.5 Isothermal Compressibility of Solids and Liquids

The isothermal compressibility factor describes the change in the volume of a substance in response to changes in pressure while keeping the temperature constant. It quantifies the compressibility of a substance under isothermal (constant temperature) conditions. It is represented by the symbol κ (kappa) and is defined as:

$$\kappa = -\frac{1}{V}\left(\frac{\partial V}{\partial P}\right)_T \tag{5.1}$$

Or, at constant T, we rearrange the terms and replace the partial with the total derivative:

$$\frac{dV}{V} = -\kappa dP \tag{5.2}$$

In integral form, we have:

$$\ln\left(\frac{V_f}{V_i}\right) = -\kappa(P_f - P_i) = -\kappa(\Delta P) \tag{5.3}$$

Alternatively, we can write Eq. (5.3) as:

$$V_f = V_i e^{-\kappa(\Delta P)} \tag{5.4}$$

where:

- κ is the isothermal compressibility factor.
- V_f is the final volume of the substance.
- V_i is the initial volume of the substance.
- P_f is the final pressure.
- P_i is the initial pressure.
- T is the temperature.

Notice that the minus sign is added on purpose to make κ a positive quantity as the partial derivative of volume with respect to pressure at constant temperature is, in general, a negative quantity.

5.1.5.1 Isothermal Compressibility of Solids

For most solids, the isothermal compressibility factor (κ) is very low because solids are relatively incompressible. This means that when pressure is applied to a solid at a constant temperature, there is a quite negligible change in volume. Solids typically have high bulk moduli (a measure of resistance to compression), which make them less responsive to changes in pressure. As a result, their κ values are very small, indicating that a significant increase in pressure causes only a negligible change in volume.

5.1.5.2 Isothermal Compressibility of Liquids

Liquids are more compressible than solids but much less compressible than gases. When pressure is applied to a liquid at a constant temperature, it causes a more significant change in volume compared with solids. The κ factor of a liquid represents its responsiveness to pressure changes. Higher values of κ indicate that the liquid is more vulnerable to compression, meaning a large change in pressure will lead to a pronounced change in volume.

Table 5.1 shows κ values for some selected solids and liquids. Notice that, except for liquid mercury, liquid materials do have κ values that are at least two orders of magnitude (10^2) larger than those of solids. In other words, liquids are more compressible than solids upon change in pressure.

Understanding the isothermal compressibility factor is crucial in various fields, including materials science, geology, and engineering. It helps in predicting the behavior of substances under different pressure conditions, understanding their responses to external forces, and designing materials or systems to withstand or utilize pressure changes effectively.

Table 5.1 Isothermal compressibility (κ) of some solids and liquids at room temperature (25°C).

Material	State	κ (GPa^{-1})
Aluminum	Solid[a]	13.8×10^{-3}
Carbon (diamond)	Solid[a]	1.8×10^{-3}
Copper	Solid[a]	7.3×10^{-3}
Gold	Solid[a]	5.77×10^{-3}
Iron	Solid[a]	5.9×10^{-3}
Water	Liquid[b]	0.458
Mercury	Liquid[b]	0.037
Ethanol	Liquid[b]	1.1
Glycerin	Liquid[b]	0.21

[a] https://www.knowledgedoor.com/2/elements_handbook/isothermal_compressibility.html.
[b] http://hyperphysics.phy-astr.gsu.edu/hbase/Tables/compress.html.

Example 5.1 Comparing the Volume Change of Aluminum and Water Under Pressure
At room temperature, a cubic centimeter (1 cm^3) of aluminum block is subjected to a pressure increase $\Delta P = 10$ MPa (10 million Pa). Using Eq. (5.2) and given κ from Table 5.1, we can calculate the fractional change in volume:

$$\Delta V/V = -\kappa \times \Delta P = -\left(13.8 \times 10^{-3}\ \text{GPa}^{-1}\right) \times (10\ \text{MPa})$$

$$= -\left(13.8 \times 10^{-3} \frac{1}{1000\ \text{MPa}}\right) \times (10\ \text{MPa}) = -0.138 \times 10^{-3}$$

$$\frac{\Delta V}{V} \times 100\% = -0.138 \times 10^{-3} \times 100\% = -0.0138\%$$

This means the volume of the aluminum block will decrease by 0.0138% (a quite negligible change).

On the other hand, from Table 5.1, water has $\kappa = 0.458\ \text{GPa}^{-1}$. Applying the same pressure increase (10 MPa) to a cubic centimeter of water at room temperature, we get:

$$\Delta V/V = -\kappa\, \Delta P = -\left(0.458\ \text{GPa}^{-1}\right)(10\ \text{MPa})$$

$$= -\left(0.458 \frac{1}{1000\ \text{MPa}}\right) \times (10\ \text{MPa}) = -0.458 \times 10^{-2}$$

$$\frac{\Delta V}{V} \times 100\% = -0.458 \times 10^{-2} \times 100\% = -0.458\%$$

Compared with aluminum, the volume of water will decrease by 0.458%, which is significantly larger due to its relatively higher compressibility. Water is (0.458%/0.0138%) = 33.2 times larger than aluminum in terms of volume change for the same applied pressure change.

Example 5.2 Estimating the Pressure Needed to Compress Glycerin by a Specific Amount
From Table 5.1, glycerin at 25°C has $\kappa = 0.21\ \text{GPa}^{-1}$. We want to reduce the volume of 100 mL of glycerin by 10%. Using Equation. (5.2), we get:

$$\Delta P = -\frac{1}{\kappa} \times \left(\frac{\Delta V}{V}\right) = -\frac{1}{0.21\ \text{GPa}^{-1}} \times \left(\frac{V_\text{f} - V_\text{i}}{V_\text{i}}\right)$$

$$= -4.762\ \text{GPa} \times \left(\frac{0.9V_\text{i} - V_\text{i}}{V_\text{i}}\right) = -4.762\ \text{GPa} \times \left(\frac{V_\text{i}(0.9 - 1)}{V_\text{i}}\right)$$

$$= -4.762\ \text{GPa} \times (-0.1) = 0.4762\ \text{GPa}$$

Therefore, a pressure of approximately 0.476 GPa (476 MPa) would be needed to achieve this 10% volume reduction in the liquid glycerin at room temperature.

The above two examples demonstrate how the isothermal compressibility factor quantifies the susceptibility of a substance to pressure changes and allows us to predict the volume response under varying pressure conditions

5.2 Temperature Effects on Densities of Solids

The effect of temperature on the densities of solids is largely governed by the substance's structure, composition, and phase transitions. Here is an elaboration on how temperature affects the densities of solids:

1) **Thermal Expansion**: Most solids expand when heated and contract when cooled. This is due to increased thermal energy at higher temperatures causing the atoms or molecules within the solid to vibrate more vigorously, leading to increased average interatomic distances. Consequently, this expansion generally results in a decrease in density as the volume of the solid increases. Conversely, as temperature decreases, solids tend to contract, leading to higher densities.
2) **Linear and Volume Expansion Coefficients**: Solids have coefficients of thermal expansion that describe how their dimensions change with temperature. Linear expansion coefficients quantify the change in length, while volume expansion coefficients describe the change in volume. These coefficients vary among different materials and crystal structures, influencing how a solid's density changes with temperature.
3) **Anomalous Behavior**: Some materials display anomalous behavior concerning density changes with temperature. The density of ice at 0°C is about 0.917 g/cm^3 (or 917 kg/m^3), compared with the liquid density of 999.87 kg/m^3 at 0°C. As liquid water is transformed into ice, it forms different ice cage structures (due to the presence of H-bonds), causing ice to be less dense than liquid water at the same temperature. Below 0°C, ice continues to contract, and its density increases.
4) **Phase Transitions**: Changes in temperature can induce phase transitions in solids, leading to alterations in their densities. For example, certain materials undergo transitions from one crystalline structure to another at specific temperature ranges, affecting their packing arrangement and resulting in changes in density.
5) **Applications**: Understanding how a solid's density changes with temperature is vital in various fields such as materials science, engineering, and geology. It is used to design materials for specific thermal conditions, predict expansion or contraction in structures, and study the behavior of substances under changing temperature environments.

5.3 Temperature Effects on Densities of Liquids

The effect of temperature on the densities of liquids is significant and is primarily influenced by the intermolecular forces and molecular arrangements within the liquid. Here is an elaboration on how temperature affects the densities of liquids:

1) **Thermal Expansion**: Similar to solids, most liquids expand when heated and contract when cooled. As temperature increases, the thermal energy causes the molecules within the liquid to move more vigorously. This increased motion tends to slightly increase the average intermolecular distances, leading to expansion and a decrease in density. Conversely, decreasing temperature reduces molecular motion, causing the liquid to contract and become denser. The fact that mercury in a thermometer rises or falls with changing temperature shows that the effect of temperature on liquid density is measurable. See Chapter 3 | Example 3.3, which demonstrates how the volume of mercury increases with increasing temperature.
2) **Phase Transitions**: Some liquids undergo phase transitions with changes in temperature. For instance, certain substances experience transitions from liquid to solid or liquid to gas at specific temperature ranges. These phase changes can significantly impact density. For example, when a liquid transitions to a gas (vaporization), the resulting gas usually has a much lower density than the liquid.

3) **Anomalous Behavior**: Similar to solids, certain liquids exhibit anomalous behavior regarding changes in density with temperature. Water exhibits an unusual property where it reaches its maximum density (1.0 g/cm^3) at approximately 4°C. As it cools further from 4 to 0°C, the liquid water density slightly drops from 1.000 down to 0.99987 g/cm^3.
4) **Critical Points and Phase Diagrams**: Some substances have critical temperatures at which they undergo abrupt changes in properties, including density. Near these critical points on a phase diagram, the behavior of liquids can become more complex due to phase changes or shifts in molecular arrangements.
5) **Applications**: Understanding how temperature affects the density of liquids is crucial in various fields such as chemistry, engineering, and environmental science. It's used to predict changes in volume in industrial processes, study the behavior of liquids in different temperature environments, and design systems where precise temperature control is necessary.

5.3.1 Isobaric Thermal Expansion of Solids and Liquids

The isobaric thermal expansion coefficient measures the relative change in volume of a substance with respect to changes in temperature at constant pressure. In simpler terms, it tells us how much a solid "grows" as we heat it up while keeping the pressure on it the same. The isobaric thermal expansion coefficient is denoted by β and is defined as:

$$\beta = \frac{1}{V}\left(\frac{\partial V}{\partial T}\right)_P \tag{5.5}$$

Or, at constant P, we re-arrange the terms and replace the partial with the total derivative:

$$\frac{\mathrm{d}V}{V} = \beta \mathrm{d}T \tag{5.6}$$

In integral form, we have:

$$\ln\left(\frac{V_\mathrm{f}}{V_\mathrm{i}}\right) = \beta(T_\mathrm{f} - T_\mathrm{i}) = \beta(\Delta \mathrm{T}) \tag{5.7}$$

Alternatively, we can write Eq. (5.7) as:

$$V_\mathrm{f} = V_\mathrm{i} \mathrm{e}^{\beta(\Delta \mathrm{T})} \tag{5.8}$$

where:

- β is the isobaric thermal expansion coefficient (1/°C).
- V_f is the final volume of the substance.
- V_i is the initial volume of the substance.
- T_f is the final temperature.
- T_i is the initial temperature.

Notice that β is a positive quantity as the partial derivative of volume with respect to temperature at constant pressure is, in general, a positive term.

5.3.1.1 Isobaric Thermal Expansion Coefficient of Solids

Solids generally have a positive but relatively low isobaric thermal expansion coefficient. When subjected to changes in temperature while maintaining constant pressure, most solids expand. However, compared to liquids and gases, the expansion of solids due to changes in temperature is typically small. The isobaric thermal expansion coefficient for solids varies depending on the material. Crystalline solids might have different expansion coefficients along different crystallographic directions. Table 5.2 shows β values for some solids.

Table 5.2 Isobaric thermal expansion coefficient, β, for some selected solids and liquids.

Material	Coefficient of Volume Expansion β(1/°C) or (1/K)
Solids	
Aluminum	75×10^{-6}
Brass	56×10^{-6}
Copper	51×10^{-6}
Gold	42×10^{-6}
Iron or steel	35×10^{-6}
Invar (nickel–iron alloy)	2.7×10^{-6}
Lead	87×10^{-6}
Silver	54×10^{-6}
Glass (ordinary)	27×10^{-6}
Glass (Pyrex®)	9×10^{-6}
Quartz	1×10^{-6}
Concrete, brick	36×10^{-6}
Marble (average)	21×10^{-6}
Liquids	
Ether	1650×10^{-6}
Ethyl alcohol	1100×10^{-6}
Petrol	950×10^{-6}
Glycerin	500×10^{-6}
Mercury	180×10^{-6}
Water	210×10^{-6}

Source: https://pressbooks.bccampus.ca/collegephysics/chapter/thermal-expansion-of-solids-and-liquids/.

5.3.1.2 Isobaric Thermal Expansion Coefficient of Liquids

Liquids also exhibit a positive but generally higher isobaric thermal expansion coefficient compared to solids. When temperature increases, most liquids expand, leading to a relatively larger change in volume compared to solids. Liquids typically have higher thermal expansion coefficients than solids, making them more responsive to temperature changes at constant pressure. Table 5.2 shows β values for some liquids.

Factors Affecting β:

- **Interatomic Forces**: Strong interatomic forces like covalent bonds in the diamond lead to lower β values, signifying less expansion due to the "tight grip" on the atoms. Weaker forces like van der Waals interactions in rubber allow for greater expansion as temperatures rise.
- **Crystal Structure**: Different crystal structures pack atoms in unique ways, affecting how much thermal vibrations can influence their positions and contribute to expansion. Densely packed structures like FCC and HCP have lower β compared to looser ones like simple cubic.

Applications of β:

- **Engineering Design**: β is critical for designing structures that can withstand temperature changes without cracking or buckling. For example, bridges and pipelines use materials with low β to minimize expansion and potential damage.
- **Precision Instruments**: Understanding β is crucial for maintaining accurate measurements in scientific instruments, as even small thermal expansions can affect readings.
- **Material Selection**: β guides the selection of materials for various applications based on their desired thermal expansion behavior. For instance, dental fillings need low β to avoid cracking upon contact with hot or cold beverages.

Example 5.3 Aluminum Rod Expanding due to Temperature change

An aluminum rod with an initial volume of 100.0 cm^3 is initially at a temperature of 20.0°C. When the temperature is increased to 120.0°C, the rod expands to a final volume, V_f. What is V_f?

$$V_f = V_i e^{\beta(\Delta T)} = 100\ cm^3 \times e^{75 \times 10^{-6}(120-20)} = 100 \times 1.0075 = 100.75\ cm^3$$

There is a change in volume by 0.75 cm^3 or a percent relative change of $(100.75 - 100)/(100) \times 100\,\% = 0.75\%$.

The change in volume of the aluminum rod is due to the thermal expansion of the aluminum. The isobaric thermal expansion coefficient of aluminum is $75 \times 10^{-6}\,°C^{-1}$, which means that for every 1°C increase in temperature, the volume of the rod will increase by 0.0075%.

Example 5.4 Water Volume Change in a Container due to Temperature Change

A container is filled with 100.0 cm^3 of water at a temperature of 0.0°C. When the temperature is increased to 100.0°C, what will be the final volume of water at the final temperature?

$$V_f = V_i e^{\beta(\Delta T)} = 100\ cm^3 \times e^{210 \times 10^{-6}(100-0)} = 100 \times 1.0214 = 102.14\ cm^3$$

There is a change in volume by 2.14 cm^3, or a percent relative change of $(102.14 - 100)/(100) \times 100\,\% = 2.14\%$. The change in volume of the water is due to the thermal expansion of the water. The isobaric thermal expansion coefficient of water is $210 \times 10^{-6}\,°C^{-1}$, which means that for every 1°C increase in temperature, the volume of the water will increase by 0.021%.

The above two examples show that the isobaric thermal expansion coefficient is an important property of materials. It can be used to predict how much a material will expand or contract when its temperature changes. This is important for many engineering applications, such as the design of bridges, buildings, and aircraft.

In general, if both temperature and pressure do change, then we can account for both effects via the following equation:

$$\frac{dV}{V} = \beta dT - \kappa dP \tag{5.9}$$

In integral form, we have:

$$\ln\left(\frac{V_f}{V_i}\right) = \beta(T_f - T_i) - \kappa(P_f - P_i) = \beta(\Delta T) - \kappa(\Delta P) \tag{5.10}$$

For a constant-volume (isochoric or isometric) process, we set $dV = 0$. Then, we have:

$$\beta dT - \kappa dP = 0 \rightarrow \frac{dT}{dP} = \frac{\kappa}{\beta} \tag{5.11}$$

The partial derivative is used here as we have, in general, three variables at a time: P, T, and V. One variable is expressed in terms of the other two variables.

$$\left(\frac{\partial T}{\partial P}\right)_V = \left(\frac{\Delta T}{\Delta P}\right)_V = \frac{\kappa}{\beta} \tag{5.12}$$

Equation (5.12) simply says that if an increase in pressure is in proportion with that of temperature such that the ratio amounts to κ/β, then the volume will be held constant. This is called the compensation effect. The thermal expansion in volume as a result of increasing T will be offset by the decrease in or compression of volume as a result of increasing P.

5.3.2 Density of a Solid Solution

Density in a solid solution, much like liquid mixtures, goes beyond simply averaging the densities of its individual components. It is a story of intricate interactions, non-ideality, and unexpected twists depending on the type of solution. Let us dive in:

Types of Solid Solutions:

1) **Substitutional Solid Solution**: Here, atoms of one element replace those of another in the crystal lattice, maintaining the overall structure.
2) **Interstitial Solid Solution**: Smaller atoms squeeze into empty spaces (interstices) within the host lattice, expanding the crystal structure slightly. Example: Carbon in steel fills gaps between iron atoms.

Factors Affecting Density:

- **Atomic Weight and Size**: Substituting a heavier atom or squeezing in a smaller one can alter the overall mass and packing efficiency, influencing density.
- **Crystal Structure**: The arrangement of atoms in the lattice determines how efficiently packing occurs, impacting density.
- **Non-ideality**: Similar to liquid mixtures, interactions between different atoms can lead to volume changes, either contraction or expansion, affecting density.

Predicting Density:

Unlike liquids, there is no one-size-fits-all method for predicting density in solid solutions. Different approaches are used depending on the type of solution:

1) **Vegard's Law (Substitutional Solutions)**: This empirical rule assumes a linear relationship between lattice parameters and composition, allowing density prediction based on individual component densities and atomic weights. See Example 5.5 below.
2) **Interstitial Solutions**: More complex models consider the expansion of the lattice due to interstitial atoms and the specific interactions between host and guest atoms.
3) **Experimental Data**: For many systems, accurate density values are available from handbooks or databases, often at different temperatures and compositions.

Applications:

Understanding the density of solid solutions plays a crucial role in various fields:

- **Materials Science**: Designing alloys with desired properties like strength, weight, and corrosion resistance.
- **Geochemistry**: Analyzing mineral compositions and understanding geological processes.
- **Semiconductors**: Tailoring doping levels to control electrical conductivity in electronic devices.
- **Biomaterials**: Developing implants and prosthetics with biocompatible and appropriate density.

Example 5.5 Vegard's Law: A Case of Gold-Silver Alloy

Consider an Ag–Au alloy with a 40% atomic fraction of silver ($AF_{Ag} = 0.4$). We want to find its theoretical density using Vegard's Law. Notice that the atomic fraction is similar to the mole fraction, but here we have atoms of both gold and silver but not molecules of Au and Ag.

Input Data:

- **Lattice Parameter of Pure Silver (a_{Ag})**: 4.086 Å
- **Lattice Parameter of Pure Gold (a_{Au})**: 4.078 Å
- **Atomic Weights**:
  - **Silver (M_{Ag})**: 107.87 g/(mol of atoms) = 107.87 g/(mol)
  - **Gold (M_{Au})**: 196.97 g/(mol of atoms) = 196.97 g/(mol)

Solution #1:

1) **Calculate the Average Lattice Parameter (a)**:

$$a = \left(1 - AF_{Ag}\right) \times a_{Au} + AF_{Ag} \times a_{Ag} = (1 - 0.4) \times 4.078\,\text{Å} + 0.4 \times 4.086\,\text{Å} = 4.0812\,\text{Å}$$

2) **Calculate the Theoretical Density (ρ)**:

Assuming a face-centered cubic (FCC) structure with four atoms per unit cell:

$$\text{Volume of a cubic unit cell} = a^3 = \left(4.0812\,\text{Å}\right)^3 = 68.0\,\text{Å}^3/\text{cell}$$

$$\frac{\text{Volume}}{\text{cell}} = \frac{68.0\,\text{A}^3}{\text{cell}} \times \left(\frac{1 \times 10^{-8}\text{cm}}{\text{A}}\right)^3 = 6.8 \times 10^{-23}\frac{\text{cm}^3}{\text{cell}}$$

$$\frac{\text{Atomic mass}}{\text{cell}} = 4\frac{\text{atom}}{\text{cell}}\left(0.4 \times 107.87\frac{\text{g}}{\text{mol}} + 0.6 \times 196.97\frac{\text{g}}{\text{mol}}\right)$$

Divide by Avogadro's number to convert from mol into atoms:

$$\frac{\text{Atomic mass}}{\text{cell}} = 4\frac{\text{atom}}{\text{cell}} \times 161.33\frac{\text{g}}{\text{mol}} \times \frac{1\ \text{mol}}{6.023 \times 10^{23}\text{atom}} = 1.071 \times 10^{-21}\frac{\text{g}}{\text{cell}}$$

$$\rho = \frac{\text{Atomic mass/cell}}{\text{Volume/cell}} = \frac{1.071 \times 10^{-21}\frac{\text{g}}{\text{cell}}}{6.8 \times 10^{-23}\frac{\text{cm}^3}{\text{cell}}} = 15.76\frac{\text{g}}{\text{cm}^3}$$

Let us convert the atomic fraction into weight fraction.

Basis: 10 mol of gold metal alloy

$$4\ \text{mol} \times \frac{107.87\ \text{g Ag}}{\text{mol}} = 431.5\ \text{g Ag}$$

$$6\ \text{mol} \times \frac{196.97\ \text{g Au}}{\text{mol}} = 1{,}181.8\ \text{g Au}$$

$$x_{\text{Au}} = \frac{1{,}181.8\ \text{g Au}}{431.5\ \text{g Ag} + 1{,}181.8\ \text{g Au}} = 0.73$$

The measured density for Ag-Au alloys with 73 wt.% gold is reported as 15.75 g/cm^3 by Kraut, J. and Stern, W.B. (2000). The density of Au–Ag–Cu alloys and its calculation from the chemical composition. Gold Bulletin 33(2): 52–55. http://doi:10.1007/BF03216580.

$$\text{PRE\%} = \frac{|\text{True Value} - \text{Predicted Value}|}{\text{True Value}} \times 100\% \tag{5.13}$$

$$\text{Percent Relative Error (PRE)\%} = \frac{(|15.76 - 15.75|)}{15.75} \times 100\% = 0.06\% \ll 10\%$$

Solution #2:

Let us apply the mass-fraction weighted formula for estimating the average density of a solid solution (See upcoming section: Density of a Liquid Mixture), given that the density of silver is 10.49 g/cm^3 and that of gold is 19.3 g/cm^3:

$$\frac{1}{\rho_{\text{solution}}} = V_{\text{solution}} = \sum_{i=1}^{N} x_i V_i \cong \sum_{i=1}^{N} x_i \frac{1}{\rho_i^o} \cong \sum_{i=1}^{N} \frac{x_i}{\rho_i^o} \cong \frac{0.27}{10.49} + \frac{0.73}{19.3} \cong 0.06356$$

$$\rho_{\text{solution}} \cong 15.73\frac{\text{g}}{\text{cm}^3}$$

The estimated solution density, based on the mass fraction and pure density of each individual component, is practically the same as the previously estimated density, and both match with the measured or experimental value.

NOTE #2

Vegard's law is an approximate method that can be used in the absence of accurate experimental data. It is more accurate when the metal alloy is made of the same crystal lattice unit type (i.e., FCC, BCC, or HCP). I tried it with zinc–copper alloy, but it gave an erroneous estimation as both metals are of different lattice units. It applies well to solid solutions with small size differences between atoms and minimal non-ideality. Deviations from linearity are often more pronounced at the extreme ends of the composition range due to the increasing likelihood of phase transitions, structural changes, and altered atomic interactions. In the intermediate compositions, where the mix of elements is more balanced, Vegard's Law tends to be more accurate. Moreover, the presence of defects or impurities can also contribute to deviations from Vegard's Law. These can disrupt the regular lattice arrangement, affecting the overall properties of the alloy.

5.3.3 Density of a Liquid Mixture

The density of a liquid mixture is an interesting topic with several layers of complexity. It goes beyond simply averaging the densities of its individual components.

5.3.3.1 Simple Calculations

1) **Mass–Volume Ratio**: At its core, density (ρ) is the mass (m) per unit volume (V): $\rho = m/V$. So, for a mixture, you can calculate the total mass (sum of individual masses) and total volume (sum of individual volumes, assuming ideal mixing). However, this is often an inaccurate oversimplification.
2) **Volume Change (Non-ideality)**: In reality, mixing liquids can induce volume changes due to interactions between molecules, especially when it comes to solute–solvent interactions. Imagine squeezing tennis balls into a container – they occupy less space together than the sum of their individual volumes. Similarly, some mixtures contract, while others expand upon mixing.

5.3.3.2 Factors Affecting Non-ideality

- **Intermolecular Forces**: Strong interactions like hydrogen bonding (e.g., ethanol–water) can lead to contraction. Conversely, weak forces or repulsion cause expansion.
- **Molecular Packing**: The size and shape of molecules influence how they fit together, impacting their final volume.
- **Concentration**: The proportions of each component can affect the degree of non-ideality.

5.3.3.3 Predicting Density

Several methods attempt to predict the density of mixtures beyond the simple average:

- **Mixing Rules**: These empirically derived equations, like the rule of mixtures or excess volume models, incorporate non-ideality using interaction parameters specific to the components.
- **Density Data Sources**: Handbooks and online databases often provide experimental density data for various mixtures at different temperatures and compositions.

- **Thermodynamic Models**: Advanced thermodynamic models account for molecular interactions and predict not only density but also other properties like excess enthalpy and entropy.

Applications:
Understanding the density of liquid mixtures is crucial in various fields:

- **Chemical Engineering**: Designing distillation columns, pipelines, and storage tanks.
- **Food Science**: Formulating recipes, optimizing mixing processes, and predicting shelf life.
- **Environmental Science**: Modeling pollutant transport and fate in water bodies.
- **Pharmaceuticals**: Developing drug delivery systems and ensuring dosage precision.

Estimating the density of a mixture of liquids or gases involves considering the densities of the individual components and their respective proportions within the mixture. The overall density of a mixture can be calculated using various methods depending on the nature of the mixture.

Weighted by the mass-fraction of each component, the solution or mixture density is given by:

$$\frac{1}{\rho_{\text{solution}}} = \frac{1}{\rho_{\text{mixture}}} = V_{\text{solution}} = \sum_{i=1}^{N} x_i V_i = \sum_{i=1}^{N} x_i \frac{1}{\rho_i} = \sum_{i=1}^{N} \frac{x_i}{\rho_i} \cong \sum_{i=1}^{N} \frac{x_i}{\rho_i^o} \tag{5.14}$$

However, the following equation can be used as a linear approximation:

$$\rho_{\text{solution}} = \sum_{i=1}^{N} x_i \rho_i = \sum_{i=1}^{N} x_i \rho_i^o \tag{5.15}$$

Notice that whether V_i or ρ_i appears in Equations (5.14) and (5.15) must be the partial specific property of the participating component; however, for lack of such data, as they are complex functions of P, T, and solution composition, we replace the partial with the pure property, with the understanding that a penalty is levied on the accuracy of the model.

NOTE #3
Equations (5.14) and (5.15) can be applied to a solid, liquid, or gas solution/mixture.

5.3.4 Real Mixtures or Solutions

Ideal solutions are theoretical models that simplify the behavior of mixtures. Real solutions, on the other hand, exhibit complexities often absent in the ideal picture. These differences manifest in various ways, including the behavior of partial molar properties like volume and enthalpy.

Here are some key features that differentiate real solutions from ideal ones, specifically regarding partial molar volume (V) and partial molar enthalpy (H):

Ideal Solutions:

- **No Intermolecular Interactions**: Ideal solutions assume no attractive or repulsive forces between molecules of different components. All interactions are like-pair interactions (e.g., A–A and B–B), leading to no energetic changes from mixing.

- **Equal Partial Molar Properties**: V and H of each component are equal to their molar volumes and enthalpies in the pure state, respectively. This means the volume or enthalpy contribution of a single molecule is independent of the solution's composition.
- **Volume Additivity**: The total volume of the solution is simply the sum of the individual volumes of its components, expressed as their mole fractions and molar volumes.
- **Zero Excess Properties**: No excess volume or excess enthalpy exists; all thermodynamic properties are directly predicted from pure component properties.

Real Solutions:

- **Intermolecular Interactions**: Real solutions involve attractions and repulsions between unlike molecules (A–B interactions) in addition to like-pair interactions. These interactions can lead to volume changes, energy changes, and deviations from ideal behavior.
- **Composition-dependent Partial Molar Properties**: V and H of each component can vary depending on the solution's composition, in addition to T and P. The presence of other components can influence the volume and energy associated with a single molecule of a specific type.
- **Non-additivity of Volume**: The total volume of the solution might not be the simple sum of individual component volumes due to volume changes upon mixing. This volume change can be positive (expansion) or negative (contraction).
- **Nonzero Excess Properties**: Excess volume and excess enthalpy are defined as the difference between the actual property value and the ideal value predicted from pure component properties. Nonzero excess properties indicate the presence of non-ideal interactions.

Example 5.6 Mixing Ethanol and Water

Using two 250 mL graduated cylinders filled precisely to the mark with water and pure ethanol, initially at room temperature, a temperature probe records their temperature. Subsequently, the temperature probe is transferred to an empty 500 mL graduated cylinder. The contents of the two smaller cylinders are then simultaneously poured into the larger cylinder for thorough mixing. Following the mixing process, the temperature of the combined solution increases by approximately 8°C. Simultaneously, the volume decreases noticeably to 480 mL, evident on the scale of the 500 mL cylinder. Reference: (https://sciencedemonstrations.fas.harvard.edu/presentations/mixing-ethanol-and-water).

In this experiment, the volume decrease upon mixing water with ethanol indicates a negative excess volume, and the observed temperature increase indicates also a negative excess enthalpy. This simply means that the ethanol–water interactions involve stronger attractive forces between the dissimilar molecules than either between similar water–water or between similar ethanol–ethanol molecules, leading to a decrease both in volume and enthalpy where the latter is manifested via the release of energy in the form of heat.

$$\Delta V_{\text{mixing}} = V_{\text{after mixing}} - V_{\text{before mixing}} = V_{\text{mixture}} - \left[V^{o}_{\text{Ethanol}} + V^{o}_{\text{Water}}\right] \tag{5.16}$$

$$\Delta V_{\text{mixing}} = n_{\text{total}}\Delta\tilde{V}_{\text{mixing}} = n_{\text{total}}\tilde{V}_{\text{mixture}} - \left[n_{\text{Ethanol}}\tilde{V}^{o}_{\text{Ethanol}} + n_{\text{Water}}\tilde{V}^{o}_{\text{Water}}\right] \tag{5.17}$$

For water – ethanol mixing, $\Delta V_{\text{mixing}} = n_{\text{total}}\Delta\hat{V}_{\text{mixing}} = 480\,\text{mL} - 500\,\text{mL} = -20\,\text{mL} < 0$.

$$\Delta H_{\text{mixing}} = H_{\text{after mixing}} - H_{\text{before mixing}} = H_{\text{mixture}} - \left[H^{o}_{\text{Ethanol}} + H^{o}_{\text{Water}}\right] \tag{5.18}$$

$$\Delta H_{\text{mixing}} = n_{\text{total}}\Delta\tilde{H}_{\text{mixing}} = n_{\text{total}}\tilde{H}_{\text{mixture}} - \left[n_{\text{Ethanol}}\tilde{H}^{o}_{\text{Ethanol}} + n_{\text{Water}}\tilde{H}^{o}_{\text{Water}}\right] \tag{5.19}$$

For water–ethanol mixing, $\Delta H_{\text{mixing}} = Q < 0$ (i. e. , Exothermic mixing).

It is worth mentioning here that both $\tilde{V}_{\text{mixture}}$ and $\tilde{H}_{\text{mixture}}$ must be expressed in terms of partial molar volume and partial molar enthalpy of participating components, respectively. Since this is an introductory course, the solution properties will not be presented here; instead, they are covered in thermodynamics of solutions in a junior-level thermodynamics course.

NOTE #4

An ideal solution is characterized by: $\Delta V_{\text{mixing}} = \Delta H_{\text{mixing}} = 0$. A real solution has the properties of $\Delta V_{\text{mixing}} \neq 0$ and also $\Delta H_{\text{mixing}} \neq 0$. Of course, ΔV_{mixing} and ΔH_{mixing} can be positive or negative for a real solution. In general, mixing two or more similar species, to form a solution, will result in an ideal liquid mixture. For example, mixing benzene with toluene will form an ideal benzene-toluene mixture. So, does this case with a methanol–ethanol mixture or *n*-heptane–octane mixture? The similarity among chemical species is exemplified in terms of equal numbers and the same type of functional groups (i.e., —OH, —OOH, —O—, —NH_2, =, ≡ ,etc.) to a large extent and of the length of the hydrocarbon chain (or tail) of the molecule, to a small extent. The deviation from an ideal liquid mixture (or the formation of a nonideal liquid mixture) arises from the mixing of chemical species characterized by different chemical activities (i.e., different functional groups) and by different sizes of homologous series or both. For example, mixing ethanol with water will form a nonideal liquid mixture, as water and ethanol are of two different categories. Mixing of ethanol with hexanol will also show some sort of deviation from ideality compared with mixing of ethanol with propanol, next to ethanol in a homologous series of alcohols.

Source: Al-Malah, K. (2022). *Aspen Plus: Chemical Engineering Applications*, 2. Wiley (December 2022). https://www.wiley.com/en-ae/Aspen+Plus:+Chemical+Engineering+Applications,+2nd+Edition-p-9781119868699.

5.4 Gases

5.4.1 Kinetic Theory of Gases

The kinetic theory of gases is a fundamental model used to explain the behavior of gases based on the motion of their constituent particles. It establishes several key principles that help in understanding how gases behave under different conditions.

Particle Motion and Temperature: According to this theory, gases consist of tiny particles (atoms or molecules) in constant, random motion. When the temperature increases, the average kinetic energy of these particles also increases. This means that at higher temperatures, gas particles move faster, colliding with each other and with the walls of the container more frequently and with greater force. The increased kinetic energy leads to the expansion of the gas, causing it to occupy more space and resulting in a decrease in density.

Effect of Intermolecular Forces: The kinetic theory also considers intermolecular forces, which are attractive forces between particles. Stronger intermolecular forces tend to pull particles closer together. In the context of gases, gases with stronger intermolecular forces will have their particles more tightly bound, partially counteracting the tendency of the particles to move apart due to their kinetic energy. This can lead to higher densities compared to gases with weaker intermolecular forces at the same temperature. For instance, water vapor, which has stronger intermolecular forces due to hydrogen bonding, will have a slightly higher density compared to gases like helium at the same temperature.

Adaptation to Understanding Liquid Densities: While the kinetic theory primarily focuses on gases, its principles can be extended to comprehend liquid densities to some extent. In liquids, particles are more closely packed than in gases. Temperature affects liquid densities similarly to gases – higher temperatures lead to increased kinetic energy, causing particles to move faster and thus slightly decreasing density. However, the impact of temperature on liquid density is not as pronounced as in gases due to the closer proximity and stronger intermolecular forces in liquids.

Pressure and Collisions: The kinetic theory goes beyond simply stating that gas particles move. It relates their motion to pressure. Imagine gas particles constantly bombarding the walls of their container like tiny balls. Each collision exerts a force, and the cumulative effect of these collisions on the container's walls is what we experience as pressure. The higher the average kinetic energy (and therefore speed) of the particles, the more frequent and forceful their collisions, leading to higher pressure. This explains why pressure increases with temperature.

Explaining Gas Laws: The kinetic theory forms the foundation for understanding several gas laws. For example, Boyle's Law (pressure and volume are inversely proportional) can be visualized as squeezing the container, forcing particles closer together, and increasing their collision frequency, therefore increasing pressure. Similarly, Charles' Law (pressure and temperature are directly proportional) can be explained by the increased kinetic energy and collision rate at higher temperatures.

Beyond Ideal Gases: While the basic principles apply to most gases, the theory initially assumes ideal gas behavior, where particles have no size or attractive forces. Real gases deviate from this ideal behavior, especially at high-pressures and low-temperatures. The theory can be refined to account for these deviations using concepts like van der Waals forces and corrections for particle size.

Applications: The kinetic theory has far-reaching applications. It guides engineers in designing gas turbines, refrigeration systems, and rocket engines. It helps chemists predict gas transport phenomena and analyze reaction rates. In meteorology, understanding gas behavior is crucial for weather prediction and climate modeling.

Further Recommended Exploration:

- **Maxwell–Boltzmann Distribution**: This describes the probability distribution of a gas's particle velocities at a given temperature, offering a deeper understanding of particle movement.
- **Mean Free Path**: This refers to the average distance a particle travels between collisions, providing insight into how frequently particles interact with each other.
- **Kinetic Theory and Other States of Matter**: The principles extend to understanding liquids and solids to some extent, highlighting the connections between different states of matter.

5.4.2 The Ideal Gas Equation of State

The ideal gas equation of state originates from the kinetic theory of gases through the presumption that gas molecules possess insignificant volume, exert no mutual forces, and undergo perfectly elastic collisions with the container walls. Typically, this equation is presented in the following format:

$$PV = nRT \text{ or } P\dot{V} = \dot{n}RT, \tag{5.20}$$

where

P = *absolute* pressure of a gas
$V(\dot{V})$ = volume (volumetric flow rate) of the gas
$n(\dot{n})$ = number of moles (molar flow rate) of the gas
R = the *gas constant,* whose value depends on the units of P, V, n, and T
T = *absolute* temperature of the gas

The equation can also be written as

$$P\tilde{V} = RT \tag{5.21}$$

where $\tilde{V} = V/n$ (or $\dot{V}/\dot{n}$) is the molar volume of the gas.

A gas whose *PVT* behavior is well described by Eq. (5.20) or (5.21) is said to behave as an ideal gas or a perfect gas. The use of this equation does not require knowledge of the gas species: **1 mol of an ideal gas at 0°C and 1 atm** *occupies* **22.415 L**, whether the gas is argon, hydrogen, nitrogen, a mixture of ethane and air, or any other single species or mixture of gases.

Therefore, ***R*** can be calculated via the same equation as:

$$R = \frac{PV}{nT} = \frac{1\,\text{atm} \times 22.414\,\text{L}}{1\,\text{mol} \times 273.15\,\text{K}} = 0.082057\frac{\text{atm} \cdot \text{L}}{\text{mol} \cdot \text{K}} = 82.057\frac{\text{atm} \cdot \text{cm}^3}{\text{mol} \cdot \text{K}} = 8.314\frac{\text{J}}{\text{mol} \cdot \text{K}} \tag{5.22}$$

NOTE #5

The ideal gas equation of state is an approximation. It works well under some conditions, generally speaking, at temperatures above 0°C and pressures below about 1 atm., but at other conditions, its use may lead to significant errors. ***A percent relative error (PRE) less than 1%*** may be expected if the quantity *RT/P* (the *ideal molar volume*) satisfies one of the following two criteria:

$$\text{PRE} < 1\% \text{if } \tilde{V}_{\text{ideal}} = \frac{RT}{P} > 5\frac{\text{L}}{\text{mol}} = 80\frac{\text{ft}^3}{\text{lb}-\text{mole}} \text{(diatomic gases)} \tag{5.23a}$$

$$\text{PRE} < 1\% \text{if } \tilde{V}_{\text{ideal}} = \frac{RT}{P} > 20\frac{\text{L}}{\text{mol}} = 320\frac{\text{ft}^3}{\text{lb}-\text{mole}} \text{(other gases)} \tag{5.23b}$$

PRE is given by Eq. (5.13):

$$\text{PRE\%} = \frac{|\text{True Value} - \text{Predicted Value}|}{\text{True Value}} \times 100\%$$

Example 5.7 The Ideal Gas Equation of State
Two hundred eighty (280) grams of nitrogen is stored in a rigid container at 27.0°C and 2.0 atm.

1) Assuming ideal gas behavior, calculate the *container volume* in liters.

$$n_{N2} = \frac{280\ \text{g}}{28\ \frac{\text{g}}{\text{mol}}} = 10.0\ \text{mol}$$

$$PV = nRT$$

$$2.0\ \text{atm} \times V = 10.0\ \text{mol} \times 0.082057 \frac{\text{atm} \cdot \text{L}}{\text{mol} \cdot \text{K}} \times 300.15\ \text{K}$$

$$V_{\text{container}} = \frac{10.0 \times 0.082057 \times 300.15}{2.0} = 123.15\ \text{L}$$

2) Verify that the ideal gas equation of state is a good approximation for the given conditions.

$$\tilde{V}_{\text{ideal}} = \frac{RT}{P} = \frac{0.082057 \frac{\text{atm} \cdot \text{L}}{\text{mol} \cdot \text{K}} \times 300.15\ \text{K}}{2.0\ \text{atm}} = 12.31 \frac{\text{L}}{\text{mol}} > 5 \frac{\text{L}}{\text{mol}}$$

Since the calculated value of $\tilde{V}_{\text{ideal}}$ exceeds the criterion value of 5 L/mol, the ideal gas equation of state should yield an error of less than 1%.

For N_2 gas at the same T_1 and P_1, $\tilde{V}_{\text{real}} = 12.314$ L/mol, using the Soave–Redlich–Kwong (SRK) Equation of State. See Section 5.5.

$$\text{PRE}\% = \frac{|12.314 - 12.31|}{12.314} \times 100\% = 0.0325\% < 1\%$$

Example 5.8 Application of the Ideal Gas Law
If the gas in Example 5.1 is *heated* to a new temperature, T_2, equal to 67°C, what will be the value of the new pressure, P_2? *Notice that* the volume of the container in this case is fixed (i.e., held constant) $V_1 = V_2$.

Solution
Since the gas is enclosed in a *container*, then the number of moles ($n_{N2} = 10.0$ mol) at *state 1* will be the same as that at *state 2*.

Apply the ideal gas law twice:

$$n_{N2} = 10.0\ \text{mol} = \frac{P_1 V_1}{RT_1}$$

$$n_{N2} = 10.0\ \text{mol} = \frac{P_2 V_2}{RT_2}$$

Since the left-hand side of both the equations is the same, then the right-hand side will be the same.

$$\frac{P_1 V_1}{RT_1} = \frac{P_2 V_2}{RT_2} \quad \rightarrow \frac{P_1}{T_1} = \frac{P_2}{T_2} \quad \rightarrow \frac{T_2}{T_1} = \frac{P_2}{P_1} \quad \rightarrow \frac{340.15\ \text{K}}{300.15\ \text{K}} = \frac{P_2}{2.0\ \text{atm}}$$

$$P_2 = 2.0\ \text{atm} \times \frac{340.15\ \text{K}}{300.15\ \text{K}} = 2.267\ \text{atm}$$

We can check the condition of ideality at state 2:

$$\tilde{V}_{\text{ideal}} = \frac{RT_2}{P_2} = \frac{0.082057\dfrac{\text{atm}\cdot\text{L}}{\text{mol}\cdot\text{K}} \times 340.15\ \text{K}}{2.267\ \text{atm}} = 12.31\frac{\text{L}}{\text{mol}} > 5\frac{\text{L}}{\text{mol}}$$

The $\tilde{V}_{\text{ideal}}$ value did not change simply because the T/P ratio is the same for both states.

5.4.3 Ideal Gas Mixtures

Assume n_A moles of substance A, n_B moles of B, n_C moles of C, and so on, are all contained in volume V at a temperature T and a total pressure P. The partial pressure p_A and pure component volume v_A of A in the mixture are defined as:

p_A: The pressure that would be exerted by n_A moles of A alone in the same total volume V at the same temperature T.

$$\text{The definition of partial pressure}: p_A V = n_A RT \tag{5.24}$$

$$PV = nRT$$

Divide the first by the second equation, side by side, we get:

$$\frac{p_A V}{PV} = \frac{n_A RT}{nRT} \quad \rightarrow \frac{p_A}{P} = \frac{n_A}{n} = y_A \quad \rightarrow p_A = y_A P \tag{5.25}$$

In words, *the partial pressure of a component in an ideal gas mixture is the mole fraction of that component times the total pressure.* Moreover, since $y_A + y_B + \cdots = 1$,

$$p_A + p_B + p_C + \ldots = y_A P + y_B P + y_C P + \ldots = (y_A + y_B + y_C + \ldots)P = (1)P \tag{5.26}$$

*The partial pressures of the components of an ideal gas mixture add up to the total pressure (**Dalton's law**).*

v_A: The volume that would be occupied by n_A moles of A alone at the total pressure P and temperature T of the mixture. A similar series of calculations can be performed for pure-component volumes:

$$Pv_A = n_A RT$$

$$\frac{Pv_A}{PV} = \frac{n_A RT}{nRT} \rightarrow \frac{v_A}{V} = \frac{n_A}{n} = y_A \quad \rightarrow v_A = y_A V$$

$$v_A + v_B + v_C + \ldots = y_A V + y_B V + y_C V + \ldots = (y_A + y_B + y_C + \ldots)V = (1)V \tag{5.27}$$

Equation (5.27) is called **Amagat's law**.

The quantity v_A/V is the volume fraction of A in the mixture, and 100 times this quantity gives the volume percentage (% v/v) of this component.

5.5 Real Gases

Nonideal gases deviate from the behavior predicted by the ideal gas law due to various factors not accounted for in the ideal gas equation. While the ideal gas law assumes certain idealized conditions, such as negligible volume of gas particles and no intermolecular forces, real gases deviate from these assumptions under certain conditions, leading to deviations from ideality.

1) **Volume of Gas Particles**: In the ideal gas law, gas particles are assumed to have negligible volume compared to the volume of the container they occupy. However, at high-pressures or low temperatures, gas particles occupy a non-negligible volume. This results in a decrease in the available space for particle movement, leading to deviations from ideal behavior.
2) **Intermolecular Forces**: Ideal gas behavior assumes that gas molecules do not exert forces upon each other. In reality, gas molecules experience intermolecular forces such as van der Waals forces, dipole–dipole interactions, or hydrogen bonding. These forces become significant at lower temperatures and higher pressures, causing gas molecules to deviate from ideal behavior by attracting each other, affecting the pressure and volume relationships within the gas.

There are different means or models by which investigators manage to account for deviation from ideality. Some are shown below.

5.5.1 Virial Equation of State

A virial equation of state expresses the quantity PV/RT as a power series in the inverse of a specific volume:

$$\frac{P\tilde{V}}{RT} = 1 + \frac{B}{\tilde{V}} + \frac{C}{\tilde{V}^2} + \frac{D}{\tilde{V}^3} + \ldots \tag{5.28}$$

where B, C, and D are functions of temperature and are known as the second, third, and fourth virial coefficients, respectively. From a statistical mechanics standpoint, this equation of state has a theoretical basis, but the procedures for estimating the coefficients are not well established, especially for C, D, etc. Note that the ideal gas equation of state is obtained if $B = C = D = \cdots = 0$. Truncating the virial equation of state after the second term yields

$$\frac{P\tilde{V}}{RT} = 1 + \frac{B}{\tilde{V}} \tag{5.29}$$

Notice that Eq. (5.29) is ***not applicable for polar compounds*** (asymmetrical compounds with a nonzero dipole moment, such as water).

The following procedure may be used to estimate V or P for a given T for a ***nonpolar*** species (one with a dipole moment close to zero, such as hydrogen and oxygen and all other molecularly symmetrical compounds).

1) Calculate $T_r = T/T_c$.
2) Look up the Pitzer acentric factor from Table 5.3, w. A parameter that reflects the geometry and polarity of a molecule.
3) Estimate B using the following equations:

$$B_0 = 0.083 - \frac{0.422}{T_r^{1.6}} \tag{5.30}$$

Table 5.3 The value of the acentric factor, w, for some selected gases.

Compound	w	Compound	w	Compound	w
Ammonia	0.25	Ethane	0.098	Oxygen	0.021
Argon	−0.004	Hydrogen Sulfide	0.1	Propane	0.152
Carbon dioxide	0.225	Methane	0.008	Sulfur dioxide	0.251
Carbon monoxide	0.049	Methanol	0.559	Water	0.344
Chlorine	0.073	Nitrogen	0.04		

$$B_1 = 0.139 - \frac{0.172}{T_r^{4.2}} \tag{5.31}$$

$$B = \frac{RT_c}{P_c}(B_0 + \omega B_1) \tag{5.32}$$

4) Substitute B in Eq. (5.29) and solve for the molar volume $\hat{V}$.

Example 5.9 The Virial Equation of State

Let us solve Example 5.7 using the virial equation of state.

Two hundred eighty (280) grams of nitrogen is stored in a rigid container at 27.0°C and 2.0 atm.

DATA: For N_2, $T_c = 126.15$ K, $P_c = 34$ bar $= 33.55$ atm, and $\omega = 0.039$

$$n_{N2} = \frac{280\ \text{g}}{28\ \frac{\text{g}}{\text{mol}}} = 10.0\ \text{mol}$$

$$T_r = \frac{T}{T_c} = \frac{300.15\ \text{K}}{126.15\ \text{K}} = 2.38$$

$$B_0 = 0.083 - \frac{0.422}{T_r^{1.6}} = 0.083 - \frac{0.422}{2.38^{1.6}} = -0.0224$$

$$B_1 = 0.139 - \frac{0.172}{T_r^{4.2}} = 0.139 - \frac{0.172}{2.38^{4.2}} = 0.1345$$

$$B = \frac{RT_c}{P_c}(B_0 + \omega B_1) = \frac{0.08206\frac{\text{atm.L}}{\text{mol}\cdot\text{K}} \times 126.15\ \text{K}}{33.55\ \text{atm}}(-0.0224 + 0.04 \times 0.1345)$$

$$= 0.30855\frac{\text{L}}{\text{mol}} \times (-0.01702) = -0.00525\frac{\text{L}}{\text{mol}}$$

$$\frac{P\tilde{V}}{RT} = 1 + \frac{B}{\tilde{V}} = 1 + \frac{(-0.00525)}{\tilde{V}} \rightarrow \frac{2.0\ \text{atm} \times \tilde{V}}{0.08206 \times 300.15} = 1 - \frac{0.00525}{\tilde{V}}$$

$$0.0812\tilde{V} + \frac{0.00525}{\tilde{V}} = 1.0 \rightarrow \frac{0.0812\tilde{V}^2 + 0.00525}{\tilde{V}} = 1.0$$
$$\rightarrow 0.0812\tilde{V}^2 - \tilde{V} + 0.00525 = 0$$

Notice that the two roots for the quadratic equation $Ax^2 + Bx + C = 0$ can be found using the quadratic formula: $x = (-B \pm \sqrt{(B^2 - 4AC)})/2A$

$$\tilde{V}_{1,2} = \frac{1 \mp \sqrt{1 - 4(0.0812)(0.00525)}}{2(0.0812)} = \frac{1 \mp 0.99915}{0.1624}$$

The first root will be: $\tilde{V}_1 = \frac{1 - 0.99915}{0.1624} = 0.00523\ \text{L/mol}$. One can say that $\tilde{V}_1 = 0.00523\ \text{L/mol}$ accounts for liquid molar volume at the given T and P. On the other hand, the second root will be: $\tilde{V}_2 = \frac{1 + 0.99915}{0.1624} = 12.31\ \text{L/mol}$which accounts for the vapor molar volume.

$$V_{\text{container}} = n \times \tilde{V}_2 = 10\ \text{mol} \times 12.31 \frac{\text{L}}{\text{mol}} = 123.1\ \text{L}$$

Compare with the ideal gas volume: $PV = nRT$

$$2.0\ \text{atm} \times V = 10.0\ \text{mol} \times 0.082057 \frac{\text{atm} \cdot \text{L}}{\text{mol} \cdot \text{K}} \times 300.15\ \text{K}$$

$$V_{\text{container}} = \frac{10.0 \times 0.082057 \times 300.15}{2.0} = 123.15\ \text{L}$$

Both estimated volumes are practically the same.

5.5.2 Cubic Equations of State (E.o.S.)

Several analytical *PVT* relationships are referred to as cubic equations of state because, when expanded, they yield third-order equations for the molar volume. In general, the additional intruded parameters are expressed in terms of the critical properties of a given species. Such parameters are meant to account for deviation from ideality as the fluid material is put under nonideal conditions, like high pressure and/or low temperature. Some cubic equations of state are shown below:

5.5.3 Van der Waals (vdW) Equation of State

The vdW equation of state is the earliest of these expressions, and it remains useful for discussing deviations from ideal gas behavior.

$$P = \frac{RT}{\tilde{V} - b} - \frac{a}{\tilde{V}^2} \tag{5.33}$$

$$a = \frac{27R^2T_c^2}{64P_c} \tag{5.34a}$$

$$b = \frac{RT_c}{8P_c} \tag{5.34b}$$

In the vdW derivation, the term $a/\tilde{V}^2$ accounts for attractive forces between molecules and b term is a correction accounting for the volume occupied by the molecules themselves.

NOTE #6

Equation (5.33) is an algebraic equation that is implicit in $\tilde{V}$ should we know ahead P and T. In general, it has three possible roots because it is a cubic equation in $\tilde{V}$.

$$\tilde{V}_{IG} = \tilde{V}_{RG} - b \qquad \rightarrow \tilde{V}_{IG} < \tilde{V}_{RG} = \tilde{V}_{vdW}$$

$$P_{IG} = P_{RG} + \frac{a}{\tilde{V}^2} \qquad \rightarrow P_{IG} > P_{RG} = P_{vdW}$$

IG stands for ideal gas, and RG for real gas.

The following code shows a demonstration of how to run MATLAB code shown in Appendix C | "C.2 MATLAB Code for Van der Waals (vdW) Equation of State" section. The **bold font** indicates the user's input parameters and generated results.

```
>> Van_Der_Waals_HC
Input "1" to search by Compound Name "2" to search by Formula: 1
Input the name of the Compound, say methane: Water
{[433]}     {'Water'}     {'H2O'}
Input the number for the chosen compound, say "1" for methane: 433
Input the absolute temperature in Kelvin 300
Input the absolute pressure in atm 1
{'The material is:'}     {'Water'}
Tr = 0.46346    Pr = 0.0045924
Vg=24.4241 L/mol
VL=0.036485 L/mol
Experimental VL = 0.01802 L/mol
PRE in VL=102.4675 %
```

Example 5.10 Concept of P_{vdW}, P_{IG}, $\tilde{V}_{vdW}$, and $\tilde{V}_{IG}$

Given the fluid to be propane, C_3H_8, and the molar volume, $\tilde{V}$, as 10 L/mol, temperature $T = 27°C$, find the pressure as given by vdW E.o.S. (Eq. 5.33), utilizing also Eqs. (5.34a) and (5.34b) for evaluating parameters a and b. Compare the pressure and molar volume, given by the ideal gas law vs. those given by vdW E.o.S.

DATA: $MW_{C3H8} = 44.1$ g/mol $T_c = 369.85$ K $P_c = 41.92$ atm.

Calculate PRE for both $\tilde{V}$ and P as predicted by the ideal gas, assuming the true value for both is given by vdW E.o.S.

Solution

NOTICE that Eq. (5.33) will give a straight forward value for P should we know ahead the value of $\tilde{V}$ and T. In other words, it becomes an algebraic equation that is explicit in P.

$$a = \frac{27R^2T_c^2}{64P_c} \qquad b = \frac{RT_c}{8P_c}$$

$$a = \frac{27R^2T_c^2}{64P_c} = \frac{27 \times \left(0.08206 \dfrac{\text{atm.L}}{\text{mol.K}}\right)^2 \times (369.85\ \text{K})^2}{64 \times 41.92\ \text{atm}} = 9.2699 \frac{\text{atm.L}^2}{\text{mol}^2}$$

$$b = \frac{RT_c}{8P_c} = \frac{0.08206 \dfrac{\text{atm.L}}{\text{mol.K}} \times 369.85\ \text{K}}{8 \times 41.92\ \text{atm}} = 0.0905 \frac{\text{L}}{\text{mol}}$$

$$P_{\text{vdW}} = \frac{RT}{\tilde{V} - b} - \frac{a}{\tilde{V}^2} = \frac{0.08206 \dfrac{\text{atm.L}}{\text{mol.K}} \times 300.15\ \text{K}}{10 \dfrac{\text{L}}{\text{mol}} - 0.0905 \dfrac{\text{L}}{\text{mol}}} - \frac{9.27 \dfrac{\text{atm.L}^2}{\text{mol}^2}}{10^2 \left(\dfrac{\text{L}}{\text{mol}}\right)^2} = 2.3928\ \text{atm}$$

$$\tilde{V}_{\text{IG}} = \tilde{V} - b = 10 \frac{\text{L}}{\text{mol}} - 0.0905 \frac{\text{L}}{\text{mol}} = 9.9095 \frac{\text{L}}{\text{mol}}$$

$$\left(\tilde{V}_{\text{IG}} = 9.9095 \frac{\text{L}}{\text{mol}}\right) < \left(\tilde{V}_{\text{RG}} = \tilde{V}_{\text{vdW}} = 10 \frac{\text{L}}{\text{mol}}\right)$$

$$P_{\text{IG}} = \frac{RT}{\tilde{V}_{\text{IG}}} = \frac{0.08206 \dfrac{\text{atm.L}}{\text{mol.K}} \times 300.15\ \text{K}}{9.9095 \dfrac{\text{L}}{\text{mol}}} = 2.4855\ \text{atm}$$

Notice that $P_{\text{IG}} = P_{\text{vdW}} + \dfrac{a}{\tilde{V}^2} = 2.3928\ \text{atm} + \dfrac{9.2699}{100} = 2.4855\ \text{atm} = 2.4855\ \text{atm}$ (from ideal gas law)

$$(P_{\text{IG}} = 2.4855\ \text{atm}) > (P_{\text{RG}} = P_{\text{vdW}} = 2.3928\ \text{atm})$$

$$\text{PRE}(\%)|_{\tilde{V}} = \left(\frac{10.0 - 9.9095}{10.0}\right) \times 100\% = 0.905\%$$

$$\text{PRE}(\%)|_{P} = \left(\frac{2.4855 - 2.3928}{2.3928}\right) \times 100\% = 3.87\%$$

```
Here is the executed MATLAB code via defining terms appearing in
equations (5.33)-(5.34a-b) in MATLAB Command prompt:
>> aterm=(27*(0.08206)^2*(369.85)^2)/(64 *41.92)
aterm = 9.2699
>> bterm=(0.08206*369.85)/(8*41.92)
bterm = 0.0905
>> VvdW=10.0;
>> PvdW = (0.08206*300.15)/(VvdW-bterm)-aterm/(VvdW^2)
PvdW = 2.3928
>> VIG=VvdW-bterm
VIG = 9.9095
>> PIG=(0.08206*300.15)/(VIG)
PIG = 2.4855
>>PIG2= PvdW + aterm/VvdW^2
PIG2 = 2.4855
>>PREV = (VvdW-VIG)/VvdW*100.0
PREV = 0.9050
>> PREP = ((PIG-PvdW)/PvdW)*100.0
PREP = 3.8741
```

Example 5.11 The vdW Cubic Equation of State

Let us solve Example 5.7 using the vdW equation of state, but we will use $P = 4\,atm$ for the sake of seeing the difference in behavior between ideal and real gas as far as the molar volume is concerned.

Two hundred eighty (280) grams of nitrogen is stored in a rigid container at 27.0°C and 4.0 atm (4.053 bar).

DATA: For N_2, $T_c = 126.15$ K, $P_c = 34$ bar $= 33.555$ atm, and $\omega = 0.039$

$$n_{N2} = \frac{280\ \text{g}}{28\ \dfrac{\text{g}}{\text{mol}}} = 10.0\ \text{mol}$$

$$T_r = \frac{T}{T_c} = \frac{300.15\ \text{K}}{126.15\ \text{K}} = 2.379 \qquad P_r = \frac{P}{P_c} = \frac{4.053\ \text{bar}}{34\ \text{bar}} = 0.1192$$

Then we have to find both a and b as given by Eqs. (5.34a) and (5.34b). After that, we have to substitute values of a and b in Eq. (5.33). However, Eq. (5.33) will be a polynomial equation of order three (3) in the variable as in the following form:

$$P\tilde{V}^3 - Pb\tilde{V}^2 - RT\tilde{V}^2 + a\tilde{V} - ab = 0 \tag{5.35}$$

If the absolute values of pressure and temperature (P and T) are given for a pure substance, then Eq. (5.35) will be *a nonlinear algebraic equation* in $\tilde{V}$ which will have, in general, three roots for the molar volume at the given pressure and temperature. MATLAB® code (Appendix C | C.2 MATLAB Code for Van der Waals (vdW) Equation of State), in the form of an *m-file*, is written to utilize the MATLAB® built-in root-finding algorithm for finding the roots for such an equation. Two roots are chosen such that the imaginary part is zero, which correspond to the volume of a gas and volume of liquid. Any additional parameters appearing in the main E.o.S, like $\boldsymbol{a}$ and $\boldsymbol{b}$, have to be defined by some additional equations other than the main equation itself. Using MATLAB code, it was found that: $\tilde{V} = 6.1415$ L/mol

Compare with the ideal gas volume: $P\tilde{V}_{IG} = RT$

NOTICE that if we set $a = b = 0$ in vdW E.o.S. (Eq. 5.33), the ideal gas law will be formed; thus, it is a simplified version of vdW E.o.S.

$$4.0\ \text{atm} \times \tilde{V}_{IG} = 0.082057 \frac{\text{atm} \cdot \text{L}}{\text{mol} \cdot \text{K}} \times 300.15\ \text{K}$$

$$\tilde{V}_{IG} = \frac{0.082057 \times 300.15}{4.0} = 6.157 \frac{\text{L}}{\text{mol}}$$

$$\text{PRE}\% = \frac{|6.157 - 6.1415|}{6.1415} \times 100\% = 0.25\% < 1\%$$

NOTE #7
Finding the roots of a cubic equation of state, like vdW and those listed below may converge to one, two identical, or two distinct real roots, depending on the selected values of P and T.

5.5.4 Peng–Robinson (P–R) Equation of State

$$P = \frac{RT}{\tilde{V} - b} - \frac{a\alpha}{\tilde{V}^2 + 2b\tilde{V} - b^2} \tag{5.36}$$

where

$$a = \frac{0.457235\, R^2 T_c^2}{P_c} \tag{5.37}$$

$$b = \frac{0.077796 R T_c}{P_c} \tag{5.38}$$

$$\alpha = \left[1 + \kappa(w) \times \left(1 - T_r^{0.5}\right)\right]^2 \tag{5.39}$$

$$\kappa(w) = 0.37464 + 1.54226 \times w - 0.26992 \times w^2 \tag{5.40}$$

See Appendix C I C.3 MATLAB Code for Peng Robinson (P-R) Equation of State, which is written to find the roots of the cubic equation (Eq. 5.36). Below is an example showing how to run MATLAB code for Peng–Robinson (P–R) Equation of State. The **bold font** signifies the user's input parameters and the generated results.

```
>> P_R
Input "1" to search by Compound Name "2" to search by Formula: 1
Input the name of the Compound, say methane: toluene
{[326]}     {'TOLUENE'}      {'C7H8'}
{[333]}     {'O-ETHYLTOLUENE'}     {'C9H12'}
{[334]}     {'M-ETHYLTOLUENE'}     {'C9H12'}
{[335]}     {'P-ETHYLTOLUENE'}     {'C9H12'}
Input the number for the chosen compound, say "1" for methane: 326
Input the absolute temperature in Kelvin 300
Input the absolute pressure in atm 1
{'The material is:'}    {'TOLUENE'}
Tr=0.50694   Pr=0.024662
Vg=23.012  L/mol
VL=0.10738  L/mol
Experimental VL=0.10539  L/mol
PRE in VL=1.8953  %
```

5.5.5 Peng–Robinson–Stryjek–Vera (P–R–S–V) Equation of State

$$P = \frac{RT}{\tilde{V} - b} - \frac{a\alpha}{\tilde{V}^2 + 2b\tilde{V} - b^2} \tag{5.41}$$

where

$$a = \frac{0.457235\, R^2 T_c^2}{P_c} \tag{5.42}$$

$$b = \frac{0.077796 R T_c}{P_c} \tag{5.43}$$

$$\alpha = \left[1 + \kappa(w) \times \left(1 - T_r^{0.5}\right)\right]^2 \tag{5.44}$$

$$\kappa(w) = \kappa_o(w) - 0.0089 \times \left(1 + T_r^{0.5}\right)(0.7 - T_r) \tag{5.45}$$

$$\kappa_o(w) = 0.378893 + 1.4897153 \times w - 0.17131848 \times w^2 + 0.0196554w^3 \tag{5.46}$$

Below is an example showing how to run MATLAB code Appendix C | C.4 MATLAB Code for Peng-Robinson-Stryjek-Vera (P-R-S-V) Equation of State. The **bold font** signifies the user's input parameters and the generated results.

```
>> P_R_S_V
Input "1" to search by Compound Name "2" to search by Formula: 1
Input the name of the Compound, say methane: n-hexane
{[6]}     {'N-HEXANE'}     {'C6H14'}
Input the number for the chosen compound, say "1" for methane: 6
Input the absolute temperature in Kelvin 300
Input the absolute pressure in atm 2
{'The material is:'}     {'N-HEXANE'}
Tr=0.59128    Pr=0.067274
Vg=10.7125   L/mol
VL=0.13104   L/mol
```

5.5.6 Redlich–Kwong (RK) Equation of State

$$P = \frac{RT}{\tilde{V} - b} - \frac{a}{\tilde{V}^2 + b\tilde{V}} \tag{5.47}$$

where

$$a = \frac{0.42748\, R^2 T_c^{2.5}}{P_c T^{0.5}} \tag{5.48}$$

$$b = \frac{0.08664 R T_c}{P_c} \tag{5.49}$$

See Appendix C | C.5 MATLAB Code for Redlich-Kwong (R-K) Equation of State, which is written to find the roots of the cubic equation (Eq. 5.47).

Below is an example showing how to run MATLAB code for the R–K Equation of State. The **bold font** signifies the user's input parameters and the generated results.

```
>> Redlich_Kwong
Input "1" to search by Compound Name "2" to search by Formula: 1
Input the name of the Compound, say methane: n-heptane
{[7]}     {'N-HEPTANE'}     {'C7H16'}
Input the number for the chosen compound, say "1" for methane: 7
Input the absolute temperature in Kelvin 300
Input the absolute pressure in atm 2
{'The material is:'}     {'N-HEPTANE'}
Tr=0.55534    Pr=0.074072
Vg=10.5197   L/mol
VL=0.17485   L/mol
```

5.5.7 Soave–Redlich–Kwong (SRK) Equation of State

$$P = \frac{RT}{\tilde{V} - b} - \frac{a}{\tilde{V}^2 + b\tilde{V}} \tag{5.50}$$

where

$$a = \frac{0.42748\, R^2 T_c^{2.0} \times \left[1 + f(w) \times \left(1 - T_r^{0.5}\right)\right]^2}{P_c} \tag{5.51}$$

$$b = \frac{0.08664 R T_c}{P_c} \tag{5.52}$$

$$f(w) = 0.48 + 1.574 \times w - 0.176 \times w^2 \tag{5.53}$$

See Appendix C | C.6 MATLAB Code for Soave-Redlich-Kwong (S-R-K) Equation of State, which is written to find the roots of the cubic equation (Eq. 5.50).

Below is an example showing how to run Appendix C: | C.6 MATLAB Code for Soave-Redlich-Kwong (S-R-K) Equation of State. The **bold font** signifies the user's input parameters and the generated results.

```
>> Soave_Redlich_Kwong
Input "1" to search by Compound Name "2" to search by Formula: 1
Input the name of the Compound, say methane: n-octane
{[8]}     {'N-OCTANE'}     {'C8H18'}
Input the number for the chosen compound, say "1" for methane: 8
Input the absolute temperature in Kelvin 350
Input the absolute pressure in atm 1
{'The material is:'}     {'N-OCTANE'}
Tr=0.6153    Pr=0.040743
Vg=26.7591   L/mol
VL=0.20153   L/mol
```

For more information on such cubic equations of state and comparison among their performance, please, refer to: Al-Malah, K.I. (2015). RK-, SRK-, & SRK-PR-Type equation of state for hydrocarbons, based on simple molecular properties. *Journal of Applied Chemical Science International* 2(2): 65–74. Retrieved from https://ikprress.org/index.php/JACSI/article/view/2808.

5.5.8 The Compressibility Factor Equation of the State

The compressibility factor of a *gaseous species* is defined as the ratio

$$Z = \frac{P\tilde{V}}{RT} \tag{5.54}$$

If the gas behaves ideally, $Z = 1$. The extent to which Z differs from 1 is a measure of the extent to which the gas is behaving non-ideally. The value of Z is found in Figure 5.1 as a function of T_r and P_r.

Example 5.12 Estimate the Molar Volume Using the Compressibility Factor, *Z*
Estimate the molar volume of N_2 at $T = 250$ K and $P = 50$ bar $= 49.35$ atm.

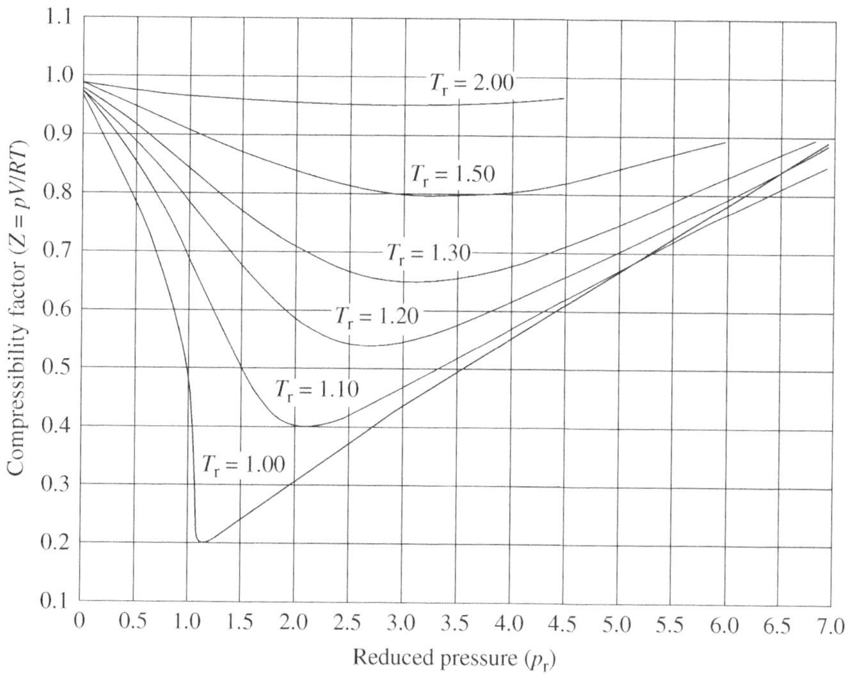

Figure 5.1 The value of Z as a function of T_r and P_r. *Source:* Wikipedia, Compressibility factor. Retrieved from https://en.wikipedia.org/wiki/Compressibility_factor.

DATA: For N_2, $T_c = 126.15$ K, $P_c = 34$ bar $= 33.555$ atm, and $\omega = 0.039$

Solution

$$T_r = \frac{T}{T_c} = \frac{250\text{ K}}{126.15\text{ K}} = 1.98 \cong 2 \qquad P_r = \frac{P}{P_c} = \frac{50\text{ bar}}{34\text{ bar}} = 1.47 \cong 1.5$$

From Figure 5.1, $Z \approx 0.96$

$$\tilde{V} = \frac{ZRT}{P} = \frac{0.96 \times 0.08206\dfrac{\text{atm.L}}{\text{mol.K}} \times 250\text{ K}}{49.35\text{ atm}} = 0.4\frac{\text{L}}{\text{mol}}$$

Using SRK Equation of State (Eq. 5.50), see Appendix C | C.6 MATLAB Code for Soave-Redlich-Kwong (S-R-K) Equation of State, here are the results after selecting the gas which is N_2 and inputting T in K and P in atm. The estimated molar volume is 0.40616 L/mol which is equal to that given by the compressibility factor approach.

```
>> Soave_Redlich_Kwong
Input "1" to search by Compound Name "2" to search by Formula: 2
```

```
Input the chemical formula, say CH4: N2
  1x3 cell array
 {[431]}     {'Nitrogen'}     {'N2'}
Input the number for the chosen compound, say "1" for methane: 431
Input the absolute temperature in Kelvin 250
Input the absolute pressure in atm 49.35
{'The material is:'}    {'Nitrogen'}
Tr=1.9818   Pr=1.4707
Vg=0.40616  L/mol
VL=0.40616  L/mol
```

5.5.9 Kay's Rule for Estimating Pseudo-critical Properties Mixtures

Kay's rule estimates *pseudo-critical properties* of mixtures as simple averages of pure component critical constants:

$$\text{Pseudo-critical Temperature: } T_c' = y_A Tc_A + y_B Tc_B + y_C Tc_C + \ldots \tag{5.55}$$

$$\text{Pseudo-critical Pressure: } P_c' = y_A Pc_A + y_B Pc_B + y_C Pc_C + \ldots \tag{5.56}$$

where $y_A, y_B, \ldots$ are mole fractions of species $A, B, \ldots$ in the mixture. Assuming that the system temperature T and pressure P are known, the pseudo-critical properties can be used to estimate the *pseudo reduced temperature and pressure* of the mixture:

$$\text{Pseudo-reduced Temperature: } T_r' = T/T_c' \tag{5.57}$$

$$\text{Pseudo-reduced Pressure: } P_r' = P/P_c' \tag{5.58}$$

Example 5.13 Application of Kay's Rule for Air as a Gas Mixture

One mole of air is contained in a tank at 100 atm and 25°C. Estimate the specific volume of the mixture in L/mol using Kay's rule.

Solution

$$\mathbf{N2}: T_c = 126.2\text{ K} \qquad P_c = 33.5\text{ atm} \qquad\qquad \mathbf{O_2}: T_c = 154.4\text{ K} \qquad P_c = 49.7\text{ atm}$$

$$T_c' = y_A Tc_A + y_B Tc_B = 0.79{*}126.2 + 0.21{*}154.4 = 99.7 + 32.4 = 132.1\text{ K}$$

$$P_c' = y_A Pc_A + y_B Pc_B = 0.79{*}33.5 + 0.21{*}49.7 = 26.46 + 10.44 = 36.9\text{ atm}$$

$$T_r' = T/T_c' = 298.15\text{ K}/132.1\text{ K} = 2.257 \qquad P_r' = P/P_c' = 100\text{ atm}/36.9\text{ atm} = 2.71$$

$$\text{Using Figure 5.1, } Z \cong 0.96\ \tilde{V} = \frac{ZRT}{P} = \frac{0.96 \times 0.08206 \times 298.15}{100\text{ atm}} = 0.235\frac{\text{L}}{\text{mol}}$$

$$\text{Using ideal gas while setting } Z = 1\text{: } \tilde{V}_{IG} = \frac{ZRT}{P} = \frac{1.00 \times 0.08206 \times 298.15}{100\text{ atm}} = 0.245\frac{\text{L}}{\text{mol}}$$

$$\text{P.R.E.} = \frac{|(0.245 - 0.235)|}{0.235} \times 100\% = 4.3\%$$

End of Chapter Problems

CHW 5.1 Which of the following properties is characteristic of solids?
A Fluidity
B Definite melting point
C Viscosity
D No fixed shape but a fixed volume

CHW 5.2 What primarily distinguishes solids from liquids?
A Moderate intermolecular forces
B Definite volume
C Compressible
D Fixed volume
E Fixed shape

CHW 5.3 Which property allows liquids to exhibit surface tension?
A High viscosity
B Cohesive forces at interfaces
C Definite volume
D Strong intermolecular forces

CHW 5.4 What property allows solids to resist changes in shape and maintain structural integrity?
A Viscosity
B Low density
C Mechanical strength
D Fluidity

CHW 5.5 Which state of matter exhibits ordered or regular arrangements in a crystal lattice structure?
A Solids
B Liquids
C Gases
D Plasma

CHW 5.6 What property allows liquids to take the shape of their container while maintaining a constant volume?
A Viscosity
B Surface tension
C Fluidity
D High density

CHW 5.7 Crystallographic data allows for calculating the density of crystalline solids because it provides information about:
A The number of electrons in the molecule
B The arrangement of atoms/molecules in the crystal lattice
C The chemical formula of the solid
D The number of atoms in a molecule

CHW 5.8 Density functional theory (DFT) focuses on:
- **A** Counting individual electrons in a molecule
- **B** The overall electron density distribution around a molecule
- **C** The specific bonds between atoms in a molecule
- **D** The mass of a molecule

CHW 5.9 Which statement BEST describes the Murnaghan equation of state?
- **A** It predicts the energy of a molecule based on its shape
- **B** It determines the ideal temperature for a chemical reaction
- **C** It calculates the force needed to break a chemical bond
- **D** It relates the volume of a solid to the pressure it experiences

CHW 5.10 The Tait equation is primarily used to:
- **A** Predict the chemical reactivity of a liquid
- **B** Describe the variation of liquid density with temperature and pressure
- **C** Determine the atomic composition of a liquid mixture
- **D** Calculate the viscosity of a liquid under flow conditions

CHW 5.11 The analogy of "bouncy balls" is used in the text to explain the Tait equation because it highlights:
- **A** The specific types of chemical bonds present in the liquid
- **B** The balance between pressure and intermolecular forces in determining density
- **C** The role of temperature in affecting the flow properties of a liquid
- **D** The importance of considering individual molecules in predicting macroscopic properties

CHW 5.12 Margules and Redlich–Kister equations are most relevant for:
- **A** Predicting the density of pure liquids at different temperatures
- **B** Understanding the behavior of solids under extreme pressure
- **C** Calculating the density of liquid mixtures based on their components
- **D** Determining the surface tension of liquids at various interfaces

CHW 5.13 The main reason liquids behave differently under high-pressure is due to:
- **A** Changes in the chemical reactivity of the liquid molecules
- **B** Increased intermolecular forces and packing arrangements
- **C** The transition of the liquid to a gas phase
- **D** The limitations of experimental techniques for measuring density

CHW 5.14 Mixing rules like the Margules equation are based on the assumption that:
- **A** The liquid components are chemically bonded to each other
- **B** The individual densities and interactions of components influence the mixed liquid density
- **C** The liquid mixture behaves ideally, like a mixture of perfect gases
- **D** The intermolecular forces between different components are negligible

CHW 5.15 PVT measurements and cubic equations of state are typically used in conjunction because:

A They provide redundant information about the same thermodynamic properties
B PVT data helps to validate the accuracy of the cubic equation of state predictions
C Cubic equations can be directly solved for density using measured PVT data
D Both methods rely on complex mathematical calculations and are unsuitable for practical applications

CHW 5.16 Which statement best describes the general effect of pressure on the density of most solids?

A Pressure significantly increases the density of all solids
B Pressure has no impact on the density of solids
C Pressure always decreases the density of solids
D The effect of pressure on density depends on the material's properties

CHW 5.17 What factor plays a key role in a solid's resistance to changes in volume under pressure?

A Its porosity
B Its chemical composition
C Its crystal structure
D Its bulk modulus

CHW 5.18 Compared to solids, how does pressure generally affect the density of liquids?

A It has a negligible impact on both
B It increases the density of both to a similar extent
C It has a more pronounced effect on liquids, increasing their density more readily
D It decreases the density of both due to expansion under pressure

CHW 5.19 What explains the increase in density of a liquid as pressure increases?

A Chemical reactions occurring between the liquid molecules
B A decrease in the average temperature of the liquid
C The liquid entering a new phase with a higher density
D The molecules packing closer together due to reduced intermolecular distances

CHW 5.20 Which statement best describes the isothermal compressibility (κ) of most solids compared to liquids?

A Solids have significantly lower κ values, emphasizing their strong resistance to compression
B Solids have much higher κ values, indicating greater compressibility than liquids
C Solids have slightly lower κ values, suggesting minor differences in compressibility
D Liquids and solids have practically the same κ values, implying similar compressibility

CHW 5.21 What factor primarily contributes to the low-isothermal compressibility of most solids?

A Their high temperature compared to liquids
B The presence of chemical bonds between atoms
C The absence of intermolecular forces within the solid
D Their high bulk modulus, a measure of resistance to volume change

CHW 5.22 Table 5.1 shows that liquid mercury has a much lower κ value than other liquids listed. What can you infer from this observation?

A Mercury is significantly less dense than other liquids due to its lower compressibility
B The compressibility of liquids is directly proportional to their temperature
C Mercury molecules are tightly packed and resist compression more effectively than other liquids
D Differences in κ values among liquids are mainly determined by their chemical composition

CHW 5.23 Compare the volume change of copper and ethanol under the same pressure change.

Initial Conditions:

- **Material 1**: Copper (solid)
- **Initial Volume (V_1)**: 10 cm^3
- **Isothermal Compressibility (κ_1)**: 7.3×10^{-3} GPa^{-1} (from Table 5.1)
- **Material 2**: Ethanol (liquid)
- **Initial Volume (V_2)**: 10 cm^3
- **Isothermal Compressibility (κ_2)**: 1.10 GPa^{-1} (from Table 5.1)
- **Applied Pressure Change (ΔP)**: 5 MPa

CHW 5.24 We want to reduce the volume of 100 mL of water by 10%. Calculate the pressure needed to achieve 10% volume reduction. Water at 25°C has $\kappa = 0.458$ GPa^{-1}.

CHW 5.25 In general, what effect does increasing temperature have on the density of most solids?

A It typically decreases the density due to thermal expansion leading to a larger volume
B It significantly increases the density due to the tighter packing of atoms
C It has minimal impact on the density as solids are almost incompressible
D It depends solely on the chemical composition of the solid

CHW 5.26 What solid's property plays a key role in determining how much a solid's density changes with temperature?

A Melting point
B Heat capacity
C Mass
D Coefficient of thermal expansion

CHW 5.27 Which scenario BEST exemplifies the anomalous behavior regarding density and temperature in solids?

A A steel bridge expands slightly in hot weather
B Iron undergoes a phase transition at high temperatures
C Water freezes into ice, which floats due to lower density
D Aluminum maintains a constant density across a wide temperature range

CHW 5.28 Which statement BEST describes the general effect of increasing temperature on the density of most liquids?

A It has minimal impact because liquids are incompressible
B It significantly increases the density due to thermal expansion
C It causes negligible changes as intermolecular forces dominate
D It typically decreases the density due to molecular motion and expansion

CHW 5.29 Water's density behavior with temperature is an example of:

A A typical liquid expansion pattern with no anomalies
B An anomalous behavior where density decreases upon cooling
C A critical point phenomenon specific to water molecules
D A standard phase transition from liquid to solid state

CHW 5.30 Brass rod expanding due to temperature change

A brass rod with an initial volume of 10.0 cm^3 is initially at a temperature of 20.0°C. When the temperature is increased to 120.0°C, the rod expands to a final volume, V_f. Calculate the final volume and the percent relative change in volume.

DATA: β for brass is 56.0×10^{-6} (1/K).

CHW 5.31 A container is filled with 100.0 cm^3 of petrol at a temperature of 0.0°C. If the temperature is increased to 60.0°C, calculate the final volume of petrol at the final temperature and the percent relative change in volume.

DATA: β for petrol is 950×10^{-6} (1/K).

CHW 5.32 If a sample of liquid ethanol is heated up from the initial condition at room temperature and pressure up to 65°C, what will be the required pressure increase to keep the volume constant? What do you conclude if you heat up an ethanol sample at constant atmospheric pressure, 1 atm?

CHW 5.33 Vegard's Law is applicable to:

A Both substitutional and interstitial solid solutions
B Only substitutional solid solutions with linear changes in lattice parameters
C Interstitial solid solutions with minimal expansion of the host lattice
D Any solid solution where the individual component densities are known

CHW 5.34 To accurately predict the density of a complex interstitial solid solution, it is best to:

A Apply Vegard's Law and assume a linear relationship
B Use experimental data from similar systems at the desired temperature and composition
C Average the densities of the individual components
D Develop a model that accounts for the expansion of the lattice due to interstitial atoms and the specific interactions between host and guest atoms

CHW 5.35 Vegard's Law Applied to a Copper–Nickel Alloy

Consider a Cu–Ni alloy with a 60% atomic fraction of copper ($AF_{Cu} = 0.6$). Calculate its theoretical density using Vegard's Law. Compare it with the calculated density based on mass fraction.

DATA:

- **Lattice Parameter of Pure Copper (a_{Cu})**: 3.615 Å (FCC)
- **Lattice Parameter of Pure Nickel (a_{Ni})**: 3.524 Å (FCC)
- **Atomic Weights**: Copper (M_{Cu}): 63.55 g/mol and Nickel (M_{Ni}): 58.69 g/mol
- $\rho^{o}_{Cu} = 8.96 \frac{g}{cm^3}$ $\rho^{o}_{Ni} = 8.908 \frac{g}{cm^3}$

CHW 5.36 Which of the following statements about the enthalpy of mixing (ΔH_{mix}) is definitely wrong for a real solution?

A ΔH_{mix} is always zero
B ΔH_{mix} can be positive or negative
C ΔH_{mix} is directly related to the intermolecular interactions between components
D ΔH_{mix} is a function of the composition of the solution

CHW 5.37 What distinguishes ideal solutions from real solutions concerning partial molar volume and enthalpy?

A Ideal solutions exhibit composition-dependent partial molar volume and enthalpy, while real solutions have equal partial molar properties
B Ideal solutions involve intermolecular interactions, unlike real solutions
C Real solutions have zero excess properties, while ideal solutions demonstrate nonzero excess volume and excess enthalpy
D Ideal solutions assume equal intermolecular interactions and thus composition-independent partial molar volume and enthalpy, while real solutions involve unequal intermolecular interactions and thus exhibit composition-dependent partial molar volume and enthalpy

CHW 5.38 In the experiment mixing ethanol and water, what does the negative excess volume indicate?

A Strong attractive forces between dissimilar molecules
B Strong repulsive forces between dissimilar molecules
C No intermolecular interactions between the components
D Perfect mixing behavior

CHW 5.39 Which statement regarding the deviation from an ideal liquid mixture is accurate?

A Deviation from an ideal liquid mixture arises from mixing chemical species characterized by different chemical activities or different sizes of homologous series but not both

B Mixing chemical species characterized by similar functional groups always results in ideal liquid mixtures

C Deviation from an ideal liquid mixture can arise from mixing chemical species characterized by different chemical activities and by different sizes of homologous series, or both

D Deviation from an ideal liquid mixture is solely dependent on the length of the hydrocarbon chain within the molecule

CHW 5.40 How does the kinetic theory explain the relationship between pressure and temperature in gases?

A Higher temperatures cause gas particles to move faster, leading to fewer collisions and lower pressure

B Elevated temperatures increase the average kinetic energy of gas particles, resulting in faster and more frequent collisions, thereby increasing pressure

C An increase in temperature results in decreased kinetic energy, reducing particle collisions and pressure

D Higher temperatures have no impact on gas pressure according to the kinetic theory

CHW 5.41 How does the kinetic theory accommodate the behavior of liquids concerning density and temperature compared with gases?

A Liquids experience a significant increase in density with higher temperatures due to increased kinetic energy

B Unlike gases, liquids do not exhibit any change in density with temperature variations

C Liquids experience a decrease in density at higher temperatures by the same magnitude of decrease found in gases

D Higher temperatures cause liquids to expand, resulting in decreased density, but this change is less pronounced compared to gases due to stronger intermolecular forces

CHW 5.42 How does the kinetic theory account for deviations in real gases from ideal gas behavior?

A Real gases deviate from ideal behavior due to larger particle sizes and stronger attractive forces, which the theory can address by considering volume corrections and van der Waals forces

B Real gases have stronger intermolecular forces, resulting in lower kinetic energy compared to ideal gases

C The kinetic theory disregards real gas behavior and cannot account for deviations

D Deviations in real gases occur solely due to higher temperatures and lower pressures, which the kinetic theory cannot explain

CHW 5.43 One gram-mole of ethane vapor is contained in a vessel at 127°C and 10 atm.

A Use the ideal gas equation of state to estimate the system volume.

B Suppose the actual volume of the vessel is 3.2 L, given by SRK E.o.S. What percent relative error arises from assuming ideal gas behavior?

CHW 5.44 The pressure gauge on a 30.0 m^3 tank of nitrogen at 27°C reads 9 bar. Estimate the mass of nitrogen in the tank using the ideal gas equation of state.

CHW 5.45 A stream of air enters a 5.0 cm ID pipe at a velocity of 80.0 m/s at 27°C and 1.90 bar (gauge). At a point downstream, the air flows through a 3.75 cm ID pipe at 67°C and 1.6 bar (gauge). What is the velocity of the gas at this point? Use the ideal gas law to estimate the inlet and outlet density of air at the given inlet and outlet pressure and temperature.

CHW 5.46 We have a container with two moles of nitrogen (N_2) and three moles of oxygen (O_2) at a total pressure of 1.5 atm and a temperature of 25°C. Both N_2 and O_2 are ideal gases.

What is the partial pressure of oxygen (O_2) in the container?

CHW 5.47 A tank contains four moles of helium (He) and six moles of neon (Ne) at a total pressure of 2 atm and a temperature of 27°C. Both He and Ne are considered ideal gases. Calculate the total volume of the gas mixture. Determine the volume fraction of helium in this mixture.

CHW 5.48 Which of the following conditions will not significantly contribute to a gas deviating from ideal behavior?

A High-pressure

B Low-temperature

C Presence of large, complex molecules

D Strong chemical reactions between gas molecules

CHW 5.49 What factor causes real gases to deviate from ideal behavior, particularly at high-pressures or low temperatures, according to the ideal gas law?

A Particle speed

B Volume of gas particles

C Type of gas molecules

D Container material

CHW 5.50 Under which conditions do intermolecular forces become significant, leading to deviations from ideal gas behavior?

A Low-pressures and low temperatures

B High-pressures and low temperatures

C Low-pressures and high temperatures

D High-pressures and high temperatures

CHW 5.51 Five-hundred sixty (560) grams of nitrogen is stored in a rigid container at 27.0°C and 20.0 atm. Estimate the molar volume of the gas, $\tilde{V}$, using the virial equation of state.

DATA: For N_2, $T_c = 126.15$ K, $P_c = 34$ bar $= 33.55$ atm, and $\omega = 0.039$

CHW 5.52 The vdW Equation of State

Given the fluid to be propane, C_4H_{10}, and the molar volume, $\tilde{V}$, as 12 L/mol, temperature $T = 27°C$, find the pressure as given by vdW E.o.S. (Eq. 5.33), utilizing also Eqs. (5.34a) and (5.34b) for evaluating parameters a and b. Compare pressure and molar volume, given by the ideal gas law vs. those given by vdW E.o.S.

DATA: $MW_{C4H10} = 58.1$ g/mol $T_c = 425.2$ K $P_c = 37.47$ atm.

Calculate PRE for both $\tilde{V}$ and P as predicted by the ideal gas, assuming the true values for both are given by vdW E.o.S.

CHW 5.53 Find the molar volume (s) of the following fluid materials at $T = 27°C$ and $P = 10$ atm. USE MATLAB SRK E.o.S. code, shown in Appendix C | C.6 MATLAB Code for SRK Equation of State. The critical data of organic hydrocarbons, water, and simple gases are provided as a Microsoft® Excel sheet (HCCrit.xlsx). A subset of compounds is selected and is shown in Table 5.4. Which materials give two distinct roots (one for liquid and another for vapor) using SRK E.o.S.?

Table 5.4 Critical properties of some selected hydrocarbons.

Num	Compound	Formula	C_{num}	H_{num}	Type	MW	T_c (K)	P_c (atm)	Z_c	W
32	Isopentane	C_5H_{12}	5	12	P	72.15	460.43	33.37	0.2700	0.2275
99	Cyclopentane	C_5H_{10}	5	10	N	70.13	511.76	44.43	0.2730	0.1943
187	Propylene	C_3H_6	3	6	O	42.08	364.76	45.52	0.2750	0.1424
283	Propadiene	C_3H_4	3	4	OD	40.06	393.15	53.98	0.2710	0.1594
314	Acetylene	C_2H_2	2	2	AC	26.04	308.32	60.59	0.2710	0.1873
325	Benzene	C_6H_6	6	6	A	78.11	562.16	48.34	0.2710	0.2108

CHW 5.54 Estimate the molar volume using the compressibility factor, Z

Estimate the molar volume of Toluene at $T = 400$ K and $P = 5$ atm.

DATA: For Toluene, $T_C = 591.79$ K, $P_C = 40.55$ atm

CHW 5.55 Kay's rule for estimating the molar volume of a gas mixture

A rigid tank contains 2 kmol of methane (CH_4) and 3 kmol of ethane (C_2H_6) at 400 K and 50 atm. Estimate the molar volume of the mixture in m^3/kmol using Kay's rule followed by the compressibility factor equation of state.

DATA: CH_4: $T_c = 190.6$ K, $P_c = 45.3$ atm C_2H_6: $T_c = 305.3$ K, $P_c = 48.1$ atm

CHW 5.56 Mono-chlorobenzene (M) is produced commercially by the direct catalytic chlorination of *benzene* (B) at 50°C and 130 kPa absolute. In the process, *dichlorobenzene* (D) is generated as a co-product:

$$C_6H_6 + Cl_2 \rightarrow C_6H_5Cl + HCl$$

$$C_6H_5Cl + Cl_2 \rightarrow C_6H_4Cl_2 + HCl$$

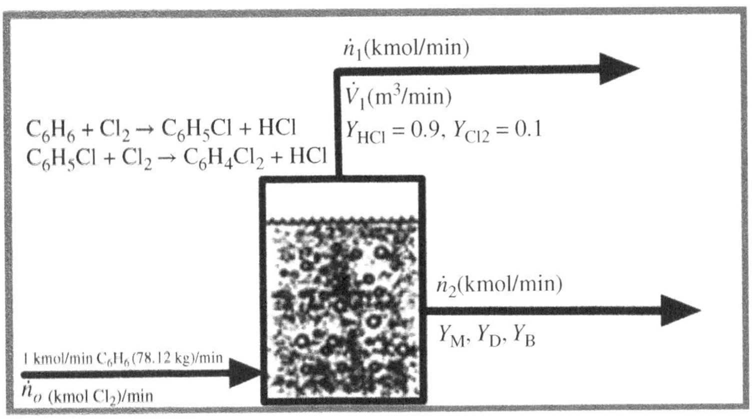

Figure CHW 5.56 A flowchart for the production of mono-chlorobenzene (M) from benzene (B) with a side reaction forming dichlorobenzene (D).

Liquid and gas streams leave the reactor. *The exit liquid stream has a mass flow rate of 100 kg/min and contains 50 wt.% M, 30% D, and the remaining 20% unreacted B.* The gas, which is sent to a treatment facility, contains 90% (v/v) HCl and 10% unreacted chlorine, as shown in Figure CHW 5.56 above.

A Convert the exit liquid mass flow rate and its mass fraction into molar flow rate ($\dot{n}_2$) and mole fraction, respectively, utilizing the molecular mass data for each species. At this step, you will be able to find $\dot{n}_2, Y_M, Y_D,$ and Y_B.

B Choose a basis of reaction as 1 kmol/min of inlet benzene and $\dot{n}^o_{Cl2}$ of chlorine gas (to be estimated later). Then, express the exit molar flow rate of each species in terms of the initial number of moles of each species and of the two extents of reaction: $\dot{\xi}_1$ and $\dot{\xi}_2$.

C Carry out component molar balance for benzene (B), mono-chlorobenzene (M), and di-chlorobenzene (D) to find both $\dot{\xi}_1$ and $\dot{\xi}_2$.

D Carry out HCl molar balance to find $\dot{n}_1$.

E Carry out Cl_2 balance to find $\dot{n}^o_{Cl2}$.

F Apply ideal gas law to find the exit gas volumetric flow rate: $\dot{V}_1$ m^3/ min.

6

Multiphase, Multicomponent Systems

All-natural resources, encompassing air, water, soil, plants, and animals, inherently exist in the forms of gaseous mixtures, liquid solutions, solid solutions, or a combination thereof. In various chemical, petrochemical, biochemical, food, and pharmaceutical processes, feed streams predominantly consist of mixtures or solutions containing multiple species. Consequently, a chemical process involves treating chemical components or different species through physical, chemical, or combined methods. The primary objective of such treatment steps is to enhance value or transform raw, inexpensive material(s) into valuable, final finished products (akin to gold). Physical treatment steps may encompass activities such as mixing, separation (de-mixing), including absorption, distillation, and extraction, as well as heating/cooling, with or without a phase change. On the other hand, the chemical treatment step encompasses singular or sets of parallel, series, or mixed reactions, leading to a change in the chemical identity of each reacting species.

Please refer to Chapter 1 | Chemical Engineering Industrial Processes section on different chemical engineering industrial processes, which involve, in general, a multiphase, multicomponent reacting, non-reacting system, or a combination of both.

Also, refer to Chapter 1 | What Chemical Engineers Can Do section to explore tasks and duties tailored to the role of a chemical engineer. To achieve such tasks and duties, characterization of multiphase, multicomponent reacting, and non-reacting systems is a must.

6.1 Equilibrium: A Balancing Act in the Universe

Equilibrium, a fundamental concept across various scientific disciplines, describes a state where a system exhibits no net change over time. It is like a delicately balanced seesaw; opposing forces perfectly counteract each other, resulting in stability. Understanding the

To my dear readers,
When copying MATLAB code or text, please right-click and attempt both the "Copy with Formatting" and "Copy" options from the context menu. This approach is effective within Adobe Acrobat. If neither option works, kindly carefully review any MATLAB error messages and compare the pasted text to the original to identify and correct any discrepancies. Alternatively, a soft copy of the MATLAB code is available in the Wiley book companion site of this book for your convenience.

MATLAB-based Computations of Chemical Engineering Principles, First Edition. Kamal I. M. Al-Malah.

Companion website: www.wiley.com/go/al-malah/matlabbasedcomputations

conditions for equilibrium and the forces that drive a system toward it is crucial in numerous fields, from physics and chemistry to ecology and economics.

Conditions for Equilibrium:
For a system to be in equilibrium, two key conditions must be met:

1) **Net Force and Torque Cancellation**: All external and internal forces and torques acting on the system must balance out, resulting in a net force and torque of zero. Imagine a ball suspended mid-air by two strings pulling with equal force in opposite directions. Its state of no net movement represents equilibrium.
2) **Constant Internal Properties**: Internal properties like temperature, pressure, concentration, and potential energy don't change over time. This implies no net flow of energy or matter within the system. For example, a closed container with a mixture of hot and cold water eventually reaches a thermal equilibrium where both liquids share the same temperature.

Driving Forces and Nonequilibrium:
Every system naturally exists in a state of minimal free energy, which acts as the driving force for attaining equilibrium. When a system is not in equilibrium, it possesses non-zero driving forces, pushing it toward a balance. These forces can arise from various factors, including:

- **Temperature Differences**: Heat flows from hotter objects to colder ones until thermal equilibrium is established.
- **Concentration Gradients**: Molecules diffuse from areas of higher concentration to lower concentration, achieving diffusion equilibrium.
- **Pressure Differences**: Gases expand to fill available space until pressure equilibrium is reached.
- **Chemical Reactions**: Reactants undergo spontaneous reactions to minimize free energy, reaching equilibrium with the products.

Examples of Driving Forces and the Journey to Equilibrium:

1) **Electrical Potential Differences**:
   - **Driving Force**: Electrons naturally flow from areas of higher potential (voltage) to lower potential, aiming for an equal distribution of charge.
   - **Equilibrium**: When we connect a battery to a light bulb, the potential difference drives the flow of current through the circuit, lighting the bulb, until the circuit reaches a steady state (equilibrium) where the energy consumption of the bulb balances the energy supplied by the battery.

2) **Gravitational Forces**:
   - **Driving Force**: Gravity pulls objects toward the center of mass, minimizing potential energy.
   - **Equilibrium**: A rolling ball on a ramp accelerates downhill due to gravity, aiming for the lowest point (minimum potential energy). After reaching the bottom and coming to rest, the ball achieves equilibrium with its surroundings.

3) **Mechanical Oscillations**:
   - **Driving Force**: Any external force or disturbance applied to an object sets it into motion (nonequilibrium).
   - **Equilibrium**: A swinging pendulum experiences friction and air resistance, gradually slowing down until it comes to rest at its equilibrium position (vertical hanging position).

4) **Biological Processes**:
   - **Driving Force**: Living organisms strive to maintain internal homeostasis, balancing essential factors like temperature, pH, and blood sugar levels.
   - **Equilibrium**: When exercising, your body temperature rises. Sweating is a response to cool you down and restore thermal equilibrium. Similarly, your body releases insulin after a meal to regulate blood sugar levels and achieve metabolic equilibrium.

5) **Thermal Systems: Hot Coffee in a Cool Room**:
   - **Driving Force**: As long as there exists a temperature difference, heat naturally flows from hotter objects (coffee) to colder objects (room) until both reach thermal equilibrium. This minimizes the imbalance in thermal energy between the two subsystems.
   - **Equilibrium**: The hot coffee gradually loses heat to the surrounding air, causing its temperature to decrease. Meanwhile, the cool air absorbs the coffee's heat, causing its temperature to increase. This exchange continues until both coffee and air reach the same temperature, achieving thermal equilibrium.

6) **Chemical Systems: Dissolving Sugar Cube in Water**:
   - **Driving Force**: As long as there exists a concentration gradient, sugar molecules exhibit random motion, spreading from areas of higher concentration (around the cube) to areas of lower concentration (bulk water). This process continues until the sugar is evenly distributed throughout the water, reaching a saturated solution and minimizing the concentration gradient.
   - **Equilibrium**: As the sugar cube enters the water, its high concentration drives sugar molecules to diffuse into the surrounding liquid. Water molecules, in turn, move in the opposite direction. This diffusion process continues until the sugar concentration throughout the water becomes uniform, establishing equilibrium.

7) **Mechanical Systems: Compressed Gas Expanding in a Container (with Valve Control)**:
   - **Driving Force**: As long as the pressure inside the container remains higher than the outside pressure, the compressed gas molecules possess kinetic energy and exert pressure on the container walls. This pressure difference creates a driving force for the gas to expand and equalize the pressures.
   - **Equilibrium**: When the valve is opened, the gas molecules, driven by their internal pressure, rush outward into the connected empty container. This expansion continues until the pressure inside the container is equal to the source pressure and then the flow of gas stops, signaling a mechanical equilibrium.

8) **Photosynthesis and Cellular Respiration in an Ecosystem**:
   - **Driving Force (Imbalance of Gases)**: Photosynthesis absorbs CO_2 and releases O_2, while cellular respiration consumes O_2 and releases CO_2. These opposing processes create a natural feedback loop, striving to maintain a stable balance of these crucial gases in the ecosystem.
   - **Equilibrium (Ecological Balance)**: Plants and algae perform photosynthesis during the day, consuming CO_2 and releasing O_2. Animals and microbes, on the other hand, engage in cellular respiration, consuming O_2 and releasing CO_2. This constant exchange regulates the levels of CO_2 and O_2 within the ecosystem, aiming for a dynamic equilibrium crucial for life and environmental stability.

Tendency Toward Equilibrium:
Nature favors stability and minimizes free energy. Therefore, any process that creates a nonequilibrium state triggers reactions or changes within the system to counteract the driving forces and ultimately reach equilibrium. This "natural flow" toward equilibrium underlies numerous phenomena, from ice melting at room temperature to ecosystems balancing predator–prey populations.

Challenges and Applications:
Understanding and manipulating equilibrium conditions has far-reaching consequences. Chemical reactions rely on achieving specific equilibria for product formation. Engineers design bridges and buildings to maintain structural equilibrium under various loads. Ecologists strive to understand the delicate equilibrium of ecosystems for managing sustainability.

However, achieving perfect equilibrium in real-world scenarios is often impossible due to continuous external influences and internal fluctuations. We can, however, use our understanding of equilibrium principles to predict system behavior, optimize processes, and maintain stability in dynamic situations.

Equilibrium Criteria:
Based on the above-listed life examples, one can say that the defining traits of a system in equilibrium include the following:

1) There's no change as time progresses.
2) The system is homogenous (uniform) or consists of multiple homogenous subsystems.
3) There is an absence of mass, heat, and work exchange, both internally and with the surrounding environment.
4) There is no overall change in the rate of chemical reactions occurring within the system.

However, if the environment enforces a mass flow in and out of the system (such as through a pumping mechanism) or a heat flow (as seen when different parts of the system experience varying temperatures), the system can only reach a time-invariant state if the flows remain constant. These time-invariant states, although distinct from equilibrium states, are referred to as steady states.

Undriven systems, exemplified by natural processes, attain equilibrium as spontaneous flows inherent in nature tend to dissipate the driving forces behind them. Consequently, the heat flow triggered by a temperature difference dissipates the temperature disparity, and the mass diffusion flux prompted by a concentration gradient leads to the establishment of a uniform concentration state. Notably, the evaluation of Gibbs free energy is employed to determine the system's equilibrium status. For a closed, non-reacting system:

$$\mathrm{d}G^t = -S^t\mathrm{d}T + V^t\mathrm{d}P \tag{6.1}$$

For such a system that exists under equilibrium, $\mathrm{d}G^t$ will be essentially zero. $\mathrm{d}G^t$ can be zero under one of the following two conditions:

First, if both $\mathrm{d}T$ and $\mathrm{d}P$ are both zero; this case represents a fixed point in T vs. P domain (or plane).

Second, if $S^t\mathrm{d}T = V^t\mathrm{d}P$; this case represents a series of points along a given trajectory (or curve) such that the pressure and temperature are correlated. In other words, perturbing the temperature by an increment or decrement will accordingly perturb the pressure in the same direction.

Furthermore, if $dG^t < 0$, the system will be considered to be in a nonequilibrium state, signifying a spontaneous situation. In such instances, the system strives to attain equilibrium, where dG^t will reach zero and all disturbing driving forces diminish to zero.

Conversely, if $dG^t > 0$, the system is also regarded as being in a nonequilibrium state, characterizing a non-spontaneous scenario. In this case, the system remains unchanged unless an external force (or means) is applied to drive it toward a specific direction, resulting in a new set of pressure and temperature.

$$\text{For an open or reacting system}: \mathrm{d}(n\tilde{G}) = -(n\tilde{S})\mathrm{d}T + (n\tilde{V})\mathrm{d}P + \sum_{i=1}^{N} \overline{G}_i \mathrm{d}n_i \qquad (6.2)$$

where $\overline{G}_i$ represents the partial molar Gibbs free energy of species i in the mixture at the given pressure and temperature of the system. The topic of partial molar properties of solutions will be handled in the thermodynamics course(s).

6.2 Thermodynamic Properties of Pure Fluids

Thermodynamic properties of pure fluids hold the key to understanding and predicting their behavior in various applications, from chemical engineering to environmental science. Let us embark on a journey to explore these fascinating properties and unlock their secrets!

A pure fluid is a substance that exists in a single phase (liquid, gas, or solid) and is chemically homogeneous. Unlike mixtures, pure fluids exhibit consistent properties throughout their entire composition. Examples include water, carbon dioxide, nitrogen, and ammonia.

Key Thermodynamic Properties:

1) **Pressure (*P*)**: The force exerted by the fluid per unit area of its container. Pressure plays a crucial role in phase transitions, like boiling and condensation.
2) **Temperature (*T*)**: A measure of the average kinetic energy of the fluid molecules. Temperature influences a variety of properties, including density, viscosity, and solubility.
3) **Volume (*V*)**: The space occupied by the fluid. The relationship between pressure, temperature, and volume is described by the equation of state, a fundamental equation in thermodynamics.
4) **Internal Energy (*U*)**: Internal energy (U) stands as a crucial thermodynamic property, representing the total energy residing within the system. It encompasses a diverse array of contributions, reflecting the microscopic activity and interactions of the fluid molecules. Here is a breakdown of what U comprises:
   a) **Kinetic Energy**: This component accounts for the movement of individual molecules within the fluid. The higher the temperature, the greater the average kinetic energy and, consequently, the higher the internal energy.
   b) **Potential Energy**: This includes various intermolecular forces influencing the relative positions and interactions of molecules. These can range from attractive forces like van der Waals interactions to repulsive forces due to electron clouds overlapping. Understanding these forces and their dependence on temperature and pressure is crucial for accurately defining U.
   c) **Zero-point Energy**: Zero-point energy refers to the minimum energy that molecules retain due to quantum mechanical effects, even at absolute zero temperature (0 K), where classical physics predicts no motion. In quantum mechanics, molecules still

exhibit residual vibrations because their energy cannot drop below a certain limit. This is due to the Heisenberg uncertainty principle, which prevents both position and momentum from being precisely zero. As a result, zero-point energy provides a constant baseline (or datum) of energy (U) that persists regardless of the temperature.

d) **Chemical Bonding Energy**: In the case of molecular fluids, internal energy also encompasses the energy stored in the chemical bonds holding the atoms together. Breaking or forming these bonds involves significant energy changes, impacting the overall U of the system.

5) **Enthalpy (*H*)**: The total heat content of a system, including both internal energy and the product of pressure and volume.
6) **Entropy (*S*)**: A measure of the system's disorder or randomness. Higher entropy indicates a greater degree of disorder.
7) **Gibbs Free Energy (*G*)**: It is a crucial thermodynamic property for understanding the spontaneity and equilibrium of processes involving pure fluids. It combines enthalpy (H), a measure of heat content, and entropy (S), a measure of disorder, into a single value. In the context of pure fluids, Gibbs free energy can be defined as follows: $G = H - TS$.
8) **Helmholtz Free Energy (*A*)**: It is another critical thermodynamic potential for analyzing pure fluids. While closely related to Gibbs free energy, it focuses on a different set of parameters, offering another perspective on the system's stability and direction of change. Helmholtz free energy is defined for a pure fluid: $A = U - TS$.

Notice that the first three properties (pressure, temperature, and volume) are fundamental and directly measurable. We can use instruments like gauges, thermometers, and measuring containers to determine these values with relative ease. These three basic properties form the foundation of classical thermodynamics, allowing us to describe the state of a system without needing knowledge of its internal structure. The remaining properties, however, delve deeper into the system's energetics and randomness. While measuring them directly can be more challenging, they can be calculated or estimated based on the readily measurable P, T, and V:

- **Internal Energy (*U*)**: While not directly measurable, we can calculate it using specific heat data at constant volume and knowledge of temperature changes in the system. Alternatively, processes like constant-volume calorimetry can indirectly measure internal energy changes.
- **Enthalpy (*H*)**: While not directly measurable, we can calculate it using specific heat data at constant pressure and knowledge of temperature changes in the system. Alternatively, processes like constant-pressure calorimetry can indirectly measure enthalpy changes.
- **Entropy (*S*)**: Directly measuring entropy is highly complex, typically involving statistical mechanics and knowledge of the microscopic state of the system. However, for pure fluids, we can relate entropy changes to heat transfer and temperature changes through relationships like the Clausius–Clapeyron equation.
- **Gibbs Free Energy (*G*)**: As mentioned above, this is a combination of enthalpy and entropy, so its calculation directly relies on knowing ***H*** and ***S***. Similarly, we can determine changes in ***G*** based on measurable parameters like heat flow and temperature changes.
- **Helmholtz Free Energy (*A*)**: Similar to ***G***, ***A*** depends on internal energy (***U***) and entropy (***S***), which again require calculation or estimation based on ***P***, ***T***, and ***V***.

In essence, the first three properties act as the "ingredients" from which the others are derived or expressed. By knowing P, T, and V, we can access a whole realm of information about the system's energetics, stability, and direction of change through these additional derived properties. Assessment of these properties helps us:

- **Predict Phase Behavior**: Determine the conditions under which a pure fluid will transition between liquid, gas, and solid phases.
- **Design and Optimize Processes**: Select appropriate materials and operating conditions for chemical reactors, heat exchangers, and other equipment.
- **Model and Simulate Fluid Behavior**: Develop computer models to predict the flow and transport of fluids in various contexts.
- **Assess Environmental Impact**: Understand the thermodynamics of emissions and pollution control.

6.3 Clapeyron Equation

Consider the *equilibrium* state of a two-phase, interacting, pure-component, and closed system. For example, liquid benzene is in equilibrium with its vapor in a closed vessel. Since at equilibrium, $\mathrm{d}(n\tilde{G}) = 0$, this implies that:

$$\mathrm{d}(n\tilde{G}) = \mathrm{d}\left(n^v\tilde{G}^v + n^l\tilde{G}^l\right) = 0 = \mathrm{d}n^v\tilde{G}^v + n^v\mathrm{d}\tilde{G}^v + \mathrm{d}n^l\tilde{G}^l + n^l\mathrm{d}\tilde{G}^l \tag{6.3}$$

Since $n = n^l + n^v$ and given that the total number of moles is fixed for a closed system, it implies that:

$$\mathrm{d}(n) = \mathrm{d}\left(n^v + n^l\right) = 0 = \mathrm{d}n^v + \mathrm{d}n^l \tag{6.4}$$

which means that $\mathrm{d}n^v = -\mathrm{d}n^l$. By replacing $\mathrm{d}n^l$ by $-\mathrm{d}n^v$ in Eq. (6.3), it reduces to:

$$\mathrm{d}n^v\tilde{G}^v - \tilde{G}^l\mathrm{d}n^v + n^v\mathrm{d}\tilde{G}^v + n^l\mathrm{d}\tilde{G}^l = 0 \tag{6.5}$$

or:

$$\left(\tilde{G}^v - \tilde{G}^l\right)\mathrm{d}n^v + n^v\mathrm{d}\tilde{G}^v + n^l\mathrm{d}\tilde{G}^l = 0 \tag{6.6}$$

Now, since the transfer of molecules from one phase to another takes place at constant temperature (i.e., $\mathrm{d}T = 0$) and constant pressure (i.e., $\mathrm{d}P = 0$), then both $\mathrm{d}\tilde{G}^l = -\tilde{S}^l\mathrm{d}T + \tilde{V}^l$ $\mathrm{d}P = 0$ and $\mathrm{d}\tilde{G}^v = -\tilde{S}^v\mathrm{d}T + \tilde{V}^v\mathrm{d}P = 0$. Hence, Eq. (6.6) is simplified to:

$$\left(\tilde{G}^v - \tilde{G}^l\right)\mathrm{d}n^v = 0 \tag{6.7}$$

Now, we have two possible scenarios:

The first is the macroscopic static equilibrium with time, where one phase is separated from another in terms of mass transfer (i.e., the presence of a barrier or film in between). In this case, $\mathrm{d}n = \mathrm{d}n^v = \mathrm{d}n^l = 0$. This is not the case in a typical gas–liquid phase equilibrium.

The second is the macroscopic dynamic equilibrium with time, where one phase can exchange molecules with another. In this case, $\mathrm{d}n = 0$; however, $\mathrm{d}n^v = -\mathrm{d}n^l \neq 0$. Molecules keep jumping from one phase into another in both directions. This case represents the typical gas–liquid phase equilibrium. Consequently, Eq. (6.7) requires that the bracketed term $\left(\tilde{G}^v - \tilde{G}^l\right)$ should reduce to zero as $\mathrm{d}n^v \neq$:

$$\tilde{G}^v = \tilde{G}^l \tag{6.8}$$

or:

$$d\tilde{G}^v = d\tilde{G}^l \tag{6.9}$$

From which we obtain:

$$-\tilde{S}^v dT + \tilde{V}^v dP = -\tilde{S}^l dT + \tilde{V}^l dP \tag{6.10}$$

Further rearrangement yields:

$$\left(\tilde{S}^v - \tilde{S}^l\right) dT = \left(\tilde{V}^v - \tilde{V}^l\right) dP \tag{6.11}$$

Thus:

$$\left.\frac{dP}{dT}\right|_{\tilde{G}^v = \tilde{G}^l} = \left(\frac{\tilde{S}^v - \tilde{S}^l}{\tilde{V}^v - \tilde{V}^l}\right) = \left(\frac{\Delta\tilde{S}_{lv}}{\Delta\tilde{V}_{lv}}\right) = \frac{\Delta\tilde{H}_{lv}}{T\Delta\tilde{V}_{lv}} \tag{6.12}$$

Equation (6.12) is called the *Clapeyron* equation. In words, Eq. (6.12) says that under *phase equilibrium*, both pressure and temperature *are not linearly independent*; Eq. (6.12) relates the slope of the P–T curve to the enthalpy and volume change at a phase transition. Since neither the entropy change nor the enthalpy change is equal to zero, for solid–liquid, liquid–vapor, or solid–vapor phase transitions, the slope of any P–T curve is not equal to zero nor is it equal to infinity (i.e., both $\Delta\tilde{V}_{lv}$ and $\Delta\tilde{H}_{lv}$ are finite). Further simplification of Eq. (6.12) results from the assumption that $\tilde{V}^v \gg \tilde{V}^l$ and $\tilde{V}^v = \frac{RT}{P}$ yields:

$$\frac{dP}{P} = d\ln P = \frac{\Delta\tilde{H}_{lv}}{RT^2} dT = -\frac{\Delta\tilde{H}_{lv}}{R} d\left(\frac{1}{T}\right) \tag{6.13}$$

or:

$$\frac{d\ln P}{d\left(\frac{1}{T}\right)} = -\frac{\Delta\tilde{H}_{lv}}{R} \tag{6.14}$$

The previous equations (up to Eq. (6.12)) are also valid for other phase changes (i.e., solid to liquid and solid to vapor).

For example, if $\Delta\tilde{H}_{lv}/R$ is assumed constant, then Eq. (6.14) can be put in the form of:

$$\frac{d\ln P}{d\left(\frac{1}{T}\right)} = A \tag{6.15}$$

Upon integration, we reach at:

$$\int d\ln P = \int A \times d\left(\frac{1}{T}\right) \quad \rightarrow \ln P = \frac{A}{T} + \text{Constant of integration} \tag{6.16}$$

or:

$$\ln P = \frac{A}{T} + B \tag{6.17}$$

Equation (6.17) is named as Clausius–Clapeyron equation. Different forms can be deduced, depending on how to express the right-hand side of Eq. (6.13) as a function of temperature.

6.4 Antoine Equation

In light of Clausius–Clapeyron equation (Eq. 6.17), a relatively simple empirical equation that correlates vapor pressure–temperature data extremely well is the Antoine equation:

$$\log_{10} P^{sat} = A - \frac{B}{T + C} \tag{6.18}$$

$$P^{sat} = 10^{\left(A - \frac{B}{T+C}\right)} \tag{6.19}$$

$$T = \frac{B}{A - \log_{10} P^{sat}} - C \tag{6.20}$$

Some textbooks may use the natural log, *ln*, instead of the base-10 logarithm, log_{10}, as in:

$$\ln P^{sat} = a - \frac{b}{T + c} \tag{6.21}$$

Notice that the Antoine equation evaluates the pressure at the saturation (boiling) condition where *T* and *P* are correlated. This equation simply says that knowing T^{sat} will enable us to evaluate P^{sat} and vice versa. Moreover, for a pure substance, there are so many boiling points, starting down from the triple up to the critical point. However, there is only one normal boiling point temperature (T_{NBP}), which gives a vapor (saturation) pressure equal to one standard atmosphere, expressed in different units.

Table 6.1 shows Antoine's equation constants using Eq. (6.18), where:

- *P* is the vapor pressure in mmHg
- *T* is the temperature in °C
- *A*, *B*, and *C* are the Antoine constants specific to each compound

Table 6.1 Antoine equation (Eq. 6.18) constants for some common substances.

Formula	Compound Name	*A*	*B*	*C*	T_{MIN} (°C)	T_{MAX} (°C)
C_3H_6O	Acetone	7.2316	1277.03	237.23	−32	77
$C_2H_4O_2$	Acetic acid	7.2996	1479.02	216.82	17	157
C_2H_3N	Acetonitrile	7.0735	1279.2	224.01	−13	117
C_6H_6	Benzene	6.90565	1211.033	220.79	−16	104
C_4H_{10}	Butane	6.80896	935.86	238.73	−78	19
CCL_4	Carbon tetrachloride	6.8941	1219.58	227.17	−20	101
C_6H_5CL	Chlorobenzene	6.9781	1431.05	217.56	47	147
C_4H_9CL	1-Chlorobutane	6.9379	1227.43	224.11	−18	112

(Continued)

Table 6.1 (Continued)

Formula	Compound Name	*A*	*B*	*C*	T_{MIN} (°C)	T_{MAX} (°C)
C_4H_9CL	2-Chlorobutane	6.9447	1195.8	226.01	−23	102
$CHCL_3$	Chloroform	6.9371	1171.2	227	−13	97
C_6H_{12}	Cyclohexane	6.8413	1201.531	222.647	6	105
C_5H_{10}	Cyclopentane	6.88676	1124.162	231.361	−40	72
$C_{10}H_{22}$	Decane	6.96375	1508.75	195.374	58	203
CH_2CL_2	Dichloromethane	7.0803	1138.91	231.46	−44	59
$C_4H_{10}O$	Ethyl ether	6.92032	1064.07	228.8	−61	20
$C_4H_8O_2$	*p*-Dioxane	7.0063	1288.5	211.01	2	137
$C_{20}H_{42}$	Eicosane	7.1522	2032.7	132.1	198	379
C_2H_6O	Ethyl alcohol (ethanol)	8.2133	1652.05	231.48	−3	96
C_8H_{10}	Ethylbenzene	6.95719	1424.255	213.206	26	163
$C_2H_6O_2$	Ethylene glycol	8.7945	2615.4	244.91	91	221
C_7H_{16}	Heptane	6.89385	1264.37	216.636	−2	123
C_6H_{14}	Hexane	6.87024	1168.72	224.21	−25	92
CH_4O	Methanol	8.0724	1574.99	238.87	−16	91
$C_4H_8O_2$	Ethyl acetate	7.0146	1211.9	216.01	−13	112
CH_3NO_2	Nitromethane	7.044	1291	209.01	5	136
C_9H_{20}	Nonane	6.9344	1429.46	201.82	39	179
C_8H_{18}	Octane	6.9094	1349.82	209.385	19	152
C_5H_{12}	Pentane	6.87632	1075.78	233.205	−50	58
C_6H_6O	Phenol	7.1345	1516.07	174.57	72	208
C_7H_8	Toluene	6.95848	1344.80	219.16	5	300
H_2O	Water	8.07131	1730.63	233.426	1	100
H_2O	Water	8.14019	1810.94	244.485	99	374

Source: https://myengineeringtools.com/Data_Diagrams/Antoine_Law_Coefficients.html, with permission of MyEngineeringTools.

Example 6.1 Benzene TNBP

Calculate the normal boiling point temperature of benzene and compare it with the experimental value $T_{NBP} = 80.1\,°C$.

Solution:

At T_{NBP}, $P = 1\,\text{atm} = 760\,\text{mmHg}$

$$\log_{10}(P^{sat}) = \log_{10}(760) = 2.8808 = 6.90565 - \frac{1211.039}{T + 220.79}$$

$$\frac{1211.04}{T + 220.8} = 6.90565 - 2.8808 = 4.02485 \rightarrow (T + 220.8) = \frac{1211.04}{4.02485} = 300.9°\text{C}$$

$T = 300.9°\text{C} - 220.8°\text{C} = 80.1°\text{C}$, which is exactly equal to the experimental value.

6.5 A Volatile Versus Less Volatile Substance

A substance's volatility refers to its tendency to change from a liquid or solid state to a gas state. We can compare the volatility of different substances using their normal boiling point and vapor pressure:

1) **Normal Boiling Point**:
   - **Volatile Substance**: A substance with a lower normal boiling point is more volatile. This means it requires less heat to transition into a gas, readily evaporating even at room temperature.
   - **Less Volatile Substance**: A substance with a higher normal boiling point is less volatile. It requires more heat to vaporize and typically evaporates slower or only at higher temperatures.
2) **Vapor Pressure**:
   - **Volatile Substance**: A substance with a higher vapor pressure at a given temperature is more volatile. This means it readily releases more gas molecules into the vapor/air, creating a higher pressure of its vapor above the liquid.
   - **Less Volatile Substance**: A substance with a lower vapor pressure at a given temperature is less volatile. It releases fewer gas molecules, creating a lower vapor pressure.

Example 6.2 Which is more volatile: benzene (B) or toluene (T)?

A typical example in chemical engineering textbooks is the use of a distillation tower to separate the binary mixture of benzene-toluene into almost pure benzene and pure toluene. The question is: Which one should constitute the substantial composition in the top (distillate) stream?

Solution:

If the normal boiling point data are given or if the vapor pressure at a given temperature is given, then the user will be able to tell which one will be at the top and the other will be at the bottom.

Alternatively, the Antoine equation can be used to tell which one is more volatile. Let us try.

1) Select a certain temperature, say $T = 80°C$.
2) Calculate the vapor pressure for both at the assigned temperature, using data given in Table 6.1.

$$\log_{10}\left(P_B^{sat}\right) = 6.90565 - \frac{1{,}211.039}{80 + 220.79} = 6.90565 - 4.02619 = 2.87946$$

$$P_B^{sat} = 10^{2.87946} = 757.6\,\text{mmHg}$$

$$\log_{10}\left(P_T^{sat}\right) = 6.95848 - \frac{1{,}344.8}{80 + 219.16} = 6.95848 - 4.4953 = 2.4632$$

$$P_T^{sat} = 10^{2.4632} = 290.5\,\text{mmHg}$$

3) Since $P_B^{sat} = 757.6\,\text{mmHg} > P_T^{sat} = 290.5\,\text{mmHg}$, then benzene (B) is more volatile than toluene (T). Beneze will be substantially in the top (distillate) stream and toluene in the bottom.

6.6 MATLAB Code for Estimating the Saturation Pressure or Temperature for a Pure Component

Appendix D | D.2 MATLAB Code (**AntoinEq.m**) for Antoine's Equation shows MATLAB code for estimating either the vapor pressure for a given saturation temperature or estimating the temperature for a given saturation pressure (i.e., vapor pressure). The database (i.e., Excel sheet) contains five thousand (5000) different species. The reference for data is: Yaws et al. [1].

NOTE #1

Since the database is too huge to pinpoint the right code number for the desired component, here are three remedies to follow up:

1) You go first and open the Microsoft Excel file and use "Find" button found in "Editing" group under "Home" tab found in the top toolbar. Enter the search text and keep looking for the right record or row. Alternatively, you may also use the shortcut key combination: Ctrl+F.
2) Manually search by the naked eye as the database compounds are alphabetically ordered from *A* to *Z*.
3) Alternatively, when you run MATLAB code, and you are prompted by a huge list of potential compounds, while in MATLAB's Command Window, right-click the mouse and select Find from the context menu, key in the search text and keep checking the right entry. Alternatively, you may also use the shortcut key combination: Ctrl+F as in Microsoft Excel sheet.
4) You can search either by 1. Name or 2. Formula. For example, searching benzene by name will populate a dataset with large records whereas by formula it will be with less records.

Please be careful when you run the code and select the entry or record number for the given species. There are different synonyms for the same species, which will be searched by MATLAB text finder to help facilitate reaching the desired component. The human language description of the algorithm is shown at the start of Appendix D. Below is a sample of running MATLAB code, where the **bold font** represents the user's input data or final results.

```
>> AntoinEq
Input "1" to search by Compound Name "2" to search by Formula: 2
Input the chemical formula, say C3H6O: C6H6
{[448]}  {'1,2,3-benzenetriol,   C.I. 76515,   C....'}  {'C6H6O3'}
{[3415]}  {'(6)annulene,   benzeen,   benzen,   b...'}  {'C6H6'}
Input the number for the chosen compound, say "1" for methane: 3415
Input "1" to calculate T or "2" to calculate P: 2
Input the temperature in oC for the chosen compound: 85
   {'The material is:'}   {'benzene'}
   {'The Molecular Formula = '}   {'C6H6'}
The estimated Psat in mm Hg = 883.1484
The estimated Psat in bar = 1.1774
The estimated Psat in atm = 1.162
```

6.7 The Gibbs Phase Rule

When two phases come into contact, there is typically a redistribution of their components through processes such as evaporation, condensation, dissolution, and precipitation. This continues until a state of equilibrium is reached, characterized by equal temperatures and pressure values in both phases, and a composition in each phase that remains constant over time.

Consider a closed vessel containing components A and B, distributed between gas and liquid phases. To precisely replicate this system, providing details such as system temperature, pressure, the mass of each phase, and the mass (or mole) fraction of one component in each phase would be adequate. However, these variables are not all independent; specifying some will inherently imply or allow the calculation of others based on the system's physical properties and constraints.

Variables describing the state of a process system can be categorized into two types: extensive variables, dependent on the system's size (e.g., mass and volume), and intensive variables, which are size-independent (e.g., temperature, pressure, density, specific volume, and mass and mole fraction of an individual system component in each phase).

Let's examine a phase, defined in this context as a homogeneous region with consistent intensive properties at equilibrium. This phase could manifest as a liquid, vapor, or solid, and may be continuous or discontinuous (dispersed). The degrees of freedom (DoF) of a system represents the number of independently specifiable intensive variables under equilibrium conditions. Essentially, the DoF indicates the number of independent intensive variables required to completely describe our system. In a state of equilibrium, specifying this number allows us to fully characterize the system, as the remaining intensive variables can be derived or deduced.

For a multi-component multiphase system without chemical reactions, the expression for DoF is:

$$\mathrm{DoF} = N + 2 - \pi \tag{6.22}$$

where

N: Number of components present in the multiphase system,
π: Number of phases present within the control volume/control mass, and
2: Accounts for the number of macroscopic equilibrium variables; namely, T and P.

If there are r independent chemical reactions at equilibrium, then DoF will be given by:

$$\mathrm{DoF} = N + 2 - \pi - r \tag{6.23}$$

Example 6.3 Use the phase rule to determine the DoF of each of the following *equilibrium* systems and give a possible set of variables that may be specified.

a) KCI crystals suspended in an aqueous KCI solution.

$$N = 2 = H_2O + KCl$$
$$\pi = 2 = \text{Solid crystal} + \text{Aqueous liquid}$$
$$\mathrm{DoF} = N + 2 - \pi = 2 + 2 - 2 = 2$$

We have *three* intensive variables: P, T, and X_{KCl}. X_{KCl} represents the mole fraction of KCl in the aqueous phase mixture. Notice that X_{H2O} will be $(1 - X_{KCl})$. Moreover, mass or weight fraction of KCl in solution is synonymous to its mole fraction in solution. Knowing one will enable the evaluation of the other based on the molecular weight data of participating compounds, namely, KCl and water. The probability of having *two* variables randomly selected out of *three* intensive variables is given by calculating the number of possible combinations:

There are $_n\mathrm{Cr} = \binom{n}{r} = \binom{3}{2} = \frac{n!}{(n-r)! \times r!} = \frac{3!}{1! \times 2!} = 3$ combinations

They are: (P, T), (T, X_{KCl}), and (P, X_{KCl}).

b) Humid air in equilibrium with condensed water (dry air may be considered a single species).

$$N = 2 = H_2O + \text{Air}$$
$$\pi = 2 = \text{Gas mixture} + \text{Liquid water}$$
$$\mathrm{DoF} = N + 2 - \pi = 2 + 2 - 2 = 2$$

We have *three* intensive variables: P, T, and Y_{H_2O}. Y_{H_2O} represents the mole fraction of H_2O in the gas phase mixture. Notice that Y_{Air} will be $(1 - Y_{H_2O})$. Notice that we will assume that liquid water is already saturated by dissolved air ($O_2 + N_2$). This means that the net diffusion of air molecules into and out of liquid water will be zero. However, water molecules can still vaporize off the liquid or condense back into water. The probability of having *two* variables randomly selected out of *three* intensive variables is given by calculating the number of possible combinations:

There are $_n\mathrm{Cr} = \binom{n}{r} = \binom{3}{2} = \frac{n!}{(n-r)! \times r!} = \frac{3!}{1! \times 2!} = 3$ combinations

They are: (P, T), (T, Y_{H_2O}), and (P, Y_{H_2O}).

c) A vapor–liquid mixture of four hydrocarbons.

$N = 4 =$ Four hydrocarbons

$\pi = 2 =$ Vapor mixture + Liquid mixture

$\text{DoF} = N + 2 - \pi = 4 + 2 - 2 = 4$

We have *eight* intensive variables: P, T, X_1, X_2, X_3, Y_1, Y_2, and Y_3. Xi represents the mole fraction of species i in the liquid phase mixture and Y_i in the vapor phase mixture. Notice that X_4 will be $(1 - (X_1 + X_2 + X_3))$ and Y_4 will be $(1 - (Y_1 + Y_2 + Y_3))$. The probability of having *four* variables randomly selected out of *eight* intensive variables is given by calculating the number of possible combinations:

There are $_n\text{Cr} = \binom{n}{r} = \binom{8}{4} = \dfrac{n!}{(n-r)! \times r!} = \dfrac{8!}{4! \times 4!} = 70$ combinations

Examples are:

(P, T, X_1, X_2), (P, T, Y_1, Y_2), (P, T, X_1, X_3), (P, T, Y_1, Y_3), (P, T, X_2, X_3), (P, T, Y_2, Y_3), (X_1, X_2, Y_1, Y_2), ... etc.

d) A gaseous mixture of N_2, O_2, and Ar.

$N = 3 = N_2 + O_2 + \text{Ar}$

$\pi = 1 =$ Gas mixture

$\text{DoF} = N + 2 - \pi = 3 + 2 - 1 = 4$

We have *four* intensive variables: P, T, Y_{N_2}, and Y_{O_2}. Yi represents the mole fraction of species i in the gas phase mixture. Notice that Y_{Ar} will be $(1 - (Y_{N_2} + Y_{O_2}))$.

The probability of having *four* variables randomly selected out of *four* intensive variables is given by calculating the number of possible combinations:

There are $_n\text{Cr} = \binom{n}{r} = \binom{4}{4} = \dfrac{n!}{(n-r)! \times r!} = \dfrac{4!}{0! \times 4!} = 1$ combination

It is only: (P, T, Y_{N_2}, Y_{O_2}).

e) A gaseous mixture of H_2, Cl_2, and HCl, given that the single reaction

$$H_2 + Cl_2 \rightleftharpoons 2HCl \quad (6.24)$$

$N = 3 = H_2 + Cl_2 + HCl$

$\pi = 1 =$ Gas mixture

$r = 1 =$ One independent reaction

$\text{DoF} = N + 2 - \pi - r = 3 + 2 - 1 - 1 = 3$

We have *four* intensive variables: P, T, Y_{H_2}, and Y_{Cl_2}. Y_i represents the mole fraction of species i in the gas phase mixture. Notice that Y_{HCl} will be $(1 - (Y_{H_2} + Y_{Cl_2}))$.

The probability of having *three* variables randomly selected out of *four* intensive variables is given by calculating the number of possible combinations:

There are $_n\text{Cr} = \binom{n}{r} = \binom{4}{3} = \dfrac{n!}{(n-r)! \times r!} = \dfrac{4!}{1! \times 3!} = 4$ combinations

They are:

(P, T, Y_{H_2}), (P, T, Y_{Cl_2}), (P, Y_{H_2}, Y_{Cl_2}), and (T, Y_{H_2}, Y_{Cl_2}).

Notice here that DoF was reduced by 1 simply because the presence of an independent equilibrium chemical reaction will give us one extra piece of information about the relationship among those four intensive variables given by:

$$\Delta G^0 = -RT\ln K_P = -RT\ln \frac{p_{HCl}^2}{p_{H_2}p_{Cl_2}} = -RT\ln K_Y = -RT\ln \frac{Y_{HCl}^2}{Y_{H_2}Y_{Cl_2}}$$

$$= -RT\ln \frac{(1-Y_{H_2}-Y_{Cl_2})^2}{Y_{H_2}Y_{Cl_2}} \tag{6.25}$$

ΔG^0 accounts for the standard molar Gibb free energy change for a given reaction at a given reference temperature, T_0. Again, as said in Chapter 4, it is either tabulated in literature for some common chemical reactions or can be deduced from the standard molar Gibbs free energy of formation for products and reactants.

$$\Delta G^0 = \sum_{i=1}^{N} \left(\Delta G_f^0\right)_{\text{Products}} - \sum_{i=1}^{N} \left(\Delta G_f^0\right)_{\text{Reactants}} \tag{6.26}$$

Notice that $K_C = K_P = K_Y$ since $\Delta n = [2 - 1 - 1] = 0$ for the reaction shown in Eq. 6.24. See Ch. 4 | 4.4.3 Chemical Equilibrium and Equilibrium Constant section for recalling the relevant topic.

Example 6.4 More examples on evaluating DoF for equilibrium systems.

a) **Solid Crystal Saturated Liquid Solution Equilibrium System**: $CaCl_2$ crystals (solid) in contact with an aqueous $CaCl_2$ solution (liquid).
- $N = 2$ ($CaCl_2$, H_2O)
- $p = 2$ (solid, liquid solution)
- DoF = 2 + 2 - 2 = 2

We have *three* intensive variables: P, T, and X_{CaCl_2}. X_{CaCl_2} represents the mole fraction of $CaCl_2$ in the aqueous phase mixture. Notice that X_{H_2O} will be $(1 - X_{KCl})$.

The probability of having *two* variables randomly selected out of *three* intensive variables is given by calculating the number of possible combinations:

There are $_n\text{Cr} = \binom{n}{r} = \binom{3}{2} = \frac{n!}{(n-r)! \times r!} = \frac{3!}{1! \times 2!} = 3$ combinations

They are: (P, T), (T, X_{CaCl_2}), and (P, X_{CaCl_2}).

b) **Vapor–Liquid Equilibrium Binary System**: A vapor mixture of methanol (M_OH) and ethanol (E_OH) in contact with a liquid mixture of both.
- $N = 2$ (M_OH, E_OH)
- $p = 2$ (vapor, liquid)
- DoF = 2 + 2 − 2 = 2

We have *four* intensive variables: P, T, Y_{M_OH}, and X_{M_OH}. Y_i represents the mole fraction of species i in the vapor phase mixture. Notice that Y_{E_OH} will be $(1 - Y_{M\_OH})$. On the other hand, X_i represents the mole fraction of species i in the liquid phase mixture. Notice that X_{E_OH} will be $(1 - X_{M\_OH})$.

The probability of having *two* variables randomly selected out of *four* intensive variables is given by calculating the number of possible combinations. There are:

$$_n\text{Cr} = \binom{n}{r} = \binom{4}{2} = \frac{n!}{(n-r)! \times r!} = \frac{4!}{2! \times 2!} = 6 \text{ combinations}$$

They are: (P, T), (P, Y_{M_OH}), (P, X_{M_OH}), (T, Y_{M_OH}), (T, X_{M_OH}), and (Y_{M_OH}, X_{M_OH}).
At this stage, we will not elaborate further, but we would like to say that each of the above six cases represents a possible scenario of vapor–liquid equilibrium (VLE) for a binary system. We will come later to what is called Raoult's law and the bubble- and dew-point calculations.

c) **Reactive Gaseous Mixture with an Independent Reaction System**: A gas mixture of CO, H_2O, CO_2, and H_2 under equilibrium condition with the following reaction:
$CO + H_2O \rightleftharpoons CO_2 + H_2$
- $N = 4$ (CO, H_2O, CO_2, and H_2)
- $p = 1$ (gas mixture)
- $r = 1$ (one independent reaction)
- $\text{DoF} = 4 + 2 - 1 - 1 = 4$

We have *five* intensive variables: P, T, Y_{CO}, Y_{H_2O}, and Y_{CO_2}. Y_i represents the mole fraction of species i in the gas phase mixture. Notice that Y_{H2} will be $(1 - (Y_{CO} + Y_{H_2O} + Y_{CO_2}))$.
The probability of having *four* variables randomly selected out of *five* intensive variables is given by calculating the number of possible combinations. There are:

$$_n\text{Cr} = \binom{n}{r} = \binom{5}{4} = \frac{n!}{(n-r)! \times r!} = \frac{5!}{1! \times 4!} = 5 \text{ combinations}$$

They are: (P, T, Y_{CO}, Y_{H_2O}), (P, T, Y_{CO}, Y_{CO_2}), $(P, T, Y_{H_2O}, Y_{CO_2})$, $(P, Y_{CO}, Y_{H_2O}, Y_{CO_2})$, and $(T, Y_{CO}, Y_{H_2O}$, and $Y_{CO_2})$.

6.8 Vapor–Liquid Equilibrium (VLE)

In chemical engineering, VLE refers to the state where a substance simultaneously exists in both vapor (gaseous) and liquid phases while maintaining a balance between the two. This equilibrium occurs at a specific temperature and pressure when the rates of evaporation (conversion of liquid into vapor) and condensation (conversion of vapor into liquid) are equal. It's a crucial concept because it helps us understand and predict how different components in a mixture distribute themselves between the vapor and liquid phases.

The following derivations are excerpted from: Al-Malah [2].

From a solution thermodynamics point of view, the nonideal liquid mixture can be described by introducing the concept of fugacity, which basically accounts for the effective "thermodynamic" presence of species i in a solution. *This means that its thermodynamic concentration not only reflects its count (or composition) but also extends to include its "social" interaction with its neighbors, the so-called activity coefficient.* Thus, the fugacity of species i in a mixture is given by:

$$\hat{f}_i^L = X_i \times \gamma_i \times f_i^o \tag{6.27}$$

where

$\hat{f}_i^L$: The fugacity of species i in a liquid mixture at the given pressure, temperature, and composition,
X_i: The mole fraction of species i in a mixture,

γ_i: The activity coefficient of species i in a solution at the given pressure, temperature, and composition, and
f_i^o: The fugacity of pure liquid species i at the given pressure and temperature.

On the other side, and in a similar fashion, the gas phase fugacity can be expressed as:

$$\hat{f}_i^V = \hat{\varphi}_i \times Y_i \times P \tag{6.28}$$

where

$\hat{f}_i^V$: The fugacity of species i in a vapor/gas mixture at the given pressure, temperature, and composition,
Y_i: The mole fraction of species i in a vapor/gas mixture,
$\hat{\varphi}_i$: The fugacity coefficient of species i in a vapor mixture at the given pressure, temperature, and composition, and
P: The system pressure.

At VLE condition, both fugacities are equal:

$$\hat{f}_i^L = X_i \times \gamma_i \times f_i^o = \hat{f}_i^V = \hat{\varphi}_i \times Y_i \times P \tag{6.29}$$

Equation (6.29) represents the general case for predicting a multiphase, multicomponent system where both liquid and vapor mixtures are nonideal solutions.

However, there are simpler versions of Eq. (6.29). Let's start with the vapor phase mixture.

At low pressure, the fugacity coefficient of species i, $\hat{\varphi}_i$, in a gas phase mixture reduces to unity, and we have:

$$\hat{f}_i^V = Y_i \times P \tag{6.30}$$

In addition, we can replace f_i^o by $P_i^{sat}(T)$, the vapor pressure of pure i at the given system temperature, T. Notice that $P_i^{sat}(T)$ is given by Antoine's equation (Eqs. 6.18–6.19).

Thus Eq. (6.29) reduces to:

$$X_i \times \gamma_i \times P_i^{sat}(T) = Y_i \times P \tag{6.31}$$

Equation (6.31) is called the modified Raoult's law for predicting the VLE relationship for a given species i.

A further simplified version of Eq. (6.31) can be deduced if we have an ideal liquid mixture, where the activity coefficient is reduced to unity. Thus, Eq. (6.31) becomes:

$$X_i \times P_i^{sat}(T) = Y_i \times P \tag{6.32}$$

Equation (6.32) is called Raoult's law, which represents the simplest case, that is, ideal liquid mixture-ideal gas mixture under vapor–liquid phase equilibrium (VLE), where both fugacities are set equal.

$$\sum_{i=1}^{N} X_i \times P_i^{sat}(T) = \sum_{i=1}^{N} Y_i \times P = P\sum_{i=1}^{N} Y_i = P = P_{\text{Bubble}} \tag{6.33}$$

Alternatively,

$$X_i = \frac{Y_i \times P}{P_i^{sat}(T)} \tag{6.34}$$

Let us take the summation for both sides:

$$\sum_{i=1}^{N} X_i = \sum_{i=1}^{N} \frac{Y_i \times P}{P_i^{sat}(T)} \tag{6.35}$$

The left side of Eq. (6.35) reduces to unity as the summation of X_i goes to 1.00. Also, P can be factored out.

$$1 = \sum_{i=1}^{N} \frac{Y_i \times P}{P_i^{sat}(T)} = P \sum_{i=1}^{N} \frac{Y_i}{P_i^{sat}(T)} \quad \rightarrow P_{Dew} = \frac{1}{\sum_{i=1}^{N} \frac{Y_i}{P_i^{sat}(T)}} = \frac{1}{\frac{Y_A}{P_A^{sat}(T)} + \frac{Y_B}{P_B^{sat}(T)} + \ldots} \tag{6.36}$$

Notice that P_{Bubble} (Eq. 6.33) is expressed in terms of vapor pressures and liquid mole fractions, whereas P_{Dew} (Eq. 6.36) is expressed in terms of vapor pressures and vapor mole fractions. In general, $P_{Bubble} \neq P_{Dew}$.

6.9 Bubble-point and Dew-point *P/T* Calculations

Referring to case #b in Example 6.4, we have shown that the six possible scenarios are: (P, T), (P, Y_{M_OH}), (P, X_{M_OH}), (T, Y_{M_OH}), (T, X_{M_OH}), and (Y_{M_OH}, X_{M_OH}). For simplicity, we will use Raoult's model (Eq. 6.32) to explore all such scenarios, and we will calculate the mole fraction of methanol in both phases with the understanding that ethanol will be the complement of methanol to 100% mole percent. Appendix E | Bubble-point and Dew-point *P/T* Calculations for Ideal, Binary Mixtures contains MATLAB code for handling any ideal, binary-component mixture selected out of the 5000-component database. The user can select any of the above-listed six possible scenarios. The solution for the methanol–ethanol VLE system is shown below for all six cases, both manually and using the MATLAB code available in Appendix E | MATLAB **Bubble_Dew.m** Code for Solving the Six Potential Scenarios Describing VLE of An Ideal, Binary Mixture.

Case #1 Given P and T, find X_{M_OH} and Y_{M_OH} in a VLE condition.

Since the normal boiling point temperature of M_OH = 64.7°C (148.5°F) and that of E_OH is 78.4°C (173°F), then will choose P = 1 atm and T somewhere between the lower and upper boiling points, say T = 70°C, to assure a two-phase system. Using MATLAB Antoine equation, here are the results for both materials: P_{M_OH} = 1.2447 bar = 1.2284 atm and P_{E_OH} = 0.73177 bar = 0.7222 atm. The database code number for methanol is **4268** and that of ethanol is **3945**.

$$P_{Bubble} = 1\ \text{atm} = X_{M\_OH}P_{M\_OH} + (1 - X_{M\_OH})P_{E\_OH}$$

$$P_{Bubble} = 1\ \text{atm} = X_{M\_OH} \times 1.2284\ \text{atm} + (1 - X_{M\_OH}) \times 0.7222\ \text{atm}$$

$$(1\ \text{atm} - 0.7222\ \text{atm}) = X_{M\_OH} \times (1.2284\ \text{atm} - 0.7222\ \text{atm})$$

$$X_{M\_OH} = \frac{(1\ \text{atm} - 0.7222\ \text{atm})}{(1.2284\ \text{atm} - 0.7222\ \text{atm})} = 0.5488$$

Solve for $X_{E\_OH} = 1 - 0.5488 = 0.4512$.

$$X_i \times P_i^{sat}(T) = Y_i \times P \quad \rightarrow 0.5488 \times 1.2284 \text{ atm} = Y_{M\_OH} \times 1.0 \text{ atm}$$

$$Y_{M\_OH} = 0.6741 \qquad Y_{E\_OH} = 1 - 0.6741 = 0.3259$$

$$P_{Dew} = \frac{1}{\frac{0.6741}{1.2284} + \frac{0.3259}{0.7222}} = 1.0 \text{ atm}$$

Below are MATLAB **Bubble_Dew.m** Case #1 results. The results for the estimated X_{M_OH} (X_1) and Y_{M_OH} (Y_1) are the same. Keep in mind that MATLAB-generated results or more precise than those manually estimated. The user may select ethanol first and methanol second but he/she has to keep track on the order they are entered. The results will show which one is which.

Please choose one of the following cases:
1. Given P and T, find X and Y for Binary Mixture
2. Given P and Y, find X and T for Binary Mixture
3. Given P and X, find Y and T for Binary Mixture
4. Given T and Y, find X and P for Binary Mixture
5. Given T and X, find Y and P for Binary Mixture
6. Given X and Y, find P and T for Binary Mixture

Enter your choice (1-6): **1**
You have chosen Case: 1. Given P and T, find X and Y for Binary Mixture
Input pressure of system in atm: **1**
Input the temperature in oC for the system: **70**
{'The material #1 is:'} {'methyl alcohol'}
{'The material #2 is:'} {'ethyl alcohol'}
The given Tsat in oC = 70
The given Psat in atm = 1
The estimated X1 = 0.54876
The estimated Y1 = 0.67411

Figure 6.1 shows a simplified version, only for the methanol and ethanol VLE system, of the general MATLAB code (Appendix F) for generating a 2-in-1 plot, representing P_{Bubble} and P_{Dew} for any binary VLE system. Figure 6.2 shows the 2-in-1 plot for the methanol–ethanol VLE system. Notice that $P_{Dew} = P_{Bubble} = P_{System} = 1.0$ atm, which is shown as a horizontal, equilibrium tie line in Figure 6.2. This horizontal line is a tie (or equilibrium) line that connects X_i with Y_i at the same P and T. Of course, $T = 70°C$.

Moreover, the dew-point pressure curve lies below the bubble-point pressure curve (i.e., line for Raoult's case). This means at high pressure for the same temperature, we should expect the presence of a liquid phase, and at low pressure, the vapor phase. The liquid and vapor phase regions are also shown in Figure 6.2. In terms of phase region location, the opposite will be true for temperature versus X_i or Y_i. The envelope area bounded by the bubble-point pressure and dew-point pressure curve represents the liquid–vapor phase region where both coexist at given T and P.

```
% Define functions
PMOH=1.2284; PEOH=0.7222; %@T=70.0°C
Pbubble = @(X) X*PMOH+(1-X)*PEOH;  % Define PBubble
Pdew = @(Y) 1./(Y/PMOH+(1.0-Y)/PEOH);  % Define PDew
% Define X and Y range
X = linspace(0, 1, 100);  % Create 100 evenly spaced points between 0 and 1
Y = linspace(0, 1, 100);  % Create 100 evenly spaced points between 0 and 1
% Calculate Y values for both functions
y1 = Pbubble(X);
y2 = Pdew(Y);
% Create a horizontal line for P=1.0 atm
y3=linspace(1, 1, 100);  % Create 100 evenly spaced points between 1 and 1
% Create the combined plot
figure;  % Create a new figure window
hold on;  % Allow multiple plots on the same axes
plot(X, y1, 'r', 'LineWidth', 2);  % Plot Pbubble(X) in red
plot(Y, y2, 'b--', 'LineWidth', 1.5);  % Plot Pdew(Y) in blue dashed line
plot(X, y3, 'k--', 'LineWidth', 1.5);  % Plot P=1.0 atm in black dashed line
% Add labels and title
xlabel('X');
ylabel('P');
title('PBubble and PDew on the Same Plot');
% Add legend
legend('PBubble(X)', 'PDew(Y)');
% Customize grid and appearance
grid on;
box on;
hold off; % Release hold on the current axes
```

Figure 6.1 MATLAB code for plotting P_{Bubble} and P_{Dew} for the methanol–ethanol VLE system @70C.

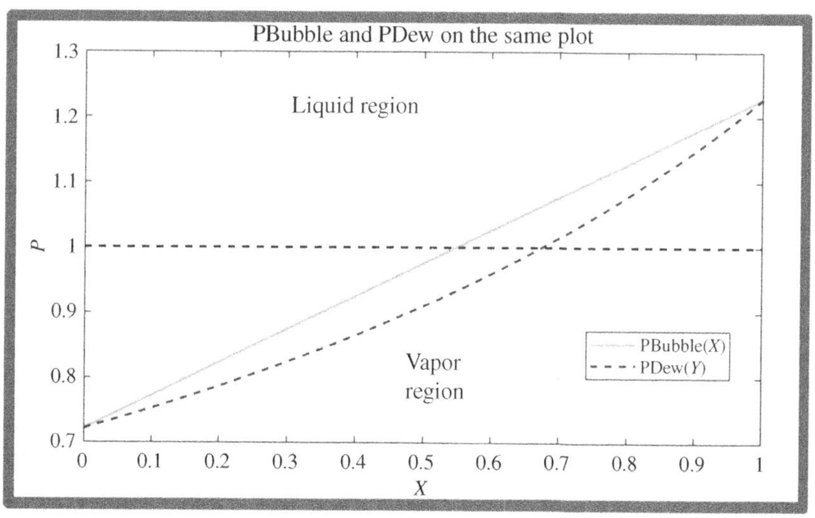

Figure 6.2 The bubble-point and dew-point pressure curve for the methanol–ethanol two-phase system, evaluated at $T = 70°C$. The dashed tie line represents $P = P_{DB} = P_{BP} = 1.0$ atm.

Appendix F | "F.2 P_BP_DP_Plot.m MATLAB Code" section generates a 2-in-1 plot for both bubble-point and dew-point pressure curves for any given ideal binary VLE system selected out of the 5000-component database. Below is an example of how to run **P_BP_DP_Plot.m** MATLAB code for acetone–benzene binary mixture. T is selected between the lower and upper normal boiling point temperatures of acetone and benzene, respectively. A system pressure of 1.0 atm is chosen to show the equilibrium tie line between liquid and vapor phases.

```
>> P_BP_DP_Plot
Search for the first component
Input "1" to search by Compound Name "2" to search by Formula: 1
Input the name of the Compound, say acetone: acetone
{[3384]} {'aceton,    acetone,    a...'}   {'C3H6O'}
Input the number for the chosen compound, say "3384" for acetone: 3384
Search for the second component
Input "1" to search by Compound Name "2" to search by Formula: 1
Input the name of the Compound, say benzene: benzene
{[3415]} {'(6)annulene,    benzeen,...'}   {'C6H6'}
Input the number for the chosen compound, say "3415" for benzene: 3415
Input system temperature in oC to create 2-in-1 plot: 65
Input system pressure in atm to create a horizontal tie line: 1.0
```

Figure 6.3 shows the generated plot for running **P_BP_DP_Plot.m** file. Again, the liquid phase resides above the bubble-point pressure curve (red color curve); the vapor phase below the dew-point pressure curve (blue color curve); and the two-phase region within the envelope. Notice that the two extreme points (corners) of the envelope represent the pure state of each compound. At the far left, this corner represents the saturation vapor pressure at $T = 65°C$ of pure benzene (i.e., $X_{\text{Acetone}} = X_1 = 0$). On the contrary, at the far right, the corner represents the vapor pressure at the same T for acetone at $X_1 = 1.0$ (pure acetone). Notice that I wrote MATLAB code considering X to be the mole fraction of the first component no matter which one is selected first.

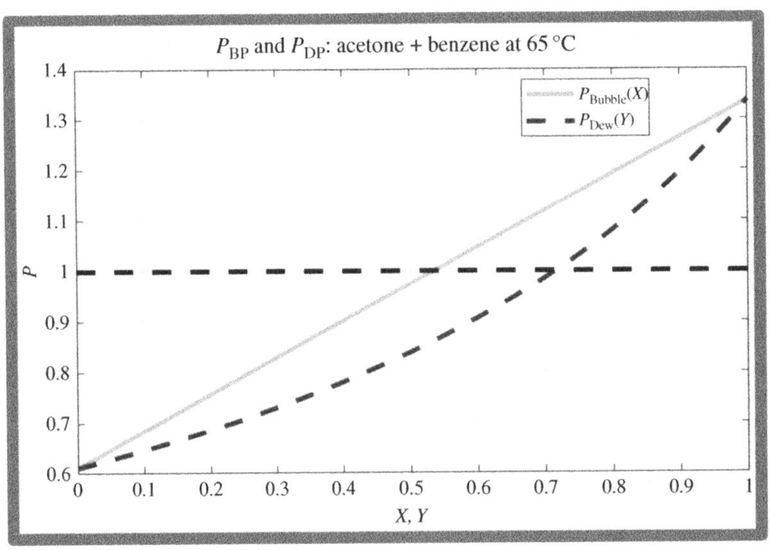

Figure 6.3 The bubble-point and dew-point pressure curve for acetone benzene mixture at 65°C, with a horizontal tie line at 1.0 atm.

Case #2 Given P and Y_{M_OH}, find X_{M_OH} and T in a VLE condition.

Let us choose $P = 1.0$ atm and $Y_{M\_OH} = 0.70$. Notice here that T is missing. Selecting the right T will dictate the right liquid composition. In other words, it is a matter of trial and error to find the right T.

Trial #1

Let us assume $T = 70°C$. The vapor pressure data for methanol and ethanol are given in Case #1. Let us use Eq. (6.34) to find X_{M_OH} and X_{E_OH}.

$$X_{M\_OH} = \frac{Y_{M\_OH} \times P}{P^{sat}_{M\_OH}(T)} = \frac{0.70 \times 1.0 \text{ atm}}{1.2284} = 0.5698$$

$$X_{E\_OH} = \frac{Y_{E\_OH} \times P}{P^{sat}_{E\_OH}(T)} = \frac{0.30 \times 1.0 \text{ atm}}{0.7222} = 0.4154$$

Check the sum of X_i: $0.5698 + 0.4154 = 0.9852 < 1.0$.

Keep trying. Where to go? You can either increase or decrease T. Since the sum = 0.9852 < 1.0, then we need to decrease T such that both P_{isat} values in the denominator will decrease accordingly; hence, the sum will get it closer and closer to 1.0.

Trial #2

Let us run the second trial:

$$T = 68°C \qquad P^{sat}_{M\_OH} = 1.1376 \text{ atm} \qquad P^{sat}_{E\_OH} = 0.6659 \text{ atm}$$

$$X_{M\_OH} = \frac{Y_{M\_OH} \times P}{P^{sat}_{M\_OH}(T)} = \frac{0.70 \times 1.0 \text{ atm}}{1.1376} = 0.61533$$

$$X_{E\_OH} = \frac{Y_{E\_OH} \times P}{P^{sat}_{E\_OH}(T)} = \frac{0.30 \times 1.0 \text{ atm}}{0.6659} = 0.45052$$

Check the sum of X_i: $0.61533 + 0.45052 = 1.0658 > 1.0$.

Trial #3

So, the third trial should lie between 68 and 70°C and be around 69.5°C:

$$T = 69.5°C \qquad P^{sat}_{M\_OH} = 1.2052 \text{ atm} \qquad P^{sat}_{E\_OH} = 0.70777 \text{ atm}$$

$$X_{M\_OH} = \frac{Y_{M\_OH} \times P}{P^{sat}_{M\_OH}(T)} = \frac{0.70 \times 1.0 \text{ atm}}{1.2052} = \mathbf{0.5808} \qquad X_{E\_OH} = \frac{Y_{E\_OH} \times P}{P^{sat}_{E\_OH}(T)}$$

$$= \frac{0.30 \times 1.0 \text{ atm}}{0.70777} = 0.4238$$

Check the sum of X_i: $0.5808 + 0.4238 = 1.0046 \approx 1.0$.

The correct temperature will be **69.5°C**.

In other words, the liquid phase with (0.5808, 0.4238) composition is in equilibrium with the vapor phase (0.7, 0.3) composition at $T = 69.5°C$ and $P = 1.0$ atm. Below is a portion of MATLAB **Bubble_Dew.m** Case #2 results:

```
Enter your choice (1-6): 2
You have chosen Case: 2. Given P and Y, find X and T for Binary Mixture
Input pressure of system in atm: 1.0
Input the vapor mole-fraction for component #1: 0.7
**Solution found after 27 iterations.
Temperature: 69.625°C
The Liquid composition is below:
 {'methyl alcohol'}   {': X1 is: '}   {'0.57805'}
 {'ethyl alcohol'}   {': X2 is: '}   {'0.42173'}
```

In addition, Appendix F | **T_BP_DP_Plot.m** MATLAB code generates a 2-in-1 plot for both bubble-point and dew-point temperature curves for any given ideal binary VLE system selected out of the 5000-component database. Below is an example of how to run **T_BP_DP_Plot.m** MATLAB code for the acetone–benzene binary mixture. P is selected as 1.0 atm. The temperature is iteratively solved such that to find the root for both of the following two functions:

```
f = @(T, X) Pbubble_target - (X.*Patm1(T) + (1-X).*Patm2(T));
f = @(T, Y) PDew_target - 1./((Y./Patm1(T)) + ((1-Y)./Patm2(T)));
```

For the first function, at the correct temperature and X, Pbubble_target=Pbubble=1.0 atm. For the second function, at the correct temperature and Y, PDew_target=PDew=1.0 atm.

NOTE #2

It is not only one single T that will give the root. There is a series of T values that will map with a series of X values for P_{BP} estimation and with a series of Y values for P_{DP} estimation.

```
>> T_BP_DP_Plot
Search for the first component
Input "1" to search by Compound Name "2" to search by Formula: 1
Input the name of the Compound, say acetone: acetone
 {[3384]}  {'aceton,   acetone,   a...'}  {'C3H6O'}
Input the number for the chosen compound, say "3384" for acetone: 3384
Search for the second component
Input "1" to search by Compound Name "2" to search by Formula: 1
Input the name of the Compound, say benzene: benzene
{[3415]} {'(6)annulene,   benzeen,...'}   {'C6H6'}
Input the number for the chosen compound, say "3415" for benzene: 3415
Input system pressure in atm to create plots: 1
```

Figure 6.4 shows the generated plot for running **T_BP_DP_Plot.m** file. Now, opposite to what is shown in either Figure 6.2 or Figure 6.3, the liquid phase resides below the bubble-point temperature curve (red color curve); the vapor phase above the dew-point temperature curve (blue color curve); and the two-phase region within the envelope. Again, notice that the

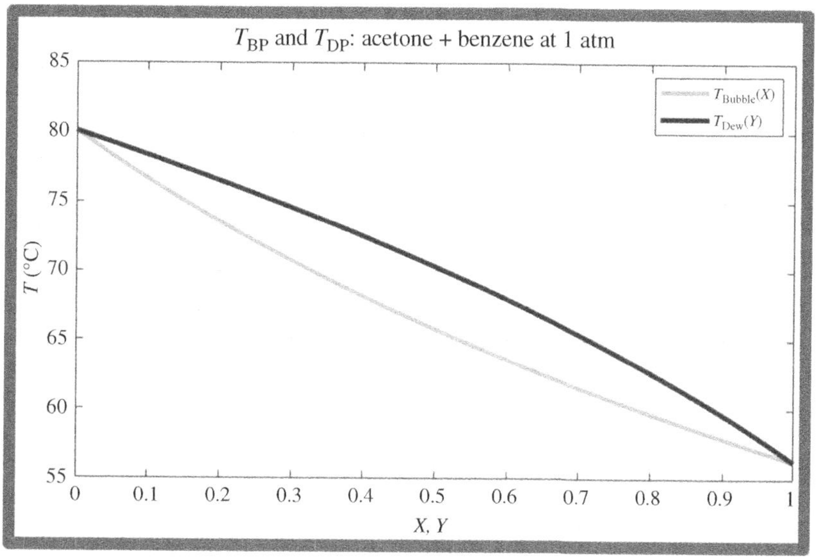

Figure 6.4 T_{BP} and T_{DP} for the acetone–benzene VLE system, evaluated at 1.0 atm. The extreme left corner is the normal boiling point T of benzene, and on the extreme right, it is the normal boiling point T of acetone.

two extreme points (corners) of the envelope represent the pure state of each compound. At the far left, this corner represents the normal boiling point temperature at $P = 1.0$ atm of pure benzene (i.e., $X_{Acetone} = X_1 = 0$). On the contrary, at the far right, the corner represents the boiling point temperature at the same P for acetone at $X_1 = 1.0$ (pure acetone). Notice that I wrote MATLAB code considering X to be the mole fraction of the first component, no matter which one comes first.

For the sake of brevity, there will be no further plots for the remaining four cases with the understanding that the user can at any time generate what he/she likes, as in figures (6.3) or (6.4).

Case #3 Given P and X_{M_OH}, find Y_{M_OH} and T in a VLE condition.

Let us choose $P = 1.0$ atm and $X_{M\_OH} = 0.60$.

Trial #1

Let us assume $T = 68°C$

The vapor pressure data for methanol and ethanol are given in Case #2 | Trial #2.

$$T = 68°C \qquad P^{sat}_{M\_OH} = 1.1376 \text{ atm} \qquad P^{sat}_{E\_OH} = 0.6659 \text{ atm}$$

$$P_{Bubble} = 1 \text{ atm} = X_{M\_OH}P_{M\_OH} + (1 - X_{M\_OH})P_{E\_OH}$$

$$P_{Bubble} = 1 \text{ atm} = X_{M\_OH} \times 1.1376 \text{ atm} + (1 - X_{M\_OH}) \times 0.6659 \text{ atm}$$

$$P_{Bubble} = 1 \text{ atm} = 0.6 \times 1.1376 \text{ atm} + 0.4 \times 0.6659 \text{ atm}$$

$$= 0.9489 \text{ atm} < P_{Given} = 1.0 \text{ atm}$$

Since the calculated is less than the given pressure, then we need to increase T value.

Trial #2

Let us assume $T = 70°C$. Case #1 has the vapor pressure data.

$P_{M\_OH} = 1.2284$ atm

and

$P_{E\_OH} = 0.7222$ atm.

$$P_{Bubble} = 1 \text{ atm} = 0.6 \times 1.2284 \text{ atm} + 0.4 \times 0.7222 \text{ atm} = 1.0259 \text{ atm} > P_{Given} = 1.0 \text{ atm}$$

Trial #3

We have to slightly decrease the temperature. Let $T = 69.5°C$. The vapor pressure data are given in Case #2 | Trial #3.

$\mathbf{T = 69.5°C}$ $P^{sat}_{M\_OH} = 1.2052$ atm $P^{sat}_{E\_OH} = 0.70777$ atm

$$P_{Bubble} = 1 \text{ atm} = 0.6 \times 1.2052 \text{ atm} + 0.4 \times 0.70777 \text{ atm} = 1.0062 \text{ atm} \approx P_{Given} = 1.0 \text{ atm}$$

We will accept and terminate the trial process. Let us make use of Eq. (6.32) and apply it twice:

$$X_{M\_OH} \times P^{sat}_{M\_OH}(T) = Y_{M\_OH} \times P$$

$$0.6 \times 1.2052 \text{ atm} = Y_{M\_OH} \times 1.0 \text{ atm}$$

$$\mathbf{Y_{M\_OH} = 0.7231}$$

$$X_{E\_OH} \times P^{sat}_{E\_OH}(T) = Y_{E\_OH} \times P \qquad 0.4 \times 0.70777 \text{ atm} = Y_{E\_OH} \times 1.0 \text{ atm}$$

$$\mathbf{Y_{E\_OH} = 0.2831}$$

Check that the sum of Y_i adds up to unity: $0.72312 + 0.2831 = 1.006 \approx 1.00$

Below is a portion of running MATLAB **Bubble_Dew.m** Case #3 results:

Enter your choice (1-6): **3**
You have chosen Case: 3. Given P and X, find Y and T for Binary Mixture
Input pressure of system in atm: **1.0**
Input the liquid mole-fraction for component #1: **0.6**
**Solution found after 33 iterations.
Temperature: 69.325°C
The Vapor composition is below:
{'methyl alcohol'} {': Y1 is: '} {'0.71828'}
{'ethyl alcohol'} {': Y2 is: '} {'0.28111'}
PBubble = 0.99939
Pdew = 1

Case #4 Given T and Y_{M_OH}, find X_{M_OH} and P in a VLE condition.

Let us choose $T = 68.0°C$ and $Y_{M\_OH} = 0.60$.

The vapor pressure data for methanol and ethanol are given in Case #2, Trial #2.

$$\mathbf{T = 68°C} \qquad P^{sat}_{M\_OH} = 1.1376 \text{ atm} \qquad P^{sat}_{E\_OH} = 0.6659 \text{ atm}$$

Let us use Eq. (6.36) to calculate P_{Dew}.

$$P_{Dew} = \frac{1}{\sum_{i=1}^{N} \frac{Y_i}{P_i^{sat}(T)}} = \frac{1}{\frac{Y_{M\_OH}}{P^{sat}_{M\_OH}(T)} + \frac{Y_{E\_OH}}{P^{sat}_{E\_OH}(T)}} = \frac{1}{\frac{0.6}{1.1376} + \frac{0.4}{0.6659}} = \mathbf{0.8864\ atm}$$

$$X_{M\_OH} = \frac{Y_{M\_OH} \times P}{P^{sat}_{M\_OH}(T)} = \frac{0.60 \times 0.8864 \text{ atm}}{1.1376 \text{ atm}} = \mathbf{0.4675}$$

$$X_{E\_OH} = \frac{Y_{E\_OH} \times P}{P^{sat}_{E\_OH}(T)} = \frac{0.40 \times 0.8864 \text{ atm}}{0.6659 \text{ atm}} = \mathbf{0.5325}$$

Notice that the sum of X_i adds up to one: $0.4675 + 0.5325 = 1.0000$.

Below is a portion of running MATLAB **Bubble_Dew.m** Case #4 results:

Enter your choice (1-6): **4**
You have chosen Case: 4. Given T and Y, find X and P for Binary Mixture
Input the vapor mole-fraction of Component #1: **0.6**
Input the temperature in oC for the system: **68**
{'The material #1 is:'} {'methyl alcohol'}
{'The material #2 is:'} {'ethyl alcohol'}
The given Tsat in oC = 68
The estimated PBubble in atm = 0.88644
The estimated PDew in atm = 0.88644
The estimated X1 = 0.46752
The sum of X1+X2 = 1
The given Y1 = 0.6

Case #5 Given T and X_{M_OH}, find Y_{M_OH} and P in a VLE condition.

Let us choose $T = 68.0°C$ and $X_{M\_OH} = 0.65$.

The vapor pressure data for methanol and ethanol are given in Case #2, Trial #2.

$T = 68°C$

$P^{sat}_{M\_OH} = 1.1376 \text{ atm} \qquad P^{sat}_{E\_OH} = 0.6659 \text{ atm}$

$$P_{Bubble} = X_{M\_OH}P_{M\_OH} + (1 - X_{M\_OH})P_{E\_OH}$$

$$P_{Bubble} = 0.65 \times 1.1376 \text{ atm} + 0.35 \times 0.6659 \text{ atm} = 0.9725 \text{ atm}$$

$$X_{M\_OH} \times P^{sat}_{M\_OH}(T) = Y_{M\_OH} \times P \qquad 0.65 \times 1.1376 \text{ atm} = Y_{M\_OH} \times 0.9725 \text{ atm}$$

$\mathbf{Y_{M\_OH} = 0.76035}$

$X_{E\_OH} \times P^{sat}_{E\_OH}(T) = Y_{E\_OH} \times P \qquad 0.35 \times 0.6659 \text{ atm} = Y_{E\_OH} \times 0.9725 \text{ atm}$

$\mathbf{Y_{E\_OH} = 0.23965}$

Notice that the sum of Y_i adds up to one: 0.76035 + 0.23965 = 1.0000
Also, double check:

$$P_{Dew} = \frac{1}{\sum_{i=1}^{N} \frac{Y_i}{P_i^{sat}(T)}} = \frac{1}{\frac{Y_{M\_OH}}{P^{sat}_{M\_OH}(T)} + \frac{Y_{E\_OH}}{P^{sat}_{E\_OH}(T)}} = \frac{1}{\frac{0.76035}{1.1376} + \frac{0.23965}{0.6659}} = 0.9725 \text{ atm}$$

Below is a portion of running MATLAB **Bubble_Dew.m** Case #5 results:

Enter your choice (1-6): **5**
You have chosen Case: 5. Given T and X, find Y and P for Binary Mixture
Input the liquid mole-fraction of Component #1: **0.65**
Input the temperature in oC for the system: **68**
{'The material #1 is:'} {'methyl alcohol'}
{'The material #2 is:'} {'ethyl alcohol'}
The given Tsat in oC = 68
The estimated PBubble in atm = 0.97252
The estimated PDew in atm = 0.97252
The estimated Y1 = 0.76035
The Sum of Y1+Y2 = 1
The given X1 = 0.65

Case #6 Given X_{M_OH} and Y_{M_OH}, Find P and T in a VLE condition.

Let us choose X_{M_OH} = **0.65** and Y_{M_OH} = **0.76035**

$$P_{Bubble} = X_{M\_OH} P^{sat}_{M\_OH}(T) + (1 - X_{M\_OH}) P^{sat}_{E\_OH}(T)$$

$$P_{Bubble} = 0.65 \times P^{sat}_{M\_OH}(T) + 0.35 \times P^{sat}_{E\_OH}(T)$$

$$P_{Dew} = \frac{1}{\frac{Y_{M\_OH}}{P^{sat}_{M\_OH}(T)} + \frac{Y_{E\_OH}}{P^{sat}_{E\_OH}(T)}} = \frac{1}{\frac{0.76035}{P^{sat}_{M\_OH}(T)} + \frac{0.23965}{P^{sat}_{E\_OH}(T)}}$$

The trial-and-error approach or algorithm is:

1. Assume T
2. Calculate both $P^{sat}_{M\_OH}(T)$ and $P^{sat}_{E\_OH}(T)$ using MATLAB Antoine equation.
3. Calculate $P_{Bubble} = 0.65 \times P^{sat}_{M\_OH}(T) + 0.35 \times P^{sat}_{E\_OH}(T)$.
4. Calculate $P_{Dew} = 1/\left(\left((0.76035)/\left(P^{sat}_{M_{OH}}(T)\right)\right) + \left((0.23965)/\left(P^{sat}_{E\_OH}(T)\right)\right)\right)$.

5. Check if $|P_{Bubble} - P_{Dew}| \leq \varepsilon = 0.00001$. If yes, terminate and accept assumed T in step #1.

6. Elseif (PBubble-PDew) < 0, then $T = T - 0.05$; % decrease T.

7. Elseif (PBubble-PDew) > 0, then $T = T + 0.05$; % increase T.

8. Go to step #2 and recalculate all parameters based on the newly assumed T.

9. Repeat steps #2–#8 until the solution converges by satisfying step #5 condition. The parameter epsilon, ε, represents the radius of convergence, which measures how precise and accurate P_{Bubble} is equal to P_{Dew}. Ideally, the exact solution should give a radius of convergence equal to zero; i.e., $P_{Bubble} = P_{Dew}$.

In our case here, this is another way of expressing Case #5. So, the solution should end up with $\mathbf{P_{Bubble} = P_{Dew} = P_{VLE} = 0.9725}$ **atm and $\mathbf{T = 68°C}$.**

Below is a portion of running MATLAB **Bubble_Dew.m** Case #6 results:

Enter your choice (1-6): **6**
You have chosen Case: 6. Given X and Y, find P and T for Binary Mixture
Input the liquid mole-fraction of Component #1: **0.65**
Input the vapor mole-fraction of Component #1: **0.76035**
**Solution found after 59 iterations.
Estimated Temperature: 68.025°C
The Liquid composition is below:
{'methyl alcohol'} {': Given X1 is: '} {'0.65'}
The Vapor composition is below:
{'methyl alcohol'} {': Given Y1 is: '} {'0.76035'}
Estimated PBubble = 0.97348 atm
Estimated Pdew = 0.97348 atm

NOTE #3

In all examined six scenarios, notice that whenever T is missing (i.e., cases: #2, #3, and #6), a trial-and-error approach is a must. The vapor pressure data are nonlinear functions in T, as shown by the Antoine equation. One can say that T is an implicit function in bubble-point and dew-point pressure calculations.

6.10 Henry's Law

Henry's law is a fundamental principle in physical chemistry that describes the relationship between the concentration of a gas in a liquid and the pressure of that gas above the liquid. It is particularly applicable to dilute solutions where the interaction between solute molecules is negligible. Henry's law is expressed as:

$$P_i = K_{P/C} C_i^{aq} \quad \text{or} \quad K_{P/C} = \frac{P_i}{C_i^{aq}} \tag{6.37a}$$

Alternatively,

$$C_i^{\text{aq}} = K_{C/P}P_i \quad \text{or} \quad K_{C/P} = \frac{C_i^{\text{aq}}}{P_i} \tag{6.37b}$$

$$P_i = K_{P/X}X_i \quad \text{or} \quad K_{P/X} = \frac{P_i}{X_i} \tag{6.37c}$$

$$C_i^{\text{aq}} = K_{C_{\text{aq}}/C_{\text{gas}}}C_i^{\text{gas}} \quad \text{or} \quad K_{C_{\text{aq}}/C_{\text{gas}}} = \frac{C_i^{\text{aq}}}{C_i^{\text{gas}}} \tag{6.37d}$$

where:

- C_i^{aq}: The concentration of the dissolved gas species i in the liquid phase (usually in mol/L or M),
- C_i^{gas}: The concentration of the gas species i in the gas above the liquid (usually in mol/L or M),
- $K_{P/C}$, $K_{C/P}$, $K_{P/X}$, and $K_{C_{\text{aq}}/C_{\text{gas}}}$: Henry's law constants, specific to a particular solute–solvent pair of units, and
- P_i is the partial pressure of the gas species i above the liquid.

Henry's law is commonly used to predict the solubility of gases in liquids, especially when the gas is not highly soluble. It is important to note that Henry's law is an empirical relationship and may not be accurate for all concentrations and conditions.

For a non-condensable gas like those commonly encountered in environmental science, such as oxygen, carbon dioxide, or volatile organic compounds, Henry's law provides insights into their behavior in water. When a gas is in contact with a liquid, the gas molecules exert pressure on the liquid surface. According to Henry's law, the concentration of the gas in the liquid is directly proportional to the partial pressure of the gas above the liquid.

The solubility of gases in liquids typically increases with increasing pressure, assuming that the temperature remains constant. This is because higher pressure results in more gas molecules being forced into the liquid phase. Henry's law constants are determined experimentally and depend on factors such as temperature and the nature of the solute and solvent.

In environmental and chemical engineering applications, Henry's law is crucial for understanding phenomena like gas dissolution in natural waters or aquatic systems, the transfer of gases across biological membranes, and the behavior of volatile compounds in various liquid phases.

Henry's law is generally applicable to solutions where the mole fraction (X_A) of component A is close to zero, indicating dilute solutions. This validity holds true, especially when A does not undergo dissociation, ionization, or reaction within the liquid phase. The law is commonly employed in scenarios involving non-condensable gases dissolved in a liquid. Perry's Chemical Engineers' Handbook (Green and Perry [3]) provides values for Henry's law constants (or gas solubility) for various gases in water on pages 2–130 through 2–133.

Appendix G | G.2 MATLAB Code for Estimating Henry's Constants as $f(T)$ section shows the code needed to estimate Henry's constants as $f(T)$ and expressed in different units, as explained in Eq. 6.37a–d. A sample result is shown below for finding Henry's K's for methane gas in water at $T = 25°\text{C}$.

```
>> Henry_K
Available gases:
Acetylene
CO2
CO
C2H6
Ethene
He
H2
CH4
N2
O2
Enter gas name: CH4
Input the temperature in K for the system: 298.15
Gas name: CH4
Temperature: 298.15 K
Mole fraction of solute: 2.5484e-05
K_{P/C} Henry Constant: 7.0703e+02
K_{C/P} Henry Constant: 1.4144e-03
K_{P/X} Henry Constant: 3.9240e+04
K_{C/C} Henry Constant: 3.4582e-02
```

NOTE #4

The units of *K*'s appearing in Eq. (6.37a–d) can be expressed in different units, depending on the units of concentration as well as on the units of pressure. Table 6.2 shows Henry's constants expressed in different units, mainly generated by MATLAB **Henry_K.m** file. References:

1) Green and Perry [3].
2) Helmenstine [4].

My way of defining Henry's constant is, for example, $K_{P/C}$ which reads as *K* sub-*P*-over-*C*; thus, it means P_i (numerator) is over C_i (denominator): $K_{P/C} = (P_i)/C_i^{aq}$, Eq. (6.37a). Another way of saying $P_i = K_{P/C}C_i^{aq}$, is P_i lies to the left side and C_i to the right side.

For dissolved gases but diluted in water, the following equations can be used:

$$P_i = PY_i \qquad C_i^{aq} = C_{total}X_i = 55.5\frac{\text{mol}}{\text{L}}*X_i \qquad C_i^{gas} = \left(\frac{P_i}{RT}\right) = \left(\frac{PY_i}{RT}\right)$$

For example, in Table 6.2:

$$\boldsymbol{K}_{P/C} = \left(\boldsymbol{K}_{P/X}/55.5\ \text{mol/L}\right)$$

$$\boldsymbol{K}_{C/P} = \left(\mathbf{1}/\boldsymbol{K}_{P/C}\right)$$

$$K_{C/C} = \frac{C_{aq}}{C_{gas}} = RT\frac{55.5\,X_i}{P_i} = 55.5RT\frac{1}{K_{P/X}} = 55.5*(0.08206*298)\frac{1}{K_{P/X}} = \frac{1.357\times10^3}{K_{P/X}}$$

All ***K*** columns in Table 6.2 can be reproduced from one another.

Table 6.2 Henry's constants for common gases in water at room temperature, 25°C.

Gas	$K_{P/C} = P_i/C_i$ (L_{Soln} · atm)/molgas	$K_{C/P} = C_i/P_i$ (molgas/(L_{Soln} · atm)	$K_{P/X} = P_i/X_i$ (atm·molsoln/molgas)	$K_{C/C} = C_{aq}/C_{gas}$ Unitless
Acetylene	24.011	4.165×10^{-2}	1.333×10^{3}	1.0183
Argon	712.7	1.4×10^{-3}	3.955×10^{4}	0.0342
C_2H_4	211.3	4.73×10^{-3}	1.173×10^{4}	0.1157
C_2H_6	529.8	1.89×10^{-3}	2.94×10^{4}	0.0461
CH_4	707.03	1.41×10^{-3}	3.924×10^{4}	0.0346
CO	1044.7	9.57×10^{-4}	5.798×10^{4}	2.340×10^{-2}
CO_2	29.46	3.39×10^{-2}	1.635×10^{3}	0.8299
H_2	1276.3	7.83×10^{-4}	7.083×10^{4}	1.916×10^{-2}
He	2575.2	3.88×10^{-4}	1.429×10^{5}	9.495×10^{-3}
N_2	1523.5	6.56×10^{-4}	8.455×10^{4}	1.605×10^{-2}
Ne	2216.2	4.5×10^{-4}	12.30×10^{4}	1.101×10^{-2}
O_2	782.83	1.28×10^{-3}	4.345×10^{4}	3.123×10^{-2}

Example 6.5
How many grams of CO_2 gas are dissolved in a 300 mL bottle of carbonated water if the manufacturer uses a pressure of 2.5 atm in the bottling process at 25°C?

Solution:

From Table 6.2, $K_{C/P}$ of CO_2 in water = 0.034 mol/(L·atm) at 25°C.

When a gas dissolves in a liquid, an equilibrium is eventually reached between the gas source and the solution. Let us apply Henry's law:

$$C_{CO_2} = K_{C/P} P_{CO_2} = 0.034 \frac{\text{mol of gas}}{L_{\text{Soln.}}.\text{atm}} \times 2.5\,\text{atm} = 0.085\,\frac{\text{mol}}{\text{L}}$$

$$n_{CO_2} = C_{CO_2} \times V_{\text{Soln.}} = 0.085\,\frac{\text{mol}}{\text{L}} \times 0.300\,\text{L} = 0.0255\,\text{mol of CO}_2$$

$$m_{CO_2} = n_{CO_2} \times \text{MW}_{CO_2} = 0.0255\,\text{mol of CO}_2 \times 44.0\,\frac{\text{g CO}_2}{\text{mol CO}_2} = 1.122\,\text{g of CO}_2$$

There are 1.122 g of CO_2 dissolved in a 0.3 L (300 mL) bottle of carbonated water from the manufacturer. Notice that prior to opening a soda can, almost all the gas present above the liquid is carbon dioxide. Upon opening the bottle, the released gas reduces the partial pressure of carbon dioxide, facilitating the liberation of dissolved gas from the solution.

This phenomenon is responsible for the effervescence or fizziness observed in soda. Moreover, the taste becomes sweet as the release of CO_2 from the solution will transform the soda solution from acidic to an almost neutral medium, giving rise to a sweeter sugar taste. That is why soda companies basically betray the drinker by masking the sugar taste.

Coke and most other soft drinks are around 10–12% sugar, thus about 30–36 g in 300 mL (https://www.cocacolaep.com/png/our-beverages/products/coca-cola-original-taste-300ml/). Keep in mind that the weight of one sugar cube can vary, but it's typically around 2–2.5 g. Thus, a 300 mL soda solution will have as low as (30 g/2.5 g/cube) = 12 sugar cubes and as high as (36 g/2.0 g/cube) = 18 sugar cubes.

6.11 Humidification/Dehumidification Systems

In such systems, we usually have water as a condensable gas while air is non-condensable. Water can exist or co-exist in two phases: liquid and vapor.

Humidification Systems:

1) **Direct Injection Humidification**:
   - In this method, water is directly injected into a gas stream, typically air.
   - Water is atomized or sprayed into the air, leading to vaporization and an increase in humidity.
2) **Adiabatic Humidification**:
   - Adiabatic humidifiers use the principle of evaporative cooling. Water is evaporated into the air, cooling it and increasing its humidity.
   - Common devices include wick-type humidifiers and spray humidifiers.
3) **Steam Injection Humidification**:
   - High-pressure saturated steam is injected into the air stream, leading to the transfer of vapor or vaporization of some condensed water and an increase in humidity of outlet air compared with inlet air.

Dehumidification Systems:

1) **Condensation Dehumidification**:
   - In this method, humid air is cooled below its dew point, causing water vapor to condense into liquid water.
   - This condensed water is then collected and removed, leaving the air drier.
2) **Absorption Dehumidification**:
   - Desiccant materials, such as silica gel or lithium chloride, absorb moisture from the air.
   - The desiccant material is regenerated by heating or passing a dry air stream through it.
3) **Refrigeration Dehumidification**:
   - This method is commonly used in HVAC systems. Humid air is cooled below its dew point, causing moisture to condense.
   - The condensed water is then drained away, leaving the air less humid.

Applications:

1) **Industrial Processes**:
   - Humidification is crucial in processes like paper manufacturing, textile production, and pharmaceuticals.
   - Dehumidification is essential in food processing, where moisture control is critical.

2) **HVAC Systems**:
   - In HVAC, humidification ensures comfort during dry winter months, while dehumidification is necessary for maintaining indoor air quality and preventing mold growth, in warm, humid climates.

3) **Preservation and Storage**:
   - Museums, archives, and storage facilities use dehumidification systems to protect artifacts and materials from moisture-related damage.

In these systems, the choice of method depends on factors such as the required humidity level, the type of material being processed, and energy efficiency considerations. The goal is to control humidity for optimal performance, comfort, and preservation of materials.

6.12 Levels of Air Saturation by Water Vapor

Several indices are used to quantify the notion of relative humidity or the degree of air saturation with water vapor. These indices help express the moisture content in the air relative to the maximum amount it can hold (or accommodate) at a given temperature. Here are some common indices related to humidity:

1) **Relative Humidity (RH)**:

The relative humidity, RH, is defined as:

$$\mathrm{RH}\ (\%) = \frac{\text{Actual Water Vapor Pressure}}{\text{Saturation Water Vapor Pressure}} \times 100\% = \frac{P_{\text{Water}}}{P_{\text{Water}}^{\text{sat}}(T)} \times 100\% \tag{6.38}$$

Relative humidity is expressed as a percentage and represents the ratio of the actual water vapor pressure in the air to the saturation water vapor pressure at a specific temperature. A relative humidity of 100% indicates that the air is saturated, while lower values indicate drier air. Simply stated, it quantifies the current amount of water vapor in the air relative to the total amount it could hold (or accommodate) at its present temperature. Warmer air has the capacity to hold more water vapor than colder air. The perceptible humidity we experience outdoors corresponds to the actual quantity of moisture (absolute humidity) present in the air. Notice that $P_{\text{Water}}^{\text{sat}}(T)$ can be calculated at a given temperature using one of the Antoine equation forms (Eq. 6.18–6.21).

2) **Dew Point**:

The dew point is the temperature at which air becomes saturated and dew (water droplets) begin to form. It is a measure of air's absolute humidity. When the air temperature equals the dew-point temperature, the air is saturated, and the relative humidity is 100%. Meteorologists routinely utilize the "dew-point" temperature, akin to absolute humidity, to assess moisture levels, particularly during spring and summer. The dewpoint temperature, indicating the actual quantity of water vapor in the air, represents the temperature at which air must be cooled to achieve saturation. Though individuals may experience weather conditions differently, as a general guideline, during spring and summer, surface dew-point temperatures in the 50s (Fahrenheit scale) are generally comfortable, in the 60s somewhat uncomfortable (humid), and in the 70s quite uncomfortable (very humid). Although extremely oppressive, dew points as high as 80 or the lower 80s are fortunately relatively rare. While the dew point provides a quick indication of air moisture, relative humidity does not, as humidity is relative to air temperature. In simpler terms, knowing the dew point alone does not determine relative humidity; one must also be aware of the actual air temperature. If the air is entirely saturated at a specific level, such as the surface, the dew-point temperature equals the actual air temperature, resulting in a relative humidity of 100%.

3) **Specific Humidity**:

The specific humidity is defined as:

$$\text{Specific Humidity} = \frac{\text{Mass of Water Vapor}}{\text{Total Mass of Moist Air}} \tag{6.39}$$

Specific humidity represents the mass of water vapor per unit mass of wet air. It is often expressed in grams of water vapor per kilogram of wet air and provides an absolute measure of humidity. From a chemical engineering point of view, it is the mass or weight fraction of water in a water vapor–air mixture. This simply means it will vary from 0 to 1.0, if it is expressed as g/g or kg/kg.

4) **Absolute Humidity (AH)**:

The absolute humidity is defined as:

$$\text{Absolute Humidity (AH)} = \frac{\text{Mass of Water Vapor}}{\text{Total Volume of Moist Air}} \tag{6.40}$$

Absolute humidity is a measure of the mass of water vapor per unit volume of humid air. It is more or less related to the density of water vapor in humid air. It is often expressed in grams of water vapor per cubic meter of humid air. It is a measure of the actual amount of water vapor (moisture) in the air, regardless of the air's temperature. The higher the amount of water vapor, the higher the absolute humidity. For example, a maximum of about 30 g of water vapor can exist in a cubic meter volume of air with a temperature in the middle 80s (Fahrenheit scale). Absolute and specific humidity are quite similar in concept. If an ideal gas behavior is assumed, the absolute humidity can be calculated using the following formula:

$$\text{AH}\left(\frac{\text{kg}}{\text{m}^3}\right) = \alpha \times \frac{P_{\text{Water}}}{T + 273.15} \tag{6.41}$$

where $\alpha = 2.167 \times 10^{-3}$ kg K/J and P_{Water} is expressed in Pascal (or J/m^3). Notice that:

$$P_{\text{Water}} = \frac{\text{RH}(\%)}{100\%} \times P_{\text{Water}}^{\text{sat}}(T) \tag{6.42}$$

5) **The Molar/Molal Humidity**:

The molar humidity is defined as:

$$h_{\text{Molar}}(\text{Dimensionless}) = \frac{\text{Moles of Water}}{\text{Moles of Dry Air}} = \frac{n_{\text{Water}}}{n_{\text{Dry Air}}} = \frac{\dot{n}_{\text{Water}}}{\dot{n}_{Dry\ Air}} = \frac{P_{Water}}{P - P_{Water}} \tag{6.43}$$

On the other hand, the molal humidity is defined as:

$$h_{\text{Molal}}\left(\frac{\text{kg Water}}{\text{kg Dry Air}}\right) = h_{\text{Molar}} \times \left(\frac{18.02\ \text{kg Water}}{\text{kmol Water}}\right) \times \left(\frac{\text{kmol Dry Air}}{29.0\ \text{kg Dry Air}}\right) \tag{6.44}$$

6) **Mixing Ratio**:

$$x\left(\frac{\text{g}}{\text{kg}}\right) = h_{\text{Molal}} \times \frac{1000\ \text{g Water}}{\text{kg Water}} \tag{6.45}$$

Alternatively, the mixing ratio (x g of water vapor/kg of dry gas) is calculated with the following formula:

$$x\left(\frac{\text{g}}{\text{kg}}\right) = \beta \times \frac{P_{\text{Water}}}{P - P_{\text{Water}}} = \beta \times h_{\text{Molar}} \tag{6.46}$$

where, $\beta = 622.0$ g/kg. The value of β depends on the gas. $\beta = 622.0$ g/kg is valid for air. P is the barometric (or applied) pressure where both P and P_{Water} must have the same unit of pressure.

6.13 Dew-point Estimation

There are several approximate equations and models for calculating the dew point of air–water systems based on temperature and relative humidity.

Magnus Equation:

The Magnus equation (Perović et al. [5]) is a widely used and relatively accurate empirical formula for calculating the dew point temperature. It is based on the relative humidity and dry bulb temperature of air. Here is the equation:

$$T_{\text{DP}}(^\circ\text{C}) = \left[\frac{243.04 * \left[\ln\left(\frac{\text{RH\%}}{100\%}\right) + \frac{17.625 * T_{\text{DB}}}{243.04 + T_{\text{DB}}}\right]}{17.625 - \ln\left(\frac{\text{RH\%}}{100\%}\right) - \frac{17.625 * T_{\text{DB}}}{243.04 + T_{\text{DB}}}}\right] \tag{6.47}$$

where:

- T_{DP} is the dew point temperature (°C),
- T_{DB} is the dry bulb temperature (°C), and
- RH is the relative humidity (%).

A simplified version ("Application Note Dew-point Calculation". https://irtfweb.ifa.hawaii.edu/tcs3/tcs3/Misc/Dewpoint_Calculation_Humidity_Sensor_E.pdf) of Eq. (6.47) can be made:

$$H = \frac{(\log_{10}(\text{RH}\%) - 2)}{0.4343} + \frac{17.62 * T_{\text{DB}}}{243.12 + T_{\text{DB}}} \tag{6.48}$$

$$T_{\text{DP}}(^\circ\text{C}) = \frac{243.12 * H}{(17.62 - H)} \tag{6.49}$$

Equation (6.47) can be used by a computer software and the set of the two equations (Eqs. 6.48 and 6.49) can be used in manual calculations.

NOTE #5
In addition to Eq. (6.47), the psychrometric charts can be used as well.

Appendix G | G.3 and G.4 show both the human language algorithm and MATLAB code based on Magnus Eq. (6.47) for expressing air/water dew-point temperature, T_{DP}, as a function of the dry bulb temperature, T_{DB}, and the percent relative humidity, RH. Below is a sample result upon executing the code.

```
>> MagnusEq
Input the dry bulb temperature, TDB: 80
Input the percent relative humidity, on 100% scale, RH%: 60
The entered dry bulb T in oC, TDB = 80 oC
The entered percent relative humidity, RH% = 60 %
The estimated dew point T, TDP in oC = 68.0171 oC
```

Example 6.6

Air and liquid water are contained at equilibrium in a closed chamber at 80°C and 760 mmHg. Calculate the molar composition of the gas phase.

Solution:

Using Appendix D | D2 MATLAB Antoine equation code, and using water as the substance, we have

```
Input "1" to search by Compound Name "2" to search by Formula: 1
Input the name of the Compound, say acetone: water
{[4999]}   {'liquid water, H2O'}   {'H2O'}
{[5000]}   {'vapor water, H2O'}   {'H2O'}
Input the number for the chosen compound, say "3384" for acetone: 4999
Input "1" to calculate T or "2" to calculate P: 2
Input the temperature in oC for the chosen compound: 80
    {'The material is:'}   {'water'} {'The Molecular Formula = '}   {'H2O'}
The estimated Psat in mm Hg = 354.5323
The estimated Psat in bar = 0.47267
The estimated Psat in atm = 0.46649
```

$$Y_{H_2O} = \frac{P^{sat}_{H_2O}}{P} = \frac{354.5\ \text{mmHg}}{760.0\ \text{mmHg}} = 0.466$$

$$Y_{\text{Dry Air}} = (1 - Y_{H_2O}) = (1 - 0.466) = 0.534$$

Example 6.7
Humid air at 85°C, 1.2 bar (1.184 atm), and 40% relative humidity is fed into a processing unit at a rate of 800 m^3/h. Assume an ideal gas mixture. Determine:

(1) The molar flow rates of water, dry air, and oxygen entering the process unit,

(2) The molal (kg Water/kg Dry Air) humidity, absolute humidity, and the mixing ratio of the air, and

(3) The dew point.

Solution:

(1) The molar flow rates of water, dry air, and oxygen entering the process unit
From Eq. (6.38), we have:

$$\text{RH}\ (\%) = 40\% = \frac{P_{\text{Water}}}{P^{sat}_{\text{Water}}(T)} \times 100\%$$

Using Appendix D | D2 MATLAB Antoine equation for water, we have $P^{sat}_{\text{Water}}(85°\text{C}) = 432.9\ \text{mmHg}$.
Thus, $P_{\text{Water}} = 0.4 \times P^{sat}_{\text{Water}}(85°\text{C}) = 173.2\ \text{mmHg}$

$$Y_{\text{water}} = \frac{P_{\text{Water}}}{P} = \frac{173.2\ \text{mmHg}}{1.2\ \text{bar} \times \dfrac{760\ \text{mmHg}}{1.01325\ \text{bar}}} = \frac{173.2\ \text{mmHg}}{900\ \text{mmHg}} = 0.1924$$

The molar flow rate of wet air can be given by the ideal gas equation of state:

$$\dot{n}_{\text{Wet Air}} = \frac{\text{P}\dot{\text{V}}}{\text{RT}} = \frac{1.184\ \text{atm} \times 800\ \dfrac{\text{m}^3}{\text{h}}}{0.08206\ \dfrac{\text{atm.m}^3}{\text{kmol.K}} \times 358.1\ \text{K}} = 32.23\ \frac{\text{kmol}}{\text{h}}$$

$$\dot{n}_{\text{Water}} = Y_{\text{Water}} \times \dot{n}_{\text{Wet Air}} = 0.1924 \times 32.23\ \frac{\text{kmol}}{\text{h}} = 6.2\frac{\text{kmol Water}}{\text{h}}$$

$$\dot{n}_{\text{Dry Air}} = 32.23 - 6.2 = 26.03\frac{\text{kmol Dry Air}}{\text{h}}$$

$$\dot{n}_{\text{Oxygen}} = Y_{O_2} \times \dot{n}_{\text{Dry Air}} = 0.21\frac{\text{kmol } O_2}{\text{kmol Dry Air}} \times 26.03\frac{\text{kmol Dry Air}}{\text{h}} = 5.47\frac{\text{kmol } O_2}{\text{h}}$$

(2) The molal (kg Water/kg Dry Air) humidity, absolute humidity, and the mixing ratio of the air

$$h_{\text{Molar}} = \frac{\text{Moles of Water}}{\text{Moles of Dry Air}} = \frac{\dot{n}_{\text{Water}}}{\dot{n}_{\text{Dry Air}}} = \frac{6.2\frac{\text{kmol Water}}{\text{h}}}{26.03\frac{\text{kmol Dry Air}}{\text{h}}} = 0.2382\frac{\text{kmol Water}}{\text{kmol Dry Air}}$$

Alternatively,

$$h_{\text{Molar}} = \frac{\text{Moles of Water}}{\text{Moles of Dry Air}} = \frac{P_{\text{Water}}}{P_{\text{Dry Air}}} = \frac{P_{\text{Water}}}{P - P_{\text{Water}}} = \frac{173.2}{900 - 173.2} = 0.2383 \frac{\text{kmol Water}}{\text{kmol Dry Air}}$$

$$h_{\text{Molal}} = 0.2382 \frac{\text{kmol Water}}{\text{kmol Dry Air}} \times \left(\frac{18.02\,\text{kg Water}}{\text{kmol Water}}\right) \times \left(\frac{\text{kmol Dry Air}}{29.0\,\text{kg Dry Air}}\right) = 0.148 \frac{\text{kg Water}}{\text{kg Dry Air}}$$

From Eq. (6.40), we have:

$$\text{Absolute Humidity (AH)} = \frac{\text{Mass of Water Vapor}}{\text{Total Volume of Moist Air}} = \frac{\dot{m}_{\text{Water}}}{\dot{V}_{\text{Wet Air}}} = \frac{\dot{n}_{\text{Water}} \times \text{MW}_{\text{Water}}}{800 \frac{\text{m}^3}{\text{h}}}$$

$$\text{Absolute Humidity (AH)} = \frac{\dot{n}_{\text{Water}} \times \text{MW}_{\text{Water}}}{800 \frac{\text{m}^3}{\text{h}}} = \frac{6.2 \frac{\text{kmol Water}}{\text{h}} \times \frac{18\,\text{kg Water}}{\text{kmol Water}}}{800 \frac{\text{m}^3}{\text{h}}}$$
$$= 0.1395 \frac{\text{kg Water}}{\text{m}^3\ \text{Wet Air}}$$

Alternatively, from Eq. (6.41), we have:

$$\text{AH}\left(\frac{\text{g}}{\text{m}^3}\right) = \alpha \times \frac{P_{\text{Water}}}{T + 273.15} = \left(0.002167 \frac{\text{kg.K}}{\text{J}}\right) \times \frac{0.1924 \times 1.2\,\text{bar} \times \frac{10^5\text{Pa}}{\text{bar}} \times \frac{\text{J/m}^3}{\text{Pa}}}{(85 + 273.15)\text{K}}$$
$$= 0.1397 \frac{\text{kg Water}}{\text{m}^3\ \text{Wet Air}}$$

From Eq. (6.45), we have:

$$x\left(\frac{\text{g}}{\text{kg}}\right) = h_{\text{Molal}} \times 1000 = 0.148 \frac{\text{kg Water}}{\text{kg Dry Air}} \times \frac{1000\,\text{g Water}}{\text{kg Water}} = 148 \frac{\text{g Water}}{\text{kg Dry Air}}$$

Alternatively, Eq. (6.46) can be used:

$$x\left(\frac{\text{g}}{\text{kg}}\right) = \beta \times \frac{P_{\text{Water}}}{P - P_{\text{Water}}} = 622.0 \frac{\text{g}}{\text{kg}} \times \frac{173.2}{900 - 173.2} = 148.2 \frac{\text{g Water}}{\text{kg Dry Air}}$$

(3) The dew point

Using Appendix G | G.4 MATLAB code for Magnus equation, the following results are obtained:

```
>> MagnusEq
Input the dry bulb temperature, TDB: 85
Input the percent relative humidity, on 100% scale, RH%: 40
The entered dry bulb T in °C, TDB = 85 °C
The entered percent relative humidity, RH% = 40 %
The estimated dew point T, TDP in °C = 63.4907 °C
```

Example 6.8 A gas mixture contains 12.0 mole% $H_2O(v)$ and 88.0 mole% N_2. The gas temperature and absolute pressure at the start of each of the three parts of this example are $T_1 = 60°C$ and $P_1 = 600$ mmHg. Ideal gas behavior can be assumed in every part of this example.

(a) If some of the gas mixture is put in a cylinder and slowly cooled at constant pressure, at what temperature would the first drop of liquid form?

Using **AntoinEq.m** file (Appendix D | D.2) while selecting water as the component and $T = 60°C$, we have:

The estimated P_W^{sat} for water is 149 mmHg.

$$P_{Water} = 0.12 \times 600 = 72 \text{ mmHg}$$

Thus, from Eq. (6.38), we have:

$$\text{RH}(\%) = \frac{P_{Water}}{P_{Water}^{sat}(T)} \times 100\% = \frac{72 \text{ mmHg}}{149 \text{ mmHg}} \times 100\% = 48.32\%$$

Using MATLAB Magnus equation (Appendix G | G.4):

```
>> MagnusEq
Input the dry bulb temperature, TDB: 60
Input the percent relative humidity, on 100% scale, RH%: 48.32
The entered dry bulb T in °C, TDB = 60 °C
The entered percent relative humidity, RH% = 48.32 %
The estimated dew point T, TDP in °C = 45.1703 °C
```

(b) If a 20.0-L flask is filled with some of the gas mixture and sealed and the water vapor in the flask is completely condensed, what volume (cm^3) would be occupied by the liquid water?

$$P_W = 72 \text{ mmHg} = 0.0947 \text{ atm}$$

Applying the Ideal Gas:

$$P_W V_1 = n_W R T_1 \Rightarrow 0.0947 \text{ atm} \times 20.0 \text{ L} = n_w \times 0.08206 \frac{\text{atm.L}}{\text{mol.K}} \times (333.15 \text{ K})$$

Solve for n_W. $n_W = 0.0693$ mol of water

The condensed water will be the same amount under complete condensation.

$$\text{Mass of condensed water} = 0.0693 \text{ mol of water} \times \frac{18.02 \text{ g}}{1 \text{ mol of water}} = 1.248 \text{ g of liquid water}$$

$$\text{Volume} = \text{Mass/Density}$$

$$\text{Liquid water volume} = 1.248 \text{ g of liquid water} \times \left(\frac{1 \text{ cm}^3}{1 \text{ g liquid water}}\right)$$
$$= 1.248 \text{ cm}^3 \text{ liquid water}$$

(c) If the gas mixture is stored in a rigid-walled cylinder and a low-pressure weather front moves in and the barometric (atmospheric) pressure drops, which of the following would change:

(A) The gas density
(B) The absolute pressure of the gas
(C) The partial pressure of water in the gas

(D) The gauge pressure of the gas
(E) The mole fraction of water in the gas
(F) The dew-point temperature of the mixture
Correct answer: (D) The gauge pressure of the gas.

6.14 Solid Solubility

6.14.1 What is Solid Solubility?

Solid solubility refers to the maximum amount of a solid substance (solute) that can dissolve in a liquid (solvent) at a specific temperature and pressure. When this limit is reached, the solution is said to be saturated.

Imagine adding sugar to a cup of tea. At first, the sugar readily dissolves in the hot water. But as we keep adding more and more sugar, it eventually stops dissolving and settles at the bottom of the cup. This is because the solution has become saturated, and there is no more solvent available to dissolve the additional sugar.

For more information on the aqueous solubility of drug-like, polar, and ionic molecules, please refer to the following articles:

1) Al-Malah [6].
2) Al-Malah [7].
3) Al-Malah [8].

6.14.2 Factors Affecting Solid Solubility

Several factors influence how much a liquid can dissolve a solid. Here are some of the most important points:

1) **Nature of the Solute and Solvent**: This is often summarized by the saying "like dissolves like." Polar solutes, such as ionic salts and sugars, tend to dissolve well in polar solvents like water because their intermolecular forces can readily break and form new interactions with the solvent molecules. Conversely, nonpolar solutes, like oils and fats, dissolve well in nonpolar solvents like gasoline, as their similar intermolecular forces allow them to mix easily.
2) **Temperature**: In general, the solubility of an inorganic solid salt increases with increasing temperature. This is because the higher kinetic energy of the solvent molecules at higher temperatures makes it easier for them to break the intermolecular forces holding the solute particles together and incorporate them into the solution. On the contrary, the solubility of gases, like HCl, NH_3, O_2, N_2, CO_2, H_2S, and SO_2 in a polar solvent does decrease with increasing temperature. Remember, gases are freedom fighters and do not like to be trapped in liquids. With increasing temperature, they experience thermal expansion effects as far as the dispersed gas bubble size is concerned. The larger the bubble size, the more difficult it will be to accommodate the gas bubbles in a liquid medium; hence, the upper limit of solubility will drop.
3) **Pressure**: For gases, the solubility increases with increasing pressure. However, for most solids in liquids, pressure has a negligible effect on solubility.

4) **Presence of Other Solutes**: Adding another solute to a solution can affect the solubility of the original solute. This can happen through various mechanisms, such as competition for solvent molecules or the formation of complexes between the solutes.
5) **Particle Size**: Smaller particles have a larger surface area relative to their volume, which allows them to interact more effectively with the solvent molecules and dissolve more readily.

Understanding these factors is essential in various fields, including:

- **Chemistry**: Predicting the behavior of chemical reactions and designing separation processes.
- **Food Science**: Developing stable and flavorful food products.
- **Pharmacology**: Formulating drugs for efficient delivery and absorption.
- **Environmental Science**: Understanding the fate and transport of pollutants in water and soil.

6.15 Salting In and Salting Out

The terms "salting in" and "salting out" refer to the complex effects that adding salt (electrolyte) can have on the solubility of other solutes (both electrolytes and nonelectrolytes) in a solution. These effects depend on various factors and can be quite intricate. Let us briefly explain the differences between these two phenomena:

Salting In:

Increased Solubility: In salting in, adding salt enhances the solubility of another solute. This typically occurs with hydrophobic solutes (molecules that repel water) and kosmotropic ions (salts that strengthen the water structure or order). In the salt-in phenomenon, the addition of certain salts increases the solubility of a substance in a solvent. The mechanism behind the salt-in effect can be explained by the following factors:

- **Water Structure Enhancement**: Kosmotropic ions strengthen the hydrogen bonds between water molecules, creating a more ordered and structured solvent. This ordered structure can better accommodate the hydrophobic moiety, promoting its solubility.
- **New Hydration Shell (Niqab or Veil)**: The salt-in molecules interact with the nonpolar solute in a way that creates a "shield" or protective layer around it. This layer prevents the nonpolar solute from directly interacting with the polar solvent environment, which it inherently dislikes. At the same time, the salt-in molecules themselves are readily accepted by the polar solvent because of their ionic nature and ability to form strong hydrogen bonds with water molecules. This essentially makes the solvent environment more "friendly" and accessible to the shielded nonpolar solute, enhancing its overall solubility. Therefore, it is more accurate to think of the salting-in process as a cooperative phenomenon rather than a competitive one. The salt-in molecules act as intermediaries, bridging the gap between the nonpolar solute and the polar solvent, thus facilitating their coexistence in the solution.

Salting Out:

Decreased Solubility: Salting out is a phenomenon in which the solubility of a substance in a solvent is reduced by the addition of a salt. The mechanisms described, including ion pairing, water structure disruption, and increased ionic strength, are commonly associated with salting out. Here is a brief explanation of each point:

1) **Ion Pairing**: When ions with opposite charges are added to a solution, they can form ion pairs with the solute ions. These ion pairs have reduced net charges compared to the individual ions, weakening the attractive forces between the solute ions and the polar solvent molecules. As a result, the solute becomes less soluble in the solution.
2) **Water Structure Disruption**: Chaotropic ions, which are ions that disrupt the structure of water, can weaken the hydrogen bonds between water molecules. This disruption makes the solvent (water) less structured and less able to accommodate the solute, leading to a decrease in solubility.
3) **Increased Ionic Strength**: The addition of salt increases the ionic strength of the solution. High ionic strength can decrease the water activity, meaning that water molecules are less available for interacting with solute molecules. This decrease in water activity makes it more difficult for the solute to dissolve in the solution.

The salt-in effect is observed with certain substances, particularly those that have specific interactions with ions present in the added salt. It is important to note that the salting-in and salting-out effects are not universal and depend on the nature of the solute, solvent, and the specific ions involved. These effects have practical applications in various fields, including biochemistry, where they are utilized in processes such as protein solubilization and stabilization. Both the salt-in and salt-out effects can be significant for a range of molecules with varying degrees of hydrophobicity or amphiphilicity.

6.16 The Maximum Solubility of a Solid Solute in a Liquid

The solubility of a solid in a liquid refers to the maximum quantity of that substance that can be dissolved in a specified volume of the liquid at equilibrium. This characteristic varies significantly between different solute–solvent combinations. Figure 6.5 shows the aqueous solubility of some inorganic salts as a function of medium temperature. Except for cerium sulfate ($Ce_2(SO_4)_3$), the other shown inorganic salts exhibit a typical behavior of solubility increase with temperature.

Perry's Chemical Engineers' Handbook (Green and Perry [3]) provides values for the solubilities of inorganic compounds in water at various temperatures. See Table 2.122 on pages 2-126–2-130.

A solution saturated with a dissolved species contains the maximum amount of that species at equilibrium. Equilibrium with a solid solute necessitates saturation with that solute; otherwise, more solute would dissolve. When a saturated solution is cooled, the solubility of the solute typically decreases. To return to equilibrium, some solute must precipitate as solid crystals. Although crystallization may proceed slowly, a metastable state can occur where the solute concentration surpasses the equilibrium value at the solution temperature. In such cases, the solution is deemed supersaturated, and the disparity between actual and equilibrium concentrations is termed supersaturation. In this textbook, we will not entertain supersaturation conditions; only equilibrium between solid and liquid phases will be considered.

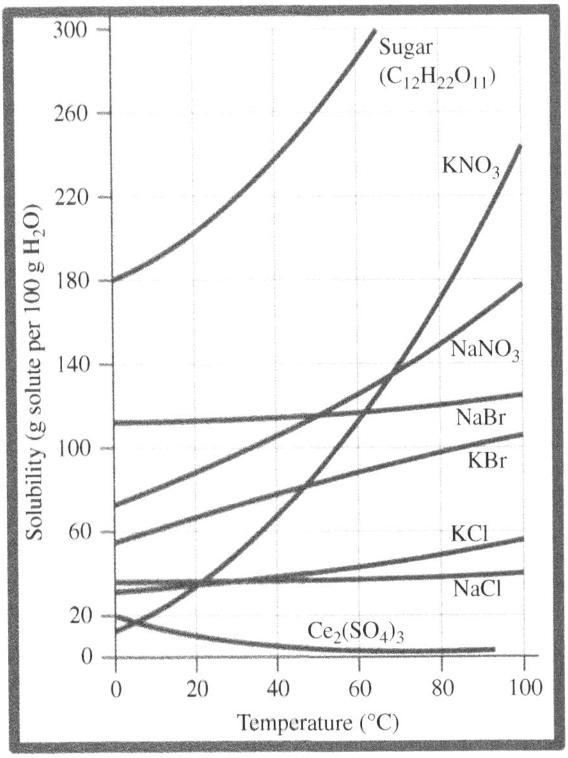

Figure 6.5 Aqueous solubility of some inorganic salts as a function of medium temperature. *Source:* Chemistry for Majors. Provided by: OpenStax. Located at: https://openstax.org/. License: CC BY: Attribution. License Terms: Access for free at https://courses.lumenlearning.com/chemistryformajors/chapter/solubility-2/.

Example 6.9 Crystallization, Filtration, and Drying of Potassium Chloride (KCl)
We have 200 kg of a saturated aqueous solution of KCl at 80°C. Upon cooling to 25°C, KCl crystals form and are separated by filtration. The remaining KCl stays in the exit solution, called filtrate, which passes through the filter. The wet filter cake contains 75% solid crystals and 25% saturated solution by mass. The remaining KCl crystals are recovered by drying the filter cake, as shown in Figure 6.6.

Calculate the percent recovery of the KCl in the feed stream, eventually recovered as dry crystals and the amount of water that must be removed in the drying stage.

DATA: Saturated Solutions:

At 80°C: 50 g KCl/100 g H_2O. At 25°C: 35 g KCl/100 g H_2O

Solution:

The saturated solution at 80°C will have 50 g KCl/100 g H_2O. Thus,

$$x_{\text{KCl}} = \frac{\text{Mass of KCl}}{\text{Mass of KCl + Mass of H2O}} = \frac{50\ \text{g}}{50\ \text{g} + 100\ \text{g}} = 0.333$$

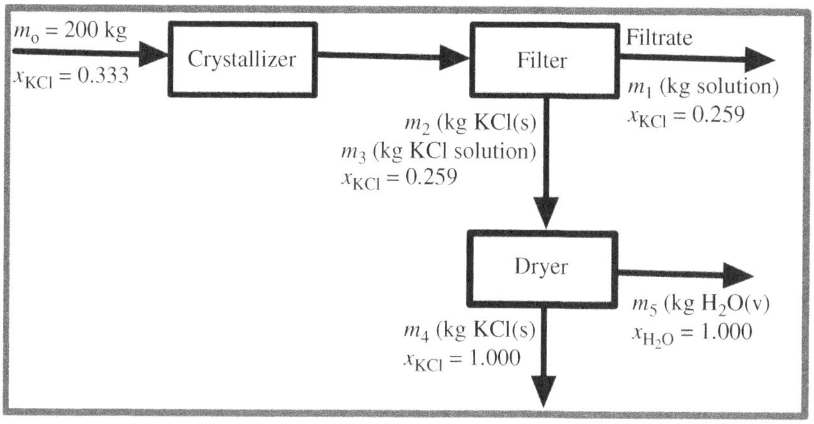

Figure 6.6 A flowchart showing the crystallization step followed by filtration and drying of the KCl solution. The mass fraction of KCl is shown on each stream with water as the complement.

On the other hand, the saturated solution at 25°C will have 35 g KCl/100 g H_2O. Thus,

$$x_{KCl} = \frac{\text{Mass of KCl}}{\text{Mass of KCl + Mass of H2O}} = \frac{35\text{ g}}{35\text{ g} + 100\text{ g}} = 0.259$$

Basis: One unit time of operation with 200 kg feed of saturated KCl solution at 80°C.

For the sake of learning, let us carry out the overall total material and KCl component balance. Total material balance around the entire (Crystallizer + Filter + Dryer) block:

$$m_0 = 200\text{ kg} = m_1 + m_4 + m_5 \qquad \text{(a)}$$

KCl component balance around the entire process:

$$m_0 x_{KCl} = m_1 x_{KCl} + m_4 x_{KCl} + m_5 x_{KCl}$$
$$200 * 0.333 = m_1 * 0.259 + m_4 * 1 + m_5 * (0.000)$$
$$66.6 = m_1 * 0.259 + m_4 \qquad \text{(b)}$$

We see that we have three unknown terms: m_1, m_4, and m_5 with only two linearly independent equations: (a) and (b). Of course, one might say why not carry out water component balance around the entire process? The answer is that water component balance is not linearly independent of equations: (a) and (b); it can be obtained by combining equations: (a) and (b). We need to go to other blocks where we have extra information (i.e., additional equation) between the unknown terms. Let us carry out the same balance, but this time around the first two blocks: Crystallizer and Filter.

Total material balance around the "Crystallizer + Filter" block:

$$m_0 = 200\text{ kg} = m_1 + m_2 + m_3 \qquad \text{(c)}$$

KCl component balance around the "Crystallizer + Filter" block:

$$m_0 x_{KCl} = m_1 x_{KCl} + m_2 x_{KCl} + m_3 x_{KCl}$$
$$200 * 0.333 = 66.6 = m_1 * 0.259 + m_2 * 1 + m_3 * (0.259) \qquad \text{(d)}$$

Again, we see that we have three unknown terms: m_1, m_2, and m_3 with only two linearly independent equations: (c) and (d). As said earlier, water component balance is not linearly independent of equations: (c) and (d) and it can be obtained by combining equations: (c) and (d).

What else do we have?

Looking back at the question statement, it says that "The wet filter cake contains 75% solid crystals and 25% saturated solution by mass." Great, it means that $m_2/m_3 = 75\,\%\,/25\,\% = 3$.

$$m_2 = 3m_3 \qquad \text{(From MATLAB code point of view : } 0{*}m_1 + m_2 - 3m_3 = 0) \qquad \text{(e)}$$

Now the DoF is zero, and the system is just specified. Three unknown terms: m_1, m_2, and m_3 with three linearly independent equations: (c), (d), and (e). Let us reduce the number of equations from three down to two by eliminating m_2.

Let us utilize Eq. (e) and replace m_2 by $3m_3$ wherever m_2 appears in Eqs. (c) and (d).

$$200\text{ kg} = m_1 + 3m_3 + m_3 = m_1 + 4m_3 \qquad (c')$$

$$66.6 = m_1 * 0.259 + 3m_3 + m_3 * (0.259) = 0.259m_1 + 3.259m_3 \qquad (d')$$

Now, we have two equations: (c′) and (d′) with two unknown terms: m_1 and m_3. Solve by either substitution or elimination. From Eq. (c′), we have:

$$m_1 = 200 - 4m_3$$

Substitute in Eq. (d′), we have:

$$66.6 = 0.259(200 - 4m_3) + 3.259m_3 = 51.8 + 2.223m_3$$

Solve for $m_3 = 6.65767$.

$$m_1 = 200 - 4m_3 = 200 - 4(6.65767) = 173.3693$$

From Eq. (e), we have: $m_2 = 3m_3 = 19.973$

For the sake of utilizing MATLAB's computational power, Figure 6.7 shows the MATLAB code for solving three simultaneous algebraic equations: (c), (d), and (e) with three unknowns: m_1, m_2, and m_3.

```
clear variables;
%Define the system of equations as a matrix of coefficients and the right-hand side vector
%Eq. (c): m1 + m2 + m3 = 200
%Eq. (d): 0.259m1 + 1.0m2 + 0.259m3=66.6
%Eq. (e): 0*m1+m2-3m3=0
A = [1 1 1; 0.259 1 0.259;0 1 -3];
b = [200; 66.6; 0];
% Solve the system using matrix inversion
x = A \ b;
% Print the solution for m1 and m3
m1 = x(1);
m2=x(2);
m3 = x(3);
fprintf('m1 = %f\n', m1);
fprintf('m2 = %f\n', m2);
fprintf('m3 = %f\n', m3);
```

Figure 6.7 MATLAB code snippet for solving three simultaneous algebraic equations.

Below are the values of m_1, m_2, and m_3 as calculated by MATLAB code:

```
m1 = 173.369321
m2 = 19.973009
m3 = 6.657670
```

From Eq. (b), we have:
$66.6 = m_1 * 0.259 + m_4 \rightarrow 66.6 = 173.369 * 0.259 + m_4 \rightarrow m_4 = 21.6973$ kg KCl(s). From Eq. (a), we have:

$$m_5 = 200 - (m_1 + m_4) = 4.933\ H_22O(v)$$

$$\text{Percent Recovery} = \frac{m4}{x_{KCl} \times m_o} \times 100\% = \frac{21.6973\text{ kg KCl}}{66.6\text{ kg KCl}} \times 100\% = 32.6\%$$

6.17 Hydrated Salts

A salt compound that contains water molecules chemically bound within its crystal structure. These water molecules are loosely held and can be removed through heating (I call it roasting) or dehydration processes. An example is magnesium sulfate heptahydrate ($MgSO_4.7H_2O$). This molecule has seven water molecules (H_2O) attached to the $MgSO_4$ unit. Thus, the water of hydration represents the water molecules that are chemically bound within the crystal structure of a hydrated salt. These water molecules play a role in the crystal structure and properties of the salt, such as color, solubility, and melting point. On the contrary, an anhydrous salt is a salt compound that does not contain any water molecules within its crystal structure. It is essentially the "dehydrated" version of a hydrated salt. An example is magnesium sulfate ($MgSO_4$). This is the anhydrous form of $MgSO_4{\cdot}7H_2O$. $MgSO_4{\cdot}7H_2O$ occurs as a mineral known as epsomite or Epsom salt. Epsomite is a white, crystalline substance that is commonly found in evaporite deposits and is often associated with other minerals such as gypsum. It can form in certain geological environments where there is a concentration of magnesium sulfate and suitable conditions for crystallization, such as in salt flats or playas. Epsomite is also known to occur in some caves and salt pans. Compared to the anhydrous form, hydrates are generally softer and more soluble in water. Different hydrates have different stabilities and can convert to each other or the anhydrous form depending on pressure and temperature conditions.

On the contrary, sodium nitrate does not form stable hydrates. There are several reasons for this:

1) **Strong Ionic Bond**: In sodium nitrate ($NaNO_3$), the ionic bond between the positively charged sodium ion (Na^+) and the negatively charged nitrate ion (NO_3^-) is very strong. This strong bond makes it difficult for water molecules to insert themselves and disrupt the crystal structure.
2) **Hydration Energy vs. Lattice Energy**: When forming a hydrate, the energy released when water molecules interact with the salt (hydration energy) needs to be greater than

the energy required to break the existing ionic bonds in the salt (lattice energy). In sodium nitrate, the lattice energy is high enough that the available hydration energy from water molecules is insufficient to overcome it and form a stable hydrate.

3) **Alternative Stability**: Instead of forming hydrates, sodium nitrate readily dissolves in water. This means it achieves a stable state by dissociating into individual Na^+ and NO_3^- ions in solution, rather than incorporating water molecules into its crystal structure.

Example 6.10

Sodium sulfate (Na_2SO_4) can form hydrates, which are compounds containing a specific number of water molecules associated with each formula unit of the salt. One common hydrated form of sodium sulfate is sodium sulfate decahydrate, often represented as $Na_2SO_4{\cdot}10H_2O$. This compound is known as Mirabilite or Glauber's salt.

An aqueous solution of sodium sulfate at 100°C containing 30 wt.% Na_2SO_4 is fed to a cooling crystallizer that operates at 25°C. The stream leaving the crystallizer is a slurry of solid sodium sulfate decahydrate particles [$Na_2SO_4{\cdot}10H_2O(s)$] suspended in a liquid solution, as shown in Figure 6.8.

Solubility at 100°C: 42.7 g/100 mL ($x_{Na_2SO_4} = 0.30$)
Solubility at 25°C: 28.1 g/100 mL ($x_{Na_2SO_4} = 0.22$)

The molecular weight of anhydrous sodium sulfate is 142.04, and that of the decahydrate salt is 322.20 kg/kmol.

References:

1) (https://www.chemeurope.com/en/encyclopedia/Sodium_sulfate.html).
2) (https://en.wikipedia.org/wiki/Sodium_sulfate).

Solution:

Basis: 1000 kg/h of [$Na_2SO_4{\cdot}10H_2O(s)$] at the exit of the crystallizer.

There are two unknowns: m_1 and m_2 and two independent component balances can be formulated: one for Na_2SO_4 and another for H_2O. Hence, the system is just specified with a zero DoF.

Total Material Balance:

$$\dot{m}_1 = \dot{m}_2 + \dot{m}_{\text{solid}} = \dot{m}_2 + 1000\,\frac{\text{kg}}{\text{h}} \tag{a}$$

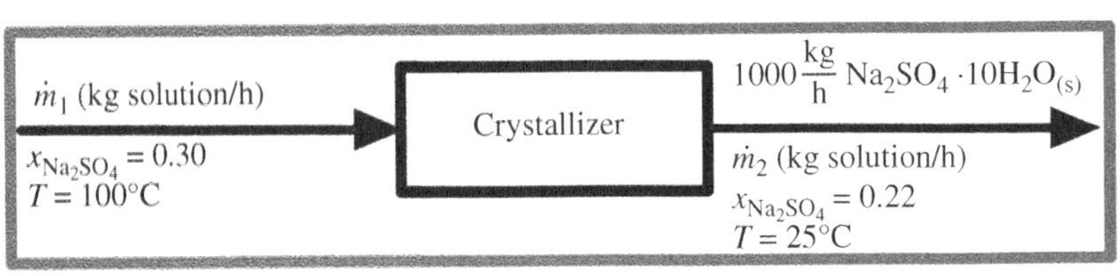

Figure 6.8 A schematic for crystallization of $Na_2SO_4{\cdot}10H_2O(s)$ out of Na_2SO_4 solution.

Na_2SO_4 Component Balance:

$$x_{Na_2SO_4} \times \dot{m}_1 = x_{Na_2SO_4} \times \dot{m}_2 + \text{mass flow rate (MFR) of } Na_2SO_4 \text{ in solid slurry} \quad \text{(b)}$$

$$\text{MFR of } Na_2SO_4 \text{ in solid slurry} = 1000\ \frac{\text{kg } Na_2SO_4 \cdot 10H_2O}{\text{h}} \times \left(\frac{1 \text{ kmol } Na_2SO_4 \cdot 10H_2O}{322.2 \text{ kg } Na_2SO_4 \cdot 10H_2O}\right)\left(\frac{142.04 \text{ kg } Na_2SO_4}{1 \text{ kmol } Na_2SO_4}\right)\left(\frac{1 \text{ kmol } Na_2SO_4}{1 \text{ kmol } Na_2SO_4 \cdot 10H_2O}\right)$$

Notice that the first two bracketed terms account for mass-mole conversion for both species: $Na_2SO_4 \cdot 10H_2O$ and Na_2SO_4, respectively. The third and last bracketed term accounts for the molar stoichiometric ratio between $Na_2SO_4 \cdot 10H_2O$ and Na_2SO_4, which says 1 mole of the first produces one mole of the second but not mass-wise.

$$\text{MFR of } Na_2SO_4 \text{ in solid slurry} = 440.84 \frac{\text{kg } Na_2SO_4}{\text{h}} \quad \text{(c)}$$

Substitute Eq. (c) in Eq. (b) and simplify, we have:

$$0.30 \times \dot{m}_1 = 0.22 \times \dot{m}_2 + 440.84 \frac{\text{kg } Na_2SO_4}{\text{h}} \quad \text{(d)}$$

Figure 6.9 shows the MATLAB code for solving two simultaneous algebraic equations: (a) and (d), with two unknowns: $\dot{m}_1$ and $\dot{m}_2$.

Below are the values of mdot1 and mdot2, as calculated by MATLAB code:

mdot1 = 2760.500000
mdot2 = 1760.500000

```
clear variables;
% Define the system of equations as a matrix of coefficients and the right-hand side vector
%Eq. (a): m1 - m2 = 1000
%Eq. (d): 0.30m1 - 0.22m2 = 440.84
A = [1 -1; 0.30 -0.22];
b = [1000; 440.84];
% Solve the system using matrix inversion
x = A \ b;
% Print the solution for m1 and m3
mdot1 = x(1);
mdot2=x(2);
fprintf('mdot1 = %f\n', mdot1);
fprintf('mdot2 = %f\n', mdot2);
```

Figure 6.9 MATLAB code snippet for solving two simultaneous algebraic equations.

6.18 Colligative Properties

Colligative properties are physical properties of a solution that depend on the concentration of solute particles, regardless of the type of solute. These properties are particularly applicable to solutions where the solute is a nonvolatile substance (doesn't easily evaporate) and does not undergo chemical changes in the solution.

The four main colligative properties are:

6.18.1 Vapor Pressure Depression

In a solution, the vapor pressure is lower than that of the pure solvent. The extent of vapor pressure lowering is directly proportional to the molality (moles of solute per kilogram of solvent) of the solute particles. Examples are shown below.

Example 6.11 Vapor Pressure Depression of Glucose Solution

Calculate the vapor pressure of a solution made by dissolving 300.0 g glucose, $C_6H_{12}O_6$, in 600 g of water at 50°C. Using Antoine equation MATLAB code, the vapor pressure of pure water is 92.3 mmHg at 50°C. The solubility of glucose at 50°C is 244 g/100 mL of water.

DATA: MW of glucose = 180.2 g/mol. MW of water = 18 g/mol

Solution:

To use Raoult's law (Eq. 6.32), we need to calculate the mole fraction of water (the solvent) in this sugar–water solution.

$$X_{\text{Water}} = \frac{\text{Moles of Water}}{\text{Moles of Water} + \text{Moles of Glucose}}$$

$$n_{\text{Water}} = \frac{\text{Mass}}{\text{MW}} = \frac{600\ \text{g}}{18\dfrac{\text{g}}{\text{mol}}} = 33.333\ \text{mol}$$

$$n_{\text{Glucose}} = \frac{\text{Mass}}{\text{MW}} = \frac{300\ \text{g}}{180.2\dfrac{\text{g}}{\text{mol}}} = 1.665\ \text{mol}$$

$$X_{\text{Water}} = \frac{33.333}{33.333 + 1.665} = 0.9524$$

$$P_{\text{Gluc.Soln.}} = X_{\text{Water}} P_{\text{Water}}^{\text{sat}}(50°\text{C}) = 0.9524 \times 92.3\ \text{mmHg} = 87.9\ \text{mmHg}$$

$$\Delta P = P_{\text{Gluc.Soln.}} - P_{\text{Water}}^{\text{sat}} = 87.9 - 92.3 = -4.4\ \text{mmHg}$$

The negative sign means that there is a pressure reduction in the vapor pressure by 4.4 mmHg. One can conclude that glucose solution is now less volatile than pure water at the same $T = 50°\text{C}$.

Example 6.12 Vapor Pressure Depression of $CaCl_2$ Solution

Calculate the vapor pressure of a solution made by dissolving 100.0 g $CaCl_2$ in 600 g of water. Using Antoine equation MATLAB code, the vapor pressure of pure water is 92.3 mmHg at 50°C. The maximum solubility of $CaCl_2$ in water is 124.4 g/100 g at (50°C).

Reference: https://en.wikipedia.org/wiki/Calcium_chloride.

DATA: The molecular mass of $CaCl_2$ is 111 g/mol.

Solution:

To use Raoult's law (Eq. 6.32), we need to calculate the mole fraction of water (the solvent) in this salt–water solution.

$$n_{\text{Water}} = \frac{\text{Mass}}{\text{MW}} = \frac{600\,\text{g}}{18\,\dfrac{\text{g}}{\text{mol}}} = 33.333\ \text{mol}$$

$$n_{\text{CaCl2}} = \frac{\text{Mass}}{\text{MW}} = \frac{100\,\text{g}}{111\,\dfrac{\text{g}}{\text{mol}}} = 0.9009\ \text{mol}$$

However, in this case, we have an ionic species that, upon dissolution, will give more than one ion:

$$\text{CaCl}_{2(s)} \Rightarrow \text{Ca}^{++}_{(aq)} + 2\text{Cl}^{-}_{(aq)} \tag{6.50}$$

This is called van't Hoff factor (i):

$$i = \frac{\text{Actual number of particles in solution after dissociation}}{\text{Number of formula units initially dissolved in solution}} = \frac{1+2}{1} = 3 \tag{6.51}$$

This means that one mole of $CaCl_2$ will give one mole of Ca^{+2} and two moles of Cl^-. The multiplication effect of dissolving $CaCl_2$ in water is tripled.

Thus,

$$X_{\text{Water}} = \frac{n_{\text{Water}}}{n_{\text{Water}} + n_{\text{Ca}++} + n_{\text{Cl}-}}$$

$$X_{\text{Water}} = \frac{33.333}{33.333 + 0.9009 + 2(0.9009)} = 0.925$$

$$P_{\text{CaCl}_2\ \text{Soln.}} = X_{\text{Water}} P^{\text{sat}}_{\text{Water}}(50^\circ\text{C}) = 0.925 \times 92.3\ \text{mmHg} = 85.4\ \text{mmHg}$$

$$\Delta P = P_{\text{CaCl}_2\ \text{Soln.}} - P^{\text{sat}}_{\text{Water}} = 85.4 - 92.3 = -6.9\ \text{mmHg}$$

Notice that the pressure reduction in the $CaCl_2$ case is larger in magnitude than that of the glucose case, although a smaller number of moles were dissolved in the $CaCl_2$ case.

6.18.2 Boiling Point Elevation

The boiling point of a solution is higher than that of a pure solvent. This increase is also directly proportional to the molality of the solute particles. This property is utilized in processes like boiling point elevation in antifreeze solutions for cars.

Table 6.3 shows the boiling point elevation constants (K_b) and freezing point depression constants (K_f) for some common solvents. Reference: ["13.8: Freezing-Point Depression and Boiling-Point Elevation of Nonelectrolyte Solutions". Reference: [https://en.wikipedia.org/wiki/List_of_boiling_and_freezing_information_of_solvents].

Table 6.3 Boiling point elevation constants (K_b) and freezing point depression constants (K_f) for some common solvents.

Solvent	Boiling Point (°C)	K_b (°C/m)	Freezing Point (°C)	K_f (°C/m)
Acetic acid	117.90	3.14	16.6	3.90
Benzene	80.1	2.65	5.5	5.12
Carbon disulfide	46.2	2.34	−111.5	3.83
Carbon tetrachloride	76.8	4.88	−22.8	29.8
Chloroform	61.2	3.88	−63.5	4.90
Cyclohexane	80.7	2.79	6.55	20.2
Ethanol	78.4	1.22	−114.6	1.99
Nitrobenzene	210.8	5.24	5.70	7.00
Water	100.00	0.51	0.00	1.86

Here are two examples.

Example 6.13 Boiling Point Elevation of Glucose Solution

A solution of 50.0 g of glucose is added to 300.0 g of water in an attempt to elevate the boiling point. What is the boiling point of the solution?

DATA: MW of glucose = 180.2 g/mol.

Solution:

Glucose does not form ionic species; hence, van't Hoff factor (i) is set to 1.0:

$$\Delta T_b = (T_{b,Soln.} - T_{b,Water}) = iK_b m \tag{6.52}$$

$$\text{Molality}, m = \frac{\text{Moles of Glucose}}{\text{kg Water}} = \frac{\left(50\,\text{g} \times \dfrac{\text{mol Glucose}}{180.2\,\text{g}}\right)}{0.3\,\text{kg Water}} = 0.925\frac{\text{mol Glucose}}{\text{kg Water}} = 0.925\text{ m}$$

From Table 6.3, $K_b = 0.51$ (°C/m) for water. Substitute in Eq. (6.52), we have:

$$\Delta T_b = (T_{b,Soln.} - 100°\text{C}) = iK_b m = (1) \times 0.51\frac{°\text{C}}{\text{m}} \times 0.925\text{ m} = 0.472°\text{C}$$

$$T_{b,Soln.} = 100°C + 0.472°\text{C} \cong 100.5°\text{C}$$

Example 6.14 Boiling Point Elevation of NaCl Solution
A solution of 50.0 g of sodium chloride is added to 300.0 g of water in an attempt to elevate the boiling point. What is the boiling point of the solution?

DATA: MW of NaCl = 58.44 g/mol.

Solution:
NaCl does form two ionic species; namely, Na^+ and Cl^-. Hence, van't Hoff factor (i) is set to 2.0:

$$\text{Molality}, m = \frac{\text{Moles of NaCl}}{\text{kg Water}} = \frac{\left(50\text{ g} \times \dfrac{\text{mol NaCl}}{58.44\text{ g}}\right)}{0.3\text{ kg Water}} = 2.852\,\frac{\text{mol NaCl}}{\text{kg Water}} = 2.852\text{ m}$$

From Table 6.3, $K_b = 0.51$ (°C/m) for water. Substitute in Eq. (6.52), we have:

$$\Delta T_b = (T_{b,\text{Soln.}} - 100°\text{C}) = iK_b m = (2) \times 0.51\frac{°\text{C}}{\text{m}} \times 2.852\text{ m} = 2.91°\text{C}$$

$$T_{b,\text{Soln.}} = 100°\text{C} + 2.91°\text{C} \cong 103.0°\text{C}$$

6.18.3 Freezing Point Depression

The freezing point of a solution is lower than that of the pure solvent. Like boiling point elevation, this property is directly proportional to the molality of the solute particles. It is commonly observed in the winter when salt is spread on icy roads to melt the ice (i.e., liquid saline solution not frozen ice).

Example 6.15 Freezing Point Depression of Glucose Solution
A solution of 50.0 g of glucose is added to 300.0 g of water in an attempt to decrease the freezing point. What is the freezing point of the solution?

DATA: MW of glucose = 180.2 g/mol.

Solution:
Glucose does not form ionic species; hence, van't Hoff factor (i) is set to 1.0:

$$-\Delta T_f = (T_{f,\text{water}} - T_{f,\text{Soln.}}) = iK_f m \tag{6.53}$$

$$\text{Molality}, m = \frac{\text{Moles of Glucose}}{\text{kg water}} = \frac{\left(50\text{ g} \times \dfrac{\text{mol Glucose}}{180.2\text{ g}}\right)}{0.3\text{ kg Water}} = 0.925\,\frac{\text{mol Glucose}}{\text{kg Water}} = 0.925\text{ m}$$

From Table 6.3, $K_f = 1.86$ (°C/m) for water. Substitute in Eq. (6.53), we have:

$$-\Delta T_f = (0°\text{C} - T_{f,\text{Soln.}}) = iK_f m = (1) \times 1.86\frac{°\text{C}}{\text{m}} \times 0.925\text{ m} = 1.72°\text{C}$$

$$T_{f,\text{Soln.}} = 0°\text{C} - 1.72°\text{C} \cong -1.72°\text{C}$$

Example 6.16 Freezing Point Depression of NaCl Solution

A solution of 50.0 g of sodium chloride is added to 300.0 g of water in an attempt to decrease the freezing point. What is the freezing point of the solution?

DATA: MW of NaCl = 58.44 g/mol.

Solution:

NaCl does form two ionic species; namely, Na^+ and Cl^-. Hence, van't Hoff factor (i) is set to 2.0:

$$\text{Molality}, m = \frac{\text{Moles of NaCl}}{\text{kg Water}} = \frac{\left(50\,\text{g} \times \frac{\text{mol NaCl}}{58.44\,\text{g}}\right)}{0.3\,\text{kg Water}} = 2.852\,\frac{\text{mol NaCl}}{\text{kg Water}} = 2.852\,\text{m}$$

From Table 6.3, $K_f = 1.86$ (°C/m) for water. Substitute in Eq. (6.53), we have:

$$-\Delta T_f = (0^\circ\text{C} - T_{f,\text{Soln.}}) = iK_f m = (2) \times 1.86\frac{^\circ\text{C}}{\text{m}} \times 2.852\,\text{m} = 10.61^\circ\text{C}$$

$$T_{f,\text{Soln.}} = 0^\circ\text{C} - 10.61^\circ\text{C} \cong -10.6^\circ\text{C}$$

6.18.4 Osmotic Pressure

Osmosis is the selective passage of solvent molecules through a porous membrane from a dilute solution to a more concentrated one. A semipermeable membrane allows the passage of solvent molecules but blocks the passage of solute molecules. Osmotic pressure π is the pressure required to stop osmosis. The osmotic pressure is directly proportional to the molarity (moles of solute per liter of solution) of the solute particles. Any applied pressure higher than π, on the more concentrated solute side, will cause the phenomenon known as reverse osmosis, which is "squeezing" or pushing back solvent molecules to the less concentrated solute side.

Example 6.17 Osmotic Pressure of Yeast Cells

The typical range of osmotic pressure for actively growing yeast cells varies between 5 and 15 atm. If we place yeast cells in a concentrated salt solution, such cells will be able to produce high internal concentrations of glycerol to counteract the osmotic pressure of the surrounding medium. Suppose that the yeast cells are placed in an aqueous solution containing 3.0% NaCl by mass; the solution density is 1.02 g/cm^3 at 25°C. Calculate the osmotic pressure of a 3.0% aqueous NaCl solution at 25°C. If the normal osmotic pressure inside a yeast cell is 7.3 atm, corresponding to a total concentration of dissolved particles of 0.30 M, what concentration of glycerol must the cells synthesize to exactly balance the external osmotic pressure at 25°C?

Solution:

Basis: 100 g solution.

The solution contains 3.0 g of NaCl per 100 g of solution. Using the molecular mass of NaCl (58.44 g/mol) and the density of the solution (1.02 g/mL), we can calculate the molarity:

$$V_{\text{solution}} = \frac{\text{Mass}}{\text{Density}} = \frac{100\,\text{g solution}}{1.02\,\dfrac{\text{g solution}}{\text{cm}^3\text{solution}}} = 98.04\,\text{cm}^3\ \text{solution}$$

$$M_{\text{NaCl}} = \frac{\text{moles NaCl}}{\text{L solution}} = \frac{\left(\dfrac{3\,\text{g NaCl}}{58.44\,\dfrac{\text{g NaCl}}{\text{mol NaCl}}}\right)}{98.04\,\text{cm}^3 \times \left(\dfrac{1\,\text{L solution}}{1000\,\text{cm}^3}\right)} = 0.5236\,\frac{\text{mol NaCl}}{\text{L solution}} = 0.5236\,\text{M}$$

NaCl does form two ionic species; namely, Na^+ and Cl^-. Hence, van't Hoff factor (i) is set to 2.0:

$$\text{Osmotic Pressure } (\pi) = i\text{MRT} \tag{6.54}$$

where M is the molarity (mol/L) of solute in solution.

$$\text{Osmotic Pressure } (\pi) = i\text{MRT} = 2 \times 0.5236\frac{\text{mol}}{\text{L}} \times 0.08206\,\frac{\text{atm.L}}{\text{mol}\cdot\text{K}}(298\,\text{K}) = 25.6\,\text{atm}$$

If the yeast cells have to exactly balance the external osmotic pressure, they must produce enough glycerol to give an additional internal pressure (or resistance) of (25.6 atm − 7.3 atm) = 18.3 atm. Glycerol is a nonelectrolyte, so we set $i = 1$. From Eq. (6.54), we have:

$$M = \frac{\pi}{\text{RT}} = \frac{18.3\,\text{atm}}{0.08206\,\dfrac{\text{atm.L}}{\text{mol}\cdot\text{K}}(298\,\text{K})} = 0.7483\,\frac{\text{mol}}{\text{L}}$$

Notice that for NaCl outside of the cell membrane, we have $M = 2 \times 0.5236 = 1.047\,\text{M}$

For glycerol inside of cell membrane, we have: 0.3 M + 0.7483 M = 1.048 M

The molarity is the same on both sides of the cell membrane. This is called an *isotonic condition*; fair and square, where the yeast cell neither shrinks nor swells due to migration of water solvent molecules out or into the cell, respectively.

6.19 Liquid–Liquid Equilibrium: Extraction

Liquid–liquid equilibrium refers to the phase behavior of a mixture of two immiscible liquids at a certain temperature and pressure. In this context, water (H_2O) and methyl isobutyl ketone (MIBK) are described as partially miscible liquids. Partial miscibility means that these liquids can mix to some extent but are not completely soluble in each other.

Let us demonstrate the miscibility issue by mixing two liquids: water and MIBK. Here are the two possible scenarios:

Single-Phase Region: For example, if the mixture contains more than either 98.0% water or 97.7% MIBK by mass, a single-phase will result. This implies that at higher concentrations of one component, the liquids are sufficiently miscible to form a homogeneous single-phase solution.

Two Liquid Phases: If the mixture contains less than 98.0% water and 97.7% MIBK, the system separates into two liquid phases. The composition of these phases is specified:

One phase contains 98.0% water and 2.0% MIBK.

The other phase contains 97.7% MIBK and 2.3% water.

The statement "Water and MIBK are examples of partially miscible liquids" emphasizes that these liquids can mix to some extent, but there are limits to their mutual solubility. Beyond these limits, phase separation occurs. In light of the above scenarios, water and MIBK liquids would be termed immiscible if one phase contained a negligible amount of water and the other a negligible amount of MIBK. Immiscibility refers to the inability of two substances to dissolve in each other, hence resulting in distinct phases.

Let me refer to my book (Al-Malah [2]) regarding the separation of water acetone mixture: "A mixture containing 50.0 wt.% acetone and 50.0 wt.% water is to be separated into two streams: one enriched in acetone and the other in water. The separation process consists of extraction of acetone from water into methyl isobutyl ketone (MIBK), which preferentially dissolves acetone but is nearly immiscible with water. The overall goal of this problem is to separate the feed stream into two streams which have greater than 95% purity of water and acetone, respectively." Thus, a water-acetone mixture can eventually be separated into pure streams of acetone and water, as well as recycling MIBK stream which results in minimization of MIBK waste or maximization of solvent recovery needed for extraction processes, in general. Let us take a simple example of water–acetone extraction process using MIBK as a third agent.

Example 6.18 Separation of Water–Acetone Mixture Using MIBK

Figure 6.10 shows that upon thoroughly mixing the 50-wt. water–acetone mixture with pure MIBK stream and then admitting the entire mixture, made of the three species, into the flash-drum tank, which acts as a single equilibrium stage, two separate phases are formed: The top stream is the organic, MIBK-based phase and the bottom is the aqueous phase which is mainly water.

Find the mass flow rate and the compositions of both exit streams.

Solution:

We have three unknown terms: $\dot{m}_{\text{top}}, x_{\text{water}}^{\text{top}}$, and $x_{\text{MIBK}}^{\text{bottom}}$. We have three independent equations: one total materials balance and two component material balances.

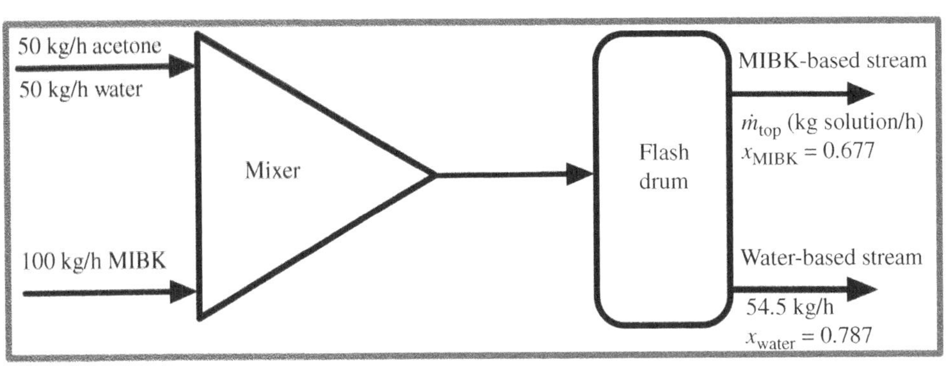

Figure 6.10 Separation of acetone–water mixture using MIBK as a third agent.

Total Material Balance:

$$\text{Sum of Mass Flow Rate In} = \text{Sum of Mass Flow Rate Out}$$

$$50\frac{\text{kg}}{\text{h}} + 50\frac{\text{kg}}{\text{h}} + 100\frac{\text{kg}}{\text{h}} = \dot{m}_{\text{top}} + 54.5\frac{\text{kg}}{\text{h}}$$

$$\dot{m}_{\text{top}} = (200 - 54.5)\frac{\text{kg}}{\text{h}} = 145.5\frac{\text{kg}}{\text{h}}$$

MIBK Balance:

$$100\frac{\text{kg}}{\text{h}} = x_{\text{MIBK}}\dot{m}_{\text{top}} + x_{\text{MIBK}}^{\text{bottom}} \times 54.5\frac{\text{kg}}{\text{h}}$$

$$100\frac{\text{kg}}{\text{h}} = 0.677 \times 145.5\frac{\text{kg}}{\text{h}} + x_{\text{MIBK}}^{\text{bottom}} \times 54.5\frac{\text{kg}}{\text{h}}$$

Solve for $x_{\text{MIBK}}^{\text{bottom}}$: $x_{\text{MIBK}}^{\text{bottom}} = \dfrac{100 - 98.5}{54.5} = 0.0275$

Water Balance:

$$50\frac{\text{kg}}{\text{h}} = x_{\text{water}}^{\text{top}}\dot{m}_{\text{top}} + x_{\text{Water}}^{\text{bottom}} \times 54.5\frac{\text{kg}}{\text{h}}$$

$$50\frac{\text{kg}}{\text{h}} = x_{\text{water}}^{\text{top}}145.5\frac{\text{kg}}{\text{h}} + 0.787 \times 54.5\frac{\text{kg}}{\text{h}}$$

Solve for $x_{\text{water}}^{\text{top}}$: $x_{\text{water}}^{\text{top}} = \dfrac{50 - 42.9}{145.5} = 0.049$

Notice that:

$$x_{\text{Acetone}}^{\text{top}} = 1 - \left(x_{\text{water}}^{\text{top}} + x_{\text{MIBK}}^{\text{top}}\right) = 1 - (0.049 + 0.677) = 0.274$$

And

$$x_{\text{Acetone}}^{\text{bottom}} = 1 - \left(x_{\text{water}}^{\text{bottom}} + x_{\text{MIBK}}^{\text{bottom}}\right) = 1 - (0.787 + 0.0275) = 0.186$$

The process is now well specified both from mass flow rate and composition point of view. Moreover, acetone has preferentially and lovely migrated toward the MIBK phase relative to the water phase. Keep in mind that acetone was originally found in water phase with 50 wt.%. Of course, as mentioned in my Aspen Plus textbook, more downstream treatment steps are needed to carry out further fractionation between acetone, water, and MIBK. The top stream is mainly MIBK simply because its density is $0.802 < 1.0\,\text{g/cm}^3$ for water. The lighter (brain) stays at the top. This is also valid in the world of politics; idiots are at the top.

Example 6.19 Separation of Water–Acetone Mixture Using Chloroform ($CHCL_3$)
Figure 6.11 shows that upon thoroughly mixing the 90-wt. water–acetone mixture with pure CHCL3 stream and then admitting the entire mixture, made of the three species, into the flash-drum tank which acts as a single equilibrium stage, two separate phases are formed: The top stream is the aqueous, water-based phase and the bottom is the organic, chloroform-based phase.

Find the mass flow rate and the compositions of both exit streams.

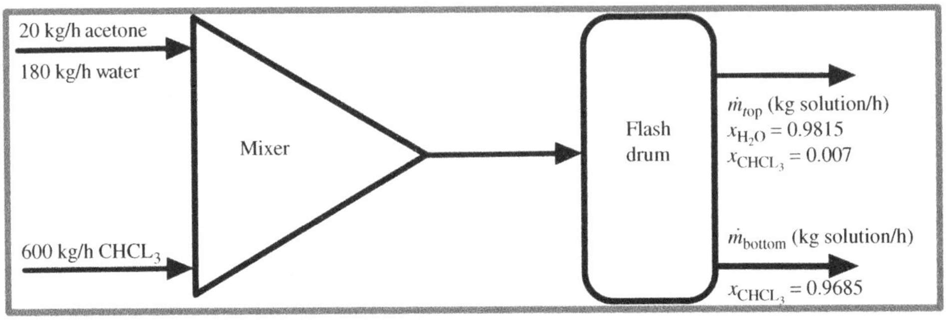

Figure 6.11 Separation of water acetone mixture using chloroform as a third agent.

Find the percent of acetone transferred into the organic phase per acetone fed to the mixing unit.

Solution:

We have three unknown terms: $\dot{m}_{\text{top}}$, $\dot{m}_{\text{bottom}}$, and $x_{\text{water}}^{\text{bottom}}$. We have three independent equations: one total materials balance and two component material balances.

Total Material Balance:

$$\text{Sum of Mass Flow Rate In} = \text{Sum of Mass Flow Rate Out}$$

$$20\frac{\text{kg}}{\text{h}} + 180\frac{\text{kg}}{\text{h}} + 600\frac{\text{kg}}{\text{h}} = \dot{m}_{\text{top}}\frac{\text{kg}}{\text{h}} + \dot{m}_{\text{bottom}}\frac{\text{kg}}{\text{h}} \tag{a}$$

$$\dot{m}_{\text{top}} = (800 - \dot{m}_{\text{bottom}})\frac{\text{kg}}{\text{h}}$$

$CHCL_3$ Balance:

$$600\frac{\text{kg}}{\text{h}} = x_{\text{CHCL}_3}\dot{m}_{\text{top}} + x_{\text{CHCL}_3}^{\text{bottom}} \times \dot{m}_{\text{bottom}}$$

$$600\frac{\text{kg}}{\text{h}} = 0.007 \times \dot{m}_{\text{top}}\frac{\text{kg}}{\text{h}} + 0.9685 \times \dot{m}_{\text{bottom}}\frac{\text{kg}}{\text{h}} \tag{b}$$

Water Balance:

$$180\frac{\text{kg}}{\text{h}} = x_{\text{water}}^{\text{top}}\dot{m}_{\text{top}}\frac{\text{kg}}{\text{h}} + x_{\text{Water}}^{\text{bottom}} \times \dot{m}_{\text{bottom}}\frac{\text{kg}}{\text{h}}$$

$$180\frac{\text{kg}}{\text{h}} = 0.9815\dot{m}_{\text{top}}\frac{\text{kg}}{\text{h}} + x_{\text{Water}}^{\text{bottom}} \times \dot{m}_{\text{bottom}}\frac{\text{kg}}{\text{h}} \tag{c}$$

Notice that Eq. (c) is nonlinear in terms of unknown variables, which makes it difficult to solve as is the case with a set of linear algebraic equations. Luckily, Eqs. (a) and (b) can be solved simultaneously.

Figure 6.12 shows the MATLAB code for solving two simultaneous algebraic equations: (a) and (b), with two unknowns: $\dot{m}_{\text{top}}$ and $\dot{m}_{\text{bottom}}$.

```
clear variables;
% Define the system of equations as a matrix of coefficients and the right-hand side vector
%Eq. (a): mdottop + mdotbottom = 800
%Eq. (b): 0.007mdottop + 0.9685mdotbottom = 600
A = [1 1; 0.007 0.9685];
b = [800; 600];
% Solve the system using matrix inversion
x = A \ b;
% Print the solution for m1 and m3
mdottop = x(1);
mdotbottom=x(2);
fprintf('mdottop = %f\n', mdottop);
fprintf('mdotbottom = %f\n', mdotbottom);
```

Figure 6.12 MATLAB code snippet for solving two simultaneous algebraic equations.

Below are the values of **mdottop** and **mdotbottom**, as calculated by MATLAB code:

mdottop = 181.799272
mdotbottom = 618.200728

From Eq. (c), we have

$$180\frac{\text{kg}}{\text{h}} = 0.9815(181.8)\frac{\text{kg}}{\text{h}} + x_{\text{Water}}^{\text{bottom}} \times (618.2)\frac{\text{kg}}{\text{h}} \tag{c}$$

Solve for $x_{\text{Water}}^{\text{bottom}}$: $x_{\text{Water}}^{\text{bottom}} = \frac{180 - 178.44}{618.2} = 0.0025$

Notice that:

$$x_{\text{Acetone}}^{\text{top}} = 1 - \left(x_{\text{water}}^{\text{top}} + x_{\text{CHCL3}}^{\text{top}}\right) = 1 - (0.9815 + 0.007) = 0.0115$$

And

$$x_{\text{Acetone}}^{\text{bottom}} = 1 - \left(x_{\text{water}}^{\text{bottom}} + x_{\text{CHCL3}}^{\text{bottom}}\right) = 1 - (0.0025 + 0.9685) = 0.029$$

The process is now well specified both from mass flow rate and composition point of view. Moreover, acetone has preferentially and lovely migrated toward the $CHCL_3$ phase relative to the water phase. Keep in mind that acetone was originally found in water phase with 10.0 wt.%. Of course, as mentioned in my Aspen Plus textbook, more downstream treatment steps are needed to carry out further fractionation between acetone, water, and $CHCL_3$. The bottom stream is mainly chloroform simply because its density is 1.48 > 1.0 g/cm^3 for water. The heavier stays at the bottom.

One way to measure the process efficiency of acetone removal from water is to evaluate what is called the percent of acetone transferred into the organic phase, which is chloroform in this case. It is given by:

$$\frac{\text{Acetone in chloroform exit stream}}{\text{acetone fed}} \times 100\% = \frac{x_{\text{Acetone}}^{\text{bottom}} \times \dot{m}_{\text{bottom}}\frac{\text{kg}}{\text{h}}}{20\frac{\text{kg}}{\text{h}}} \times 100\%$$

$$\text{Percent Transferred} = \frac{0.029 \times 618.2\frac{\text{kg}}{\text{h}}}{20\frac{\text{kg}}{\text{h}}} \times 100\% = \frac{17.93\frac{\text{kg}}{\text{h}}}{20\frac{\text{kg}}{\text{h}}} \times 100\% = 89.6\%$$

Nevertheless, the overall performance of a chemical process is governed by the process economics, which includes both the annual fixed and operating costs.

6.20 Adsorption Equilibrium Isotherms

Adsorption equilibrium isotherms describe the relationship between the amount of adsorbate (substance being adsorbed) and the concentration of the adsorbate at a constant temperature. These isotherms provide valuable insights into the adsorption process, helping researchers and engineers understand how adsorption occurs on a particular adsorbent material under specific conditions. For more information on adsorption equilibrium and isotherms, please refer to my PhD dissertation: Al-Malah [9].

There are several classical adsorption equilibrium isotherms, each characterized by its own mathematical expression. Some of the most commonly used isotherms include:

6.20.1 Langmuir Model

The Langmuir isotherm is a simple model that describes the adsorption of molecules onto a solid surface. It assumes that adsorption occurs at specific sites on the surface and that each site can only hold one molecule.

Let us consider a gas/solid phase with molecules A and a solid surface with N adsorption sites.

$$A + N \rightleftarrows A$$

At equilibrium, the rate of molecules sticking to the surface (k_1) equals the rate of molecules leaving the surface ($k - 1$):

$$k_1[A][N] = k_{-1}[AN] \tag{6.55}$$

Here, $[A]$ represents the concentration of gas phase molecules, $[N]$ represents the number of available sites on the solid surface, and $[AN]$ represents the number of adsorbed molecules on the solid surface. We can rearrange this equation to solve for the concentration of adsorbed molecules $[AN]$:

$$[AN] = \frac{k_1[A][N]}{k_{-1}} \tag{6.56}$$

Let us define the equilibrium constant K as:

$$K = \frac{k_1}{k_{-1}} \tag{6.57}$$

$$[AN] = K[A][N] \tag{6.58}$$

This equation gives the concentration of adsorbed molecules at equilibrium. However, it is more common to express the Langmuir isotherm in terms of the fraction of surface coverage (θ):

$$\theta = \frac{[AN]}{[N] + [AN]} = \frac{K[A][N]}{[N] + [AN]} = \frac{K[A][N]}{[N] + K[A][N]} = \frac{K[A]}{1 + K[A]} = \frac{[A]}{\frac{1}{K} + [A]} \tag{6.59}$$

Figure 6.13 shows the MATLAB code for plotting the Langmuir model for adsorption isotherm, as shown in Figure 6.14:

Figure 6.14 shows a plot of Langmuir monolayer surface coverage, θ versus the bulk concentration, $[A]$. Notice that the Langmuir-type isotherm shows a steep initial slope followed by the attainment of a plateau at high bulk concentrations.

```
clear variables;
A=0:10/100:10;
K=10.0;
theta=A./(1/K+A);
plot(A,theta)
title("Plot of \theta versus bulk concentration");
ylabel("\theta");
xlabel("Bulk Concentration, [A]");
```

Figure 6.13 MATLAB code for creating a plot of Langmuir model adsorption isotherm.

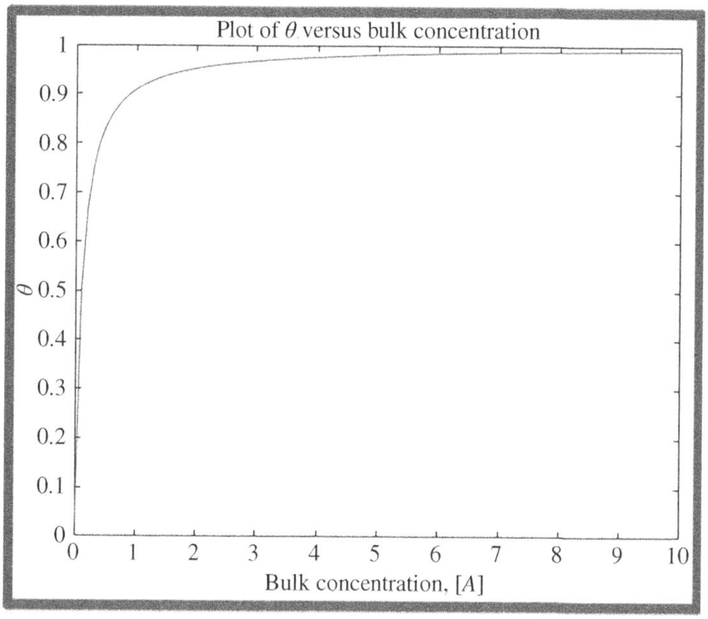

Figure 6.14 Langmuir isotherms for a monolayer surface coverage, θ, as a function of bulk concentration, $[A]$.

6.20.2 Freundlich Model

On the other hand, the Freundlich model adsorption isotherm is described by the following equation:

$$\theta = a[A]^b \tag{6.60}$$

where a and b are function constants that define the functionality of θ versus $[A]$ at equilibrium. The Freundlich isotherm shows a monotonic increase in θ with $[A]$. Upon examining the pertinent literature, one recognizes that the interpretation of data with Langmuir-type isotherms is extremely popular. Figure 6.15 shows the MATLAB code for plotting the Freundlich model for adsorption isotherm.

Figure 6.16 shows the plot of the surface coverage versus the bulk concentration under equilibrium conditions. The fact that θ is larger than one tells us we have a multilayer, not only a single monolayer surface coverage.

```
clear variables;
a=1.0;
b=0.5;
A=0:10/100:10;
theta=a*A.^b;
plot(A,theta)
title("Freundlich plot of \theta versus [A]");
xlabel("Bulk Concentration, [A]");
ylabel("\theta");
```

Figure 6.15 MATLAB code for creating a plot of the Freundlich model for adsorption isotherm.

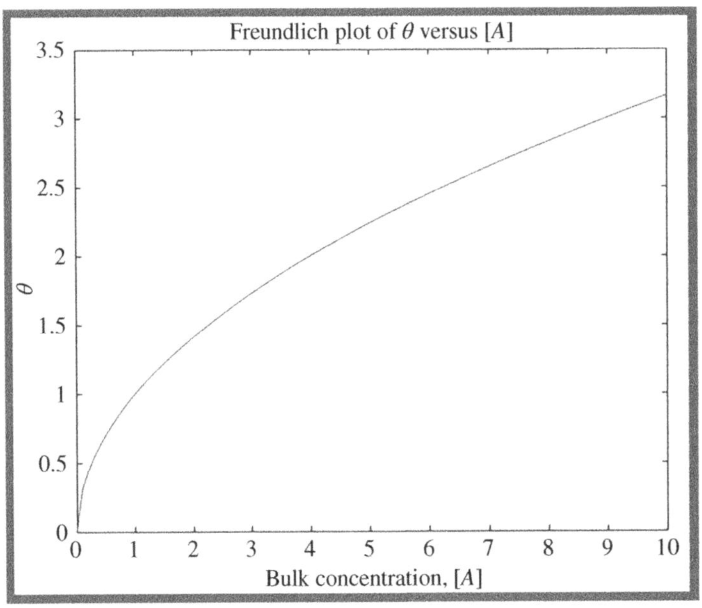

Figure 6.16 Freundlich multilayer surface coverage, θ, as a function of $[A]$.

6.20.3 Applications of Adsorption Isotherms

Adsorption isotherms play a crucial role in various chemical industry processes by providing insights into the interaction between adsorbents and adsorbates. Here are some key examples:

1) **Design and Optimization of Separation Processes**:
   - **Purification**: Removing impurities from gases and liquids, like removing CO_2 from natural gas or water purification using activated carbon. Isotherms help select the right adsorbent and predict its capacity for target impurities at different concentrations.
   - **Chromatography**: Separating different components in a mixture based on their adsorption affinities. Isotherms help optimize column design, eluent selection, and predict separation efficiency.

- **Drying**: Removing moisture from gases and liquids. Isotherms help determine the required amount of adsorbent and operating conditions for efficient drying.

2) **Catalyst Development and Optimization**:
   - **Supported Catalysts**: Understanding the adsorption of reactants and products on catalyst supports helps design catalysts with optimal surface area, pore size, and activity.
   - **Catalyst Poisoning**: Predicting the extent of adsorption of poisoning species on catalyst surfaces, aiding in developing strategies to mitigate their effects.

3) **Environmental Remediation**:
   - **Removal of Pollutants from Air and Water**: Isotherms help in selecting and designing adsorbents for air and water purification, capturing pollutants like heavy metals, organic compounds, and dyes.
   - **Soil Remediation**: Understanding the adsorption of contaminants onto soil particles helps in designing strategies for soil decontamination.

4) **Food and Beverage Industry**:
   - **Decaffeination of Coffee and Tea**: Isotherms guide the selection of adsorbents and operating conditions for efficient caffeine removal while preserving flavor and aroma.
   - **Sugar Refining**: Using activated carbon for color removal and deodorization of sugar syrups. Isotherms help optimize this process for desired purity and color.

5) **Other Applications**:
   - **Gas Storage and Separation**: Isotherms help design materials for efficient storage and separation of gases like hydrogen and methane.
   - **Sensor Development**: Utilizing the selective adsorption of specific molecules for developing sensitive and specific sensors.

End of Chapter Problems

CHW 6.1 What are the two key conditions that must be met for a system to be in equilibrium?

A Net force and torque cancellation, and constant external properties
B Net force and torque cancellation, and constant internal properties
C Energy flow and matter transfer, and constant external properties
D Temperature and pressure equilibrium

CHW 6.2 What is the driving force for attaining equilibrium in a system?

A Maximum entropy
B Minimal free energy
C Maximum energy consumption
D Constant temperature

CHW 6.3 In a closed electrical circuit, what is the driving force for electrons?

A Higher resistance
B Higher voltage
C Higher current
D Higher capacitance

CHW 6.4 What happens when a compressed gas in a chamber flows into an empty chamber to achieve mechanical equilibrium?

A The gas molecules rush inward
B The pressure inside the second chamber decreases
C The gas temperature decreases
D The gas flows until both pressures equalize

CHW 6.5 Which of the following is NOT a defining characteristic of a system in equilibrium?

A No change over time
B Homogeneous composition
C No chemical reactions
D No mass, heat, or work exchange

CHW 6.6 In the equation: $\mathbf{d}\boldsymbol{G}^t = -\boldsymbol{S}^t\mathbf{d}\boldsymbol{T} + \boldsymbol{V}^t\mathbf{d}\boldsymbol{P}$, what does a value of $\mathrm{d}G^t > 0$ indicate for a closed, non-reacting system?

A Equilibrium state
B Nonequilibrium state with spontaneous change
C Nonequilibrium state requiring external intervention
D Steady state with constant mass/heat flow

CHW 6.7 Which of the following is NOT directly measurable for a pure fluid?

A Pressure
B Temperature
C Volume
D Internal energy

CHW 6.8 Referring to Table 6.1, what is the normal boiling point temperature (T_{NBP}), expressed in °C, of *n*-hexane?

A Use Eq. (6.20)
B Use Appendix D | D.2 MATLAB Code (AntoinEq.m) for Antoine's equation

CHW 6.9 Which is slightly more volatile: benzene (B) or cyclohexane (C-H)? Use a comparison between boiling points for the same pressure or the saturation pressure for the same temperature. Search the Internet sources and try to explain the behavior in light of London dispersion forces (a type of van der Waals forces) for both molecules.

CHW 6.10 Which is more volatile: *n*-hexane (N-H) or cyclohexane (C-H)? Use a comparison between boiling points for the same pressure or the saturation pressure for the same temperature. Search the Internet sources and try to explain the behavior in light of London dispersion forces (a type of van der Waals forces) for both molecules versus their molecular packing density or efficiency.

CHW 6.11 For the following cases, carry out the DoF analysis using the Gibbs phase rule and show possible variables and scenarios for each case.

A A solution of NaCl crystals in water
B Equilibrium between liquid ethanol and its vapor
C A gas mixture of N_2 and O_2 gas
D A gaseous mixture of H_2, N_2, and NH_3, given that the single reaction: $3H_2 + N_2 \rightleftharpoons 2\ NH_3$
E Solid mixture of two salts in equilibrium with their saturated solutions

CHW 6.12 [Benzene (B)–toluene (T)] VLE system

We have a liquid mixture of benzene (B) and toluene (T) in equilibrium with its vapor mixture. As far as the DoF analysis and number of intensive variables are concerned, they will be the same as those of methanol (M_OH) and ethanol (E_OH) VLE system explained in 6.9. Bubble-point and Dew-point P/T Calculations.

For each of the following six cases, find the required and missing variables. In all given six cases, X_T and Y_T will be the complement of X_B and Y_B, respectively. The databases code for benzene (B) = 3415 and that for toluene (T) = 4785.

A **Case #1**: Given $P = 1.0$ atm and $T = 95°C$, find X_B and Y_B in a VLE condition
B **Case #2**: Given $P = 1.0$ atm and $Y_B = 0.7$, find X_B and T in a VLE condition
C **Case #3**: Given $P = 1.0$ atm and $X_B = 0.6$, find Y_B and T in a VLE condition
D **Case #4**: Given $T = 95°C$ and $Y_B = 0.6$, find X_B and P in a VLE condition
E **Case #5**: Given $T = 95°C$ and $X_B = 0.4$, find Y_B and P in a VLE condition
F **Case #6**: Given $X_B = 0.3$ and $Y_B = 0.5$, find P and T in a VLE condition

CHW 6.13 Henry's law for calculation of aqueous solubility of atmospheric O_2

Calculate the solubility of atmospheric gaseous oxygen in water at a temperature of 298 K when the partial pressure exerted by O_2 is 0.21 atm.

CHW 6.14 Henry's law for calculation of CO_2 partial pressure

At what partial pressure would the CO_2 gas have a solubility (in water at 25°C) of $2*10^{-3}$ M?

CHW 6.15 The water vapor pressure (P_{Water}), at 50°C, is 80 mmHg. Calculate:

A The percent relative humidity of air, RH
B The dew point
C The absolute humidity (AH)
D The molar humidity, h_{Molar}
E The molal humidity, h_{Molal}
F The mixing ratio, x

CHW 6.16 The mass of water vapor is 12 g, and the total mass of moist air is 100 g. What is the specific humidity of air? What is the mass fraction of water in wet air mixture?

CHW 6.17 Humid Air Compression

An air–water mixture enters the compression unit at a flow rate of 100 mol/min with water vapor mole fraction of 0.12. The compressor elevates the pressure to 8000 mmHg and with an exit temperature of 95°C, as shown in Figure CHW 6.17. Calculate the outlet flow rates and their compositions in terms of water and dry air.

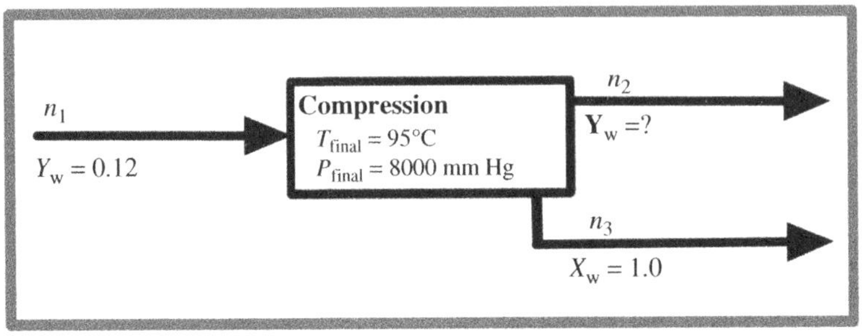

Figure CHW 6.17 Compression of wet air, which results in partial condensation of inlet water vapor.

CHW 6.18 Crystallization, filtration, and drying of silver nitrate ($AgNO_3$).

Two hundred kilograms of a saturated aqueous solution of $AgNO_3$ at l00°C is cooled down to 20°C, thereby forming $AgNO_3$ crystals, which are filtered from the remaining solution. The wet filter cake, which contains 85% solid crystals and 15% saturated solution by mass, passes to a dryer in which the remaining water is vaporized, as shown in Figure CHW 6.18.

Calculate the percent recovery of the $AgNO_3$ in the feed stream, eventually recovered as dry crystals, and the amount of water that must be removed in the drying stage.

DATA: Saturated solutions:

At 100°C: 952 g $AgNO_3$/100 g solution. At 20°C: 222 g $AgNO_3$/100 g solution

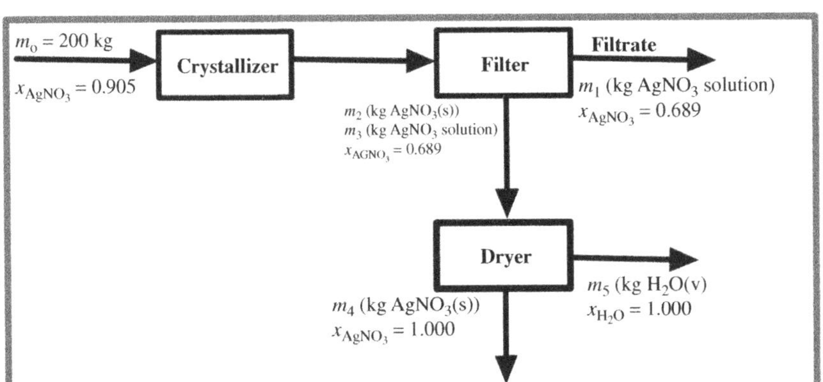

Figure CHW 6.18 A flowchart showing the crystallization step followed by filtration and drying of $AgNO_3$.

CHW 6.19 Crystallization of magnesium sulfate heptahydrate

An aqueous solution of magnesium sulfate at 100°C containing 35 wt.% $MgSO_4$ is fed to a cooling crystallizer that operates at 10°C. The stream leaving the crystallizer is a slurry of solid magnesium sulfate heptahydrate particles [$MgSO_4{\cdot}7H_2O(s)$] suspended in a liquid solution, as shown in Figure CHW6.19.

The solubility of magnesium sulfate heptahydrate [$MgSO_4(s)$] in water at 10°C is approximately 30.0 g per 100 g of water (x_{MgSO4} = 0.231). The molecular weight of anhydrous magnesium sulfate is 120.4, and that of the heptahydrate salt is 246.5 tonne/tonne mol. Notice that 1 tonne mol is 1,000 kmol.

Determine the rate at which solution must be fed to the crystallizer to produce 1 metric ton (1,000 kg) of magnesium sulfate heptahydrate per hour.

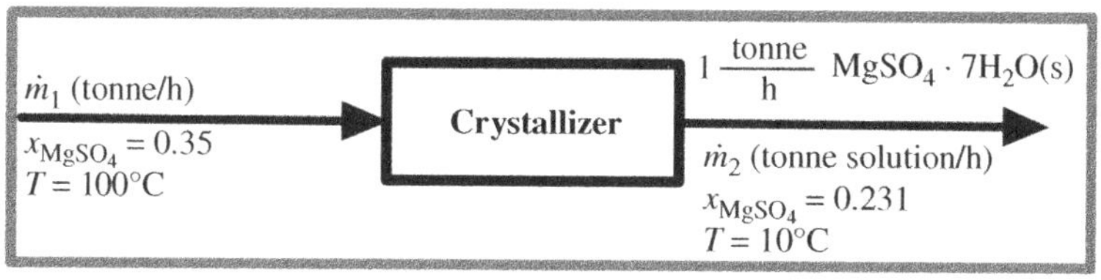

Figure CHW 6.19 A schematic for crystallization of $MgSO_4 \cdot 7H_2O(s)$ out of $MgSO_4$ solution.

CHW 6.20 Vapor pressure depression.

Calculate the vapor pressure of a solution made by dissolving 166.0 g KI in 500 g of water. Using the Antoine equation MATLAB code, the vapor pressure of pure water is 149.0 mmHg at 60°C. The maximum solubility of KI in water is 176.0 g/100 g at 60°C.

DATA: The molecular mass of KI is 166 g/mol.

Reference: https://en.wikipedia.org/wiki/Potassium_iodide.

CHW 6.21 Boiling point elevation of $FeCl_3$ solution.

A solution of 100.0 g of ferric chloride is added to 300.0 g of water in an attempt to elevate the boiling point. What is the boiling point of the solution?

The maximum solubility at 25°C is 91.2 g/100 g water. (https://en.wikipedia.org/wiki/Iron(III)_chloride).

DATA: MW of $FeCl_3$ = 162.2 g/mol.

CHW 6.22 Molecular mass and formula determination via freezing point depression

A 15.75-g sample of a compound with the empirical formula C_5H_4 is dissolved in 300 g of benzene.

The freezing point of the solution is 2.1°C below that of pure benzene. What are the molar mass and molecular formula of this compound?

CHW 6.23 Molecular mass determination via osmotic pressure of hemoglobin (Hb) solution.

A solution is prepared by dissolving 42.0 g of hemoglobin (Hb) in enough water to make up 1 L in volume. If the osmotic pressure of the solution is found to be 12.0 mmHg at 25°C, calculate the molar mass of hemoglobin.

CHW 6.24 Separation of water–acetone mixture using benzene (C_6H_6).

Figure CHW 6.24 shows that upon thoroughly mixing the 90-wt. water–acetone mixture with pure benzene stream and then admitting the entire mixture, made of the three species, into the flash-drum tank, which acts as a single equilibrium stage, two separate phases are formed: The top stream is the organic, benzene-based phase and the bottom is the aqueous, water-based phase.

Find the mass flow rate and the compositions of both exit streams.

Find the percent of acetone transferred into the organic phase per acetone fed to the mixing unit.

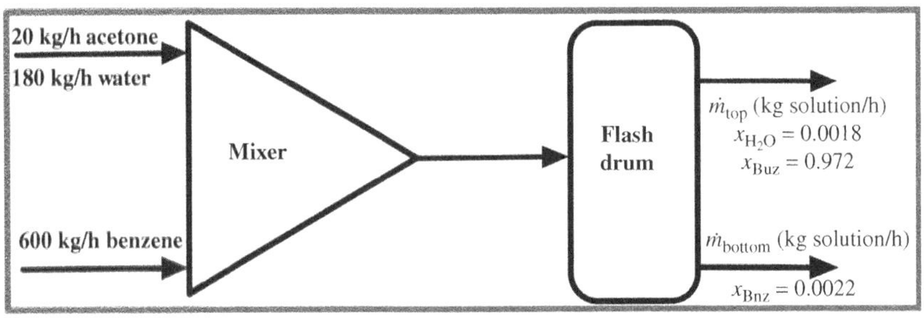

Figure CHW 6.24 Separation of water acetone mixture using benzene as a third agent.

CHW 6.25 In the Langmuir model, what does the equilibrium constant K represent?

A The rate of molecules leaving the surface
B The number of available sites on the surface
C The concentration of gas phase molecules
D The ratio of the rate of molecules sticking to the surface to the rate of molecules leaving the surface

CHW 6.26 In the Langmuir isotherm equation, what does θ represent?

Given that:

$$\theta = \frac{[AN]}{[N] + [AN]} = \frac{K[A][N]}{[N] + [AN]} = \frac{K[A][N]}{[N] + K[A][N]} = \frac{K[A]}{1 + K[A]} = \frac{[A]}{\frac{1}{K} + [A]}$$

Answer the following questions:

A The rate of adsorption
B The number of available surface sites
C The concentration of gas phase molecules
D The fraction of surface coverage by adsorbed molecules

CHW 6.27 If $K = 10$ and $[A]$ is 5, what will be θ?

A 5.0
B 10.0
C 0
D 0.33
E 0.98
F 1.02

CHW 6.28 If $K = 1$ and $[A]$ is 5, what will be θ?

A 5.0
B 6.0
C 0
D 0.83
E 1.0
F 1.2

CHW 6.29 If the value of θ is 0.6 and $K = 2$, what will be $[A]$?

A 1.0
B 1.67
C 1.33
D 0.50
E 2.0
F 0.75

CHW 6.30 In the low concentration range, what happens to θ as the concentration of gas phase molecules ($[A]$) increases?

A θ decreases
B θ remains constant
C θ increases
D θ becomes negative

CHW 6.31 What does the Freundlich equation $\theta = a[A]^b$ describe?

A The rate of desorption of adsorbed molecules
B The relationship between θ and the concentration of gas phase molecules ([A])
C The number of available surface sites
D The equilibrium constant (K) for adsorption

CHW 6.32 How does the Freundlich isotherm typically behave as the concentration of gas phase molecules ($[A]$) increases?

A It shows a monotonic decrease in θ
B It shows a monotonic increase in θ
C It remains constant regardless of [A]
D It becomes negative

CHW 6.33 If θ is larger than one according to the Freundlich model, what does it suggest about the adsorption process?

A The adsorption process is not occurring
B The adsorption process is highly efficient
C The adsorption process involves multiple layers of adsorbed molecules
D The adsorption process is irreversible

References

1 Yaws, C., Narasimhan, P., and Gabbula, C. (2005). Yaw's Handbook of Antoine Coefficients for Vapor Pressure (Electronic Edition). Knovel. ISBN: 1-59124-879-5.

2 Al-Malah, K. (2022). Aspen Plus: Chemical Engineering Applications, 2e. Wiley. https://www.wiley.com/en-ae/Aspen+Plus:+Chemical+Engineering+Applications,+2nd +Edition-p-9781119868699.

3 Green, D.W. and Perry, R.H. (2008). Perry's Chemical Engineers' Handbook, 8e. New York: McGraw-Hill, Inc. https://doi.org/10.1036/0071422943.

4 Helmenstine, T. (2020). Henry's law example problem. https://www.thoughtco.com/henrys-law-example-problem-609500 (accessed 27 February 2020.

5 Perović, B., Klimenta, D., Jevtić, M., and Milovanović, M. (2019). The effect of different sky temperature models on the accuracy in the estimation of the performance of a photovoltaic module. *Journal of the Technical University of Gabrovo* 59: 78–82.

6 Al-Malah, K. "Prediction of aqueous solubility of organic solvents as a function of selected molecular properties". *Journal of Pharmaceutics & Drug Delivery Research*, 1:2 (2012). https://doi.org/10.4172/2325-9604.1000106

7 Al-Malah, K. (2011). Aqueous solubility of a simple single-carbon organic molecule as a function of its size & dipole moment. *Journal of Molecular Modeling* 17 (5): 1029–1034. https://doi.org/10.1007/s00894-010-0800-y.

8 Al-Malah, K. (2011). Aqueous solubility of a diatomic molecule as a function of its size & electronegativity difference. *Journal of Molecular Modeling* 17 (2): 325–331. https://doi.org/10.1007/s00894-010-0729-1.

9 Al-Malah, K. (1993). A macroscopic model for apparent protein adsorption equilibrium at hydrophobic solid/liquid interfaces. Ph.D. dissertation. Oregon State University, Corvallis. 127 pages.

7

Energy and Energy Balance

7.1 Importance of Energy Balance

The energy balance is a fundamental and critical concept in chemical engineering applications, playing a pivotal role in the design, optimization, and operation of processes. The energy balance is based on the principle of conservation of energy, stating that energy cannot be created or destroyed but only transferred or converted from one form into another. In chemical processes, understanding and maintaining this balance is essential to ensure that the total energy entering a system equals the total energy leaving the system. Here are some key applications that demonstrate the importance of the energy balance in chemical engineering:

1) **Process Design and Optimization**: In the design phase of a chemical process, engineers must establish the energy requirements for different unit operations. The energy balance helps in sizing equipment such as heat exchangers, reactors, and distillation columns. It also aids in optimizing the overall process to minimize energy consumption and maximize efficiency.
2) **Heat Integration**: Energy balance is crucial for heat integration in a process. Heat exchangers and other heat transfer devices are strategically placed based on the energy balance to recover and reuse heat within the system. This not only reduces energy costs but also enhances the overall sustainability of the process.
3) **Temperature and Pressure Control**: Maintaining the energy balance is essential for controlling temperatures and pressures within a system. Deviations from the energy balance can lead to unsafe operating conditions, inefficient processes, or even equipment failure. Proper temperature and pressure instrumentation and control are vital for ensuring product quality and process safety.
4) **Economic Considerations**: Energy is a significant cost factor in chemical processes. An accurate energy balance helps in evaluating the economic feasibility of a process by

---

To my dear readers,
When copying MATLAB code or text, please right-click and attempt both the "Copy with Formatting" and "Copy" options from the context menu. This approach is effective within Adobe Acrobat. If neither option works, kindly carefully review any MATLAB error messages and compare the pasted text to the original to identify and correct any discrepancies. Alternatively, a soft copy of the MATLAB code is available in the Wiley book companion site of this book for your convenience.

MATLAB-based Computations of Chemical Engineering Principles, First Edition. Kamal I. M. Al-Malah.

Companion website: www.wiley.com/go/al-malah/matlabbasedcomputations

estimating energy requirements and, hence, costs. Engineers can identify opportunities to improve energy efficiency and reduce operational expenses.

5) **Environmental Impact**: Efficient energy use in chemical processes not only reduces costs but also minimizes environmental impact. Processes with optimized energy balances typically have lower greenhouse gas emissions, contributing to sustainability goals and meeting regulatory requirements, both locally and globally.
6) **Start-up and Shut-down Procedures**: Understanding the energy balance is crucial during the start-up and shut-down phases. Proper procedures ensure that equipment is brought to operating conditions efficiently, minimizing energy waste and preventing thermal shocks that can damage equipment.
7) **Reaction Kinetics**: Energy considerations are essential in understanding and optimizing chemical reactions. The energy balance helps determine the heat generated or absorbed during a reaction, influencing reactor design and process control strategies.

7.2 What is Internal Energy ($\overline{U}$)?

Internal energy is a property of a bulk amount of matter that reflects the combined effects of the microscopic motions and interactions of its particles. It encompasses three main components:

1) **Inter-molecular Potential Energy (P.E.)**: This represents the P.E. stored due to the forces between molecules. It's like a stretched spring holding P.E. that can be converted into other forms. Think of it as the "stickiness" among molecules.
   a) **Van der Waals Forces**: These are weak attractive forces that exist among all molecules. They exist in three types:
      i) **Permanent Dipole-permanent Dipole**: Imagine two magnets attracting or repelling each other based on their orientations, similar to how polar molecules with permanent electric dipoles can attract or repel each other.
      ii) **Permanent Dipole-induced Dipole**: A permanent dipole (like a charged magnet) can induce a temporary dipole (like a weakly magnetized object) in a nearby molecule, leading to attraction.
      iii) **Induced Dipole-induced Dipole**: Fluctuations (or what is called the transient asymmetry) in the electron distribution of one molecule or atom can create a temporary dipole, which can then induce a temporary dipole in another neighboring molecule, causing attraction.
   b) **Repulsive Forces**: As molecules get too close, love (attraction) will start diminishing. Their electron clouds start to repel (hate) each other, creating a strong repulsive force that prevents them from completely collapsing onto each other. This is analogous to pushing two balloons together, where they eventually resist further compression.
2) **Molecular Kinetic Energy (K.E.)**: This represents the energy of motion of the individual molecules. Think of gas molecules bouncing around like tiny balls, each with a certain speed and mass. The faster and heavier the molecules, the higher their K.E.
3) **Intra-molecular Energy**: This refers to the energy associated with the internal structure and motions within a molecule itself. It consists of three components:
   a) **Electronic Energy**: This is the energy of the electrons orbiting the nucleus and spinning on their own axes. Think of it as the energy levels of electrons in different orbitals.
   b) **Vibrational Energy**: This is the energy associated with the stretching and bending of the bonds among atoms within a molecule. Think of a molecule like a spring that can vibrate at different frequencies.

c) **Rotational Energy**: This is the energy associated with the rotation of the entire molecule. Picture a molecule spinning around an axis like a child's spinning toy, or a gyroscope, rotating on its axis.

To sum up, imagine a container of water vapor. The internal energy of the water vapor is a combination of:

1) The attractive and repulsive forces between water molecules (inter-molecular P.E.).
2) The random movement of individual water molecules at different speeds (molecular K.E.).
3) The energy of the electrons within each water molecule, the stretching and bending of the O—H bonds, and the rotation of the entire molecule (intra-molecular energy).

If we divide the internal energy (U) of a material by its mass, then we will get the specific internal energy: $\overline{U} = \frac{\text{Total Internal Energy}}{\text{Total Mass}} = \frac{U(\text{J})}{m(\text{kg})} = \overline{U}\left(\frac{\text{J}}{\text{kg}}\right)$.

On the other hand, if we divide the internal energy (U) of a material by its moles, then we will get the molar internal energy: $\tilde{U} = \frac{\text{Total Internal Energy}}{\text{Total Moles}} = \frac{U(\text{J})}{n(\text{mol})} = \tilde{U}\left(\frac{\text{J}}{\text{mol}}\right)$.

From a classical (macroscopic) thermodynamics point of view, the molar internal energy, $\tilde{U}$ is expressed as a function of intensive variables.

For a single-phase pure substance, $\tilde{U}$ can be expressed as: $\tilde{U} = f(P, T), f(P, \tilde{V})$, or $f(T, \tilde{V})$, where $\tilde{V}$ is the molar volume of the pure substance.

For a single-phase multicomponent solution, $\tilde{U}$ can be expressed as:

$$\tilde{U} = f(P, T, X_1, X_2, \ldots, X_{N-1}),$$

where X_i is the mole fraction of species i and N is the number of participating components.

7.3 Total K.E. ($m(v^2/2)$)

It is the macroscopic energy that a moving body possesses because of its motion. Notice the difference between the microscopic molecular K.E. and the macroscopic K.E., where the latter is attributed to the *macroscopic motion* of the entire system or control volume (CV) being examined or monitored. Of course, a frame of reference must be arbitrarily selected so that the observer comprehends what it means a static object or zero-velocity object means.

If we divide the total kinetic energy (K.E.) of a material by its mass, then we will get the specific kinetic energy: $\text{k.e.} = \frac{\text{Total Kinetic Energy}}{\text{Total Mass}} = \frac{\text{K.E.}}{\text{m}} = \frac{m\frac{v^2}{2}}{m} = \frac{v^2}{2}\left(\frac{\text{J}}{\text{kg}}\right)$.

7.4 Total Potential Energy (mgZ)

The potential to do work or to acquire kinetic energy is called potential energy, by virtue of its elevation relative to a datum. Again, the datum must be arbitrarily defined so that the potential energy at such chosen datum is set to zero.

If we divide the total potential energy (P.E.) of a material by its mass, then we will get the specific potential energy: $\text{p.e.} = \frac{\text{Total Potential Energy}}{\text{Total Mass}} = \frac{\text{P.E.}}{\text{m}} = \frac{mg(Z - Zo)}{m} = g(Z - Zo)\left(\frac{\text{J}}{\text{kg}}\right)$. Z_o is arbitrarily chosen to be zero as the datum.

7.5 Relativity of Energy Terms

While we often treat the previous three forms of common energy as absolute values, internal energy, macroscopic kinetic energy, and macroscopic potential energy are all relative quantities. This means they are not inherent properties of a system but depend on a reference point or reference state.

Macroscopic internal energy is the sum of all microscopic kinetic and potential energies within a system. While it seems absolute, the specific values depend on the chosen reference state. Consider two identical containers of gas at two different temperatures. The first container is at room temperature, 25°C, and the second at 60°C. We can arbitrarily choose the internal energy of the first container to be zero (reference state). The internal energy of the second container will then be a relative value compared to the first one, even though the reference container still contains a mass associated with the nonzero energy of molecules bouncing in a ceaseless motion, yet it is arbitrarily assigned zero value of energy.

Macroscopic kinetic energy refers to the energy of motion of a macroscopic object, like a car or a ball. It appears absolute, but it is relative to the observer's reference frame. Consider a car moving at 60 km/h relative to the ground. From the perspective of someone standing on the ground, this is the car's kinetic energy. However, for someone sitting in another car moving at 40 km/h in the same direction, the observed kinetic energy of the first car would be different (20 km/h relative).

Macroscopic potential energy represents the energy an object possesses due to its position relative to a chosen reference point. Similar to kinetic energy, it is not absolute. Consider a book resting on a table. Its potential energy is due to its position relative to the ground (reference point). However, the specific value depends on the chosen reference level. If we choose the floor as the reference, the potential energy value is different compared to choosing the table as the reference.

Therefore, when dealing with these quantities, it is critical to remember:

1) They are relative and depend on a defined reference point or state.
2) Absolute internal energy, absolute macroscopic kinetic energy, and absolute potential enrgy do not literally exist.
3) We must specify the reference point when stating the values of these quantities to avoid ambiguity and ensure clarity.

7.6 System (Control Mass) and Control Volume

Both systems (control mass) and control volumes are important concepts in thermodynamics, but they represent different approaches to analyzing energy interactions.

7.6.1 System (Control Mass)

1) **Definition**: A system is a defined portion of matter that we choose to focus on for our analysis. It has a fixed mass but it can be with a variable volume.
2) **Boundaries**: The system is surrounded by a real boundary that separates it from its surroundings. The boundaries can be stationary or moving.

3) **Focus**: The analysis focuses on the changes in the total energy (internal energy, kinetic energy, and potential energy) of the system itself by considering heat transfer and work interactions across the boundary.
4) **Examples**: A closed system with a fixed mass, like a sealed container of a gas.

7.6.2 Control Volume (CV)

1) **Definition**: A control volume is a fixed region in space through which matter can flow. It can be visualized as a box with defined inlet and outlet boundaries.
2) **Boundaries**: The boundaries of a control volume are not necessarily physical but can extend to include a conceptual boundary. Matter can cross these boundaries in the form of mass flow.
3) **Focus**: The analysis focuses on the flow of energy across the boundaries of the control volume in the form of heat transfer, work, and energy associated with mass in and out. The total mass within the control volume itself is not necessarily conserved.
4) **Examples**: A pipe carrying moving fluid, a turbine, a pump, a compressor, or an open system with varying mass, like a rocket engine burning fuel.

Figure 7.1 shows a schematic demonstrating the difference between the boundaries of a control mass (left) and those of a control volume (right). The boundaries of a control mass are real; they can be rigid as in rigid walls of a container, movable as in a piston-cylinder arrangement, or stretchable as in an expandable/contractable balloon. On the other hand, the boundaries of a control volume can be both real and imaginary.

7.6.3 System and Surroundings

To understand energy transfer, we define a specific boundary around the region of interest (our system). Everything outside is part of the surroundings. Thus, we only see and measure heat and work when energy crosses that boundary. If the system is hotter than the surroundings, heat will flow out of the system and into the surroundings. This is observed as a change in the temperature of both system and surroundings at the boundary. On the other hand, when we apply an external force on a system (compressing a gas), we are doing work on it. This work becomes evident as a change in the system's properties (e.g., increased pressure) or a change in the position of the boundary. For both heat and work, we can't talk about heat or work without specifying a system and its surroundings. The boundary is where energy

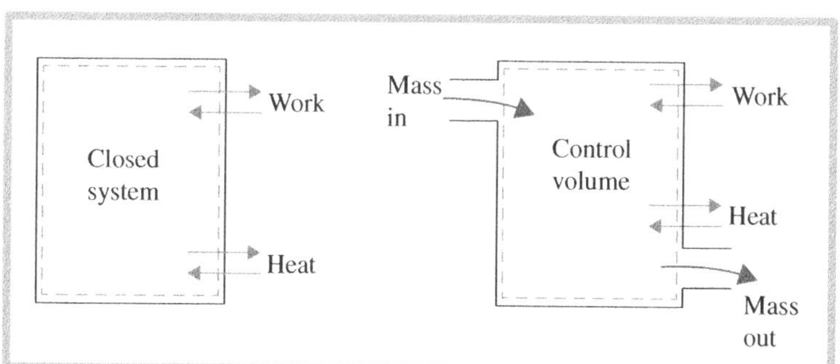

Figure 7.1 Boundaries of a control mass (left) and those of a control volume (right).

transitions from being "in the system" to "into the surroundings" or vice versa. For example, to boil water, heat must transfer *into* the system (the water) from a heat source (the stove), causing a change in its internal energy (increasing temperature). Another example is when a gas expands in a piston-cylinder setup, it does work *on* the surroundings, and we observe this as a displacement of the piston boundary in the outward direction.

7.6.4 A Closed, Open, and Isolated System

A system, in general, can be open, closed, or insulated.

For an open system, the mass may cross the boundaries of a control mass or control volume. The beaker of hot liquid coffee transfers energy to the surroundings, and it loses heat as it cools. Matter is also transferred in the form of water vapor.

For a closed system, the mass is not allowed to cross the boundaries of either the control mass or control volume. The flask of hot liquid coffee transfers energy (heat) to the surroundings as it cools. Because the flask is sealed (or stoppered), no water vapor escapes and no matter is transferred.

For an isolated system, neither mass nor energy is allowed to cross the boundaries of the control mass or control volume. Hot liquid coffee in a highly insulated container approximates a perfectly isolated system. No water vapor escapes, and, for a long time at least, very little or no heat is transferred to the surroundings. Obviously, the liquid coffee in the container cools down to room temperature, so this system is not completely isolated.

Figure 7.2 shows the three possible scenarios for a system made of a hot liquid coffee placed inside a container.

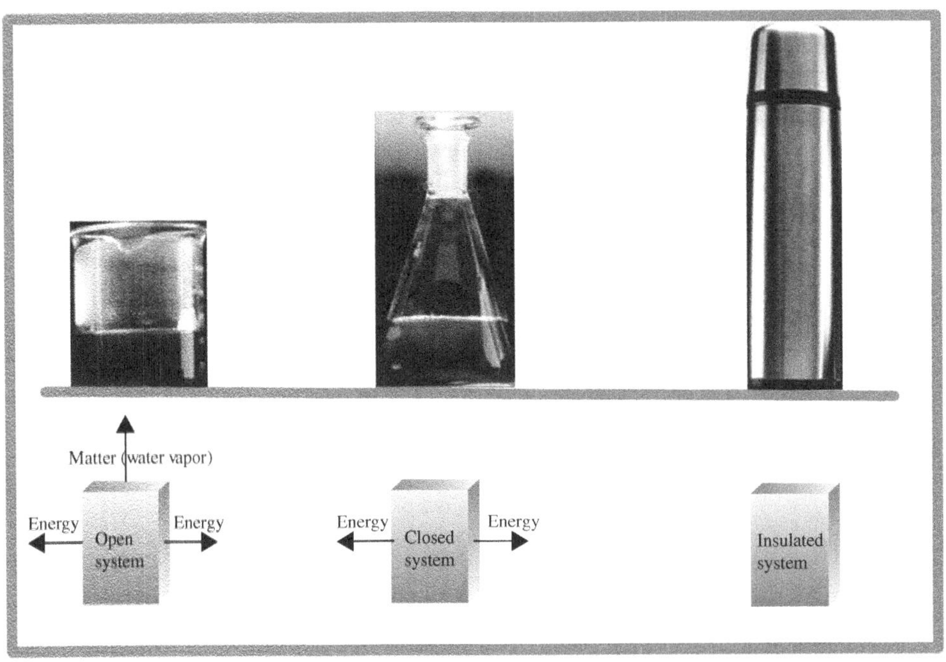

Figure 7.2 (Left), an open system with mass and energy exchange; (middle), a closed system where energy exchange is allowed; and (right), an insulated system living alone with no mass or energy exchange.

7.6.5 Heat and Work: Energy on the Move

Heat and work are not stored properties. Unlike internal energy, heat and work cannot be stored within a system. They represent energy that is being transferred into or out of a system as a process occurs. There are specific mechanisms of energy transfer:

Heat: This is energy transfer driven by a temperature difference between a system and its surroundings. Think of a hot cup of coffee cooling down. The energy flows from the hot coffee to the surroundings due to this temperature difference.

Work: This involves a force acting through a distance. Imagine compressing a gas in a piston. The work done on the gas transfers energy into the system. Work, a form of energy, flows in response to any driving force other than temperature difference, such as a force, a torque, or a voltage.

Electric current can be said as charge in transit and is driven by voltage difference across a conducting wire. Rain is water in transit that moves from the source (i.e., clouds) down to the Earth where it becomes again a surface water and is driven by gravity force.

That is why we say heat and work are "in transit," because they are only observable as they cross the boundary between a system and its surroundings.

7.7 The First Law of Thermodynamics (Closed System)

The first law of thermodynamics, also known as the law of energy conservation, states that energy cannot be created or destroyed in an isolated system. The total energy of a closed system remains constant. It can undergo various transformations between different forms, such as heat transfer, work done, and internal energy changes, but the total energy within the system remains constant. The energy of the system will be restricted to three common types of energy used in chemical engineering applications:

In total form:

$$E = U + \text{K.E.} + \text{P.E.} = m\overline{U} + m\frac{v^2}{2} + mgZ \tag{7.1a}$$

In specific form:

$$e = \overline{U} + \frac{v^2}{2} + gz \tag{7.1b}$$

The first law of thermodynamics can be expressed in different forms.

In absolute differential form:

$$\delta Q + \delta W = (\mathrm{d}E)_{\text{Sys}} = \mathrm{d}(U + \text{K.E.} + \text{P.E.})_{\text{Sys}} \tag{7.2a}$$

In absolute integral form:

$$Q_{12} + W_{12} = (\Delta E)_{\text{Sys}} = (\Delta U + \Delta\text{K.E.} + \Delta\text{P.E.})_{\text{Sys}} \tag{7.2b}$$

For open systems, we will later express the energy conservation principle in terms of rate forms.

In the absence of kinetic and potential energy changes, we have:

$$\delta Q + \delta W = d(U)_{Sys} \tag{7.3a}$$

$$Q_{12} + W_{12} = (\Delta U)_{Sys} = U_2 - U_1 \tag{7.4a}$$

where Q_{12} is net heat exchanged between the system and its surroundings as it travels from state 1 to state 2. W_{12} is the net amount of work developed by or on the system as it travels from state 1 to state 2. A state of a system is characterized by a sufficient number of intensive variables. For example, we can say that the system, upon applying heat and/or work, moves from state 1 (P_1, T_1) to state 2 (P_2, T_2).

The displacement (expansion/contraction) work is demonstrated in Figure 7.3. The pressure force is acting upward against the external pressure force and the weight force of the piston. If the piston is displaced a differential distance, dy, in the vertical direction by the pressure force, then we can define the expansion/contraction work in terms of the pertinent variables. In physics, the work is the dot vector multiplication of the force vector, $\vec{F}$ and the displacement vector, $\overrightarrow{dy}$. If the area of the piston is A, then $dV = Ady$.

$$\text{dWork} = \overrightarrow{\text{Force}} \cdot \overrightarrow{\text{Displacenet}} = \vec{F}.\overrightarrow{dy} = \overrightarrow{PA}.\overrightarrow{dy} = |PAdy| = \mp |PdV|$$

$$\text{Work} = \mp \int_{V1}^{V2} PdV$$

See notes below on the sign selection for defining the work term.

Back to Eq. (7.4a) which is written with a plus sign for the work, W_{12}. If the work term is preceded by a positive sign, then:

$$W_{12} = -\int_{V1}^{V2} PdV \tag{7.5a}$$

NOTE #1

Using the negative sign notation appearing in Eq. (7.5a), **the work** will be **negative** for a **volume expansion** process (work done by the system on the surroundings) and positive for a volume compression process (work done on the system by the surroundings).

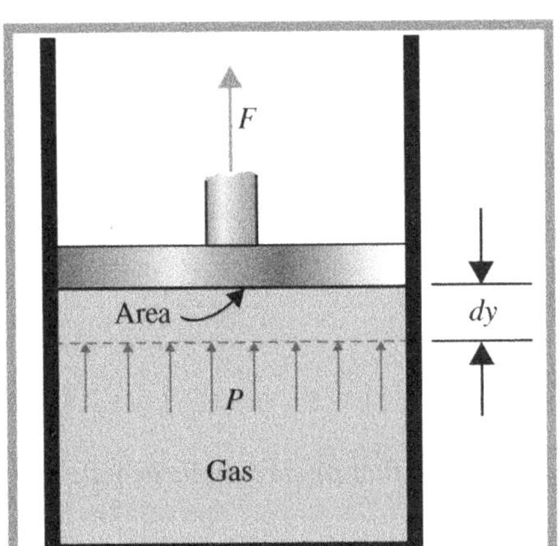

Figure 7.3 A Piston-cylinder setup showing how to express PdV work in terms of pressure force, area, and displacement.

On the other hand, if we express the first law of thermodynamics as:

$$\delta Q - \delta W = d(U)_{Sys} \tag{7.3b}$$

$$Q_{12} - W_{12} = (\Delta U)_{Sys} = U_2 - U_1 \tag{7.4b}$$

Notice that Eq. (7.4b) is written with a minus sign preceding the work, W_{12}. If the work term is preceded by a minus sign, then:

$$W_{12} = \int_{V1}^{V2} PdV \tag{7.5b}$$

NOTE #2

Using the positive sign notation appearing in Eq. (7.5b), **the work** will be **positive** for a **volume expansion** process (work done by the system on the surroundings) and negative for a volume compression process (work done on the system by the surroundings).

However, both expressions will yield the same effect on the energy balance as a result of either volume expansion or contraction. That means if we assign the minus sign to the PdV term, as in Eq. (7.5a), then δW (or W_{12}) term will be preceded by a positive sign (Eqs. 7.3a and 7.4a).

On the contrary, if we assign the plus sign to the PdV term, as in Eq. (7.5b), then δW (or W_{12}) the term will be preceded by a negative sign (Eqs. 7.3b and 7.4b). Some textbooks use Eqs. (7.3a) through (7.5a) combination; others use Eqs. (7.3b) through (7.5b) combination.

NOTE #3

δQ, the heat is **positive**; when it is transferred **from surroundings** to the system; vice versa, it will be **negative** if heat is transferred **from the system** to the surroundings. If the system is insulated, then the amount of heat, δQ, will be zero and the process is called **adiabatic**. The unit of Q will be calorie or joule. 1 cal = 4.184 J.

Example 7.1 Kinetic Energy Transported by a Flowing Stream

Water flows into a process unit through a 3-cm ID pipe at a rate of 3.6 m^3/h.

Calculate the kinetic energy for this stream in J/s (W).

Solution:

$$\dot{V} = 3.6\frac{m^3}{h} \times \frac{1\,h}{3600\,s} = 0.001\frac{m^3}{s}$$

$$\dot{V} = 0.001\,\frac{m^3}{s} = v \times A = v \times \pi\frac{(0.03\,m)^2}{4} = v \times 7.0686 \times 10^{-4}\,m^2$$

$$v = \frac{\dot{V}}{A} = \frac{0.001\,\frac{m^3}{s}}{7.0686 \times 10^{-4}\,m^2} = 1.4147\frac{m}{s}$$

$$\text{K.E.} = \frac{1}{2}\dot{m}v^2 = \frac{1}{2}\rho\dot{V}v^2 = \frac{1}{2}\left(1000\frac{\text{kg}}{\text{m}^3}\right)\left(0.001\frac{\text{m}^3}{\text{s}}\right)1.4147^2\frac{\text{m}^2}{\text{s}^2}$$

$$\text{K.E.} = 1.0007\frac{\text{kg.m}^2}{\text{s}^3} \times \frac{\text{N.m}}{\frac{\text{kg.m}^2}{\text{s}^2}} = 1.0007\frac{\text{J}}{\text{s}} = 1.0007\,\text{W}$$

Example 7.2 Minimum Power Requirement for Crude Oil Elevation

A crude oil is pumped at a rate of 15 kg/s from a point 220 m below earth's surface to a point 20 m above ground level. Calculate the rate of increase of potential energy.

Solution:

The ground surface is arbitrarily assigned zero datum.

$$\Delta(\dot{m}gZ) = \dot{m}g\Delta Z = 15\frac{\text{kg}}{\text{s}} \times 9.8\frac{\text{m}}{\text{s}^2} \times (Z_2 - Z_1) = 15\frac{\text{kg}}{\text{s}} \times 9.8\frac{\text{m}}{\text{s}^2} \times (20 - (-220))\text{m}$$

$$\Delta(\dot{m}gZ) = 15\frac{\text{kg}}{\text{s}} \times 9.8\frac{\text{m}}{\text{s}^2} \times (240)\text{m} = 35{,}280\frac{\text{N.m}}{\text{s}} = 35{,}280\frac{\text{J}}{\text{s}} = 35.28\frac{\text{kJ}}{\text{s}} = 35.28\,\text{kW}$$

As we all know that a fluid cannot move upward by itself, against the gravity field. Hence, it requires a pump which at least can handle the pumping requirement by supplying power to the fluid to raise its energy content to overcome potential energy change. Of course, there are other things to overcome, like friction within the pump itself, friction of the fluid along the pipeline, and friction of pipe fittings installed along the pipeline from the source to the delivery point. In addition, the pump power should account for the kinetic energy change if the crude oil is pumped from a storage tank.

7.8 Heat Capacity of a Solid or Liquid

The specific heat capacity of a **solid or liquid** is characterized as the amount of heat needed to elevate the temperature of one unit mass of the substance by one degree. This relationship can be expressed through the following equation:

$$Q_{12} = m\overline{C}\Delta T = m\overline{C}\,(T_2 - T_1) \tag{7.6}$$

where,

Q_{12} = Heat supplied to a substance
m = Mass of the substance
$\overline{C}$ = Specific heat capacity of either a solid or liquid
ΔT = Temperature rise, or temperature change

Example 7.3 Heating Iron Metal vs. Water

Calculate the amount of heat needed to heat up 1 kg of water from room temperature to 100°C and the same for 1 kg of iron.

DATA: Specific heat for iron = 0.451 J/(g °C) Specific heat for water = 4.184 J/(g °C)

Solution:

For Water:

$$(Q_{12})_{\text{Water}} = m\overline{C}\Delta T = 1000\,\text{g} \times 4.184\frac{\text{J}}{\text{g}\cdot^{\circ}\text{C}}(100-25)^{\circ}\text{C} = 313{,}800\,\text{J} = 313.8\,\text{kJ}$$

For Iron:

$$(Q_{12})_{\text{Iron}} = m\overline{C}\Delta T = 1000\,\text{g} \times 0.451\frac{\text{J}}{\text{g}\cdot^{\circ}\text{C}}(100-25)^{\circ}\text{C} = 33{,}825\,\text{J} = 33.825\,\text{kJ}$$

Notice that the specific heat essentially reflects a material's ability to store thermal energy. In a car engine, the goal is to maintain a specific operating temperature range to ensure optimal performance and avoid overheating. This requires efficient removal of excess heat generated by combustion. Given that water has a high specific heat, it means that it takes a significant amount of heat energy to raise the temperature of water by 1°C. Hence, water is preferred for the following reasons:

- **Higher Heat Absorption**: Due to its high specific heat, water can absorb a larger amount of heat from the hot engine components with a relatively smaller increase in its own temperature. This allows water to effectively carry away excess heat from the engine; $(Q_{12})_{\text{Water}} = 9.3 \times (Q_{12})_{\text{Iron}}$.
- **Efficient Circulation**: Water is a liquid at room temperature, making it easy to pump and circulate throughout the engine block and radiator using a water pump. This continuous circulation ensures continuous heat removal.
- **Readily Available**: Water is a readily available and inexpensive coolant compared with other substances with high specific heat.

7.8.1 MATLAB for Sensible Heat Calculation

Appendix H shows both the human language algorithm and MATLAB code for calculating the sensible heat of solids and liquids within the applicable temperature range. The database contains 301 compounds, which can be liquids or solids. For each compound, the specific heat and the acceptable temperature range are stated. You can search by entering the keyword and the program will look up for all permutations (or occurrences) containing the keyword. Your duty then is to select the material of your choice. The heat duty is calculated based on the assumption that $T_2 > T_1$. If interested in the cooling process (i.e., $T_2 < T_1$), then do the heating step and take the negative value for the calculated heat duty. The reason for restricting $T_2 > T_1$ is to assure that the user will not enter temperature values below or above the acceptable/applicable temperature range which is dictated by the melting and/or boiling point values of the given substance, in general.

There are points to consider for calculating the sensible heat, using MATLAB code, for either cooling or heating a solid or liquid substance. The state of the substance is indicated in the third column as either liquid or solid. The specific heat has units of kJ/(kg·°C) and temperatures are expressed in °C.

7.8.1.1 Solids

There is an upper limit for the melting point, but there is no lower limit. This means we can go down to 0 K (−273.15°C); however, given that the specific heat for practically all substances is reported at either room temperature (25°C) or within the applicable temperature range of liquids, one has to be precautious about evaluation of the heat duty for severely low and high temperatures.

Moreover, at low temperatures, some materials might undergo a solid phase transition or transformation, which will result in a geometrical molecular restructuring, which will definitely affect the specific heat value. For example, polymers like nylon-6 exhibit gradual changes in their mechanical properties with decreasing temperature. They become stiffer and less flexible as it gets colder, but there's no single point of transition from a completely solid state to a brittle fracture-prone state.

The same thing can be said about the effect of severely high temperatures where again other solid phase transformations may occur. Not only that but also other desired/undesired phenomena may occur as well. For example, thermal softening, degradation, and decomposition may occur, which may alter the identity of the material itself; hence, the specific heat value.

Consequently, for such severe cases of temperature range, one has to look up the proper literature pertaining to the material under investigation. Alternatively, temperature-dependent heat capacity data should be sought to have a better evaluation of the amount of heat extracted or added upon cooling or heating, respectively.

For the literature-cited melting points, I take the lowest value to be on the safe side that we still have a solid-phase.

7.8.1.2 Liquids

There are two limits for the applicable temperature range: The melting/freezing point and boiling/condensation point. Both are evaluated at 1 standard pressure. Keep in mind that the saturation boiling point is a function of the applied pressure. We select here the normal boiling point temperature at 1 atm. The applicable liquid phase temperature range is confined between the melting lower and normal boiling upper limit.

Example 7.4 MATLAB for Water-Sensible Heat Calculation

Using Appendix H | H.2 MATLAB code (**SpecHeat.m**), calculate the required heat duty for heating 1 kg of liquid water from 25 until 100°C.

Solution:

The heat duty results are shown below with different units, like: kJ, kcal, BTU, and lbf-ft.

```
>> SpecHeat
Enter a search keyword: water
Matching entries:
  - water, fresh (Row 283)
  - water, pure liquid (20°C) (Row 284)
  - water, sea 36 oF (Row 285)
Enter the row number for the compound: 284
Tmin: 0°C
Tmax: 100°C
Enter T1 (must be higher than or equal to Tmin): 25
Enter T2 (must be lower than or equal to Tmax): 100
Enter the mass of the compound in kg: 1
```

```
Sensible heat calculation for liquid phase: ...
{'The compound names is: '} {'water, pure liquid (20°C)'}
QkJ: 313.8 kJ
Qkcal: 75 kcal
Qbtu: 297.425 BTU
Qlb_ft: 231447 lbf-ft
```

Example 7.5 MATLAB for Ethanol Sensible Heat Calculation
Using Appendix H | H.2 MATLAB code, calculate the required heat duty for heating 7 kg of liquid ethanol from 25 until 60°C.

Solution:

The heat duty results are shown below with different units, like: kJ, kcal, BTU, and lbf-ft.

```
>> SpecHeat
Enter a search keyword: ethanol
Matching entries:
  - alcohol, ethyl 104 oF of (ethanol) (Row 5)
  - alcohol, ethyl 32 oF of (ethanol) (Row 6)
Enter the row number for the compound: 5
Tmin: -114°C
Tmax: 78.4°C
Enter T1 (must be higher than or equal to Tmin): 25
Enter T2 (must be lower than or equal to Tmax): 60
Enter the mass of the compound in kg: 7
Sensible heat calculation for liquid phase: ...
{'The compound names is: '}{'alcohol, ethyl 104 oF (ethanol)'}
QkJ: 666.4 kJ
Qkcal: 159.2734 kcal
Qbtu: 631.6252 BTU
Qlb_ft: 491511.4111 lbf-ft
```

7.8.2 Heat Capacity of a Gas or Vapor

On the other hand, there are two definitions for vapors and gases:

1) Specific heat capacity at constant pressure, which is given by:

$$\bar{C}_P = \left(\frac{\partial \bar{H}}{\partial T}\right)_P \tag{7.7a}$$

On a molar heat capacity is given by:

$$\tilde{C}_P = \left(\frac{\partial \tilde{H}}{\partial T}\right)_P \tag{7.7b}$$

At constant P, we have:

$$\partial\overline{H} = \overline{C}_P\partial T \rightarrow \text{Integrate with respect to } T: \Delta\overline{H} = \int_{T1}^{T2} \overline{C}_P dT$$

The total enthalpy can be given by multiplying the mass of a gas by the specific enthalpy:

$$\Delta H = m\Delta\overline{H} = m\int_{T1}^{T2} \overline{C}_P dT \tag{7.8a}$$

On a molar basis, we have:

$$\partial\tilde{H} = \tilde{C}_P\partial T \rightarrow \Delta\tilde{H} = \int_{T1}^{T2} \tilde{C}_P\partial T$$

The total enthalpy can be given by multiplying the moles of a gas by the molar enthalpy:

$$\Delta H = n\Delta\tilde{H} = n\int_{T1}^{T2} \tilde{C}_P dT \tag{7.8b}$$

2) Specific heat capacity at a constant volume, which is given by:

$$\overline{C}_V = \left(\frac{\partial\overline{U}}{\partial T}\right)_V \tag{7.9a}$$

On a molar heat capacity is given by:

$$\tilde{C}_V = \left(\frac{\partial\tilde{U}}{\partial T}\right)_V \tag{7.9b}$$

In a similar fashion as we did with enthalpy difference, we can apply it here.
At constant V, we have:

$$\partial\overline{U} = \overline{C}_V\partial T \rightarrow \text{Integrate with respect to } T: \Delta\overline{U} = \int_{T1}^{T2} \overline{C}_V dT$$

The total internal energy can be given by multiplying the mass of a gas by the specific internal energy:

$$\Delta U = m\Delta\overline{U} = m\int_{T1}^{T2} \overline{C}_V dT \tag{7.9c}$$

On a molar basis, we have:

$$\partial\tilde{U} = \tilde{C}_V\partial T \rightarrow \Delta\tilde{U} = \int_{T1}^{T2} \tilde{C}_V\partial T$$

The total internal energy can be given by multiplying the moles of a gas by the molar internal energy:

$$\Delta U = n\Delta\tilde{U} = n\int_{T1}^{T2} \tilde{C}_V \mathrm{d}T \tag{7.9d}$$

NOTE #4
For an ideal gas, both the internal energy and enthalpy are functions of temperature only. This stems from the hypothetical nature of an ideal gas that its molecules are infinitely separated from one another to the point that compression/expansion (i.e., pressure–volume change) effects will not affect the microscopic, intermolecular potential and kinetic energy of molecules. If T is held constant, so do internal energy and enthalpy. This simplification does not extend to other thermodynamic properties, like entropy, Helmholtz free energy, and Gibbs free energy.

7.9 Internal Energy and Enthalpy Relationship

For any substance, the relationship between enthalpy and internal energy is given by:

$$H = U + (PV) \tag{7.10a}$$

On a mass basis, we have:

$$\overline{H} = \overline{U} + (P\overline{V}) \tag{7.10b}$$

On a molar basis, we have:

$$\tilde{H} = \tilde{U} + (P\tilde{V}) \tag{7.10c}$$

The absolute change in U and H is given by:

$$\Delta H = \Delta U + \Delta(PV) \tag{7.10d}$$

On a mass basis, the change is given by:

$$\Delta\overline{H} = \Delta\overline{U} + \Delta(P\overline{V}) \tag{7.10e}$$

On a molar basis, the change is given by:

$$\Delta\tilde{H} = \Delta\tilde{U} + \Delta(P\tilde{V}) \tag{7.10f}$$

7.9.1 Heat Capacities of an Ideal Gas

The following relations can be used for a value of $\tilde{C}_V$ of an ideal gas

$$\text{For a mono atomic ideal gas, we have } \tilde{C}_V = \frac{3}{2}R = 12.47\frac{\text{J}}{\text{mol.K}} \tag{7.11a}$$

$$\text{For a diatomic ideal gas, we have } \tilde{C}_V = \frac{5}{2}R = 20.79\frac{\text{J}}{\text{mol.K}} \tag{7.11b}$$

For a polyatomic ideal gas, we have $\tilde{C}_V = 3R = 24.94 \dfrac{\text{J}}{\text{mol.K}}$ (7.11c)

Reference: [University Physics Volume 2. Chapter 3. The First Law of Thermodynamics. 3.5 Heat Capacities of an Ideal Gas. https://pressbooks.online.ucf.edu/osuniversityphysics2/chapter/heat-capacities-of-an-ideal-gas/].

It can be shown that for a perfect (ideal) gas:

$$\tilde{C}_P = \tilde{C}_V + R \tag{7.12}$$

where R is the universal gas constant.

The ratio $(\tilde{C}_P/\tilde{C}_V)$, has been given the symbol γ,

$$\gamma = \frac{\tilde{C}_P}{\tilde{C}_V} \tag{7.13}$$

Notice that γ is always greater than unity.

7.9.2 A Closed System Involving Ideal Gas Heating

Example 7.6 Heating a Mono-atomic Ideal Gas

When 2000 J of heat is slowly added to 10 mol of an ideal mono-atomic gas, its temperature rises by 10°C. What is the work done by the gas? Assume a constant-heat capacity.

Solution:

Let us apply Eq. (7.4a) and use Eq. (7.9d):

$$Q_{12} + W_{12} = (\Delta U)_{\text{Sys}}$$

$$2000\,\text{J} + W_{12} = n\tilde{C}_V(T_2 - T_1) = n\tilde{C}_V(\Delta T)$$

From Eq. (7.11a), for a mono-atomic ideal gas, we have $\tilde{C}_V = (3/2)R = 12.47\ (\text{J}/(\text{mol.K}))$

$$2000\,\text{J} + W_{12} = 10\,\text{mol} \times 12.47 \frac{\text{J}}{\text{mol.K}}(10°\text{C}) = 1247\,\text{J}$$

$$W_{12} = 1247\,\text{J} - 2000\,\text{J} = -753\,\text{J}$$

From Eq. (7.5a), we have:

$$W_{12} = -\int_{V1}^{V2} P\text{d}V = -753\,\text{J}$$

Work is done by the gas, and dV must be positive (i.e., volume expansion).

NOTE #5
Nothing is mentioned about how the gas moved from state 1 to state 2. Yet, we can find the amount of work done by the gas as it represents the difference between heat (Q_{12}) and internal energy difference (ΔU). However, we can conclude that the process is neither iso-thermal nor is it iso-choric (iso-metric) process. It is more likely that it is iso-baric.

Example 7.7 Heating an Ideal Gas in a Cylinder at Constant Volume (Iso-choric or Iso-metric) Process

Ten moles of a diatomic ideal gas are contained in a cylinder fitted with a piston. The initial gas temperature is 25°C, and the initial pressure is 1 atm.

The cylinder is placed in boiling water with the piston held in a fixed position. Heat in the amount of 3 kcal is transferred to the gas, which reaches a higher pressure and temperature, as shown in Figure 7.4.

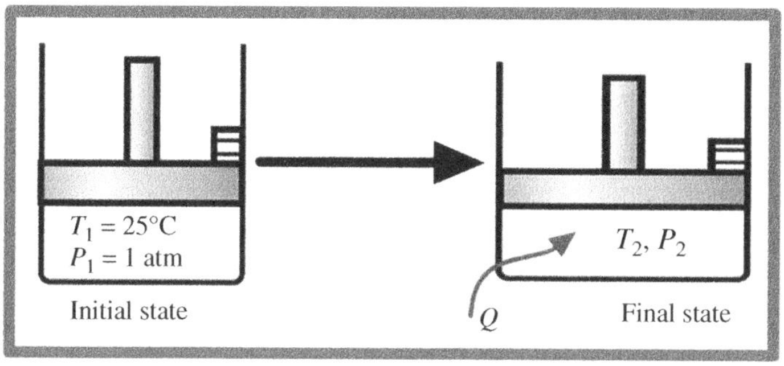

Figure 7.4 Constant-volume heating process ($T_2 > T_1$ and $P_2 > P_1$) of an ideal di-atomic gas in a piston-cylinder setup.

1) Calculate the work, W_{12}, in J.
2) Calculate ΔU
3) Calculate ΔT and thus T_2
4) Calculate ΔH
5) Calculate P_2.

Solution:

1) Let us apply Eq. (7.4a) and use Eq. (7.9d):

$$Q_{12} + W_{12} = (\Delta U)_{\text{Sys}} = U_2 - U_1 = n\int_{T1}^{T2} \tilde{C}_V \mathrm{d}T$$

Since the piston is fixed at its initial position, then the volume change is zero; hence, $W_{12} = 0$.

$$Q_{12} = (\Delta U)_{\text{Sys}}$$

2) $$Q_{12} = 3\,\text{kcal} \times \frac{4.184\,\text{kJ}}{\text{kcal}} = 12.552\,\text{kJ} = 12{,}552\,\text{J} = \Delta U$$

This is called a constant-volume heating process.

3) If we assume a constant $\tilde{C}_V$ (Eq. 7.11b), then we can factor it out from the integral term:

$$12{,}552\,\text{J} = n\int_{T1}^{T2} \tilde{C}_V \mathrm{d}T = n\tilde{C}_V(T_2 - T_1) = 10\,\text{mol} \times 20.79\frac{\text{J}}{\text{mol.K}} \times (\Delta T)$$

Solve for ΔT. $\Delta T = 60.38°C = T_2 - T_1$. Hence, $T_2 = \Delta T + T_1 = 85.38°C$.

4) Let us use Eq. (7.8b) with $\tilde{C}_P = \tilde{C}_V + R = 20.79 + 8.314 = 29.1 \dfrac{J}{mol.K}$

$$\Delta H = n\Delta\tilde{H} = n \times \tilde{C}_P(\Delta T) = 10 \text{ mol} \times 29.1 \frac{J}{mol.K} \times (60.38)K = 17{,}571 \text{ J}$$

An alternative way to calculate ΔH is to use Eq. (7.10d):
From ideal gas law: $PV = nRT$
Apply delta (differential operator) on both sides, we have:

$$\Delta(PV) = \Delta(nRT) = nR\Delta(T)$$

NOTE #6

Δ means differential operator.

$\Delta U = Q_{12}$ (From step#1)

$\Delta H = \Delta U + \Delta(PV)$ (From Eq.7.10d)

Substitute for $\Delta(PV)$ and $\Delta(U)$, we have:

$$\Delta H = \Delta U + \Delta(PV) = Q_{12} + \Delta(nRT) = Q_{12} + nR\Delta(T) =$$

$$\Delta H = 12{,}552 \text{ J} + 10 \text{ mol} \times 8.314 \frac{J}{mol.K}(60.38)K = 12{,}552 \text{ J} + 5020 \text{ J} = 17{,}572 \text{ J}$$

Notice both enthalpy-change values are the same. Moreover, for a constant-volume process, the heat duty will be equal to the change in internal energy, not to the change in enthalpy.

5) Let us apply the ideal gas law twice:

$$P_1V_1 = n_1RT_1$$

$$P_2V_2 = n_2RT_2$$

Divide side by side, we get:

$$\frac{P_1V_1}{P_2V_2} = \frac{n_1T_1}{n_2T_2}$$

Notice that $V_1 = V_2$ and $n_1 = n_2$. We have:

$$\frac{P_1}{P_2} = \frac{T_1}{T_2} \rightarrow \frac{1 \text{ atm}}{P_2} = \frac{298.15 \text{ K}}{358.5 \text{ K}}.$$ Solve for P_2. $P_2 = 1.2$ atm.

7.9.3 MATLAB Code for an Iso-choric Ideal Gas Process

NOTE #7
For all MATLAB code calculations, the difference between manual and MATLAB calculations lies in round off errors done in manual calculations.

Figure 7.5 shows MATLAB code (**IG_IsoChoric.m**) for calculating thermodynamics properties, like T_2, P_2, ΔU, ΔH, Q_{12}, and W_{12} for a constant-volume process. The user has to input some parameters and decide on the nature of the ideal gas: Mono-atomic, di-atomic, or poly-atomic. One can run the code for different input parameters, including the heat duty and the type of the ideal gas. Notice that n, T_{1C}, and P_1 are entered as priori input parameters; however, the user must select one more input out of three possibles scenarios: These three scenarios include heat duty, Q_{12}, the final pressure, P_2, and the final temperature, T_{2C}. Now, the system will be specified, and the remaining unknown thermodynamic properties can be calculated.

```
clear variables;
% Get user input for initial moles, initial volume, pressure, final temperature, and final volume
n = input('Enter the initial number of moles: ');
T1C = input('Enter the initial temperature (oC): ');
P1 = input('Enter the initial pressure (atm): ');
% User selection for gas type (modify prompts as needed)
gasType = input('Select gas type (A: Mono-atomic, B: Di-atomic, C: Poly-atomic): ', 's');
% Calculate gas constant based on user selection
switch upper(gasType)
  case 'A'
    R = 8.314; % J/mol*K (Mono-atomic)
    Cv = (3/2)*R; % J/mol*K (Specific heat at constant volume)
    Cp = Cv + R; % J/mol*K (Specific heat at constant pressure)
  case 'B'
    R = 8.314; % J/mol*K (Di-atomic)
    Cv = (5/2)*R; % J/mol*K
    Cp = Cv + R; % J/mol*K
  case 'C'
    R = 8.314; % J/mol*K (Poly-atomic)
    Cv = (3.0)*R; % J/mol*K
    Cp = Cv + R; % J/mol*K
  otherwise
    error('Invalid gas type selection.');
end
FinalCondition = input('Known Final Condition (A: Heat duty, Q12, J, B: Final P, P2, atmC: Final
T, T2C, oC): ', 's');
switch upper(FinalCondition)
```

```
  case 'A'
    Q12 = input('Enter the heat duty, Q12, (J): ');
    Rgas=0.08206;
    T1=273.15+T1C;
    V1 = ((n*Rgas*T1)/P1);
    T2=T1+(Q12)/(n*Cv);
    T2C=T2-273.15;
    DU=n*Cv*(T2-T1);
    W12=0;
    DH=n*Cp*(T2-T1);
    P2=(T2/T1)*P1;
  case 'B'
    P2 = input('Enter the final Pressure (atm): ');
    T1=273.15+T1C;
    Rgas=0.08206;
    V1 = (n*Rgas*T1)/P1;
    T2=(P2/P1)*T1;
    T2C=T2-273.15;
    DU=n*Cv*(T2-T1);
    W12=0.0;
    Q12=DU-W12;
    DH=n*Cp*(T2-T1);
  case 'C'
    T2C = input('Enter the final temperature (oC): ');
    T1=273.15+T1C;
    T2=273.15+T2C;
    Rgas=0.08206;
    V1 = ((n*Rgas*T1)/P1);
    DU=n*Cv*(T2-T1);
    W12=0;
    Q12=DU-W12;
    DH=n*Cp*(T2-T1);
    P2=(T2/T1)*P1;
  otherwise
    error('Invalid gas type selection.');
end
% Display results
disp('Results:');
disp(['Number of Moles: ', num2str(n), ' mole']);
disp(['Initial Temperature: ', num2str(T1C), ' oC']);
disp(['Final Temperature: ', num2str(T2C), ' oC']);
disp(['Initial Volume: ', num2str(V1), ' L']);
disp(['Final Pressure: ', num2str(P2), ' atm']);
disp(['Work done: ', num2str(W12), ' J']);
disp(['Change in internal energy, DU: ', num2str(DU), ' J']);
disp(['Q12 Duty done: ', num2str(Q12), ' J']);
disp(['Change in enthalpy, DH: ', num2str(DH), ' J']);
```

Figure 7.5 MATLAB code (**IG_IsoChoric.m**) for an ideal gas heating/cooling process calculations under iso-choric conditions.

Sample calculations are shown below for Example 7.7.

```
>> IG_IsoChoric
Enter the initial number of moles: 10
Enter the initial temperature (°C): 25
Enter the initial pressure (atm): 1
Select gas type (A: Monoatomic, B: Diatomic, C: Polyatomic): B
Known Final Condition (A: Heat duty, Q12, J, B: Final P, P2, atm C: Final T, T2C, °C): A
Enter the heat duty, Q12, (J): 12552
Results:
Number of Moles: 10 mole
Initial Temperature: 25°C
Final Temperature: 85.3897°C
Initial Volume: 244.6619 L
Final Pressure: 1.2025 atm
Work done: 0 J
Change in internal energy, DU: 12,552 J
Q12 Duty done: 12552 J
Change in enthalpy, DH: 17,572.8 J
```

Example 7.8 Heating an Ideal Gas in a Cylinder at Constant Pressure (Iso-baric) Process

Ten moles of a diatomic ideal gas are contained in a cylinder fitted with a freely moving piston. The initial gas temperature is 25°C, and the initial pressure is 1 atm.

The cylinder is placed in boiling water. Heat in the amount of 3 kcal is transferred to the gas, which reaches a higher temperature; however, as the piston slowly expands upward in response to the heating effect, it keeps a constant pressure process until reaching the final volume (V_2) and temperature (T_2), as shown in Figure 7.6.

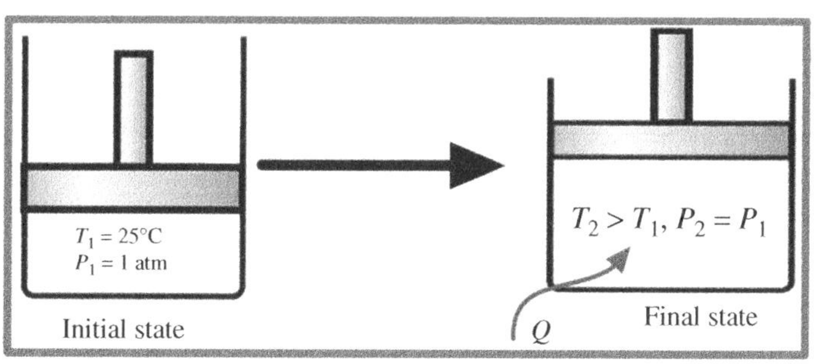

Figure 7.6 Constant-pressure heating process with expansion ($V_2 > V_1$) of an ideal diatomic gas in a piston-cylinder setup.

1) Calculate ΔH, ΔT, and thus T_2
2) Calculate ΔU
3) Calculate V_2
4) The work done by the system, W_{12}

Solution:

1) Let us apply Eq. (7.4a):

$$Q_{12} + W_{12} = (\Delta U)_{Sys} \tag{A}$$

Since the piston is freely moving, we have a constant-pressure expansion process. The work is given by Eq. (7.5a):

$$W_{12} = -\int_{V1}^{V2} PdV = -P\Delta V \tag{B}$$

$$Q_{12} + W_{12} = Q_{12} - P\Delta V = \Delta U \tag{C}$$

From an ideal gas law, we have: $V = (nRT)/P$

Hence, ΔV can be expressed in terms of ΔT by applying differentiation on both sides. nR/P is factored out as it is constant in this case.

$$\Delta V = \Delta\left(\frac{nRT}{P}\right) = \frac{nR}{P}\Delta T \tag{D}$$

Equation (C) becomes:

$$Q_{12} - P\frac{nR}{P}\Delta T = \Delta U \rightarrow Q_{12} - nR\Delta T = \Delta U \quad \rightarrow Q_{12} = nR\Delta T + \Delta U = nR\Delta T + n\tilde{C}_V(\Delta T)$$

$$Q_{12} = nR\Delta T + n\tilde{C}_V(\Delta T) = n(\Delta T)(R + \tilde{C}_V) = n(\Delta T)(\tilde{C}_P) = \Delta H \tag{E}$$

Equation (E) says that Q_{12} is equal to $n(\Delta T)(\tilde{C}_P)$ which is equal to ΔH. The process is coined as constant-pressure heating.

$$\tilde{C}_P = \tilde{C}_V + R = 20.79 + 8.314 = 29.1\ \frac{\text{J}}{\text{mol.K}} \tag{F}$$

$$(\Delta T) = \frac{Q_{12}}{n(\tilde{C}_P)} = \frac{12{,}552\,J}{10\ \text{mol} \times 29.1\ \frac{\text{J}}{\text{mol.K}}} = 43.13°\text{C} \tag{G}$$

$T_2 = 25°\text{C} + 43.13°\text{C} = 68.13°\text{C}$

Alternatively, we can rearrange the terms in Eq. (C), to have:

$$Q_{12} = \Delta U + P(\Delta V) \tag{H}$$

From Eq. (7.10d), we have:

$$\Delta H = \Delta U + \Delta(PV) = \Delta U + P(\Delta V) + V(\Delta P) = \Delta U + P(\Delta V) \tag{I}$$

Notice that $V(\Delta P) = 0$ because it is an iso-baric process, which means $(\Delta P) = 0$

The right-hand side of both Eqs. (H) and (I) is the same, thus

$$Q_{12} = \Delta H \tag{J}$$

Again, this is called the constant-pressure heating process. Let us invoke Eq. (7.8b)

$$Q_{12} = \Delta H = n\int_{T1}^{T2} \tilde{C}_P dT = n\tilde{C}_P(T_2 - T_1) = n\tilde{C}_P(\Delta T)$$

$$\text{From Eq.(7.12)} : \tilde{C}_P = \tilde{C}_V + R = 20.79 + 8.314 = 29.1\ \frac{\text{J}}{\text{mol.K}}$$

$$Q_{12} = 12{,}552\ \text{J} = 10\ \text{mol} \times 29.1\ \frac{\text{J}}{\text{mol.K}} \times (\Delta T)$$

$$(\Delta T) = 43.13^\circ\text{C} \qquad T_2 = 25^\circ\text{C} + 43.13^\circ\text{C} = 68.13^\circ\text{C (same answer)}$$

2) If we assume a constant $\tilde{C}_V$, then we can factor it out from the integral term:

$$\Delta U = n\int_{T1}^{T2} \tilde{C}_V dT = n\tilde{C}_V(T_2 - T_1) = 10\ \text{mol} \times 20.79\frac{\text{J}}{\text{mol.K}} \times (43.13) = 8966\ \text{J} = 8.966\ \text{kJ}$$

3) Let us apply the ideal gas law twice:

$$P_1V_1 = n_1RT_1$$

$$P_2V_2 = n_2RT_2$$

Divide side by side, we get:

$$\frac{P_1V_1}{P_2V_2} = \frac{n_1T_1}{n_2T_2}$$

Notice that $P_1 = P_2$ and $n_1 = n_2$. We also have:

$$V_1 = \frac{nRT_1}{P_1} = \frac{10\ \text{mol} \times 0.08206\frac{\text{atm.L}}{\text{mol.K}}(298.15\ \text{K})}{1\ \text{atm}} = 244.66\ \text{L}$$

$$\frac{V_1}{V_2} = \frac{T_1}{T_2} \rightarrow \frac{244.66\ \text{L}}{V_2} = \frac{298.15\ \text{K}}{341.3\ \text{K}}.$$

Solve for $V_2 : V_2 = 280.1$ L. $(V_2 > V_1)$.

4) From Eq. (B):

$$W_{12} = -\int_{V1}^{V2} P dV = -P\Delta V = -1.0\ \text{atm} \times \frac{101{,}325\ \text{Pa}}{\text{atm}}(280.1 - 244.7)\text{L} \times \frac{\text{m}^3}{1000\ \text{L}}$$

$$W_{12} = -3587\ \text{J} = -3.587\ \text{kJ}$$

The work is done by the system. See NOTE #1.
Alternatively, from Eq. (D)

$$\Delta V = \Delta\left(\frac{nRT}{P}\right) = \frac{nR}{P}\Delta T \rightarrow W_{12} = -P\Delta V = -P \times \frac{nR}{P}\Delta T = -nR\Delta T$$

$$W_{12} = -10\ \text{mol} \times 8.314\frac{\text{J}}{\text{mol.K}}(43.13)\text{K} = -3586\ \text{J} = -3.586\ \text{kJ}$$

7.9.4 MATLAB Code for an Iso-baric Ideal Gas Process

Figure 7.7 shows MATLAB code (**IG_IsoBaric.m**) for calculating thermodynamics properties, like T_2, V_2, ΔU, ΔH, Q_{12}, and W_{12} for a constant-pressure process. The user has to input some parameters and decide on the nature of the ideal gas: Mono-atomic, di-atomic, or poly-atomic. One can run the code for different input parameters, including the heat duty and the type of the ideal gas. Notice that n, T_{1C}, and P_1 are entered as priori input parameters; howsever, the user must select one more input out of three possibles scenarios: These three scenarios include heat duty, Q_{12}, the final volume, V_2, and the final temperature, T_{2C}. Now, the system will be specified, and the remaining unknown thermodynamic properties can be calculated.

```
% Get user input for initial moles, initial volume, pressure, final temperature, and final volume
clear variables;
n = input('Enter the initial number of moles: ');
T1C = input('Enter the initial temperature (oC): ');
P1 = input('Enter the initial pressure (atm): ');
% User selection for gas type (modify prompts as needed)
gasType = input('Select gas type (A: Mono-atomic, B: Di-atomic, C: Poly-atomic): ', 's');
% Calculate gas constant based on user selection
switch upper(gasType)
  case 'A'
    R = 8.314; % J/mol*K (Mono-atomic)
    Cv = (3/2)*R; % J/mol*K (Specific heat at constant volume)
    Cp = Cv + R; % J/mol*K (Specific heat at constant pressure)
  case 'B'
    R = 8.314; % J/mol*K (Di-atomic)
    Cv = (5/2)*R; % J/mol*K
    Cp = Cv + R; % J/mol*K
  case 'C'
    R = 8.314; % J/mol*K (Poly-atomic)
    Cv = (3.0)*R; % J/mol*K
    Cp = Cv + R; % J/mol*K
  otherwise
    error('Invalid gas type selection.');
end
FinalCondition = input('Known Final Condition (A: Heat duty, Q12, J, B: Final V, V2, L C: Final
T, T2C, oC): ', 's');
switch upper(FinalCondition)
  case 'A'
    Q12 = input('Enter the heat duty, Q12, (J): ');
    Rgas=0.08206;
    T1=273.15+T1C;
    V1 = ((n*Rgas*T1)/P1);
    T2=T1+(Q12)/(n*Cp);
    T2C=T2-273.15;
    DU=n*Cv*(T2-T1);
    W12=DU-Q12;
    V2 = ((n*Rgas*T2)/P1);
    DH=n*Cp*(T2-T1);
```

```
  case 'B'
    V2 = input('Enter the final volume (L): ');
    T1=273.15+T1C;
    Rgas=0.08206;
    V1 = (n*Rgas*T1)/P1;
    T2=(V2/V1)*T1;
    T2C=T2-273.15;
    DU=n*Cv*(T2-T1);
    W12=-n*R*(T2-T1);
    Q12=DU-W12;
    DH=n*Cp*(T2-T1);
  case 'C'
    T2C = input('Enter the final temperature (oC): ');
    T1=273.15+T1C;
    T2=273.15+T2C;
    Rgas=0.08206;
    V1 = ((n*Rgas*T1)/P1);
    V2 = ((n*Rgas*T2)/P1);
    DU=n*Cv*(T2-T1);
    W12=-n*R*(T2-T1);
    Q12=DU-W12;
    DH=n*Cp*(T2-T1);
  otherwise
    error('Invalid gas type selection.');
end
% Display results
disp('Results:');
disp(['Number of Moles: ', num2str(n), ' mole']);
disp(['Initial Temperature: ', num2str(T1C), ' oC']);
disp(['Final Temperature: ', num2str(T2C), ' oC']);
disp(['Initial Volume: ', num2str(V1), ' L']);
disp(['Final Volume: ', num2str(V2), ' L']);
disp(['Work done: ', num2str(W12), ' J']);
disp(['Change in internal energy, DU: ', num2str(DU), ' J']);
disp(['Q12 Duty done: ', num2str(Q12), ' J']);
disp(['Change in enthalpy, DH: ', num2str(DH), ' J']);
```

Figure 7.7 MATLAB code (**IG_IsoBaric.m**) for an ideal gas heating/cooling process calculations under iso-baric conditions.

Sample calculations are shown below for Example 7.8.

>> **IG_IsoBaric**
Enter the initial number of moles: **10**
Enter the initial temperature (°C): **25**
Enter the initial pressure (atm): **1**
Select gas type (A: Monoatomic, B: Diatomic, C: Polyatomic): **B**
Known Final Condition (A: Heat duty, Q12, J, B: Final V, V2, L C: Final T, T2C, °C): **A**
Enter the heat duty, Q12, (J): **12,552**

Results:
Number of Moles: 10 mole
Initial Temperature: 25°C
Final Temperature: 68.1355°C
Initial Volume: 244.6619 L
Final Volume: 280.0589 L
Work done: -3586.2857 J
Change in internal energy, DU: 8965.7143 J
Q12 Duty done: 12,552 J
Change in enthalpy, DH: 12,552 J

7.9.5 MATLAB Code for an Iso-thermal Ideal Gas Process

Example 7.9 Heating an Ideal Gas in a Cylinder at Constant Temperature (Iso-thermal) Process

Ten moles of a diatomic ideal gas are contained in a cylinder fitted with a freely moving piston. The initial gas temperature is 25°C, and the initial pressure is 2 atm.

The cylinder is placed in a hot water bath. An external dead weight is differentially decreased, which will slightly decompress or expand the gas and slightly decrease its temperature. Heat is then absorbed in proportion with the differential decrease in temperature to level off the decrease in temperature and maintain the T constant.

The cycle of continuously decreasing the dead weight and allowing heat absorption after that will terminate once the desired level of expansion is achieved. We can then report the final pressure, the final volume, or the heat duty. The heat duty is reported here. Overall, an amount of 3 kcal is transferred to the gas until reaching the final volume (i.e., $V_2 > V_1$) while keeping the same temperature ($T_2 = T_1$), as shown in Figure 7.8. So, the cylinder is placed in a heating water bath for allowing heat transfer from the surroundings to the slightly colder gas as a result of volume expansion, pointed above.

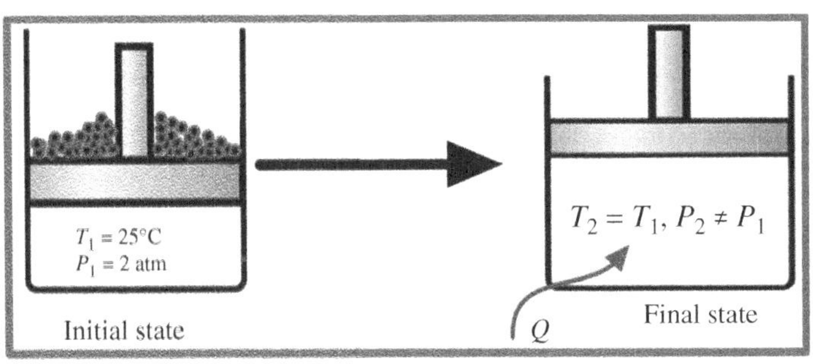

Figure 7.8 Iso-thermal expansion of an ideal gas from V_1 to V_2 ($V_2 > V_1$) and ($P_2 < P_1$).

1) Calculate ΔU and ΔH
2) Calculate the work done by the system, W_{12}
3) Calculate V_2
4) Calculate P_2

Solution:

1) Since T is constant, then $\Delta T = 0$. This implies that:

$$\Delta U = n\tilde{C}_V(\Delta T) = 0 \qquad \Delta H = n\tilde{C}_P(\Delta T) = 0$$

2) Let us apply Eq. (7.4a):

$$Q_{12} + W_{12} = (\Delta U)_{\text{Sys}} = 0$$

Rearrange to get:

$$W_{12} = -Q_{12} = -12{,}552\,\text{J} \tag{A}$$

Work is done by the system on its surroundings. See NOTE #1.

3) The work is given by Eq. (7.5a):

$$W_{12} = -\int_{V1}^{V2} P\text{d}V = -\int_{V_1}^{V_2} \frac{nRT}{V}\text{d}V = -nRT\ln\left(\frac{V_2}{V_1}\right) \tag{B}$$

Thus, from Eqs. (A) and (B), we have:

$$-12{,}552\,\text{J} = W_{12} = -nRT\ln\left(\frac{V_2}{V_1}\right) \tag{C}$$

$$\ln\left(\frac{V_2}{V_1}\right) = \frac{-12{,}552\,\text{J}}{-10\,\text{mol} \times 8.314\dfrac{\text{J}}{\text{mol.K}} \times (298.15)\text{K}} = 0.5064 \tag{D}$$

$$\left(\frac{V_2}{V_1}\right) = \text{e}^{0.5064} = 1.6593 \tag{E}$$

From an ideal gas law, we have:

$$V_1 = \frac{nRT1}{P1} = \frac{10\,\text{mol} * 0.08206\dfrac{\text{atm.L}}{\text{mol.K}} * 298.15\,\text{K}}{2\,\text{atm}} = 122.33\,\text{L}$$

$$V_2 = 1.6593 * V_1 = 1.6593 * 122.33 = 202.984\,\text{L}$$

4)
$$P_2 = \frac{nRT_2}{V_2} = \frac{10\,\text{mol} * 0.08206\,\dfrac{\text{atm.L}}{\text{mol.K}} * 298.15\,\text{K}}{202.984\,\text{L}} = 1.2053\,\text{atm}$$

NOTE #8
For an ideal gas under iso-thermal conditions: $P_1V_1 = P_2V_2 = nRT_1$ = Constant. $\Delta U = \Delta H = 0$

Figure 7.9 shows MATLAB code (**IG_IsoTherm.m**) for calculating thermodynamics properties, like P_2, V_2, ΔU, ΔH, Q_{12}, and W_{12} for a constant-temperature process. The user has to input some parameters and decide on the nature of the ideal gas: Mono-atomic, di-atomic, or

poly-atomic. One can run the code for different input parameters, including the heat duty and the type of the ideal gas. Notice that n, T_{1C}, and P_1 are entered as priori input parameters; however, the user must select one more input out of three possibles scenarios: These three scenarios include heat duty, Q_{12}, the final volume, V_2, and the final pressure, P_2. Now, the system will be specified, and the remaining unknown thermodynamic properties can be calculated.

```
% Get user input for initial moles, initial volume, pressure, final temperature, and final volume
clear variables;
n = input('Enter the initial number of moles: ');
T1C = input('Enter the initial temperature (oC): ');
P1 = input('Enter the initial pressure (atm): ');
% User selection for gas type (modify prompts as needed)
gasType = input('Select gas type (A: Mono-atomic, B: Di-atomic, C: Poly-atomic): ', 's');
% Calculate gas constant based on user selection
switch upper(gasType)
  case 'A'
    R = 8.314; % J/mol*K (Mono-atomic)
    Cv = (3/2)*R; % J/mol*K (Specific heat at constant volume)
    Cp = Cv + R; % J/mol*K (Specific heat at constant pressure)
  case 'B'
    R = 8.314; % J/mol*K (Di-atomic)
    Cv = (5/2)*R; % J/mol*K
    Cp = Cv + R; % J/mol*K
  case 'C'
    R = 8.314; % J/mol*K (Poly-atomic)
    Cv = (3.0)*R; % J/mol*K
    Cp = Cv + R; % J/mol*K
  otherwise
    error('Invalid gas type selection.');
end
FinalCondition = input('Known Final Condition (A: Heat duty, Q12, J, B: Final V, V2, L C:
Final P, P2, atm): ', 's');
switch upper(FinalCondition)
  case 'A'
    Q12 = input('Enter the heat duty, Q12, (J): ');
    T1=273.15+T1C;
    Rgas=0.08206;
    lnVterm=(Q12)/(n*R*T1);
    V1 = ((n*Rgas*T1)/P1);
    V2=V1*exp(lnVterm);
    P2=(n*Rgas*T1)/V2;
    W12=-Q12;
    DU=0.0;
    DH=0.0;
  case 'B'
    V2 = input('Enter the final Volume (L): ');
    T1=273.15+T1C;
    Rgas=0.08206;
    V1 = (n*Rgas*T1)/P1;
    lnVterm=log(V2/V1);
```

```
    Q12=(n*R*T1)*lnVterm;
    P2=(n*Rgas*T1)/V2;
    W12=-Q12;
    DU=0.0;
    DH=0.0;
  case 'C'
    P2 = input('Enter the final pressure (atm): ');
    T1=273.15+T1C;
    Rgas=0.08206;
    V1 = (n*Rgas*T1)/P1;
    V2=(n*Rgas*T1)/P2;
    lnVterm=log(V2/V1);
    Q12=(n*R*T1)*lnVterm;
    W12=-Q12;
    DU=0.0;
    DH=0.0;
  otherwise
    error('Invalid gas type selection.');
end
% Display results
disp('Results:');
disp(['Number of Moles: ', num2str(n), ' mole']);
disp(['Initial Temperature: ', num2str(T1C), ' oC']);
disp(['Final Temperature: ', num2str(T1C), ' oC']);
disp(['Initial Volume: ', num2str(V1), ' L']);
disp(['Final Volume: ', num2str(V2), ' L']);
disp(['Initial Pressure: ', num2str(P1), ' atm']);
disp(['Final Pressure: ', num2str(P2), ' atm']);
disp(['Work done: ', num2str(W12), ' J']);
disp(['Change in internal energy, DU: ', num2str(DU), ' J']);
disp(['Q12 Duty done: ', num2str(Q12), ' J']);
disp(['Change in enthalpy, DH: ', num2str(DH), ' J']);
```

Figure 7.9 MATLAB code (**IG_IsoTherm.m**) for an ideal gas heating/cooling process calculations under iso-thermal conditions.

Sample calculations are shown below for Example 7.9.

>> **IG_IsoTherm**
Enter the initial number of moles: **10**
Enter the initial temperature (°C): **25**
Enter the initial pressure (atm): **2**
Select gas type (A: Monoatomic, B: Diatomic, C: Polyatomic): **B**
Known Final Condition (A: Heat duty, Q12, J, B: Final V, V2, L C: Final P, P2, atm): **A**
Enter the heat duty, Q12, (J): **12,552**
Results:
Number of Moles: 10 mole
Initial Temperature: 25°C
Final Temperature: 25°C

Initial Volume: 122.3309 L
Final Volume: 202.9785 L
Initial Pressure: 2 atm
Final Pressure: 1.2054 atm
Work done: -12552 J
Change in internal energy, DU: 0 J
Q12 Duty done: 12,552 J
Change in enthalpy, DH: 0 J

7.9.6 MATLAB Code for an Adiabatic Ideal Gas Process

Example 7.10 An Ideal Gas Expansion in a Cylinder Under Adiabatic Process
Ten moles of a diatomic ideal gas are contained in a cylinder fitted with a freely moving piston. The initial gas temperature is 300°C, and the pressure is 10 atm. The piston-cylinder setup is well insulated from surroundings such that we can assume an adiabatic condition (i.e., $Q_{12} = 0$).

As shown in Figure 7.10, the piston is allowed to slowly expand upward in response to an external decremental dead weight mechanism, where the external pressure (weight/area) is slightly decremented until reaching the final pressure P_2, which is 1.3 atm. Notice here that all three measurable variables do change with expansion: P, T, and V. The only fixed parameters are: n and R. The process that is both adiabatic and reversible (infinitesimal driving force for change in both directions) is called isentropic, which means a constant-entropy process. The scope of entropy will be fully covered in thermodynamics courses/textbooks.

Find T_2, V_2, ΔU, ΔH, and W_{12}.

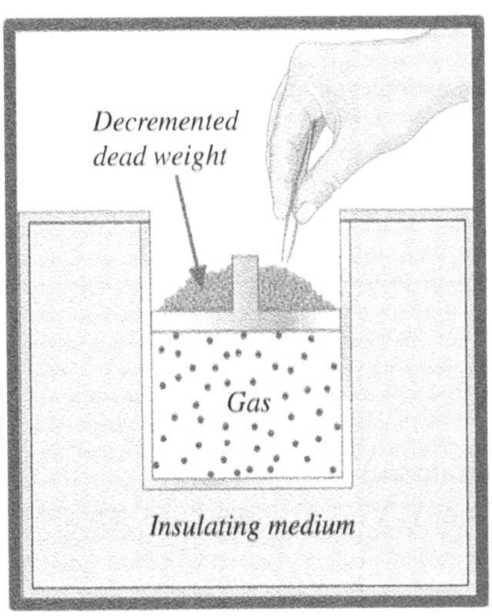

Figure 7.10 Insulated gas while the dead weight is decrementally removed to have an adiabatic and reversible expansion process.

Solution:

1) Let us apply Eq. (7.3a):

$$dQ_{12} + dW_{12} = (dU)_{Sys} \tag{A}$$

Notice that dQ_{12} is zero.

$$-PdV = n\tilde{C}_V dT \rightarrow -\frac{nRT}{V}dV = n\tilde{C}_V dT \tag{B}$$

Rearrange terms to reach at:

$$\frac{dV}{V} = -\frac{\tilde{C}_V}{R}\frac{dT}{T} \rightarrow \ln\left(\frac{V_2}{V_1}\right) = \ln\left(\frac{T_2}{T_1}\right)^{-\frac{\tilde{C}_V}{R}} \tag{C}$$

$$\frac{V_2}{V_1} = \left(\frac{T_2}{T_1}\right)^{-\frac{\tilde{C}_V}{R}} \tag{D}$$

$$\tilde{C}_P = \tilde{C}_V + R$$

Divide by $\tilde{C}_V$, and recall that $\gamma = (\tilde{C}_P)/(\tilde{C}_V)$, we have:

$$\gamma = 1 + \frac{R}{\tilde{C}_V} \qquad \rightarrow \frac{R}{\tilde{C}_V} = \gamma - 1 \qquad \rightarrow \frac{\tilde{C}_V}{R} = \frac{1}{\gamma - 1} \tag{E}$$

Equation (D) becomes:

$$\frac{V_2}{V_1} = \left(\frac{T_2}{T_1}\right)^{-\frac{1}{\gamma-1}} = \left(\frac{T_2}{T_1}\right)^{\frac{1}{1-\gamma}} \qquad \left(\frac{T_2}{T_1}\right) = \left(\frac{V_2}{V_1}\right)^{1-\gamma} \tag{F}$$

Let us apply the ideal gas law twice:

$$P_1V_1 = n_1RT_1$$

$$P_2V_2 = n_2RT_2$$

Divide side by side, we get:

$$\frac{P_1V_1}{P_2V_2} = \frac{T_1}{T_2} \qquad \left(\frac{T_2}{T_1}\right) = \frac{P_2}{P_1} \times \frac{V_2}{V_1} \tag{G}$$

The right-hand side of Eqs. (F) and (G) must be equal:

$$\left(\frac{V_2}{V_1}\right)^{1-\gamma} = \frac{P_2}{P_1} \times \frac{V_2}{V_1} \qquad \rightarrow \left(\frac{V_2}{V_1}\right)^{1-\gamma}\left(\frac{V_2}{V_1}\right)^{-1} = \frac{P_2}{P_1}$$
$$\rightarrow \frac{P_2}{P_1} = \left(\frac{V_2}{V_1}\right)^{-\gamma} = \left(\frac{V_1}{V_2}\right)^{\gamma} \tag{H}$$

Eq. (H) says that:

$$P_2V_2^{\gamma} = P_1V_1^{\gamma} = PV^{\gamma} = \text{Constant} \rightarrow P = \frac{C}{V^{\gamma}} \tag{I}$$

Also, utilizing Eqs. (F) and (H), we can reach at:

$$\left(\frac{T_2}{T_1}\right) = \left(\frac{P_2}{P_1}\right)^{\frac{\gamma-1}{\gamma}} \quad \text{or} \quad \left(\frac{P_2}{P_1}\right) = \left(\frac{T_2}{T_1}\right)^{\frac{\gamma}{\gamma-1}} \tag{J}$$

NOTE #9

Depending on what is given ahead, one can use Eqs. (F), (H), or (J) to find the missing final condition.

Back to Eq. (B), we have:

$$-PdV = n\tilde{C}_V dT \rightarrow -\frac{C}{V^\gamma}dV = n\tilde{C}_V dT \tag{K}$$

$$\int_{V_1}^{V_2} -\frac{\text{Constant}}{V^\gamma}dV = \int_{T_1}^{T_2} n\tilde{C}_V dT \tag{L}$$

Integration and rearrangement of terms, we reach at:

$$W_{12} = -\text{Constant}\left[\frac{V_2^{1-\gamma} - V_1^{1-\gamma}}{1-\gamma}\right] = -P_1V_1^\gamma\left[\frac{V_2^{1-\gamma} - V_1^{1-\gamma}}{1-\gamma}\right] = n\tilde{C}_V(T_2 - T_1) = \Delta U \tag{M}$$

Alternatively, we can utilize Eq. (H) to reach at:

$$W_{12} = -\left[\frac{P_2V_2 - P_1V_1}{1-\gamma}\right] = -\left[\frac{nRT_2 - nRT_1}{1-\gamma}\right] = n\tilde{C}_V(T_2 - T_1) = \Delta U \tag{N}$$

Back to Example 7.10, for a di-atomic molecule, we have:

$$\gamma = \frac{C_P}{C_V} = \frac{C_V + R}{C_V} = \frac{\frac{5}{2}R + R}{\frac{5}{2}R} = \frac{7}{5} = 1.4$$

$$\left(\frac{T_2}{T_1}\right) = \left(\frac{P_2}{P_1}\right)^{\frac{\gamma-1}{\gamma}} \rightarrow \quad T_2 = T_1\left(\frac{P_2}{P_1}\right)^{\frac{\gamma-1}{\gamma}} = 573.15\,K \times \left(\frac{1.3\text{ atm}}{10\text{ atm}}\right)^{\frac{1.4-1}{1.4}} = 319.97\text{ K}$$

$$\Delta U = n\tilde{C}_V(T_2 - T_1) = 10\text{ mol} \times \frac{5}{2} \times 8.314\,\frac{\text{J}}{\text{mol.K}} \times (319.97 - 573.15\,)\text{K} = -52{,}623.5\text{ J}$$

$$W_{12} = \Delta U = n\tilde{C}_V(T_2 - T_1) = -52{,}623.5\text{ J}$$

$$\Delta H = n\tilde{C}_P(T_2 - T_1) = 10\text{ mol} \times \frac{7}{2} \times 8.314\frac{\text{J}}{\text{mol.K}} \times (319.97 - 573.15\,)\text{K} = -73{,}672.8\text{ J}$$

Alternatively,

$$\Delta H = \Delta U + \Delta(PV) = \Delta U + (P_2V_2 - P_1V_1) = \Delta U + (nRT_2 - nRT_1)$$

$$\Delta H = n\tilde{C}_V(T_2 - T_1) + nR(T_2 - T_1) = n(\tilde{C}_V + R)(T_2 - T_1) = n\tilde{C}_P(T_2 - T_1) = -73{,}672.8\text{ J}$$

One can also find V_2 and V_1 via applying ideal gas law both at the final and initial state and then substitute for P_2V_2 and P_1V_1, respectively. Of course, n, P_2, T_2, P_1, and T_1 are either given or calculated earlier. See the MATLAB code below.

Figure 7.11 shows MATLAB code (**IG_Adiabatic.m**) for calculating thermodynamics properties, like T_2, V_2, ΔU, ΔH, and W_{12} for an adiabatic process. The user has to input some parameters and decide on the nature of the ideal gas: Mono-atomic, di-atomic, or poly-atomic. One can run the code for different input parameters, including the heat duty and the type of the ideal gas. Notice that n, T_{1C}, and P_1 are entered as priori input parameters; however, the user must select one more input out of three possibles scenarios: These three scenarios include the final pressure, P_2, the final volume, V_2, and the final temperature, T_2. Now, the system will be specified, and the remaining unknown thermodynamic properties can be calculated.

```
clear variables;
%Get user input for initial moles, initial volume, pressure, final temperature, and final volume
n = input('Enter the initial number of moles: ');
T1C = input('Enter the initial temperature (oC): ');
P1 = input('Enter the initial pressure (atm), P1: ');
% User selection for gas type (modify prompts as needed)
gasType = input('Select gas type (A: Mono-atomic, B: Di-atomic, C: Poly-atomic): ', 's');
% Calculate gas constant based on user selection
switch upper(gasType)
  case 'A'
    R = 8.314; % J/mol*K (Mono-atomic)
    Cv = (3/2)*R; % J/mol*K (Specific heat at constant volume)
    Cp = Cv + R; % J/mol*K (Specific heat at constant pressure)
    gama=Cp/Cv;
  case 'B'
    R = 8.314; % J/mol*K (Di-atomic)
    Cv = (5/2)*R; % J/mol*K
    Cp = Cv + R; % J/mol*K
    gama=Cp/Cv;
  case 'C'
    R = 8.314; % J/mol*K (Poly-atomic)
    Cv = (3.0)*R; % J/mol*K
    Cp = Cv + R; % J/mol*K
    gama=Cp/Cv;
  otherwise
    error('Invalid gas type selection.');
end
FinalCondition = input('Known Final Condition (A: Final P, P2, atm, B: Final V, V2, L C:
Final T, T2C, oC): ', 's');
switch upper(FinalCondition)
  case 'A'
    P2 = input('Enter the final Pressure, P2, (atm): ');
    Rgas=0.08206;
    T1=273.15+T1C;
    T2=T1*(P2/P1)^((gama-1)/gama);
    T2C=T2-273.15;
    DU=n*Cv*(T2-T1);
    V1 = ((n*Rgas*T1)/P1);% V1 in L
    V2 = ((n*Rgas*T2)/P2); %V2 in L
    W12=-(P2*V2-P1*V1)*101.325/(1-gama);% W12 calculation via D(PV)
    W12b=-((n*R*T2-n*R*T1)/(1-gama));%Another way of W12 calculation via DT
    DH=n*Cp*(T2-T1);
    Q12=0.0; %Adaiabatic Process
  case 'B'
    V2 = input('Enter the final volume (L): ');
    T1=273.15+T1C;
    Rgas=0.08206;
    V1 = (n*Rgas*T1)/P1;
    T2=T1*(V2/V1)^(1-gama);
    T2C=T2-273.15;
    P2=n*Rgas*T2/V2;
    DU=n*Cv*(T2-T1);
    W12=-(P2*V2-P1*V1)*101.325/(1-gama);% W12 calculation via D(PV)
```

```
    W12b=-((n*R*T2-n*R*T1)/(1-gama));%Another way of W12 calculation via DT
    DH=n*Cp*(T2-T1);
    Q12=0.0; % Adiabatic Process
  case 'C'
    T2C = input('Enter the final temperature (oC): ');
    T1=273.15+T1C;
    T2=273.15+T2C;
    Rgas=0.08206;
    V1 = (n*Rgas*T1)/P1;
    P2=P1*(T2/T1)^(gama/(gama-1));
    V2 = (n*Rgas*T2)/P2;
    DU=n*Cv*(T2-T1);
    W12=-(P2*V2-P1*V1)*101.325/(1-gama);% W12 calculation via D(PV)
    W12b=-((n*R*T2-n*R*T1)/(1-gama));%Another way of W12 calculation via DT
    DH=n*Cp*(T2-T1);
    Q12=0.0; % Adiabatic Process
  otherwise
    error('Invalid gas type selection.');
end
% Display results
disp('Results:');
disp(['Number of Moles: ', num2str(n), ' mole']);
disp(['Initial Temperature: ', num2str(T1C), ' oC']);
disp(['Final Temperature: ', num2str(T2C), ' oC']);
disp(['Initial Volume: ', num2str(V1), ' L']);
disp(['Final Volume: ', num2str(V2), ' L']);
disp(['Initial Pressure: ', num2str(P1), ' atm']);
disp(['Final Pressure: ', num2str(P2), ' atm']);
disp(['Work done via D(PV): ', num2str(W12), ' J']);
disp(['Work done via DT: ', num2str(W12b), ' J']);
disp(['Change in internal energy, DU: ', num2str(DU), ' J']);
disp(['Change in enthalpy, DH: ', num2str(DH), ' J']);
disp(['Heat Duty for Adiabatic Process, Q12: ', num2str(Q12), ' J']);
```

Figure 7.11 MATLAB code (**IG_Adiabatic.m**) for an ideal gas heating/cooling process calculations under adiabatic ($Q_{12} = 0$) condition.

Sample calculations are shown below for Example 7.10. The work is done by the system. See NOTE #1.

```
>> IG_Adiabatic
Enter the initial number of moles: 10
Enter the initial temperature (°C): 300
Enter the initial pressure (atm), P1: 10
Select gas type (A: Monoatomic, B: Diatomic, C: Polyatomic): B
Known Final Condition (A: Final P, P2, atm, B: Final V, V2, L C: Final T, T2C, °C): A
Enter the final Pressure, P2, (atm): 1.3
```

Results:
Number of Moles: 10 mole
Initial Temperature: 300°C
Final Temperature: 46.82°C
Initial Volume: 47.0327 L
Final Volume: 201.9749 L
Initial Pressure: 10 atm
Final Pressure: 1.3 atm
Work done via D(PV): -52,628.0814 J
Work done via DT: -52,623.464 J
Change in internal energy, DU: -52,623.464 J
Change in enthalpy, DH: -73,672.8496 J
Heat Duty for Adiabatic Process, Q12: 0 J

7.9.7 The Latent Heat of a Substance

So far, we have covered what is called the sensible heat. On the one hand, sensible heat is the heat we can sense with a thermometer. It is the kind of heat that makes a substance feel warmer as we add heat energy. For example, when we heat a cup of water on the stove, the water gets hotter (temperature increases) as it absorbs sensible heat. On the other hand, latent heat is hidden heat. It is the energy absorbed or released by a substance during a phase change (solid to liquid, liquid to gas, solid to vapor, or vice versa) without a change in temperature.

Think of it like this: the added heat isn't going toward making the particles move faster (which is what increases temperature) but rather toward overcoming the intermolecular forces holding the substance in its current state. Here are some examples to illustrate the difference:

Ice Melting: As we add heat to ice, its temperature stays at 0°C (32°F) until all the ice melts. That is because the added heat is being utilized as latent heat to break the bonds between water molecules in the ice, changing it to liquid water. Only after all the ice melts will the temperature of the water start to rise with further heating (sensible heat).

Water Boiling: When water reaches 100°C (212°F), at 1 standard atmosphere, it starts to boil. The added heat is being used as latent heat to convert liquid water into steam, not to raise the temperature.

Wax Melting: As solid wax is heated, it absorbs latent heat to break the bonds between wax molecules, changing its state to liquid wax without a temperature increase until completely melted.

Notice that specific heat capacity does not account for the absorption of heat energy during a change of state, as transitioning between states constitutes a distinct physical process. Figure 7.12 illustrates the variation in the temperature of a substance in response to the application of heat.

In Figure 7.12, the temperature exhibits a steady increase until a state change is encountered. During melting or boiling, despite the continuous addition of heat energy, no temperature change occurs. This concealed (nonsensible) energy is termed latent heat. When

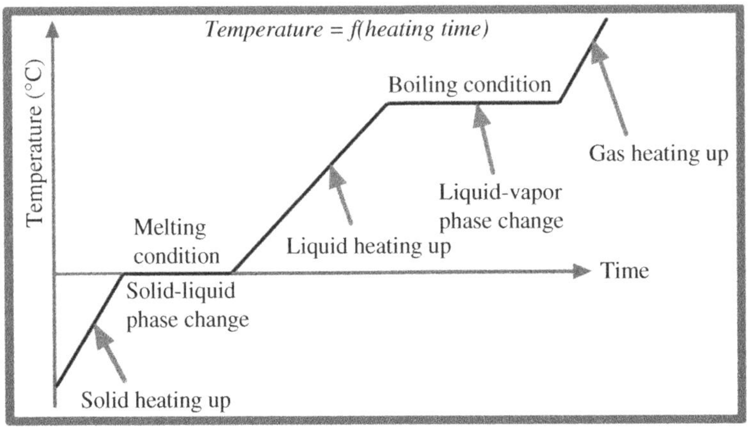

Figure 7.12 The temperature profile versus heating time of a sub-cooled ice sample (bottom left) until it becomes superheated vapor (top right).

heating a sample of a substance, heat energy is employed to elevate the temperature by augmenting the average kinetic energy of the molecules, a concept covered by specific heat capacity calculations. On the other hand, when changing the state of matter (i.e., phase transformation), the heat energy is utilized to disrupt molecular bonds, facilitating a state change, and this is accounted for through latent heat calculations. The quantity of energy necessary to break these bonds, referred to as latent heat, varies among elements and depends on the specific change of state taking place:

Latent Heat of Fusion: The energy required to melt 1 kg of a solid into a liquid.
Latent Heat of Vaporization: The energy needed to boil 1 kg of a liquid into a vapor.

It is important to note that to reverse these processes, such as freezing a liquid or condensing a gas, an equivalent amount of heat energy must be extracted from the material. In the absence of kinetic energy and potential energy changes and for constant-pressure heating, we can write:

Constant-pressure heating for liquid-vapor phase change (vaporization):

The initial phase is liquid and the final phase will be vapor. Using Eqs. (7.3a) and (7.5a), we have:

$$Q_{12} + W_{12} = \Delta U \qquad Q_{12} - P\Delta V = \Delta U$$

$$Q_{12} = P\Delta V + \Delta U = (U_V - U_L) + P(V_V - V_L) \tag{7.14}$$

Using Eq. (7.10d) for a constant-pressure process, we also have:

$$\Delta H = \Delta U + P\Delta V = (U_V - U_L) + P(V_V - V_L) \tag{7.15}$$

The right-hand side of Eq. (7.14) and of Eq. (7.15) is the same.

This implies that:

$$Q_{12} = \Delta H = (H_V - H_L) = m\Delta\overline{H} = m\left(\overline{H}_V - \overline{H}_L\right) = m\overline{H}_{LV} \tag{7.16}$$

Q_{12}: Heat energy supplied (J)
m: mass of the closed-system (kg)
$\overline{H}_{LV}$: The latent heat of vaporization (J/kg) > 0

Notice that both Q_{12} (heat absorbed or received by the system) and $\overline{H}_{LV}$ are positive values as $\overline{H}_V > \overline{H}_L$. The enthalpy of vapor is higher than that of liquid at the same temperature and pressure.

The opposite will be true if we switch the process from constant-pressure heating to cooling. The initial phase is vapor, and the final phase will be liquid. We have:

$$Q_{12} = \Delta H = (H_L - H_V) = m\Delta\overline{H} = m(\overline{H}_L - \overline{H}_V) = m\overline{H}_{VL} = -m\overline{H}_{LV} \tag{7.17}$$

$\overline{H}_{VL}$: The latent heat of condensation (J/kg) < 0

Notice here that both Q_{12} (heat leaving the closed system because of cooling) and $\overline{H}_{VL}$ must be negative quantities as $(\overline{H}_L < \overline{H}_V)$.

$$\overline{H}_{VL} = -\overline{H}_{LV} = -(\overline{H}_V - \overline{H}_L) \tag{7.18}$$

The same thing can be said about the other phase-change-related latent heat of melting (or fusion)/solidification (or freezing) and latent heat of sublimation/deposition.

$$\text{Steam Quality, } x = \frac{(\overline{H}_{\text{Sys}} - \overline{H}_L)}{(\overline{H}_V - \overline{H}_L)}$$

NOTE #10

Keep in mind, in general, as we move from $S \rightarrow L \rightarrow V$, we have to supply external heat; conversely, as we move from $V \rightarrow L \rightarrow S$, we have to cool or extract heat. The latent heat related to phase change will be either positive as in the former direction or negative as in the latter direction. There are tables made for a specific substance where it shows different thermodynamic properties, like molar/specific volume, molar/specific enthalpy, molar/specific internal energy, and molar/specific entropy for the two phases on both sides of phase change. Steam and refrigerant tables are typical examples.

Steam quality, x, expressed as a value between zero and one, describes the portion of a steam mixture that exists as vapor. Zero indicates pure saturated liquid (all liquid water), while one signifies completely dry saturated vapor (pure steam). Values in between represent a wet steam mixture containing both liquid water and vapor, with the proportion of vapor increasing as the quality approaches one.

Example 7.11 Heat Duty for Melting Ice

If the latent heat value of melting for water is 330 kJ/kg, calculate how much ice can be melted into liquid water by supplying 1200 kJ?

Solution:

From Eq. (7.16), we have:

$$Q_{12} = 1200\,\text{kJ} = m\overline{H}_{LV} = m \times 330\frac{\text{kJ}}{\text{kg}} \qquad m = \frac{1200\,\text{kJ}}{330\frac{\text{kJ}}{\text{kg}}} = 3.64\,\text{kg}$$

Example 7.12 Steam Quality
A saturated steam at $P_1 = 5.0\,\text{atm}$ and $T_1 = 152.33°\text{C}$ has the following specific enthalpy values:

$$\overline{H}_L = 642.3324\,\text{kJ/kg} \qquad \overline{H}_V = 2748.6938\,\text{kJ/kg}$$

Given that 1 kg of steam initially exists as a saturated liquid (i.e. $x = 0$) at P_1 and T_1, what will be the steam quality x, if an amount of heat is added to the closed system equal to 1500 kJ.

Solution:
We have a saturated liquid steam at $P_1 = 5.0\,\text{atm}$ and $T_1 = 152.33°\text{C}$. This means $\overline{H}_1 = \overline{H}_L = 642.3324\,\text{kJ/kg}$.

$$Q_{12} = 1500\,\text{kJ} = m\Delta\overline{H} = m\left(\overline{H}_2 - \overline{H}_1\right) = 1\,\text{kg} \times \left(\left(\overline{H}_2 - 642.3324\right)\right)\frac{\text{kJ}}{\text{kg}}$$

$$1500\frac{\text{kJ}}{\text{kg}} = \left(\left(\overline{H}_2 - 642.3324\right)\right)\frac{\text{kJ}}{\text{kg}}$$

$$\text{Solve for } \left(\overline{H}_2\right): \ \overline{H}_2 = \overline{H}_1 + 1500\frac{\text{kJ}}{\text{kg}} = 642.3324\frac{\text{kJ}}{\text{kg}} + 1500\frac{\text{kJ}}{\text{kg}} = 2142.3\frac{\text{kJ}}{\text{kg}}$$

$$\text{Steam Quality, } x = \frac{\left(\overline{H}_{\text{Sys}} - \overline{H}_L\right)}{\left(\overline{H}_V - \overline{H}_L\right)} = \frac{\left(\overline{H}_2 - \overline{H}_L\right)}{\left(\overline{H}_V - \overline{H}_L\right)} = \frac{(2143.3 - 642.3)}{2748.7 - 642.3} = 0.713$$

Or steam quality percent $= x * 100\% = 71.3\%$.

7.9.8 MATLAB Code for Water/Steam Thermodynamic Properties Calculations

Appendix H: | "H.4 MATLAB Code for Evaluating Liquid Water/Steam Properties at different T and P" section shows MATLAB code for evaluating thermodynamic properties of water in the form of subcooled, saturated liquid, two-phase liquid-vapor, saturated vapor, and superheated vapor. The temperature range falls within 0 up to 2000°C for pressure values up to 600 MPa (6000 bar), which is exceptionally very high pressure for steam-based applications. If the pressure is above 600 MPa up to 1000 MPa, the lower applicable temperature value is 10°C or above to have a valid liquid region. The linear interpolation is carried out for a temperature that lies between two given temperatures for a given listed pressure; otherwise, the warning message: "Temperature or pressure is outside the provided range" will be fired. The thermodynamic properties are specific volume (m^3/kg), specific internal energy (kJ/kg), specific enthalpy (kJ/kg), and specific entropy (kJ/(kg·K))

Figure 7.13 shows three examples of different scenarios for calculating water/steam thermodynamic properties, using **StreamProps.m** file.

```
>> SteamProps
Enter T between [0 2000] °C): 100
Enter P between [0.01 1000] MPa): 1
Temperature matches one of the provided values exactly.
Performing calculations for exact temperature...
Entered Temperature: 100°C
Entered Pressure: 1 MPa
    {'Phase @(P,T): '}    {'liquid'}
Specific volume: 0.001043 m3/kg
Internal energy: 418.8 kJ/kg
Enthalpy: 419.8 kJ/kg
Entropy: 1.307 kJ/(kg.K)

>> SteamProps
Enter T between [0 2000] °C): 107
Enter P between [0.01 1000] MPa): 1
Interpolating properties between two temperatures...
Interpolated properties at 107°C
Interpolated properties at 1 MPa
    {'Phase @(T,P): '}    {'liquid'}
Specific volume: 0.0010486 m3/kg
Internal energy: 448.34 kJ/kg
Enthalpy: 449.34 kJ/kg
Entropy: 1.385 kJ/(kg.K)

>> SteamProps
Enter T between [0 2,000] °C): 99.61
Enter P between [0.01 1,000] MPa): 0.1
Enter steam quality (x): 0.8
Performing calculations based on steam quality...
Thermodynamic properties at T:99.61 °C:
Thermodynamic properties at P: 0.1 MPa:
You are trapped in the two-phase region at a steam quality: 0.8
Specific volume: 1.3554 m3/kg
Internal energy: 2088.28 kJ/kg
Enthalpy: 2223.5 kJ/kg
Entropy: 6.1478 kJ/(kg.K)
```

Figure 7.13 Important thermodynamic properties $(\overline{V}, \overline{U}, \overline{H}, and\, \overline{S})$ of either liquid water or steam are evaluated at three different values of input pressure and temperature.

Example 7.13 Heat Duty for Heating Water

Let us make use of the steam table via utilizing MATLAB **SteamProps.m**. 0.10 kg/s of water stream is to be heated from 1 bar and 25°C up to $P_2 = 1.4$ bar and $T_2 = 152$°C as vapor. Calculate the total heat needed to carry out the mission, that is, to generate a low-pressure steam utility.

Solution:

We have a steady-state heating with negligible potential and kinetic energy changes and no shaft work (See Eq. 7.38):

$$\dot{Q} = \dot{m}\left(\Delta\overline{H}\right) = \dot{m}\left(\overline{H}_2 - \overline{H}_1\right)$$

$$\overline{H}_1@(T_1 = 25°\text{C}, P_1 = 1\text{ bar}) \qquad \overline{H}_2@(T_2 = 152°\text{C}, P_2 = 1.4\text{ bar})$$

Using **SteamProps.m**, we have:

```
Enter T between [0 2000] °C): 25
Enter P between [0.01 1000] MPa): 0.1
Temperature matches one of the provided values exactly.
Performing calculations for exact temperature...
Entered Temperature: 25 °C
Entered Pressure: 0.1 MPa
    {'Phase @(P,T): '}   {'liquid'}
Specific volume: 0.001003 m3/kg
Internal energy: 104.8 kJ/kg
Enthalpy: 104.9 kJ/kg
Entropy: 0.3672 kJ/(kg.K)
```

```
Enter T between [0 2,000] °C): 152
Enter P between [0.01 1,000] MPa): 0.14
Interpolating properties between two temperatures...
Interpolated properties at 152 °C
Interpolated properties at 0.14 MPa
    {'Phase @(T,P): '}   {'vapor'}
Specific volume: 1.3858 m3/kg
Internal energy: 2583.8 kJ/kg
Enthalpy: 2778 kJ/kg
Entropy: 7.4636 kJ/(kg.K)
```

$$\dot{Q} = \dot{m}(\Delta\overline{H}) = \dot{m}(\overline{H}_2 - \overline{H}_1) = 0.10\frac{\text{kg}}{\text{s}} * \left(2778\frac{\text{kJ}}{\text{kg}} - 104.9\frac{\text{kJ}}{\text{kg}}\right) = 267.31\frac{\text{kJ}}{\text{s}} = 267.31\text{ kW} = 0.2673\text{ MW}$$

7.10 General Energy Balance

$$\text{Energy}|_{\text{in}} - \text{Energy}|_{\text{out}} = \text{Accumulation} \tag{7.19}$$

$$\left[\left(\overline{U} + gZ + v^2/2\right)_{\text{in}} dm_{\text{in}} + \delta Q_{\text{in}} + \delta W_{\text{in}}\right] - \left[\left(\overline{U} + gZ + v^2/2\right)_{\text{out}} dm_{\text{out}} + \delta Q_{\text{out}} + \delta W_{\text{out}}\right] = d\left[m\left(\overline{U} + gz + v^2/2\right)\right]_{\text{C.V.}} \tag{7.20}$$

Let us denote:

$$\delta Q = \delta Q_{\text{in}} - \delta Q_{\text{out}} \tag{7.21}$$

$$\delta W = (\delta W_{\text{in}} - \delta W_{\text{out}}) \tag{7.22}$$

Equation (7.19) then becomes:

$$\left[\left(\overline{U} + gZ + v^2/2\right)_{\text{in}} dm_{\text{in}}\right] - \left[\left(\overline{U} + gZ + \frac{v^2}{2}\right)_{\text{out}} dm_{\text{out}}\right] + \delta Q + \delta W = d\left[m\left(\overline{U} + gZ + v^2/2\right)\right]_{\text{C.V.}} \tag{7.23}$$

Notice that Eq. (7.23) is written in absolute form. Alternatively, it can be expressed in rate form as:

$$\left[\left(\bar{U} + gZ + v^2/2\right)_{\text{in}} \frac{dm_{\text{in}}}{dt}\right] - \left[\left(\bar{U} + gZ + v^2/2\right)_{\text{out}} \frac{dm_{out}}{dt}\right] + \frac{\delta Q}{dt} + \frac{\delta W}{dt} = \frac{d\left[m\left(\bar{U} + gZ + v^2/2\right)\right]_{\text{C.V.}}}{dt} \quad (7.24)$$

Equation (7.23) or 7.24 represents energy balance in its most general form.

7.10.1 Common Types of Work in Chemical Applications

1) ***PdV* (Expansion/Contraction work)**: $dW = \mp \vec{F}.d\vec{x} = \mp PA.dx = \mp PdV$
2) **Shaft Work**: The work done on or by the system as a result of a shaft protruding into or out of the system (or control volume).
3) **Injection (Flow) Work**: The work due to the injection of a fluid element into the system or purging it out of the system. This type of work is *associated with mass flow*, in or out, *across* the boundaries of a given control volume.

For the *first type* of work, please refer to the previous section for a closed system. As open systems primarily involve the flow of fluids, the first type will not be addressed here due to the negligible impact of expansion/contraction effects in fluid flow. This is because we typically handle rigid containers and conduits in such situations.

For the *second type*, the shaft work is important in open (flow) systems as it reflects either the amount of work added to the system as is the case with pumps and compressors, or the amount of work extracted out of the system as is the case with turbines.

For the *third type*, the flow work is accounted for by virtue of fluid flow into or out of the control volume.

Hence:

$$\delta W = (\delta W_{\text{flow}})_{\text{in}} - (\delta W_{\text{flow}})_{\text{out}} + (\delta W_{\text{shaft}})_{\text{in}} - (\delta W_{\text{shaft}})_{\text{out}} \quad (7.25)$$

$$\delta W_{\text{saft}} = (\delta W_{\text{shaft}})_{\text{in}} - (\delta W_{\text{shaft}})_{\text{out}} \quad (7.26)$$

$$dV = \bar{V}dm = \frac{1}{\rho}dm \quad (7.27)$$

where $\bar{V}$ is the specific volume of the process fluid (flowing fluid), which is equal to the inverse of its density. Thus:

$$(\delta W_{\text{flow}})_{\text{in}} = P_{\text{in}}dV_{\text{in}} = P_{\text{in}}\bar{V}_{\text{in}}dm_{\text{in}} = \frac{P_{\text{in}}}{\rho_{\text{in}}}dm_{\text{in}} \quad (7.28)$$

$$(\delta W_{\text{flow}})_{\text{out}} = P_{\text{out}}dV_{\text{out}} = P_{\text{out}}\bar{V}_{\text{out}}dm_{\text{out}} = \left(\frac{P}{\rho}\right)_{\text{out}} dm_{\text{out}} \quad (7.29)$$

Thus, Eq. (7.25) becomes:

$$\delta W = P_{\text{in}}\bar{V}_{\text{in}}dm_{\text{in}} - P_{\text{out}}\bar{V}_{\text{out}}dm_{\text{out}} + \delta W_{\text{shaft}} \quad (7.30)$$

Substitute Eq. (7.30) into Eq. (7.23) and rearrange to reach at:

$$\left[\left(\bar{U} + gZ + v^2/2 + P\bar{V}\right)_{\text{in}} dm_{\text{in}}\right] - \left[\left(\bar{U} + gZ + v^2/2 + P\bar{V}\right)_{\text{out}} dm_{\text{out}}\right] + \delta Q + \delta W_{\text{shaft}} = d\left[m\left(\bar{U} + gZ + v^2/2\right)\right]_{\text{C.V.}} \tag{7.31}$$

Or, in rate form:

$$\left[\left(\bar{U} + gZ + v^2/2 + P\bar{V}\right)_{\text{in}} \frac{dm_{\text{in}}}{dt}\right] - \left[\left(\bar{U} + gZ + v^2/2 + P\bar{V}\right)_{\text{out}} \frac{dm_{\text{out}}}{dt}\right] + \frac{\delta Q}{dt} + \frac{\delta W_{\text{shaft}}}{dt} = \frac{d\left[m\left(\bar{U} + gZ + v^2/2\right)\right]_{\text{C.V.}}}{dt} \tag{7.32}$$

Given that: $\bar{H} = \bar{U} + P\bar{V}$, then Eq. (7.31) can also be written as:

$$\left[\left(\bar{H} + gZ + v^2/2\right)_{\text{in}} dm_{\text{in}}\right] - \left[\left(\bar{H} + gZ + v^2/2\right)_{\text{out}} dm_{\text{out}}\right] + \delta Q + \delta W_{\text{shaft}} = d\left[m\left(\bar{U} + gZ + v^2/2\right)\right]_{\text{C.V.}} \tag{7.33a}$$

In rate form:

$$\left[\left(\bar{H} + gZ + v^2/2\right)_{\text{in}} \frac{dm_{in}}{dt}\right] - \left[\left(\bar{H} + gZ + v^2/2\right)_{\text{out}} \frac{dm_{\text{out}}}{dt}\right] + \frac{\delta Q}{dt} + \frac{\delta W_{\text{shaft}}}{dt} = \frac{d\left[m\left(\bar{U} + gZ + v^2/2\right)\right]_{\text{C.V.}}}{dt} \tag{7.33b}$$

7.11 Simple Forms of General Energy Balance

Equation (7.33b) can be simplified under specific conditions:

1) ***Steady-State*** **Flow**:

$$\left[\left(\bar{H} + gZ + v^2/2\right)_{\text{out}} \dot{m}\right] - \left[\left(\bar{H} + gZ + v^2/2\right)_{\text{in}} \dot{m}\right] = \dot{Q} + \dot{W}_{\text{shaft}} \tag{7.34}$$

$$\Delta\left[\left(\bar{H} + gZ + v^2/2\right)\dot{m}\right] = \dot{Q} + \dot{W}_{\text{shaft}} \tag{7.35}$$

2) ***Steady-State*** **Flow with** ***Negligible Potential and Kinetic Energy Changes***:

$$\left[\left(\bar{H}\right)_{\text{out}} \dot{m}\right] - \left[\left(\bar{H}\right)_{\text{in}} \dot{m}\right] = \dot{Q} + \dot{W}_{\text{shaft}} \tag{7.36}$$

$$\Delta\left[\left(\bar{H}\right)\dot{m}\right] = \left[\Delta\dot{H}\right] = \left[\dot{H}_{\text{out}} - \dot{H}_{\text{in}}\right] = \dot{Q} + \dot{W}_{\text{shaft}} \tag{7.37}$$

3) ***Steady-State*** **Heating/Cooling with** ***Negligible Potential and Kinetic Energy Changes and no Shaft Work***:

$$\left[\left(\bar{H}\right)_{\text{out}} \dot{m}\right] - \left[\left(\bar{H}\right)_{\text{in}} \dot{m}\right] = \left[\Delta\dot{H}\right] = \left[\dot{H}_{\text{out}} - \dot{H}_{\text{in}}\right] = \dot{Q} \tag{7.38}$$

Notice that $\dot{m}$, in Eqs. (7.34) through (7.38), represents the *steady-state mass flow rate* in and out of the control volume.

For a closed system with negligible potential and K.E. changes, Eq. (7.4a) can be modified to include the shaft work, in addition to the PdV work.

$$Q + W_{\text{shaft}} + W_{PdV} = \Delta U = U_2 - U_1 \tag{7.39}$$

Notice the PdV type of work may arise in closed-system applications, in addition to the shaft work. Obviously, the *flow work is zero for a closed system*, unlike the open system.

Notice that the units of Eq. (7.33b) or any of its simplified versions will be power units.

7.11.1 Bernoulli's Equation (Energy Balance)

Example 7.14 Water Exit (or Jet Velocity) from a Hole of a Filled Tank

Water is flowing into the top of a tank at a rate of 600 L/min. The tank is 45 cm inside diameter and has an 8 cm inside diameter hole in the bottom, through which the water flows out. If the inflow rate is adjusted to match the outflow rate, what will the height of the water be in the tank if friction is negligible?

Solution:

Let us start with Eq. (7.35). In the absence of heat (i.e., $\dot{Q} = 0$) and shaft work (i.e., $\dot{W}_{\text{shaft}} = 0$) as well as negligible temperature change between the fluid inside the tank and the exit fluid (i.e., $\Delta\overline{H} = 0$), we have:

$$\Delta\left[\left(gZ + v^2/2\right)\dot{m}\right] = 0 \qquad \Delta\left[\left(gZ + v^2/2\right)\right] = 0 \qquad \left[\left(g\Delta Z + \Delta v^2/2\right)\right] = 0 \tag{7.40}$$

$$\frac{\Delta v^2}{2} = \frac{1}{2}\left(v_{\text{Hole}}^2 - v_{\text{Tank}}^2\right) = -g\Delta Z = -g(Z_2 - Z_1) = gh \tag{7.41}$$

NOTE #11
Δ always refers to: (final condition – initial condition) of a property.

where h is the depth of water measured from the free surface inside the tank down to the hole where water exits the tank. Notice that $h = -z$ and $\Delta h = -\Delta Z$. v_{Tank} is water velocity inside the tank as the free surface moves down and v_{Hole} is the exit or jet velocity of water issuing out of the hole.

Applying continuity equation between the free surface and the exit jet.

$$\dot{m} = \rho A_{\text{Tank}} v_{\text{Tank}} = \rho A_{\text{Hole}} v_{\text{Hole}} \qquad [\rho \text{ is constant; incompressible}]$$

ρ can be canceled out. Thus, we have:

$$A_{\text{Tank}} v_{\text{Tank}} = A_{\text{Hole}} v_{\text{Hole}} \qquad \frac{\pi}{4} D_{\text{Tank}}^2 v_{\text{Tank}} = \frac{\pi}{4} D_{\text{Hole}}^2 v_{\text{Hole}} \tag{7.42}$$

$$(0.45\ \text{m})^2 v_{\text{Tank}} = (0.08\ \text{m})^2 v_{\text{Hole}} \qquad v_{\text{Tank}} = 0.032 v_{\text{Hole}} \tag{7.43}$$

Substitute the value of v_{Tank} in Eq. (7.41), we have:

$$\frac{1}{2}\left(v_{\text{Hole}}^2 - 0.001 v_{\text{Hole}}^2\right) = gh \ \frac{1}{2}(0.999)\, v_{\text{Hole}}^2 \cong \frac{1}{2}\, v_{\text{Hole}}^2 = gh \tag{7.44}$$

$$v_{\text{Hole}} = \sqrt{2gh} \tag{7.45}$$

Equation (7.45) gives the velocity of the exit stream as a function of fluid height inside the tank in the absence of friction and energy forms other than potential and kinetic energy. Equation (7.45) is called Bernoulli's equation for a flow system. Its applications will be covered in the fluid mechanics course and also with characterization of frictional energy terms.

Applying the total material balance around the tank, we have:

$$\rho\dot{V}_{\text{in}} - \rho\dot{V}_{\text{out}} = \rho\frac{dV}{dt} = 0 \quad \text{[incompressible fluid, steady – state flow]} \tag{7.46}$$

$$\dot{V}_{\text{in}} = 600\frac{\text{L}}{\text{min}} \times \left(\frac{\text{m}^3}{1000\,L}\right) \times \left(\frac{1\text{ min}}{60\text{ s}}\right) = \dot{V}_{\text{out}} = A_{\text{Hole}} * v_{\text{Hole}} = \left(\frac{\pi}{4}D_{\text{Hole}}^2\right)\sqrt{2gh} \tag{7.47}$$

$$0.01\left(\frac{\text{m}^3}{\text{s}}\right) = 0.005\text{ m}^2\sqrt{2*9.8\frac{\text{m}}{\text{s}^2}*h\,[\text{m}]} \quad \rightarrow 2\left(\frac{\text{m}}{\text{s}}\right) = \left(\sqrt{2*9.8*h}\right)\frac{\text{m}}{\text{s}}$$

Solve for h: $h = 0.204\text{ m}$. Also, $v_{\text{Hole}} = \sqrt{2*9.8*0.204} = 2\frac{\text{m}}{\text{s}}$.

7.11.2 Applications of Energy Balance on Flow Systems

Example 7.15 Calculation of Enthalpy Flow Rate

The molar internal energy of helium, $\tilde{U}$, at 300 K and 1 atm is 3,800 J/mol, and the molar volume at the same temperature and pressure is $\tilde{V} = 24.63\text{ L/mol}$. Calculate the molar enthalpy of helium at the given temperature and pressure, and the rate at which enthalpy is transported by a stream of helium at 300 K and 1 atm with a molar flow rate of 5 kmol/min.

Solution:

Let us use Eq. (7.10c):

$$\tilde{H} = \tilde{U} + \left(P\tilde{V}\right) = 3{,}800\frac{\text{J}}{\text{mol}} + 1.0\text{ atm} \times \left(\frac{101{,}325\text{ Pa}}{\text{atm}}\right) \times 24.63\frac{\text{L}}{\text{mol}} \times \frac{1\text{ m}^3}{1{,}000\text{ L}}$$

$$= 3{,}800\frac{\text{J}}{\text{mol}} + 2{,}496\frac{\text{J}}{\text{mol}} = 6{,}296\frac{\text{J}}{\text{mol}}$$

$$\dot{H} = \dot{n}\tilde{H} = 5\,\frac{\text{kmol}}{\text{min}} \times \frac{1\text{ min}}{60\text{ s}} \times \frac{1{,}000\text{ mol}}{\text{kmol}} \times 6{,}296\frac{\text{J}}{\text{mol}}$$

$$= 5.247 \times 10^5\text{ W} = 0.5247\text{ MW}$$

NOTE #11

For liquids and solids, when we apply Eq. (7.10e): $\Delta\overline{H} = \Delta\overline{U} + \Delta\left(P\overline{V}\right)$, the $\left(P\overline{V}\right)$ change is quite negligible compared with either $\Delta\overline{H}$ *or* $\Delta\overline{U}$. Of course, for vapors and gases, the $\left(P\overline{V}\right)$ change will be quite significant. Unlike gases, liquids and solids are quite incompressible, hence:

$$\Delta\overline{H} \cong \Delta\overline{U} \rightarrow \overline{C}_P dT \cong \overline{C}_V dT \cong \overline{C} dT \tag{7.48}$$

See Eq. (7.6) for solids and liquids and Eq. (7.12) for gases.

Example 7.16 Crude Oil Pumping Requirement

Crude oil is pumped at a rate of 20 kg/s from a point 150 m below the earth's surface to a point 50 m above ground level. The diameter of the pipe carrying oil is 25 cm. Calculate the change in potential energy, kinetic energy, and how much work will be required to pump the oil. Assuming there is no transfer of heat and no change in internal energy (specific gravity of oil = 0.59).

Solution:

Crude oil will be sucked from a crude oil pool where the velocity is essentially zero to the final destination or delivery point where the velocity is nonzero and will be calculated via mass flow rate, density, and area of the pipe.

Let us use the steady-state simplified version (Eq. 7.35).

$$\Delta\left[\left(\overline{H} + gZ + v^2/2\right)\dot{m}\right] = \dot{Q} + \dot{W}_{\text{shaft}} \tag{A}$$

From Eq. (7.48), we have:

$$\Delta\overline{H} \cong \Delta\overline{U} = 0 \qquad \text{(no change in internal energy)}$$

$$\dot{Q} = 0 \qquad \text{(no transfer of heat)}$$

Equation (A) reduces to:

$$\Delta\left[\left(gZ + v^2/2\right)\dot{m}\right] = \dot{W}_{\text{shaft}} \qquad \rightarrow \left[\left(g\Delta Z + \Delta v^2/2\right) \times 20\,\frac{\text{kg}}{\text{s}}\right] = \dot{W}_{\text{shaft}} \tag{B}$$

$$g\Delta Z = 9.8\frac{\text{m}}{\text{s}^2}(Z_2 - Z_1) = 9.8\frac{\text{m}}{\text{s}^2}(50\text{ m} - (-150\text{ m})) = 9.8\frac{\text{m}}{\text{s}^2}(200\text{ m}) = 1960\,\frac{\text{m}^2}{\text{s}^2}$$

$$\frac{\Delta v^2}{2} = \left(\frac{v_2^2}{2} - \frac{v_1^2}{2}\right) = \frac{v_2^2}{2} \qquad (v_1 = 0 \text{ surface of crude oil pool})$$

$$v_2 = \frac{\dot{m}}{\rho A} = \frac{20\dfrac{\text{kg}}{\text{s}}}{590\,\dfrac{\text{kg}}{\text{m}^3} \times \dfrac{\pi}{4}(0.25\text{ m})^2} = 0.69\frac{\text{m}}{\text{s}}$$

$$\frac{v_2^2}{2} = \frac{(0.69)^2}{2} = 0.2384\,\frac{\text{m}^2}{\text{s}^2}$$

Substitute in Eq. (B), we have:

$$\left[\left(g\Delta Z + \Delta v^2/2\right) \times 20\frac{\text{kg}}{\text{s}}\right] = \left[\left(1960\frac{\text{m}^2}{\text{s}^2} + 0.2384\frac{\text{m}^2}{\text{s}^2}\right) \times 20\frac{\text{kg}}{\text{s}}\right]$$
$$= 3.92 \times 10^4\text{ W} = \dot{W}_{\text{shaft}}$$

$$\dot{W}_{\text{shaft}} = 3.92 \times 10^4\text{ W} = 39.2\text{ kW}$$

Example 7.17 Steam Turbine Power Generation

600 kg/h of steam drives a turbine. The steam enters the turbine at 45 atm and 500°C at a linear velocity of 50 m/s and leaves at appoint 6 m below the turbine inlet at atmospheric pressure, temperature of 250°C, and a velocity of 350 m/s. The overall heat loss from the turbine is estimated to be 10,000 kcal/h. Calculate the shaft work generated by the turbine, as shown in Figure 7.14.

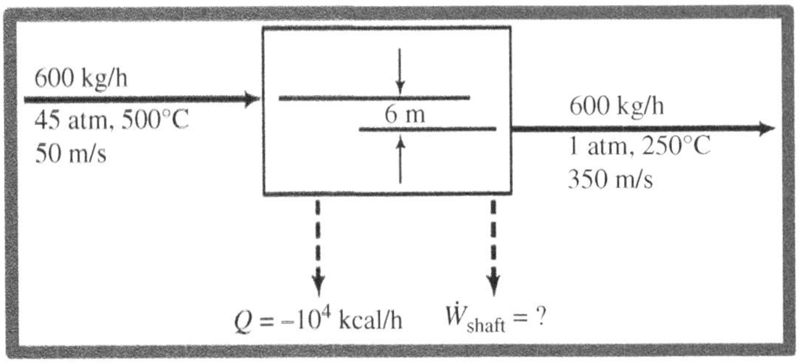

Figure 7.14 Generation of electric power using a steam turbine with known inlet and outlet steam conditions.

DATA: $\overline{H}_1$ (45 atm, 500°C) = $3439.5\frac{\text{kJ}}{\text{kg}}$ $\overline{H}_2$ (1 atm, 250°C) = $2974.5\frac{\text{kJ}}{\text{kg}}$

Solution:

Let us use the steady-state simplified version (Eq. 7.35).

$$\Delta\left[\left(\overline{H} + gZ + v^2/2\right)\dot{m}\right] = \dot{Q} + \dot{W}_{\text{shaft}} \tag{A}$$

$$\Delta\overline{H} = \left(\overline{H}_2 - \overline{H}_1\right) = (2974.5 - 3439.5)\frac{\text{kJ}}{\text{kg}} = -465\frac{\text{kJ}}{\text{kg}} = -465{,}000\frac{\text{J}}{\text{kg}}$$

$$g\Delta Z = 9.8\frac{\text{m}}{\text{s}^2}(Z_2 - Z_1) = 9.8\frac{\text{m}}{\text{s}^2}(-6\,\text{m} - (0\,\text{m})) = 9.8\frac{\text{m}}{\text{s}^2}(-6\,\text{m}) = -58.8\,\frac{\text{m}^2}{\text{s}^2}$$

$$\frac{\Delta v^2}{2} = \left(\frac{v_2^2}{2} - \frac{v_1^2}{2}\right) = \frac{(350)^2}{2} - \frac{(50)^2}{2} = 60{,}000\frac{\text{m}^2}{\text{s}^2}$$

$$\dot{m} = 600\frac{\text{kg}}{\text{h}} \times \frac{1\,\text{h}}{3600\,\text{s}} = 0.1667\frac{\text{kg}}{\text{s}}$$

$$\dot{Q} = -10{,}000\frac{\text{kcal}}{\text{h}} \times \frac{4.184\,\text{kJ}}{1\,\text{kcal}} \times \frac{1000\,\text{J}}{\text{kJ}} \times \frac{1\,\text{h}}{3600\,\text{s}} = -1.162 \times 10^4\,\text{W}$$

Substitute in Eq. (A), we have:

$$[(-465{,}000 - 58.8 + 60{,}000)0.1667] = -1.162 \times 10^4\,\text{W} + \dot{W}_{\text{shaft}}$$

$$\left[-6.7523 \times 10^4\right]\text{W} = -1.162 \times 10^4\,\text{W} + \dot{W}_{\text{shaft}}$$

$$\dot{W}_{\text{shaft}} = \left[-5.5901 \times 10^4\right]\text{W} = -55.9\,\text{kW}$$

The rate of shaft work is negative, indicating a generation of electric power.

7.11.3 MATLAB Code for Ideal Gas Pipe Flow with Heating

Example 7.18 Ideal Gas Exit Temperature in a Pipe Flow

Air gas enters a horizontal 10 cm-diameter pipe at a constant rate of 0.20 kg/s, with temperature $T_1 = 27$°C, and pressure $P_1 = 1.2$ atm. The pipe is wrapped with a 30-kW heating coil covered with an external thick layer of insulation. At the point where the gas is discharged, the pressure is $P_2 = 1.1$ atm. What is the gas temperature (T_2) at the discharge point, assuming it to be ideal with a MW of 29 and a C_p of 1 kJ/(kg °C)?

Solution:
Starting with Eq. (7.35), we have:

$$\Delta\left[\left(\overline{H} + gZ + v^2/2\right)\dot{m}\right] = \dot{Q} + \dot{W}_{\text{shaft}} \qquad \text{(A)}$$

Notice that:

$\Delta[(gZ)] = 0 \qquad$ [Horizontal line; $Z_1 = Z_2$]

$\dot{W}_{\text{shaft}}$ [No shaft work is imposed between inlet and exit point]

Equation (A) reduces to:

$$\Delta\left[\left(\overline{H}\right)\dot{m}\right] + \Delta\left[\left(v^2/2\right)\dot{m}\right] = \dot{m}\left[\overline{H}_2 - \overline{H}_1\right] + \frac{1}{2}\dot{m}\left[v_2^2 - v_1^2\right] = \dot{Q} = 30\frac{\text{kJ}}{\text{s}} \qquad \text{(B)}$$

From the Continuity equation, we have the following:

$$\dot{m} = \rho_1 v_1 A_1 = \rho_2 v_2 A_2 \qquad [A_1 = A_2;\text{pipe with constant cross} - \text{sectional area}] \qquad \text{(C)}$$

$$v_2 = \frac{\rho_1}{\rho_2}v_1 = \frac{\dfrac{P_1\text{MW}}{RT_1}}{\dfrac{P_2\text{MW}}{RT_2}}v_1 = \left(\frac{P_1}{P_2}\right)\left(\frac{T_2}{T_1}\right)v_1 = \left(\frac{1.2\text{ atm}}{1.1\text{ atm}}\right)\left(\frac{T_2\text{ K}}{300\text{ K}}\right)v_1 = 0.00364T_2v_1 \qquad \text{(D)}$$

Assuming air with a constant heat capacity, we have:

$$\overline{H}_1 = \overline{C}_pT_1 \qquad \overline{H}_2 = \overline{C}_pT_2$$

Substitute in Eq. (B), while replacing v_2 with its equivalence from Eq. (D), we have:

$$0.20\frac{\text{kg}}{\text{s}} \times 1.0\frac{\text{kJ}}{\text{kg}\cdot{}^\circ\text{C}}[T_2 - 300\text{ K}] + \frac{1}{2} \times 0.2\frac{\text{kg}}{\text{s}}\left[(0.00364T_2v_1)^2 - v_1^2\right] = 30\frac{\text{kJ}}{\text{s}} \qquad \text{(E)}$$

$$v_1 = \frac{0.2\dfrac{\text{kg}}{\text{s}}}{\rho_1 A_1} = \frac{0.2\dfrac{\text{kg}}{\text{s}}}{\dfrac{1.2\text{ atm} \times \left(\dfrac{101.325\text{ kJ/m}^3}{1\text{ atm}}\right) \times 29\dfrac{\text{kg}}{\text{kmol}}}{8.314\dfrac{\text{kJ}}{\text{kmol.K}} \times 300\text{ K}} \times \dfrac{\pi(0.1)^2}{4}\text{m}^2} = 18.013\frac{\text{m}}{\text{s}}$$

Substitute the value of v_1 in Eq. (E), we have:

$$0.2\frac{\text{kg}}{\text{s}} \times 1.0\frac{\text{kJ}}{\text{kg}\cdot{}^\circ\text{C}}[T_2 - 300\text{ K}] + \frac{0.2}{2}\frac{\text{kg}}{\text{s}}\left[(0.00364T_2 \times 18.013)^2 - 18.013^2\right] = 30\frac{\text{kJ}}{\text{s}} \qquad \text{(F)}$$

Simplify, we have:

$$0.2T_2 - 60 + 0.10 * \left[0.004298 * T_2^2 - 324.47\right] = 30 \qquad \text{(G)}$$

$$0.2T_2 - 122.447 + \left[0.00043 * T_2^2\right] = 0 \qquad \text{(H)}$$

Equation (H) is a quadrature with two possible roots for T_2.

Figure 7.15 shows MATLAB code (**IG_Flow_Heat.m**) for setting up the previous steps and also evaluating T_2 via finding the roots of the generated quadrature, shown in Eq. (H).

```
clear variables;
% Get user input for initial moles, initial volume, pressure, final temperature, and final volume
mdot = input('Enter the mass flow rate, (kg/s), mdot: ');
T1C = input('Enter the inlet temperature (oC), T1: ');
P1 = input('Enter the inlet pressure (atm), P1: ');
P2 = input('Enter the outlet pressure (atm), P2: ');
MW = input('Enter the gas molecular weight (kg/kmol), MW: ');
Qdot = input('Enter the heat duty (kJ/s), Qdot: ');
PID = input('Enter the pipe inside diameter (cm), PID: ');
% User selection for gas type (modify prompts as needed)
gasType = input('Select gas type (A: Mono-atomic, B: Di-atomic, C: Poly-atomic): ', 's');
% Calculate gas constant based on user selection
switch upper(gasType)
   case 'A'
      R = 8.314; % J/mol*K (Mono-atomic)
      Cv = (3/2)*R; % J/mol*K (Specific heat at constant volume)
      Cp = Cv + R; % J/mol*K (Specific heat at constant pressure)
      Cpm=Cp/MW;
   case 'B'
      R = 8.314; % J/mol*K (Di-atomic)
      Cv = (5/2)*R; % J/mol*K
      Cp = Cv + R; % J/mol*K
      Cpm=Cp/MW;
   case 'C'
      R = 8.314; % J/mol*K (Poly-atomic)
      Cv = (3.0)*R; % J/mol*K
      Cp = Cv + R; % J/mol*K
      Cpm=Cp/MW;
   otherwise
      error('Invalid gas type selection.');
end
T1=T1C+273.15;
row1=(P1*101.325*MW)/(8.314*300);
A1=0.25*pi*(PID/100)^2;
v1=mdot/(row1*A1);
v2coeff=(P1/P2)*(v1/T1);
T2coeff=(mdot/2)*(v2coeff)^2;
aterm=T2coeff;% Coefficient of T2^2
bterm=mdot*Cpm;% Coefficient of T2
cterm=-mdot*Cpm*T1-(mdot/2)*(v1)^2-Qdot;% Constant term;
% Find the roots using the 'roots' function
roots_of_polynomial = roots([aterm bterm cterm]);
% Display the roots
disp('Roots of the polynomial which give two T2 values, K:');
disp(roots_of_polynomial);
T2=max(roots_of_polynomial);
T2C=T2-273.15;
disp(['The positive root is set to T2: ', num2str(T2), ' K']);
disp(['T2 in oC: ', num2str(T2C), ' oC']);
```

Figure 7.15 MATLAB code (**IG_Flow_Heat.m**) for setting up the ideal gas pipe flow with a heating element and solving for the exit temperature, T_2.

MATLAB results are shown below:

```
>> IG_Flow_Heat
Enter the mass flow rate (kg/s), mdot: 0.2
Enter the inlet temperature (°C), T1: 27
Enter the inlet pressure (atm), P1: 1.2
Enter the outlet pressure (atm), P2: 1.1
Enter the gas molecular weight (kg/kmol), MW: 29
Enter the heat duty (kJ/s), Qdot: 30
Enter the pipe inside diameter (cm), PID: 10.0
Select gas type (A: Monoatomic, B: Diatomic, C: Polyatomic): B
Roots of the polynomial which give two T2 values, K:
  -818.1050
  349.8757
The positive root is set to T2: 349.8757 K
T2 in °C: 76.7257 °C
```

NOTE #12
If both roots are positive, then one should select the root that is more realistic.

7.12 Power and Power Units

Imagine you are lifting a box. The amount of effort (work) you exert in a certain amount of time determines how powerful you are in that moment. Power quantifies how quickly you can do or deliver work. More formally, power is the rate of work done or delivered. It represents the amount of work performed per unit time. The SI unit of power is the watt (W), which is equivalent to one joule per second (J/s).

Another common unit, especially for engines, is horsepower (hp).

$$1.0\,\text{hp} = 550\,\text{ft} \cdot \text{lb}_\text{f}/\text{s} = 0.746\,\text{kW} = 746\,\text{W} = 2686\,\text{kJ/h}$$

Examples of power and units are:

- A light bulb rated at 60 watts (W) consumes 60 J of energy every second.
- A car engine generating 200 horsepower (hp) produces 149.2 kW (1 kW = 1.34 hp) of power.
- A compressor using 5 kilowatts (kW) of power consumes 18 MJ (megajoules) of energy per hour (1 kW = 3.6 MJ/h).

Devices like car engines and turbines "do work" on their surroundings, producing power. They convert one form of energy (e.g., chemical, thermal) into mechanical work and ultimately into other forms of energy. The car engine converts fuel into mechanical energy, which is used to propel the vehicle forward. As the engine burns fuel and produces force, it performs work on the car's components, such as the transmission, drive shaft, and, ultimately, the wheels. This work results in the car moving and overcoming resistance forces such as friction and air resistance.

On the contrary, devices like compressors, pumps, and mixers are types of machinery that play crucial roles in various industrial processes.

1) **Compressors**:
   - **Function**: Compressors are designed to increase the pressure of a gas, typically air or other gases.
   - **Power Consumption**: Compressors consume energy in the form of electrical or mechanical power to operate. This power is used to compress the gas, resulting in an increase in its pressure.
2) **Pumps**:
   - **Function**: Pumps are devices used to transfer liquids from one place to another by increasing their kinetic energy.
   - **Power Consumption**: Pumps require energy input, either in the form of electrical power or mechanical power, to overcome resistance within the system and propel the fluid through pipes or other conduits.
3) **Mixers**:
   - **Function**: Mixers are employed to blend or combine different substances, ensuring uniformity in the composition of a mixture.
   - **Power Consumption:** Mixers use power, either electrical or mechanical, to drive rotating blades or other mixing elements. This mechanical energy is then transferred to the substances being mixed, causing them to blend thoroughly.

Many devices, like turbines, compressors, and pumps, are steady-flow devices. In these devices, power is often transmitted through a rotating shaft. The power transmitted by the shaft is simply referred to as the shaft power for that device.

Example 7.19 Shaft Power as a function of Torque (T) and Revolution per Minute (RPM)

A motor shaft exerts a torque (T) of 1200 Nm (Newton meters) on a pump. The shaft rotates at a speed of 150 RPM (revolutions per minute). What is the shaft work expressed in kW?

Solution:

$$\text{Power}, P\,[W] = T \cdot \omega$$

We need to convert the rotational speed from RPM to angular velocity (ω) in radians per second (rad/s) for the calculation. Here is the conversion formula:

$$\omega\,(\text{rad/s}) = \text{RPM}^{*}\ (2\pi)/60$$

$$\omega = 150\,\frac{\text{revolution}}{\text{min}} \times \left(\frac{2\pi\ \text{rad}}{1\ \text{revolution}}\right) \times \left(\frac{1\ \text{min}}{60\ \text{s}}\right) = 15.71\,\frac{\text{rad}}{\text{s}}$$

$$\text{Shaft Work}, \dot{W}_{\text{shaft}} = T \times \omega = 1200\ \text{N.m} \times 15.71\,\frac{\text{rad}}{\text{s}} = 18{,}852\,\frac{\text{J}}{\text{s}} = 18{,}852\ \text{W} = 18.85\ \text{kW}$$

End of Chapter Problems

CHW 7.1 What is the MAIN principle behind the importance of energy balance in chemical engineering?

A Energy can be freely created and destroyed
B Energy can only be transferred or converted, not created or destroyed (This adheres to the principle of conservation of energy)
C Chemical processes require a constant influx of new energy
D Energy can be lost due to friction

CHW 7.2 How does energy balance help in process design?

A It determines the color of the final product
B It dictates the specific chemicals used in the process
C It helps size equipment and optimize the process for efficient energy use
D It helps determine the extent of reaction not the heat of reaction

CHW 7.3 What is the role of energy balance in heat integration?

A It helps strategically place equipment to recover and reuse heat within the system
B It eliminates the need for heat exchangers altogether
C It focuses solely on maximizing the temperature of the process
D None of the above

CHW 7.4 How does achieving an energy balance impact the environment?

A It has no significant effect on environmental factors
B Processes with optimized energy balances typically have lower greenhouse gas emissions
C It helps minimize what is called thermal pollution (i.e., dumping of hot/cold streams)
D $b + c$

CHW 7.5 What does the internal energy ($\bar{U}$) of a material primarily represent?

A Only the molecular kinetic energy
B Only the intermolecular potential energy
C Only the intramolecular energy
D All of the above

CHW 7.6 Which of the following is NOT a type of intramolecular energy?

A Electronic
B Vibrational
C Repulsive
D Rotational

CHW 7.7 How is specific internal energy ($\bar{U}$) defined?

A $\bar{U}$ = Total Internal Energy/Total Moles
B $\bar{U}$ = Total Internal Energy/Total Mass
C $\bar{U}$ = Total Mass/Total Internal Energy
D $\bar{U}$ = Total Moles/Total Internal Energy

CHW 7.8 What does molar internal energy ($\tilde{U}$) represent?

A The internal energy per mole of a substance
B The internal energy per unit mass of a substance
C The total internal energy of a substance
D The internal energy per unit volume of a substance

CHW 7.9 In classical thermodynamics, how is molar internal energy ($\tilde{U}$) expressed for a single-phase pure substance?

A $\tilde{U} = f(P, T)$
B $\tilde{U} = f(P, \tilde{V})$
C $\tilde{U} = f(T, \tilde{V})$
D All of the above

CHW 7.10 What is the total kinetic energy attributed to a moving body?

A Microscopic molecular motion
B Macroscopic motion of the entire system
C Both a) and b)
D None of the above

CHW 7.11 What is the formula for specific kinetic energy?

A $\upsilon^{2/2}$
B $m\upsilon^2/2$
C $m\upsilon^2$
D υ

CHW 7.12 What does potential energy depend on?

A Mass
B Height
C Speed
D Both a) and b)

CHW 7.13 What is a system (control mass)?

A A region in space through which matter can flow
B A defined portion of matter with a fixed mass
C A conceptual boundary for analyzing energy interactions
D A box with defined inlet and outlet boundaries

CHW 7.14 Which of the following is true regarding the boundaries of a control volume (CV)?

A The boundaries are always physical
B The boundaries can be stationary only
C The boundaries can include a conceptual boundary
D The boundaries are immovable

CHW 7.15 What is the main focus of analysis in a system (control mass)?

A Flow of energy across boundaries
B Changes in the total energy of the system

C Mass flow in and out of the system
D The variation of volume within the system

CHW 7.16 Which of the following is NOT an example of a control volume (CV)?
A A sealed container of gas
B A pipe carrying moving fluid
C A turbine
D A pump

CHW 7.17 A beaker of hot coffee losing heat and water vapor to the surroundings is an example of a/an:
A Closed system
B Open system (Mass and energy transfer occur)
C Isolated system
D None of the above

CHW 7.18 A sealed flask containing hot coffee that cools down but does not lose any water vapor is a/an:
A Open system
B Closed system (Only energy transfer occurs)
C Isolated system
D All of the above

CHW 7.19 A container with perfectly insulated coffee that maintains its temperature for a very long time is:
A Open system
B Closed system
C Completely isolated system
D Nearly isolated system (Ideally no transfer of mass or energy, but practically some minimal energy transfer might occur)

CHW 7.20 In which system is the total mass ALWAYS constant?
A Open system
B Closed system (Mass cannot enter or leave)
C Semi-closed system
D Semi-open system

CHW 7.21 Which statement is TRUE about a completely isolated system?
A Energy can freely enter and leave the system
B Mass can be exchanged with the surroundings
C Neither mass nor energy can be transferred across the boundary
D The system experiences a continuous change in temperature

CHW 7.22 Which of the following best describes heat?
A Energy stored within a system
B Energy transfer driven by a temperature difference
C Energy transfer driven by a force acting through a distance
D Energy transfer driven by a voltage difference

CHW 7.23 Expansion/contraction (*PdV/Vdp*) work involves:

A Energy storage within a system
B Energy transfer driven by a temperature difference
C Energy transfer driven by a force acting through a distance
D Energy transfer driven by a voltage difference

CHW 7.24 Water flows into a process unit through a 4-cm ID pipe at a rate of 4.8 m^3/h. Calculate the kinetic energy for this stream in J/s (W).

CHW 7.25 A crude oil is pumped at a rate of 20 kg/s from a point 230 m below earth's surface to a point 15 m above ground level. Calculate the rate of increase of potential energy.

CHW 7.26 Calculate the final temperature (T_2) of 1 kg of water at room temperature subject to 280 kJ and the same for 1 kg of iron.

DATA: Specific heat for iron = 0.451 J/(g·°C) Specific heat for water = 4.184 J/(g·°C)

CHW 7.27 Using Appendix H | H.2 MATLAB code (**SpecHeat.m**), calculate the heat duty, Q_{12}, for heating 1 kg of each of the following liquids or solids for the indicated temperature range.

A Liquid acetone ($T_1 = -50°C$ and $T_2 = 50°C$)
B Solid aluminum bronze ($T_1 = -50°C$ and $T_2 = 800°C$)
C Liquid asphalt ($T_1 = 50°C$ and $T_2 = 300°C$)
D Solid Beeswax ($T_1 = -50°C$ and $T_2 = 60°C$)
E Hard brick ($T_1 = -50°C$ and $T_2 = 1500°C$)
F Olive oil ($T_1 = -5°C$ and $T_2 = 150°C$)

CHW 7.28 Heating a poly-atomic ideal gas

When 7000 J of heat is slowly added to 10 mol of an ideal poly-atomic gas, its temperature rises by 20°C. What is the work done by the gas?

CHW 7.29 Cooling a poly-atomic ideal gas

When −7000 J of heat is slowly removed from 10 mol of an ideal polyatomic gas, its temperature drops by 20°C. What is the work done on the gas?

CHW 7.30 Cooling an ideal gas in a cylinder at constant volume (iso-choric or iso-metric) process

Ten moles of a di-atomic ideal gas are contained in a cylinder fitted with a piston. The initial gas temperature is 125°C, and the initial pressure is 10 atm. The cylinder is placed in a cooling water bath with the piston held in a fixed position. Heat is transferred from the gas, which reaches a lower pressure and temperature, as shown in Figure CHW 7.30. The final reached pressure, P_2, is 8 atm.

Using both manual calculations and MATLAB **IG_IsoChoric.m** code, carry out the following steps:

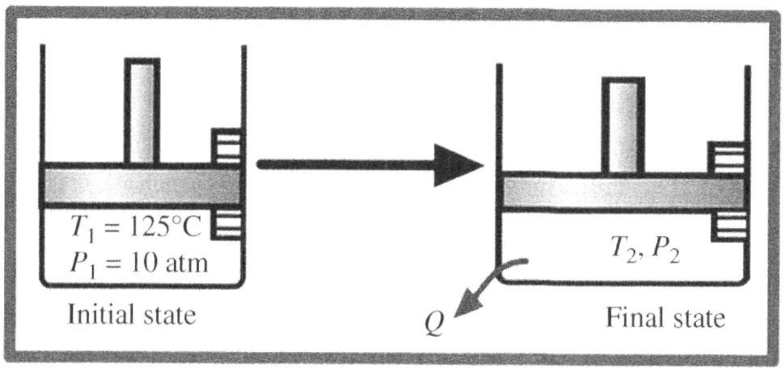

Figure CHW 7.30 Constant-volume cooling process of a 10-mol ideal diatomic gas in a piston-cylinder setup. ($T_2 < T_1$ and $P_2 < P_1$).

A Find T_2
B Calculate ΔU
C What is W_{12}?
D Find Q_{12}
E Calculate ΔH

CHW 7.31 Cooling an ideal gas in a cylinder at constant pressure (iso-baric) process

Ten moles of a diatomic ideal gas are contained in a cylinder fitted with a freely moving piston. The initial gas temperature is 125°C, and the initial pressure is 1 atm.

The cylinder is placed in a cooling water bath. Heat is transferred from the gas, which causes both a pressure and temperature decrease; however, as the piston slowly contracts downward in response to the cooling effect, it keeps a constant pressure process until reaching the final volume (V_2) and temperature (T_2), as shown in Figure CHW 7.31. The final volume, V_2, is reported as 300 L.

Using both manual calculations and MATLAB **IG_IsoBaric.m** code, carry out the following steps:

A Calculate T_2
B Calculate ΔU
C Calculate W_{12}
D Calculate Q_{12}.
E Calculate ΔH

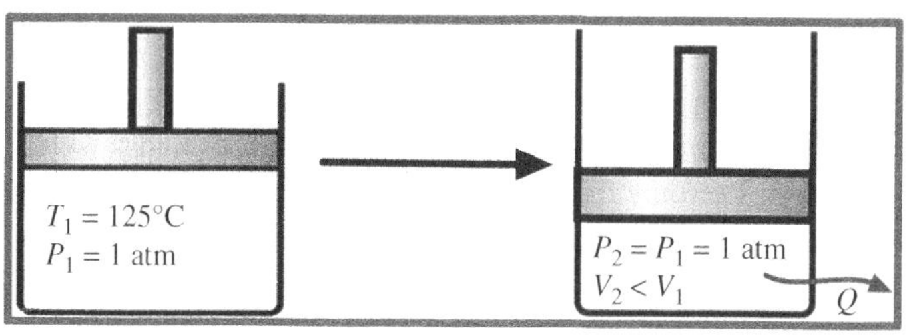

Figure CHW 7.31 Constant-pressure cooling process of an ideal di-atomic gas in a piston-cylinder setup.

CHW 7.32 Compressing an ideal gas in a cylinder at a constant temperature (iso-thermal) process

Ten moles of a diatomic ideal gas are contained in a cylinder fitted with a freely moving piston. The initial gas temperature is 125°C, and the initial pressure is 10 atm. The cylinder is placed in a cooling water bath. An external dead weight is differentially increased which will slightly compress the gas and slightly increase its temperature. Heat is released in proportion with the differential increase in temperature to level off the increase in temperature and maintain the T constant.

The cycle of continuously increasing the dead weight and allowing heat release after that will terminate once the desired level of compression is achieved. We can then report the heat duty, Q_{12}, the final volume, or the final pressure. The final pressure is reported here which is equal to $P_2 = 16$ atm, as shown in Figure CHW 7.32.

Using both manual calculations and MATLAB **IG_IsoTherm.m** code, carry out the following steps:

A Calculate V_2

B Calculate the work done on the system, W_{12}

C Calculate ΔU and ΔH

D Calculate the heat duty, Q_{12}

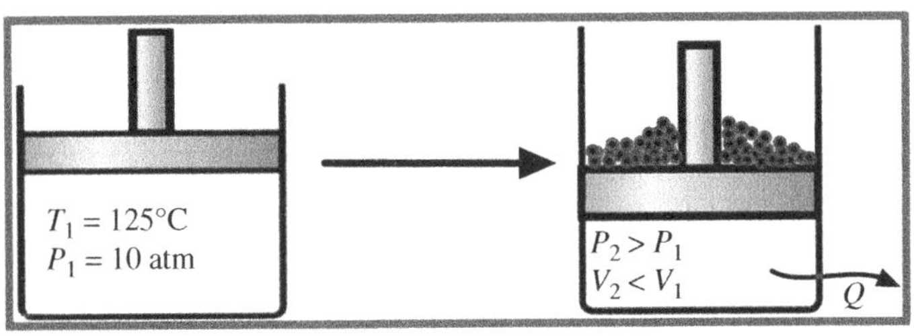

Figure CHW 7.32 Isothermal compression of an ideal gas from V_1 to V_2 ($V_2 < V_1$).

CHW 7.33 Compression of an ideal gas in a cylinder under an isentropic process

Ten moles of a di-atomic ideal gas are contained in a cylinder fitted with a freely moving piston. The initial gas temperature is 100°C, and the pressure is 2 atm. The piston-cylinder setup is well insulated from surroundings such that we can assume an adiabatic condition (i.e., $Q_{12} = 0$).

The piston is allowed to slowly compress downward in response to an external incremental dead weight mechanism, where the external pressure (weight/area) is slightly incremented until reaching the final volume, V_2, which is $V_2 = (1/2)$ V_1. Notice here that all three measurable variables do change with compression: P, T, and V. The only fixed parameters are: n and R. The process that is both adiabatic and reversible (infinitesimal driving force for change in both directions) is called isentropic, which means a constant-entropy process.

Find V_2, T_2, P_2, ΔU, Q_{12}, W_{12}, and ΔH

CHW 7.34 Heat Duty Calculations

Calculate the amount of heat to completely evaporate 1 kg of liquid water from room temperature and pressure (1 bar) until it becomes a completely saturated vapor at 1 bar, using MATLAB code **SteamProps.m** code

CHW 7.35 Fluid Exit Velocity from a Tank

Water is flowing into the top of a tank at a rate equal to the exit flow rate. If the tank holds a liquid inside such that the height above the exit hole is 1.0 m. Assume negligible energy changes except for potential and kinetic energies. The diameter of the hole is much smaller than that of the tank $((A_{\text{Hole}}/A_{\text{Tank}}) \approx 0)$. Calculate the exit velocity assuming no friction.

CHW 7.36 The Exit Mixture Temperature

To determine the flow rate of hot coal/oil slurry in a pipeline, a method involves introducing a small side stream of cooler oil and observing the resultant temperature change downstream. Initially, the slurry temperature is 160°C with a specific heat of 3.0 J/(g.°C). In the absence of the side stream, the downstream temperature after mixing exists at 158°C. However, with a side cold stream at 20°C flowing at 0.4 kg/s, the temperature at the mixing point drops to 155°C. The side cold stream has a specific heat of 2.6 J/(g.°C). Determine the mass flow rate of the slurry.

CHW 7.37 Ideal Gas Exit Temperature in a Pipe Flow

A gas enters a horizontal 10 cm-diameter pipe at a constant rate of 0.30 kg/s, with temperature of 27°C, and pressure of 1.3 atm. aThe pipe is wrapped with a 50 kW heating coil covered with a thick layer of insulation. At the point where the gas is discharged, the pressure is 1.1 atm. What is the gas temperature at the discharge point, assuming it to be ideal with a MW of 29 and a C_p of 1 kJ/(kg.°C)?

CHW 7.38 A motor shaft exerts a torque (T) of 20 N m (Newton meters) on a pump. The shaft rotates at a speed of 1500 RPM (revolutions per minute). What is the shaft work, expressed in W?

Appendix A

Units Conversion Tables

Table A.1 Length conversion from metric to imperial system.

Metric Unit	Multiplier	Imperial Equivalent
Millimeter (mm)	0.0393701	0.03937 in./mm
Centimeter (cm)	0.393701	0.393701 in./cm
Decimeter (dm)	3.93701	3.93701 in./dm
Meter (m)	3.28084	3.28084 ft/m
Kilometer (km)	0.621371	0.621371 miles/km

Table A.2 Length conversion from imperial to metric system.

Imperial Unit	Multiplier	Metric Equivalent
Inch (in)	25.4	25.4 mm/in.
Foot (ft)	0.3048	0.3048 m/ft
Yard (yd)	0.9144	0.9144 m/yd
Mile (mi)	1.60934	1.60934 km/mile

Table A.3 Length conversion factors within the imperial system.

Unit	Multiplier to Convert to Base Unit (Inch)	Base Unit Equivalent
Inch (in)	1	1 in. = 1 in. (Base Unit)
Foot (ft)	12	1 ft = 12 in.
Yard (yd)	36	1 yd = 36 in. = 3 ft
Mile (mi)	63,360	1 mi = 63,360 in. = 5280 ft = 1760 yd

MATLAB-based Computations of Chemical Engineering Principles, First Edition. Kamal I. M. Al-Malah.

Companion website: www.wiley.com/go/al-malah/matlabbasedcomputations

Table A.4 Mass conversion between imperial and metric systems.

Imperial Unit	Multiplier	Metric Equivalent
Ounce (oz)	28.3505	28.3505 g/oz
Pound (lbm)	0.453592	0.453592 kg/lbm
Ton (short)	907.185	907.185 kg/short ton
Ton (long)	1016.05	1016.05 kg/long ton

Table A.5 Volume conversion between metric and imperial systems.

Metric Unit	Multiplier	Imperial Equivalent
Cubic meter (m^3)	264.172	264.172 US gallon
Cubic meter (m^3)	219.969	219.969 UK gallon/m^3
Cubic meter (m^3)	35.3147	35.3147 ft^3/m^3
1 L	0.035315	0.035315 ft^3/L

Table A.6 Volume conversion between imperial and metric systems.

Imperial Unit	Multiplier	Metric Equivalent
Gallon (US)	3.78541	3.78541 L/US gal
Gallon (UK)	4.54609	4.54609 L/UK gal
1 ft^3	0.028317	0.028317 m^3/1 ft^3

Table A.7 Power inter-conversion from and to the metric system.

Metric Unit	Multiplier	Imperial Equivalent
Kilowatt (kW)	1.34102	1.34102 horsepower (hp)
Megawatt (MW)	1341.02	1341.02 horsepower (hp)
Imperial Unit	**Multiplier**	**Metric Equivalent**
Horsepower (hp)	745.7	745.7 watts (W)
British thermal unit per second (Btu/s)	1055.06	1055.06 watts (W)

Table A.8 Pressure conversion from metric into imperial system.

Metric Unit	Imperial Unit	Conversion Factor
Pascal (Pa)	pound-force per square inch (psi)	1 Pa = 0.000145 psi
Kilopascal (kPa)	psi	1 kPa = 0.145 psi
Megapascal (MPa)	psi	1 MPa = 145 psi
Bar	psi	1 bar = 14.5 psi
Atmosphere (atm)	psi	1 atm = 14.7 psi
Torr	1 mmHg	1 Torr = 1 mmHg
Torr (mmHg)	Pa atm psi	1 Torr = 133.3224 Pa 1 Torr = 9.80665×10^{-3} atm 1 Torr = 0.0193366 psi

Appendix B

Liquid and Solid Densities + Conversion of Mass/Mole Basis

B.1 Density of Liquids

Densities of common liquids are shown in Table B.1.

Reference: https://www.engineeringtoolbox.com/liquids-densities-d_743.html.

B.2 Density of Solids

Densities of common solids are shown in Table B.2.

Reference: https://www.engineeringtoolbox.com/density-solids-d_1265.html.

B.3 MATLAB Code for Searching Tables B.1 and B.2 and Reporting Solid/Liquid Density

Figure B.1 shows MATLAB code called **SLD_LIQ_DENS.m** file which searches Tables B.1 and B.2 for a user-input search keyword, which represents the name of liquid or solid material. The list is populated in light of the available occurrences of the search key word. Then the user has to select the row number of the material of interest and MATLAB will provide the density both in SI system as kg/m^3 and in AE system as lb_m/ft^3.

To my dear readers,
When copying MATLAB code or text, please right-click and attempt both the "Copy with Formatting" and "Copy" options from the context menu. This approach is effective within Adobe Acrobat. If neither option works, kindly carefully review any MATLAB error messages and compare the pasted text to the original to identify and correct any discrepancies. Alternatively, a soft copy of the MATLAB code is available in the Wiley book companion site of this book for your convenience.

MATLAB-based Computations of Chemical Engineering Principles, First Edition. Kamal I. M. Al-Malah.

Companion website: www.wiley.com/go/al-malah/matlabbasedcomputations

Table B.1 Densities of common liquids.

Liquid	T (°C)	ρ (kg/m³)	Liquid	T (°C)	ρ (kg/m³)
2,3-Butandione	18	981	Isopropyl Alcohol	20	785
2-Butanone	25	800	Isopropyl Myristate	20	853
Acetaldehyde	18	783	Isopropylbenzene hydroperoxide	20	1030
Acetic Acid	25	1049	Kapok seed oil	15	926
Acetone	25	784.6	Kerosene	15.6	820.1
Acetonitrile	20	783	Linolenic Acid	25	897
Acrolein	20	840	Linseed oil	25	924
Acrolonitrile	25	801	*m*-Xylene	20	864
Alcohol, ethyl (ethanol)	25	785.1	Machine oil	20	910
Alcohol, methyl (methanol)	25	786.5	Mango seed oil	15	912
Alcohol, propyl	25	800	Menhaden oil	15	920
Almond kernel oil	25	910	Mercury	25	13,595
Alyllamine	20	758	Methane	−164	465
Ammonia (aqua)	25	823.5	Methanol	20	791
Aniline	25	1019	Methyl Ethyl Ketone	20	805
Anisole	20	994	Methyl Isoamyl Ketone	20	888
Apricot kernel oil	25	910	Methyl Isobutyl Ketone	20	801
Argan seed oil	20	912	Methyl *n*-Propyl Ketone	20	808
Automobile oils	15	880–940	Methyl *t*-Butyl Ether	20	741
Avacado pulp oil	25	912	Methylamine	25	656
Babassu palm oil	25	914	Milk	15	1020–1050
Beef tallow (land animals)	25	902	Moringa peregrina seed oil	24	903
Beer (varies)	10	1010	Mustard seed oil	20	913
Benzaldehyde	25	1040	Mutton tallow	15	946
Benzene	25	873.8	*N,N*-Dimethylformamide	20	949
Benzil	15	1230	*n*-Butyl Acetate	20	880
Blackcurrant oil	20	923	*n*-Butyl Alcohol (Butanol)	20	810
Borneo tallow	100	855	*n*-Butylhloride	20	886
Brine	15	1230	*N*-Methylpyrrolidone	20	1030
Bromine	25	3120	Naphtha	15	665
Butanal	20	802	Naphtha, wood	25	960
Butane	25	599	Napthalene	25	820
Butterfat (land animals)	15	934	Neem oil	30	912
Butyric Acid	20	959	Niger seed oil	15	924
Cameline oil	15	924	Nitric acid	0	1560
Canola rapeseed oil	20	915	*o*-Dichlorobenzene	20	1306
Caproic acid	25	921	*o*-Xylene	20	880

Table B.1 (Continued)

Liquid	T (°C)	ρ (kg/m³)	Liquid	T (°C)	ρ (kg/m³)
Carbolic acid (phenol)	15	956	Oat been oil	25	904
Carbon disulfide	25	1261	Oat oil	25	917
Carbon tetrachloride	25	1584	Ocimene	25	798
Carene	25	857	Octane	15	698.6
Cashew nut oil	15	914	Oil of resin	20	940
Castor oil	25	952	Oil of turpentine	20	870
Cherry kernel oil	25	918	Oil, lubricating	20	900
Chicken fat	15	918	Oiticica oil	20	972
Chinese vegetable tallow	25	887	Olive oil	20	911
Chloride	25	1560	Oxygen (liquid)	−183	1140
Chlorobenzene	20	1106	*p*-Xylene	20	861
Chlorodifluoromethane refrigerant R-22	25	1194	Palm kernel oil	15	922
Chloroform	20	1489	Palm oil	15	914
Chloroform	25	1465	Palm olein	40	910
Citric acid, 50% aqueous solution	15	1220	Palm stearin	60	884
Cocoa butter	25	974	Palmitic Acid	25	851
Coconut oil	40	930	Paraffin		800
Cod liver oil	15	924	Paraldehyde	20	994
Cohune nut oil	25	914	Peanut oil	20	914
Corn oil	20	919	Pentane	20	626
Coriander seed oil	25	908	Pentane	25	625
Cotton seed oil	20	920	Perchlor ethylene	20	1620
Crambe oil	25	906	Perilla oil	25	924
Creosote	15	1067	Petrol, natural	15.6°C	711
Cresol	25	1024	Petrol, Vehicle	15.6°C	737
Crude oil, 32.6o API	15.6	862	Petroleum Ether	20	640
Crude oil, 35.6o API	15.6	847	Phenol (carbolic acid)	25	1072
Crude oil, 40o API	15.6	825	Phosgene	0	1378
Crude oil, 48o API	15.6	790	Phulwara butter	100	862
Crude oil, California	15.6	915	Phytadiene	25	823
Crude oil, Mexican	15.6	973	Pine nut oil	15	919
Crude oil, Texas	15.6	873	Pinene	25	857
Cumene	25	860	Poppy seed oil	25	916
Cyclohexane	20	779	Pork lard	20	898
Cyclopentane	20	745	Propanal	25	866
Decane	25	726.3	Propane	−40	493.5
Dichlorodifluoromethane refrigerant R-12	25	1311	Propane, R-290	25	494
Dichloromethane	20	1326	Propanol	25	804

(*Continued*)

Table B.1 (Continued)

Liquid	T (°C)	ρ (kg/m³)	Liquid	T (°C)	ρ (kg/m³)
Dichloromethane	20	1326	Propylamine	20	717
Diesel fuel oil 20–60	15	820–950	Propylene	25	514.4
Diethanolamine	20	1097	Propylene glycol	25	1036
Diethyl ether	20	714	Propylenearbonate	20	1201
Diethyl ether	20	714	Pyridine	25	979
Diethylene glycol	15	1120	Pyrrole	25	966
Diethylene glycol diethyl ether	20	906	Rape seed oil	20	920
Diisopropyl ether	25	719	Resorcinol	25	1269
Dimethyl acetamide	20	942	Rice bran oil	25	916
Dimethyl sulfate	20	1332	Rosin oil	15	980
Dimethyl sulfide	20	848	Salmon oil	15	924
Dimethyl sulfoxide	20	1100	Sardine oil	25	915
Dodecane	25	754.6	Sea water	25	1025
Ethane	−89	570	Seaflower seed oil	15	924
Ether	25	713.5	Shark liver oil	25	917
Ethyl Acetate	20	901	Sheanut butter	100	863
Ethyl Alcohol (Ethanol, pure alcohol, grain alcohol or drinking alcohol)	20	789	Silane	25	718
Ethyl Ether	20	713	Silicone oil	25	965–980
Ethylamine	16	681	Sodium Hydroxide (caustic soda)	15	1250
Ethylene Dichloride	20	1253	Sorbaldehyde	25	895
Ethylene glycol	25	1097	Soybean oil	20	920
Euphorbia lagascae seed oil	25	952	Stearic Acid	25	891
Formaldehyde	45	812	Stillinga seed kernel oil	25	937
Formic acid 10% concentration	20	1025	Styrene	25	903
Formic acid 80% concentration	20	1221	Sugar solution 68 brix	15	1338
Fuel oil	15.6	890	Sulfur dichloride		1620
Furan	25	1416	Sulfuric Acid 95% concentration	20	1839
Furforal	25	1155	Sulfurous acid	−20	1490
Gas oils	15.6°C	890	Sulfuryl chloride		1680
Gasoline, natural	15.6°C	711	Sunflower seed oil	20	919
Gasoline, Vehicle	15.6	737	Tall oil	25	969
Glucose	15.6	1350–440	Terpinene	25	847
Glycerin	25	1259	Tetrahydrofuran	20	888
Glycerol	25	1126	Toluene	20	867
Grape seed oil	20	923	Trichlor ethylene	20	1470

Table B.1 (Continued)

Liquid	T (°C)	ρ (kg/m^3)	Liquid	T (°C)	ρ (kg/m^3)
Hazelnut oil	25	909	Trichlorofluoromethane refrigerant R-11	25	1476
Heating oil	20	920	Triethylamine	20	728
Hempseed oil	25	921	Trifluoroacetic Acid	20	1489
Heptane	25	679.5	Tung oil	25	912
Herring oil	20	914	Turpentine	25	868.2
Hexane	25	654.8	Ucuhuba butter oil	100	870
Hexanol	25	811	Vernonia seed oil	30	901
Hexene	25	671	Walnut oil	25	921
Hexylamine	20	766	Water – pure	4	1000
Hydrazine	25	795	Water – sea	25	1022
Illipe mowrah butter	100	862	Water, heavy	11.6	1105
Ionene	25	932	Whale oil	15	925
Iso-Octane	20	692	Wheatgerm oil	25	926
Isobutyl Alcohol	20	802			

Table B.2 Densities of common solids.

Solid	Density (10^3 kg/m^3)	Solid	Density (10^3 kg/m^3)
ABS – copolymer of acrylonitrile, butadiene and styrene	1.06	Hardboard	1
Acetals	1.42	Hematite	4.9–5.3
Agate	2.5–2.7	Hornblende	3
Acrylic	1.19	Ice	0.917
Agate	2.6	Iron, cast	7.0–7.4
Alabaster carbonate	2.7–2.8	Iodine	4.95
Alabaster sulfate	2.3	Iridium	22.5
Alum, lumpy	0.881	Ivory	1.8–1.9
Alum, pulverized	0.752	Kaolin	2.6
Alumina (aluminium oxide)	3.95–4.1	Lead	11.35
Aluminum	2.7	Leather, dry	0.86
Aluminum Bronze	7.7	Lime, slaked	1.35
Albite	2.6–2.65	Limestone	2.7–2.8
Amber	1.06–1.1	Linoleum	1.2
Amphiboles	2.9–3.2	Lithium	0.53
Andesite, solid	2.77	Magnesia	3.2–3.6
Anorthite	2.74–2.76	Magnesium	1.74
Antimony, cast	6.7	Magnetite	4.9–5.2
Arsenic	4.7	Malachite	3.7–4.1
Artificial Wool	1.5	Manganese	7.43
Asbestos	2.0–2.8	Marble	2.6–2.8
Asbestos, shredded	0.35	Meerschaum	1–1.3

(*Continued*)

Table B.2 (Continued)

Solid	Density (10^3 kg/m^3)	Solid	Density (10^3 kg/m^3)
Asbestos, solid	2.45	Mica	2.6–3.2
Ashes	0.65	Mineral wool quilt	0.05
Asphalt, compacted	2.36	Molybdenum	10.2
Asphalt, crushed	0.72	Muscovite	2.8–3
Bakelite	1.36	Nickel	8.9
Baking powder	0.72	Nylon 6	1.12–1.17
Balsa Wood	0.13	Nylon 6,6	1.13–1.15
Barite, crushed	2.89	Oak	0.72
Barium	3.78	Ochre	3.5
Bark, wood refuse	0.24	Opal	2.2
Barytes	4.5	Osmium	22.48
Basalt	2.4–3.1	Palladium	12
Bauxite, crushed	1.28	Paper	0.7–1.15
Beeswax	0.96	Paraffin	0.9
Beryl	2.7	Peat blocks	0.85
Beryllia	3	Phenolic cast resin	1.24–1.32
Beryllium	1.85	Phosphorbronce	8.8
Biotite	2.7–3.1	Phosphorus	1.82
Bismuth	9.8	Pinchbeck	8.65
Boiler scale	2.5	Pitch	1.1
Bone	1.7–2.0	Pit coal	1.35
Bone, pulverized	0.88	Plaster board	0.8
Borax, fine	0.85	Platinum	21.5
Brasses	8.47–8.75	Plywood	0.54
Bronzes	8.74–8.89	Polyacrylonitrile	1.16–1.18
Brown iron ore	5.1	Polyamides	1.15–1.25
Brick	1.4–2.4	PC – poly carbonate	1.2
Brick, fire	2.3	PBT – poly butylene terephthalate	1.35
Brick, hard	2	LDPE – low-density poly ethylene	0.91
Brick, pressed	2.2	HDPE – (PEH) – high-density poly ethylene	0.96
Brickwork in cement	1.8	PET – poly ethylene terephthalate	1.35
Brickwork in mortar	1.6	PMMA – poly methyl methacrylate	1.2
Butter	0.86–0.87	POM – poly oxy methylene	1.4
Cadmium	8.64	PP – poly propylene	0.91–0.94
Calamine	4.1–4.5	PPO – poly penylene ether	1.1
Calcium	1.55	PS – poly styrene	1.03

Table B.2 (Continued)

Solid	Density (10^3 kg/m^3)	Solid	Density (10^3 kg/m^3)
Calcspar	2.6–2.8	PTFE – poly tetra fluoro ethylene, Teflon	2.28–2.30
Camphor	1	PU – poly urethane foam	0.03
Carbon	3.51	PVDF – poly vinylidene fluoride	1.76
Caoutchouc	0.9–1	Porcelain	2.3–2.5
Cardboard	0.7	Porphyry	2.6–2.9
Cast Iron	7.2	Potassium	0.86
Celluloid	1.4	Pressed wood, pulp board	0.19
Cellulose, cotton, wood pulp, regenerated	1.48–1.53	PVC – poly vinyl chloride	1.39–1.42
Cellulose acetat, moulded	1.22–1.34	Pyrex	2.25
Cellulose acetat, sheet	1.28–1.32	Pyrite	4.9–5.1
Cellulose nitrate, celluloid	1.35–1.4	Quartz	2.65
Chlorinated polyether	1.4	Radium	5
Cement, set	2.7–3	Red lead	8.6–9.1
Cement, Portland	1.5	Red metal	8.8
Cerium	6.77	Resin	1.07
Chalk	1.9–2.8	Rhenium	21.4
Charcoal, oak	0.6	Rhodium	12.3
Charcoal, pine	0.3–0.4	Rock salt	2.2
Chromium	7.1	Rock wool	0.22–0.39
Chrom oxide	5.21	Rosin	1.07
Cinnabar	8.1	Rubber, hard	1.2
Clay	1.8–2.6	Rubber, soft commercial	1.1
Coal, anthracite	1.4–1.8	Rubber, pure gum	0.91–0.93
Coal, bituminous	1.2–1.5	Rubber, foam	0.07
Cobalt	8.8	Rubidium	1.52
Cocoa, butter	0.9	Sand, dry	1.4–1.6
Coke	1–1.7	Sandstone	2.1–2.4
Concrete, lightweight	0.45–1.0	Sapphire	3.98
Concrete, medium	1.3–1.7	Selenium	4.4
Concrete, dense	2.0–2.4	Serpentine	2.5–2.65
Constantan	8.89	Silica, fused transparent	2.2
Copal	1–1.15	Silica, translucent	2.1
Copper	8.79	Silicium carbide	3.16
Cork	0.2–0.25	Silicon	2.33
Cork, linoleum	0.55	Silver	10.5
Corundum	4	Slag	2–3.9

(Continued)

Table B.2 (Continued)

Solid	Density (10^3 kg/m^3)	Solid	Density (10^3 kg/m^3)
Cotton	0.08	Slate	2.6–3.3
CPVC – Chlorinated poly vinyl chloride	1.6	Snow	0.1
Lead Crystal	3.1	Soapstone	2.6–2.8
Diamond	3–3.5	Sodium	0.98
Dolomite	2.8	Soil	2.05
Duralium	2.8	Solder	8.7–9.4
Earth, loose	1.2	Soot	1.6–1.7
Earth, rammed	1.6	Spermaceti	0.95
Ebonite	1.15	Starch	1.5
Emery	4	Steatite	2.6–2.7
Electron	1.8	Steel	7.82
Epidote	3.2–3.5	Stone	2.3–2.8
Epoxy cast resin	1.11–1.4	Sulfur, cryst.	2
Epoxy glass fibre	1.5	Sugar	1.6
Expanded polystyrene	0.015–0.03	Talc	2.7–2.8
Feldspar	2.6–2.8	Tallow, beef	0.95
Fire brick	1.8–2.2	Tallow, mutton	0.95
Flint	2.6	Tantalum	16.6
Fluorite	3.2	Tar	1.05
Galena	7.3–7.6	Teflon	2.2
Gallium	5.9	Tellurium	6.25
Gamboge	1.2	Thoria	4.16
Garnet	3.2–4.3	Thorium	11.7
Gas carbon	1.9	Tin	7.28
Gelatin	1.3	Titanium	4.5
Germanium	5.32	Topaz	3.5–3.6
Glass, common	2.4–2.8	Tourmaline	3–3.2
Glass, flint	2.9–5.9	Tungsten	19.2
Glass, Pyrex	2.21	Tungsten carbide	14.0–15.0
Glass-wool	0.025	Uranium	19.1
Glue	1.3	Urethane foam (Urea formaldehyde foam)	0.08
Gneiss	2.69	Vanadium	6.1
Gold	19.29	Vermiculite	0.12
Granite	2.6–2.8	Wax, sealing	1.8
Graphite	2.3–2.7	White metal	7.5–10
Gum arabic	1.3–1.4	Wood wool slab	0.5–0.8
Gypsum	2.3	Zinc	7.12

```
clear variables;
try
   % Attempt to read the Excel file first
   SLD_Dens = readtable('Density.xlsx','Sheet','SldDens','Format','auto');
   LIQ_Dens = readtable('Density.xlsx','Sheet','LiqDens','Format','auto');
   catch ME
   % If reading the Excel file fails, try loading the MAT file
   disp('Failed to read Excel file; it will continue with the saved MATLAB file.');
   try
     load('Density.mat', 'SLD_Dens');
     load('Density.mat', 'LIQ_Dens');
     disp('Loaded Density from MAT file.');
   catch ME2
     % If loading the MAT file also fails, display an error message
     disp('Failed to read Density MAT file.');
   end
end
try

% Prompt the user to enter a phase
prompt = 'Enter s or S for solid material, l or L for liquid material: ';
phase = input(prompt, 's');
% Check the input and display corresponding message
if strcmpi(phase, 's')
disp('You selected the solid material.');
searchKeyword = lower(input('Enter a search keyword: ', 's'));
SLD_Dens.Solid = lower(SLD_Dens.Solid);
  % Find rows containing the search keyword in the 'Compound' column
matchRows = find(contains(SLD_Dens.Solid, searchKeyword));

if isempty(matchRows)
     disp('No entries found matching the search keyword.');
     return;
else
     % Display matching compounds and row numbers (looping approach)
     disp('Matching entries:');
     for i = 1:length(matchRows)
     disp(['  - ', SLD_Dens.Solid{matchRows(i)}, ' (Row: ', num2str(matchRows(i)), ')']);
     end
   end
rowNumber = input('Enter the row number for the compound: ');
  % Check for NAN input
  if isnan(rowNumber)
     disp('Terminating program');
     return;
   end
```

Figure B.1 SLD_LIQ_DENS.m file for searching and presenting Tables B.1 and B.2 for solid and liquid density. The density is expressed both in SI system as kg/m^3 and in AE system as lb$_m$/ft^3.

```
    selectedCompound = SLD_Dens(rowNumber,:);
    SldDensity = 1000.0*selectedCompound.Density;
    cmpname=selectedCompound.Solid;
    disp(['The compound names is: ',cmpname]);
    disp(['Solid Density: ', num2str(SldDensity), ' kg/m3']);
    SldDensity2=SldDensity*(62.4/1000.0);
    disp(['Solid Density: ', num2str(SldDensity2), ' lbm/ft^3']);
elseif strcmpi(phase, 'l')
    disp('You selected the liquid material.');
    searchKeyword = lower(input('Enter a search keyword: ', 's'));
    LIQ_Dens.Liquid = lower(LIQ_Dens.Liquid);

    % Find rows containing the search keyword in the 'Compound' column
    matchRows = find(contains(LIQ_Dens.Liquid, searchKeyword));
    if isempty(matchRows)
        disp('No entries found matching the search keyword.');
        return;
    else
      % Display matching compounds and row numbers (looping approach)
      disp('Matching entries:');
      for i = 1:length(matchRows)
          disp(['  - ', LIQ_Dens.Liquid{matchRows(i)}, ' Temperature is: ', ...
          LIQ_Dens.Temperature{matchRows(i)},' (Row: ', num2str(matchRows(i)),')']);
      end
    end
rowNumber = input('Enter the row number for the compound: ');

    % Check for NAN input
    if isnan(rowNumber)
      disp('Terminating program');
      return;
    end
selectedCompound = LIQ_Dens(rowNumber,:);
    LiqDensity = selectedCompound.Density;
    cmpname=selectedCompound.Liquid;
    disp(['The compound names is: ',cmpname]);
    disp(['Liquid Density: ', num2str(LiqDensity), ' kg/m3']);
    LiqDensity2=LiqDensity*(62.4/1000.0);
    disp(['Liquid Density: ', num2str(LiqDensity2), ' lbm/ft^3']);
else
    disp('Error: Invalid input. Please enter s or S for solid material, l or L for liquid material.');
end
catch ME
disp(['Error: ', ME.message]);
end
```

Figure B.1 (Continued)

B.4 MATLAB Code for Conversion from Mass % to Molar % Composition for a Multi-component Mixture

Figure B.2 shows MATLAB code called **Mass2Mole.m** file for carrying out conversion from mass % to molar % composition. Here is the algorithm made of the following steps:

Step #1: Prompt the user to enter number of components which can be 2, 3, 4, or 5.
Step #2: Based on number of components in Step #1, the user must enter the mass percent composition of each component. We will call the first component A, second B, third C,

```
clear variables;
% Step 1: Prompt user for the number of components
num_components = input('Enter the number of components (2, 3, 4, or 5): ');
% Step 2: Prompt user for mass percent composition of each component
mass_composition = zeros(1, num_components);
for i = 1:num_components
   component = char('A' + i - 1);
   mass_composition(i) = input(['Enter the mass percent of component ' component ': ']);
end

% Step 3: Prompt user for molecular masses of each component
molecular_masses = zeros(1, num_components);
for i = 1:num_components
   component = char('A' + i - 1);
   molecular_masses(i) = input(['Enter the molecular mass of component ' component ': ']);
end
% Step 4: Calculate number of moles of each component
n = mass_composition ./ molecular_masses;
% Step 5: Calculate total number of moles
ntotal = sum(n);
% Step 6: Calculate mole fractions
X = (n ./ ntotal) * 100;
% Step 7: Calculate molecular weight average
MWavg = sum(X .* molecular_masses) / 100;
% Step 8: Display original mass composition
disp('Original mass composition:');
for i = 1:num_components
   component = char('A' + i - 1);
   fprintf('m%s = %.2f\n', component, mass_composition(i));
end
% Step 9: Display calculated molar composition
disp('Calculated molar composition:');
for i = 1:num_components
   component = char('A' + i - 1);
   fprintf('X%s = %.2f%%\n', component, X(i));
end
% Step 10: Display molecular weight average
disp(['Molecular weight average (MWavg) = ', num2str(MWavg)]);
```

Figure B.2 **Mass2Mole.m** file for carrying out mass % to mole % composition for a multi-component mixture.

fourth D, and fifth E. So, the user will enter mA, mB, mC, mD, and mE, depending on the selected number of components in Step #1.

Step #3: The user will also enter molecular mass of A, MWA, of B MWB, of C MWC, of D MWD, and of E MWE, depending on the selected number of components in Step #1.

Step #4: nA=mA/MWA; nB=mB/MWB; nC=mC/MWC; nD=mD/MWD; and nE=mE/MWE; depending on the selected number of components in Step #1.

Step #5: ntotal=nA+nB+nC+nD+nE; depending on the selected number of components in Step #1.

Step #6: XA=(nA/ntotal)*100; XB=(nB/ntotal)*100; XC=(nC/ntotal)*100; XD=(nD/ntotal)*100; and XE=(nE/ntotal)*100; depending on the selected number of components in Step #1.

Step #7: MWavg=(XA/100)*MWA+(XB/100)*MWB+(XC/100)*MWC+(XD/100)*MWD +(XE/100)*MWE; depending on the selected number of components in Step #1.

Step #8: display the original mass % composition: mA, mB, mC, mD, and mE, depending on the selected number of components in Step #1.

Step#9: display the calculated molar % composition: XA, XB, XC, XD, and XE, depending on the selected number of components in Step #1.

Step #10: display molecular weight average, MWavg, calculated in Step #7

B.5 MATLAB Code for Conversion from Molar % to Mass % Composition for a Multi-component Mixture

Figure B.3 shows MATLAB code called **Mole2Mass.m** file for carrying out conversion from mole % to mass % composition. Here is the algorithm made of the following steps:

Step #1: Prompt the user to enter number of components which can be 2, 3, 4, or 5.

Step #2: Based on number of components in Step #1, the user must enter the mole percent composition of each component. We will call the first component A, second B, third C, fourth D, and fifth E. So, the user will enter nA, nB, nC, nD, and nE, depending on the selected number of components in Step #1.

Step #3: The user will also enter molecular mass of A, MWA, of B MWB, of C MWC, of D MWD, and of E MWE, depending on the selected number of components in Step #1.

Step #4: mA=nA*MWA; mB=nB*MWB; mC=nC*MWC; mD=nD*MWD; and mE=nE*MWE; depending on the selected number of components in Step #1.

Step #5: mtotal=mA+mB+mC+mD+mE; depending on the selected number of components in Step #1.

Step #6: xA=(mA/mtotal)*100; xB=(mB/mtotal)*100; xC=(mC/mtotal)*100; xD=(mD/mtotal)*100; and mE=(mE/ntotal)*100; depending on the selected number of components in Step #1.

Step #7: MWavg=(nA/100)*MWA+(nB/100)*MWB+(nC/100)*MWC+(nD/100)*MWD +(nE/100)*MWE; depending on the selected number of components in Step #1.

Step #8: display the original molar composition: nA, nB, nC, nD, and nE, depending on the selected number of components in Step #1.

Step#9: display the calculated mass composition: xA, xB, xC, xD, and xE, depending on the selected number of components in Step #1.

Step #10: display molecular weight average, MWavg, calculated in Step #7

```
clear variables;
% Step 1: Prompt user for the number of components
num_components = input('Enter the number of components (2, 3, 4, or 5): ');
% Step 2: Prompt user for mole percent composition of each component
mole_composition = zeros(1, num_components);
for i = 1:num_components
    component = char('A' + i - 1);
    mole_composition(i) = input(['Enter the mole percent of component ' component ': ']);
end
% Step 3: Prompt user for molecular masses of each component
molecular_masses = zeros(1, num_components);
for i = 1:num_components
    component = char('A' + i - 1);
   molecular_masses(i) = input(['Enter the molecular mass of component ' component ': ']);
end
% Step 4: Calculate masses of each component
masses = mole_composition .* molecular_masses;
% Step 5: Calculate total mass
mtotal = sum(masses);
% Step 6: Calculate mass fractions
x = (masses ./ mtotal) * 100;
% Step 7: Calculate molecular weight average
MWavg = sum(mole_composition ./ 100 .* molecular_masses);
% Step 8: Display original molar composition
disp('Original molar composition:');
for i = 1:num_components
    component = char('A' + i - 1);
    fprintf('n%s = %.2f\n', component, mole_composition(i));
end
% Step 9: Display calculated mass composition
disp('Calculated mass composition:');
for i = 1:num_components
    component = char('A' + i - 1);
    fprintf('x%s = %.2f%%\n', component, x(i));
end
% Step 10: Display molecular weight average
disp(['Molecular weight average (MWavg) = ', num2str(MWavg)]);
```

Figure B.3 **Mole2Mass.m** file for carrying out molar % to mass % composition for a multi-component mixture.

Appendix C

MATLAB Code for Cubic Equations of State (E.o.S.)

C.1 Human Language Description

The following is a description of utilizing MATLAB's computational power to find the roots of a cubic equation like van der Waals'. It applies also for the other cubic equations of state. Thus, there is no need to repeat the description

1) **Load Data**:
   Attempt to read data from an Excel file named 'HCCrit.xlsx'.
   If that fails, try loading a MAT file named 'PureComp2.mat'.
   If both fail, display an error message.
2) **Search for Compound**:
   Ask the user if they want to search by compound name or formula.
   Get the search term (name or formula) from the user.
   Find matching compounds in the loaded data.
   Display a list of matching compounds and their indices.
3) **Select Compound**:
   Ask the user to select the compound number from the list.
4) **Get Temperature and Pressure**:
   Ask the user to input the absolute temperature in Kelvin.
   Ask the user to input the absolute pressure in atm.
5) **Retrieve Compound Properties**:
   Retrieve critical pressure (Pcr), critical temperature (Tcr), molecular weight (MW), and specific gravity at 60°F (SpGr60F) for the selected compound.
6) **Calculate Experimental Liquid Volume**:
   Calculate the experimental liquid volume (ExpVL) using SpGr60F and MW.

To my dear readers,
When copying MATLAB code or text, please right-click and attempt both the "Copy with Formatting" and "Copy" options from the context menu. This approach is effective within Adobe Acrobat. If neither option works, kindly carefully review any MATLAB error messages and compare the pasted text to the original to identify and correct any discrepancies. Alternatively, a soft copy of the MATLAB code is available in the Wiley book companion site of this book for your convenience.

MATLAB-based Computations of Chemical Engineering Principles, First Edition. Kamal I. M. Al-Malah.

Companion website: www.wiley.com/go/al-malah/matlabbasedcomputations

7) **Calculate Constants**:
   Calculate the constants **aterm** and **bterm** using **Rgas** (gas constant), **Pcr**, and **Tcr**.
8) **Solve for Volumes**:
   Solve a cubic equation to find all possible volumes (Vols).
   Identify the largest real root as the gas volume (**GasVol**).
   Identify the smallest real root as the liquid volume (**LiqVol**).
9) **Calculate Percent Relative Error (PRE)**:
   Calculate the PRE between the calculated liquid volume (LiqVol) and the experimental liquid volume (ExpVL).
10) **Calculate Reduced Temperature and Pressure**:
    Calculate the reduced temperature (Tr) and reduced pressure (Pr) using Tabs, Pabs, Tcr, and Pcr.
11) **Display Results**:
    Display the material name, Tr, Pr, GasVol, LiqVol, ExpVL, and PRE.

C.2 MATLAB Code for Van der Waals (vdW) Equation of State

```
clear variables;
try
  % Attempt to read the Excel file first
  PureComp = readtable('HCCrit.xlsx','Format','auto');
catch ME
  % If reading the Excel file fails, try loading the MAT file
  disp('Failed to read Excel file; it will continue with the saved MATLAB
file.');
  try
    load('PureComp2.mat', 'PureComp');
    disp('Loaded PureComp from MAT file.');
  catch ME2
    % If loading the MAT file also fails, display an error message
    disp('Failed to read MAT file.');
  end
end
myvar=PureComp.Compound;
myvar2=PureComp.Formula;
SearchKey= str2double(input('Input "1" to search by Compound Name "2"
to search by Formula: ', 's'));
if SearchKey==1
CompName= input('Input the name of the Compound, say methane: ', 's');
matchStart = regexpi(myvar, CompName);
elseif SearchKey==2
FormName=input('Input the chemical formula, say CH4: ', 's');
matchStart = regexpi(myvar2, FormName);
else
disp('You did not enter 1 or 2');
return
```

```
end
tempindex=1;
flag=0;
myindx(1,1)=1;
for idx=1:size(matchStart,1)
   if cell2mat(matchStart(idx,1))>=1
      myindx(tempindex,1)=idx;
      tempindex=tempindex+1;
      flag=1;
      end
end
if flag==0
   disp ('No Entry was Found!; Good Luck and Try Again.')
   return
end
for idex2=1:size(myindx,1)
   value=myindx(idex2,1);
   display([value,PureComp.Compound(value),PureComp.Formula(value)])
end
CompNum= str2double(input('Input the number for the chosen compound,
say "1" for methane: ', 's'));
Tabs= str2double(input('Input the absolute temperature in Kelvin ', 's'));
Pabs= str2double(input('Input the absolute pressure in atm ', 's'));
Pcr=PureComp.Pcr(CompNum);
Tcr=PureComp.Tcr(CompNum);
MW=PureComp.MW(CompNum);
SpGr60F=PureComp.SpGr60F(CompNum);
ExpVL=(1/SpGr60F)*(MW/1000.0);
Rgas=0.082057;
aterm=(27.0/64.0)*Rgas^2*Tcr^2/Pcr;
bterm=Rgas*Tcr/(8.0*Pcr);
Vols=roots([Pabs, -(Pabs*bterm+Rgas*Tabs), aterm, -aterm*bterm]); % Finds
all roots
GasVol=max(Vols(imag(Vols) == 0)); % finds largest real root for the gas
volume
LiqVol=min(Vols(imag(Vols) == 0)); % finds smallest real root for the
liquid volume
PRE=(abs(LiqVol-ExpVL)/ExpVL)*100;
Tr=Tabs/Tcr;
Pr=Pabs/Pcr;
display(['The material is:',PureComp.Compound(CompNum)]);
display(['Tr=',num2str(Tr),'    Pr=',num2str(Pr)])
disp(['Vg=',num2str(GasVol),'    L/mol'])
disp(['VL=',num2str(LiqVol),'    L/mol'])
disp(['Experimental VL=',num2str(ExpVL),'    L/mol'])
disp(['PRE in VL=',num2str(PRE),'    %'])
```

C.3 MATLAB Code for Peng Robinson (P-R) Equation of State

```
clear variables;
try
  % Attempt to read the Excel file first
  PureComp = readtable('HCCrit.xlsx','Format','auto');
catch ME
  % If reading the Excel file fails, try loading the MAT file
  disp('Failed to read Excel file; it will continue with the saved MATLAB file.');
  try
    load('PureComp2.mat', 'PureComp');
    disp('Loaded PureComp from MAT file.');
  catch ME2
    % If loading the MAT file also fails, display an error message
    disp('Failed to read MAT file.');
  end
end
myvar=PureComp.Compound;
myvar2=PureComp.Formula;
SearchKey= str2double(input('Input "1" to search by Compound Name "2" to
search by Formula: ', 's'));
if SearchKey==1
CompName= input('Input the name of the Compound, say methane: ', 's');
matchStart = regexpi(myvar, CompName);
elseif SearchKey==2
FormName=input('Input the chemical formula, say CH4: ', 's');
matchStart = regexpi(myvar2, FormName);
else
disp('You did not enter 1 or 2');
return
end
tempindex=1;
flag=0;
myindx(1,1)=1;
for idx=1:size(matchStart,1)
  if cell2mat(matchStart(idx,1))>=1
    myindx(tempindex,1)=idx;
    tempindex=tempindex+1;
    flag=1;
  end
end
if flag==0

  disp ('No Entry was Found!; Good Luck and Try Again.')
  return
end
```

```
for idex2=1:size(myindx,1)
   value=myindx(idex2,1);
   display([value,PureComp.Compound(value),PureComp.Formula(value)])
end
CompNum= str2double(input('Input the number for the chosen compound, say "1"
for methane: ', 's'));
Tabs= str2double(input('Input the absolute temperature in Kelvin ', 's'));
Pabs= str2double(input('Input the absolute pressure in atm ', 's'));
Pcr=PureComp.Pcr(CompNum);
Tcr=PureComp.Tcr(CompNum);
% Hnum=PureComp.Hnum(CompNum);
MW=PureComp.MW(CompNum);
w=PureComp.w(CompNum);
SpGr60F=PureComp.SpGr60F(CompNum);
ExpVL=(1/SpGr60F)*(MW/1000.0);
Rgas=0.082058;
Tr=Tabs/Tcr;
Pr=Pabs/Pcr;
aterm=(0.457235)*Rgas^2*Tcr^2/Pcr;
bterm=0.077796*Rgas*Tcr/(Pcr);
kw=0.37464 + 1.54226*w-0.26992*w^2;
alpha=(1.0+kw*(1-sqrt(Tr)))^2.0;
vterm=-(3*bterm^2*Pabs+2.0*bterm*Rgas*Tabs-alpha*aterm);
constterm=(bterm^3*Pabs+Rgas*Tabs*bterm^2-alpha*aterm*bterm);
Vols=roots([Pabs, (Pabs*bterm-Rgas*Tabs), vterm, constterm]); % Finds all roots
GasVol=max(Vols((imag(Vols) == 0))); % finds largest real root for the gas
volume
LiqVol=min(Vols((imag(Vols) == 0))); % finds smallest real root for liquid volume
PRE=(abs(LiqVol-ExpVL)/ExpVL)*100;
display(['The material is:',PureComp.Compound(CompNum)]);
display(['Tr=',num2str(Tr),'    Pr=',num2str(Pr)])
disp(['Vg=',num2str(GasVol),'  L/mol'])
disp(['VL=',num2str(LiqVol),'  L/mol'])
disp(['Experimental VL=',num2str(ExpVL),'  L/mol'])
disp(['PRE in VL=',num2str(PRE),'  %'])
```

C.4 MATLAB Code for Peng–Robinson–Stryjek–Vera (P–R–S–V) Equation of State

```
clear variables;
try
   % Attempt to read the Excel file first
   PureComp = readtable('HCCrit.xlsx','Format','auto');
catch ME
```

```
  % If reading the Excel file fails, try loading the MAT file
  disp('Failed to read Excel file; it will continue with the saved MATLAB
file.');
  try
    load('PureComp2.mat', 'PureComp')
    disp('Loaded PureComp from MAT file.');
  catch ME2
    % If loading the MAT file also fails, display an error message
    disp('Failed to read MAT file.');
  end
end

myvar=PureComp.Compound;
myvar2=PureComp.Formula;
SearchKey= str2double(input('Input "1" to search by Compound Name "2"
to search by Formula: ', 's'));
if SearchKey==1
CompName= input('Input the name of the Compound, say methane: ', 's');
matchStart = regexpi(myvar, CompName);
elseif SearchKey==2
FormName=input('Input the chemical formula, say CH4: ', 's');
matchStart = regexpi(myvar2, FormName);
else
disp('You did not enter 1 or 2');
return
end
tempindex=1;
flag=0;
myindx(1,1)=1;
for idx=1:size(matchStart,1)
  if cell2mat(matchStart(idx,1))>=1
    myindx(tempindex,1)=idx;
    tempindex=tempindex+1;
    flag=1;
  end
end
if flag==0
  disp ('No Entry was Found!; Good Luck and Try Again.')
  return
end
for idex2=1:size(myindx,1)
  value=myindx(idex2,1);
  display([value,PureComp.Compound(value),PureComp.Formula(value)])
end
CompNum= str2double(input('Input the number for the chosen compound,
say "1" for methane: ', 's'));
Tabs= str2double(input('Input the absolute temperature in Kelvin ', 's'));
```

```
Pabs= str2double(input('Input the absolute pressure in atm ', 's'));
Pcr=PureComp.Pcr(CompNum);
Tcr=PureComp.Tcr(CompNum);
MW=PureComp.MW(CompNum);
w=PureComp.w(CompNum);
SpGr60F=PureComp.SpGr60F(CompNum);
ExpVL=(1/SpGr60F)*(MW/1000.0);
Rgas=0.082058;
Tr=Tabs/Tcr;
Pr=Pabs/Pcr;
aterm=(0.457235)*Rgas^2*Tcr^2/Pcr;
bterm=0.077796*Rgas*Tcr/(Pcr);
kw1=0.0089*(1.0+sqrt(Tr))*(0.7-Tr);
kw=0.378893 + 1.4897153*w-0.17131848*w^2+0.0196554*w^3-kw1;
alpha=(1.0+kw*(1-sqrt(Tr)))^2.0;
vterm=-(3*bterm^2*Pabs+2.0*bterm*Rgas*Tabs-alpha*aterm);
constterm=(bterm^3*Pabs+Rgas*Tabs*bterm^2-alpha*aterm*bterm);
Vols=roots([Pabs, (Pabs*bterm-Rgas*Tabs), vterm, constterm]); % Finds all
roots
GasVol=max(Vols((imag(Vols) == 0))); % finds largest real root for the gas
volume
LiqVol=min(Vols((imag(Vols) == 0))); % finds smallest real root for the
liquid volume
display(['The material is:',PureComp.Compound(CompNum)]);
display(['Tr=',num2str(Tr),'    Pr=',num2str(Pr)])
disp(['Vg=',num2str(GasVol),'  L/mol'])
disp(['VL=',num2str(LiqVol),'  L/mol'])
```

C.5 MATLAB Code for Redlich–Kwong (R–K) Equation of State

```
clear variables;
try
  % Attempt to read the Excel file first
  PureComp = readtable('HCCrit.xlsx','Format','auto');
catch ME
  % If reading the Excel file fails, try loading the MAT file
  disp('Failed to read Excel file; it will continue with the saved MATLAB
file.');
  try
    load('PureComp2.mat', 'PureComp');
    disp('Loaded PureComp from MAT file.');
  catch ME2
    % If loading the MAT file also fails, display an error message
    disp('Failed to read MAT file.');
  end
```

```
end
myvar=PureComp.Compound;
myvar2=PureComp.Formula;
SearchKey= str2double(input('Input "1" to search by Compound Name "2"
to search by Formula: ', 's'));
if SearchKey==1
CompName= input('Input the name of the Compound, say methane: ', 's');
matchStart = regexpi(myvar, CompName);
elseif SearchKey==2
FormName=input('Input the chemical formula, say CH4: ', 's');
matchStart = regexpi(myvar2, FormName);
else
disp('You did not enter 1 or 2');
return
end
tempindex=1;
flag=0;
myindx(1,1)=1;
for idx=1:size(matchStart,1)
   if cell2mat(matchStart(idx,1))>=1
      myindx(tempindex,1)=idx;
      tempindex=tempindex+1;
      flag=1;
   end
end
if flag==0
   disp ('No Entry was Found!; Good Luck and Try Again.')
   return
end
for idex2=1:size(myindx,1)
   value=myindx(idex2,1);
   display([value,PureComp.Compound(value),PureComp.Formula(value)])

end
CompNum= str2double(input('Input the number for the chosen compound,
say "1" for methane: ', 's'));
Tabs= str2double(input('Input the absolute temperature in Kelvin ', 's'));
Pabs= str2double(input('Input the absolute pressure in atm ', 's'));
Pcr=PureComp.Pcr(CompNum);
Tcr=PureComp.Tcr(CompNum);
Tr=Tabs/Tcr;
Pr=Pabs/Pcr;
Rgas=0.082057;
aterm=(0.42748*Rgas^2*Tcr^2.5)/(Pcr*Tabs^0.5);
bterm=0.08664*Rgas*Tcr/Pcr;
Vols=roots([Pabs, - Rgas*Tabs, (aterm - Pabs*bterm*bterm -
Rgas*Tabs*bterm), -aterm*bterm]); % Finds all roots
```

```
GasVol=max(Vols(imag(Vols) == 0)); % finds largest real root for the gas
volume
LiqVol=min(Vols(imag(Vols) == 0)); % finds smallest real root for the
liquid volume
display(['The material is:',PureComp.Compound(CompNum)]);
display(['Tr=',num2str(Tr),'     Pr=',num2str(Pr)])
disp(['Vg=',num2str(GasVol),'  L/mol'])
disp(['VL=',num2str(LiqVol),'  L/mol'])
```

C.6 MATLAB Code for Soave–Redlich–Kwong (SRK) Equation of State

```
clear variables;
try
  % Attempt to read the Excel file first
  PureComp = readtable('HCCrit.xlsx','Format','auto');
catch ME
  % If reading the Excel file fails, try loading the MAT file
   disp('Failed to read Excel file; it will continue with the saved MATLAB
file.');
  try
    load('PureComp2.mat', 'PureComp');
    disp('Loaded PureComp from MAT file.');
  catch ME2
    % If loading the MAT file also fails, display an error message
    disp('Failed to read MAT file.');
  end
end
myvar=PureComp.Compound;
myvar2=PureComp.Formula;
SearchKey= str2double(input('Input "1" to search by Compound Name "2"
to search by Formula: ', 's'));
if SearchKey==1
CompName= input('Input the name of the Compound, say methane: ', 's');
matchStart = regexpi(myvar, CompName);
elseif SearchKey==2
FormName=input('Input the chemical formula, say CH4: ', 's');
matchStart = regexpi(myvar2, FormName);
else
disp('You did not enter 1 or 2');
return
end
tempindex=1;
flag=0;
```

```
myindx(1,1)=1;
for idx=1:size(matchStart,1)
  if cell2mat(matchStart(idx,1))>=1
    myindx(tempindex,1)=idx;
    tempindex=tempindex+1;
    flag=1;
  end
end
if flag==0
  disp ('No Entry was Found!; Good Luck and Try Again.')
  return
end
for idex2=1:size(myindx,1)
  value=myindx(idex2,1);
  display([value,PureComp.Compound(value),PureComp.Formula(value)])
end
CompNum= str2double(input('Input the number for the chosen compound,
say "1" for methane: ', 's'));
Tabs= str2double(input('Input the absolute temperature in Kelvin ', 's'));
Pabs= str2double(input('Input the absolute pressure in atm ', 's'));
Pcr=PureComp.Pcr(CompNum);
Tcr=PureComp.Tcr(CompNum);
w=PureComp.w(CompNum);
fw=0.48 + 1.574*w-0.176*w^2;
Rgas=0.082058;
Tr=Tabs/Tcr;
Pr=Pabs/Pcr;
Numer1=(1+fw*(1-sqrt(Tr)))^2;
aterm=0.42748*Rgas^2*(Tcr^2.0)*Numer1/(Pcr);
bterm=0.08664*Rgas*Tcr/Pcr;
Vols=roots([Pabs, - Rgas*Tabs, (aterm - Pabs*bterm*bterm -
Rgas*Tabs*bterm), -aterm*bterm]); % Finds all roots
GasVol=max(Vols(imag(Vols) == 0)); % finds largest real root for the gas
volume
LiqVol=min(Vols(imag(Vols) == 0)); % finds smallest real root for liquid
volume
display(['The material is:',PureComp.Compound(CompNum)]);
display(['Tr=',num2str(Tr),'     Pr=',num2str(Pr)])
disp(['Vg=',num2str(GasVol),'   L/mol'])
disp(['VL=',num2str(LiqVol),'   L/mol'])
```

Appendix D

Antoine's Equation

D.1 Human Language Description

Here is a description of the MATLAB code in human language, focusing on the main steps:

1) **Load Data**:
   - Attempt to read data from an Excel file named '**AntoineEq.xlsx**'.
   - If that fails, try loading a MAT file named '**PurComp.mat**'.
   - If both fail, display an error message.
2) **Search for Compound**:
   - Ask the user if they want to search by compound name or formula.
   - Get the search term (name or formula) from the user.
   - Find matching compounds in the loaded data.
   - Display a list of matching compounds and their indices.
3) **Select Compound**:
   - Ask the user to select the compound number from the list.
   - Retrieve the corresponding A, B, and C coefficients for the selected compound.
4) **Calculation**:
   - Ask the user if she/he wants to calculate temperature (T) or pressure (P).
   - If T is selected:
     - Get the pressure in bar from the user.
     - Convert pressure to mmHg.
     - Calculate the saturation temperature (Tsat) using the Antoine equation.
     - Display the material, formula, and estimated Tsat.

To my dear readers,
When copying MATLAB code or text, please right-click and attempt both the "Copy with Formatting" and "Copy" options from the context menu. This approach is effective within Adobe Acrobat. If neither option works, kindly carefully review any MATLAB error messages and compare the pasted text to the original to identify and correct any discrepancies. Alternatively, a soft copy of the MATLAB code is available in the Wiley book companion site of this book for your convenience.

MATLAB-based Computations of Chemical Engineering Principles, First Edition. Kamal I. M. Al-Malah.

Companion website: www.wiley.com/go/al-malah/matlabbasedcomputations

- If P is selected:
  - Get the temperature in °C from the user.
  - Check if the temperature is within valid limits for the compound.
  - If not, display an error message and return.
  - Calculate the saturation pressure (Psat) using the Antoine equation.
  - Convert pressure to bar.
  - Display the material, formula, and estimated Psat.
- If neither 1 nor 2 is selected, display an error message.

D.2 MATLAB Code (AntoinEq.m) for Antoine's Equation

NOTE

Since the database is so huge to pinpoint at the right code number for the desired component, here are three remedies to follow up:

1) You go first and open the Microsoft Excel file and use "Find" button found in "Editing" group under "Home" tab found in top toolbar. Enter the search text and keep looking for the right record or row. Alternatively, you may also use the shortcut key combination: Ctrl +F.
2) Manually search by the naked eye as the database compounds are alphabetically ordered from A to Z.
3) Alternatively, when you run MATLAB code and you are prompted by a huge list of potential compounds, while in MATLAB's Command Window, right click the mouse and select Find from the context menu, key in the search text and keep checking the right entry. Alternatively, you may also use the shortcut key combination: Ctrl +F as in Microsoft Excel sheet.

```
clear variables;
try
   % Attempt to read the Excel file first
   PureComp = readtable('AntoineEq.xlsx','Format','auto');
catch ME
   % If reading the Excel file fails, try loading the MAT file
   disp('Failed to read Excel file; it will continue with the saved MATLAB
file.');
   try
      load('PurComp.mat', 'PureComp');
      disp('Loaded PureComp from MAT file.');
   catch ME2
      % If loading the MAT file also fails, display an error message
      disp('Failed to read MAT file.');
   end
end
```

```
myvar=PureComp.Compound;
myvar2=PureComp.Formula;
SearchKey= str2double(input('Input "1" to search by Compound Name "2"
to search by Formula: ', 's'));
if SearchKey==1
CompName= input('Input the name of the Compound, say acetone: ', 's');
matchStart = regexpi(myvar, CompName);
elseif SearchKey==2
FormName=input('Input the chemical formula, say C3H6O: ', 's');
matchStart = regexpi(myvar2, FormName);
else
disp('You did not enter 1 or 2');
return
end
tempindex=1;
flag=0;
myindx(1,1)=1;
for idx=1:size(matchStart,1)
  if cell2mat(matchStart(idx,1))>=1
    myindx(tempindex,1)=idx;
    tempindex=tempindex+1;
    flag=1;
  end
end
if flag==0
  disp ('No Entry was Found!; Good Luck and Try Again.')
  return
end
for idex2=1:size(myindx,1)
  value=myindx(idex2,1);
  display([value,PureComp.Compound(value),PureComp.Formula(value)])
end
CompNum= str2double(input('Input the number for the chosen compound,
say "3384" for acetone: ', 's'));
Aterm= PureComp.A(CompNum);
Bterm= PureComp.B(CompNum);
Cterm= PureComp.C(CompNum);
SelKey= str2double(input('Input "1" to calculate T or "2" to calculate P: ', 's'));
if SelKey==1
Pbar= str2double(input('Input the pressure in absolute bar for the chosen
compound: ', 's'));
PmmHg=(760.0/1.01325)*Pbar;
Tsat=(Bterm/(Aterm-log10(PmmHg)))-Cterm;
disp(['The material is:',PureComp.Material(CompNum)]);
disp(['The Molecular Formula = ', PureComp.Formula(CompNum)]);
disp(['The estimated Tsat in oC = ',num2str(Tsat)]);
```

```
elseif SelKey==2
Tin= str2double(input('Input the temperature in oC for the chosen
compound: ', 's'));
Tmin=PureComp.Tmin_C(CompNum);
Tmax=PureComp.Tmax_C(CompNum);
if isnan(Tmax)
  Tmax = max(PureComp.Tmax_C);
end
if Tin < Tmin
  disp (['Re-run but enter T in oC greater than or Equal to: ', num2str
(Tmin),'  oC'])
return
elseif Tin>Tmax
  disp (['Re-run but enter T in oC less than or Equal to: ', num2str
(Tmax),'  oC'])
return
end
Psat=10.0^(Aterm-Bterm/(Tin+Cterm));
Pbar2=Psat*(1.01325/760.0);
Pbar3=Psat/760.0;
disp(['The material is:',PureComp.Material(CompNum)]);
disp(['The Molecular Formula = ', PureComp.Formula(CompNum)]);
disp(['The estimated Psat in mm Hg = ',num2str(Psat)]);
disp(['The estimated Psat in bar = ',num2str(Pbar2)]);
disp(['The estimated Psat in atm = ',num2str(Pbar3)]);
else
disp('You did not enter 1 or 2');
return
end
```

Appendix E

Bubble-point and Dew-point P/T Calculations for Ideal, Binary Mixtures

E.1 MATLAB Bubble_Dew.m Code for Solving the Six Potential Scenarios Describing VLE of An Ideal, Binary Mixture

NOTE

The algorithmic solution for each of the six cases is clearly demonstrated in Chapter 6 | Bubble-point and Dew-point Pressure/Temperature Calculations section. There are 5000 different compounds that make up the database, where the user can search and then select from. Antoine Equation constants will be calculated and populated for the selected two components making up the ideal binary VLE system. Of course, the saturation temperature can be either fixed (i.e., given) or will be iteratively solved based on a certain convergence criterion. Once the saturation temperature is fixed or be solved, other intensive VLE variables will be evaluated accordingly, using the relevant VLE or phase relationships.

```
clear variables;
try
   % Attempt to read the Excel file first
   PureComp = readtable('AntoineEq.xlsx','Format','auto');
catch ME
   % If reading the Excel file fails, try loading the MAT file
   disp('Failed to read Excel file; it will continue with the saved MATLAB
file.');
   try
      load('PurComp.mat', 'PureComp');
      disp('Loaded PureComp from MAT file.');
```

To my dear readers,
When copying MATLAB code or text, please right-click and attempt both the "Copy with Formatting" and "Copy" options from the context menu. This approach is effective within Adobe Acrobat. If neither option works, kindly carefully review any MATLAB error messages and compare the pasted text to the original to identify and correct any discrepancies. Alternatively, a soft copy of the MATLAB code is available in the Wiley book companion site of this book for your convenience.

MATLAB-based Computations of Chemical Engineering Principles, First Edition. Kamal I. M. Al-Malah.

Companion website: www.wiley.com/go/al-malah/matlabbasedcomputations

```
   catch ME2
      % If loading the MAT file also fails, display an error message
      disp('Failed to read MAT file.');
   end
end
% Get information for the first component
disp('Search for the first component');
myvar = PureComp.Compound;
myvar2 = PureComp.Formula;
SearchKey= str2double(input('Input "1" to search by Compound Name "2" to search by
Formula: ', 's'));
if SearchKey==1
CompName= input('Input the name of the Compound, say acetone: ', 's');
matchStart = regexpi(myvar, CompName);
elseif SearchKey==2
FormName=input('Input the chemical formula, say C3H6O: ', 's');
matchStart = regexpi(myvar2, FormName);
else
disp('You did not enter 1 or 2');
return
end
tempindex=1;
flag=0;
myindx(1,1)=1;
for idx=1:size(matchStart,1)
   if cell2mat(matchStart(idx,1))>=1
      myindx(tempindex,1)=idx;
      tempindex=tempindex+1;
      flag=1;
   end
end
if flag==0
   disp ('No Entry was Found!; Good Luck and Try Again.')
   return
end
for idex2=1:size(myindx,1)
   value=myindx(idex2,1);
   display([value,PureComp.Compound(value),PureComp.Formula(value)])
end
CompNum1= str2double(input('Input the number for the chosen compound, say "3384" for
acetone: ', 's'));
Aterm1= PureComp.A(CompNum1);
Bterm1= PureComp.B(CompNum1);
Cterm1= PureComp.C(CompNum1);
% Get information for the second component
disp('Search for the second component');
SearchKey2= str2double(input('Input "1" to search by Compound Name "2" to search
by Formula: ', 's'));
if SearchKey2==1
CompName2= input('Input the name of the Compound, say benzene: ', 's');
matchStart = regexpi(myvar, CompName2);
elseif SearchKey2==2
FormName2=input('Input the chemical formula, say C6H6: ', 's');
matchStart = regexpi(myvar2, FormName2);
else
disp('You did not enter 1 or 2');
return
end
```

```
tempindex=1;
flag=0;
myindx(1,1)=1;
for idx=1:size(matchStart,1)
   if cell2mat(matchStart(idx,1))>=1
      myindx(tempindex,1)=idx;
      tempindex=tempindex+1;
      flag=1;
   end
end
ifflag==0
   disp ('No Entry was Found!; Good Luck and Try Again.')
   return
end
for idex2=1:size(myindx,1)
   value=myindx(idex2,1);
   display([value,PureComp.Compound(value),PureComp.Formula(value)])
end
CompNum2= str2double(input('Input the number for the chosen compound, say "3415" for
benzene: ', 's'));
Aterm2= PureComp.A(CompNum2);
Bterm2= PureComp.B(CompNum2);
Cterm2= PureComp.C(CompNum2);
% Display options to the user
disp('Please choose one of the following cases:');
descrip1='1. Given P and T, find X and Y for Binary Mixture';
disp(descrip1);
descrip2='2. Given P and Y, find X and T for Binary Mixture';
disp(descrip2);
descrip3='3. Given P and X, find Y and T for Binary Mixture';
disp(descrip3);
descrip4='4. Given T and Y, find X and P for Binary Mixture';
disp(descrip4);
descrip5='5. Given T and X, find Y and P for Binary Mixture';
disp(descrip5);
descrip6='6. Given X and Y, find P and T for Binary Mixture';
disp(descrip6);
% Get user input
userChoice = input('Enter your choice (1-6): ');
% Validate user input
while userChoice < 1 || userChoice > 6
   disp('Invalid choice. Please enter a number between 1 and 6.');
   userChoice = input('Enter your choice (1-6): ');
end
% Based on the user's choice, proceed to the specific case code
switch userChoice
   case 1
     descrip=descrip1;% Code for Case 1
     disp(['You have chosen Case: ', descrip]);
     P1= str2double(input('Input pressure of system in atm: ', 's'));
     T1= str2double(input('Input the temperature in oC for the system: ', 's'));
     Tmin1=PureComp.Tmin_C(CompNum1);
     Tmax1=PureComp.Tmax_C(CompNum1);
     Tmin2=PureComp.Tmin_C(CompNum2);
     Tmax2=PureComp.Tmax_C(CompNum2);
     if isnan(Tmax1)
     Tmax1 = max(PureComp.Tmax_C);
     elseifisnan(Tmax2)
```

```
    Tmax2=max(PureComp.Tmax_C);
    end
    if T1 < Tmin1
    disp(['Re-run but enter T in oC greater than or Equal to: ', num2str(Tmin1),' oC'])
    return
    elseif T1 < Tmin2
    disp(['Re-run but enter T in oC greater than or Equal to: ', num2str(Tmin2),' oC'])
    return
    elseif T1>Tmax1
    disp(['Re-run but enter T in oC less than or Equal to: ', num2str(Tmax1),' oC'])
    return
    elseif T1>Tmax2
    disp(['Re-run but enter T in oC less than or Equal to: ', num2str(Tmax2),' oC'])
    end
    Psat1=10.0^(Aterm1-Bterm1/(T1+Cterm1));
    Psat2=10.0^(Aterm2-Bterm2/(T1+Cterm2));
    Patm1=Psat1/760.0;
    Patm2=Psat2/760.0;
    X1=(P1-Patm2)./(Patm1-Patm2);
    Y1=(X1*Patm1)./P1;
    disp(['The material #1 is:',PureComp.Material(CompNum1)]);
    disp(['The material #2 is:',PureComp.Material(CompNum2)]);
    disp(['The Molecular Formula for #1 = ', PureComp.Formula(CompNum1)]);
    disp(['The Molecular Formula for #2 = ', PureComp.Formula(CompNum2)]);
    disp(['The given Tsat in oC = ',num2str(T1)]);
    disp(['The given Psat in atm = ',num2str(P1)]);
    disp(['The estimated X1 = ',num2str(X1)]);
    disp(['The estimated Y1 = ',num2str(Y1)]);
  case 2
    descrip=descrip2;
    disp(['You have chosen Case: ', descrip]);
    P2= str2double(input('Input pressure of system in atm: ', 's'));
    Y1= str2double(input('Input the vapor mole-fraction for component #1: ', 's'));
    Tmin1=PureComp.Tmin_C(CompNum1);
    Tmax1=PureComp.Tmax_C(CompNum1);
    Tmin2=PureComp.Tmin_C(CompNum2);
    Tmax2=PureComp.Tmax_C(CompNum2);
    if isnan(Tmax1)
    Tmax1 = max(PureComp.Tmax_C);
    elseif isnan(Tmax2)
    Tmax2=max(PureComp.Tmax_C);
    end
    % Initial guess for temperature
    Tmin=max(Tmin1,Tmin2);
    Tmax=min(Tmax1,Tmax2);
    T2=(Tmin+Tmax)/2.0;
    if T2 < Tmin
    T2 = Tmin;
    elseif T2 > Tmax
    T2 = Tmax;
    end  % X will be 5
    % Tolerance for sum of mole fractions
    tol = 1.0e-3;
    % Iteration counter and flag
    iteration = 1;
    found_solution = false;
while ~found_solution && iteration < 10000
  % Interpolate vapor pressures for current temperature
  Psat1=10.0^(Aterm1-Bterm1/(T2+Cterm1));
```

```
  Patm1=Psat1/760.0;
  Psat2=10.0^(Aterm2-Bterm2/(T2+Cterm2));
  Patm2=Psat2/760.0;
  % Calculate vapor mole fractions
  X1 = Y1 * P2 / (Patm1);
  X2 = (1.0-Y1)*P2/(Patm2);
  % Check sum of mole fractions
  sum_Xi = X1 + X2;
  % Update temperature based on sum of mole fractions
  if sum_Xi < 1 - tol
     T2 = T2 - 0.050; % decrease T if sum < 1
  elseif sum_Xi > 1 + tol
     T2 = T2 + 0.050; % increase T if sum > 1
  else
     found_solution = true;
  end
  iteration = iteration + 1;
end
if found_solution
   disp(['**Solution found after ', num2str(iteration), ' iterations.']);
   disp(['Temperature: ', num2str(T2), '°C']);
   disp('The Liquid composition is below: ');
   disp([PureComp.Material(CompNum1),': X1 is: ', num2str(X1)]);
   disp([PureComp.Material(CompNum2),': X2 is: ', num2str(X2)]);
else
   disp('Solution not found after 10000 iterations. Please adjust search parameters.');
end
   case 3
     descrip=descrip3;
     disp(['You have chosen Case: ', descrip]);
     P3= str2double(input('Input pressure of system in atm: ', 's'));
     X1= str2double(input('Input the liquid mole-fraction for component #1: ', 's'));
     Tmin1=PureComp.Tmin_C(CompNum1);
     Tmax1=PureComp.Tmax_C(CompNum1);
     Tmin2=PureComp.Tmin_C(CompNum2);
     Tmax2=PureComp.Tmax_C(CompNum2);
     if isnan(Tmax1)
     Tmax1 = max(PureComp.Tmax_C);
     elseif isnan(Tmax2)
     Tmax2=max(PureComp.Tmax_C);
     end
     % Initial guess for temperature
     Tmin=max(Tmin1,Tmin2);
     Tmax=min(Tmax1,Tmax2);
     T3=(Tmin+Tmax)/2.0;
     if T3 < Tmin
     T3 = Tmin;
     elseif T3 > Tmax
     T3 = Tmax;
     end
     % Tolerance for sum of mole fractions
     tol = 1.0e-3;
     % Iteration counter and flag
     iteration = 1;
     found_solution = false;
while ~found_solution && iteration < 10000
  % Interpolate vapor pressures for current temperature
  Psat1=10.0^(Aterm1-Bterm1/(T3+Cterm1));
  Patm1=Psat1/760.0;
```

```
  Psat2=10.0^(Aterm2-Bterm2/(T3+Cterm2));
  Patm2=Psat2/760.0;
  % Calculate vapor mole fractions
  Y1 = X1 * Patm1 / P3;
  Y2 = (1.0-X1)*Patm2/P3;
  PBubble=X1*Patm1+(1-X1)*Patm2;
  PDew=1.0/((Y1/Patm1)+(Y2/Patm2));
  % Update temperature based on sum of mole fractions
  if PBubble < P3 - tol
     T3 = T3 + 0.050; % increase T
  elseif PBubble > 1 + tol
     T3 = T3 - 0.050; % decrease T
  else
     found_solution = true;
  end
  iteration = iteration + 1;
end
   if found_solution
   disp(['**Solution found after ', num2str(iteration), ' iterations.']);
   disp(['Temperature: ', num2str(T3), '°C']);
   disp('The Vapor composition is below: ');
   disp([PureComp.Material(CompNum1),': Y1 is: ', num2str(Y1)]);
   disp([PureComp.Material(CompNum2),': Y2 is: ', num2str(Y2)]);
   disp(['PBubble = ',num2str(PBubble)]);
   disp(['Pdew = ',num2str(PDew)]);
   else
   disp('Solution not found after 10000 iterations. Please adjust search parameters.');
   end
 case 4
    descrip=descrip4;
    disp(['You have chosen Case: ', descrip]);
    Y1= str2double(input('Input the vapor mole-fraction of Component #1: ', 's'));
    T4= str2double(input('Input the temperature in oC for the system: ', 's'));
    Tmin1=PureComp.Tmin_C(CompNum1);
    Tmax1=PureComp.Tmax_C(CompNum1);
    Tmin2=PureComp.Tmin_C(CompNum2);
    Tmax2=PureComp.Tmax_C(CompNum2);
    if isnan(Tmax1)
    Tmax1 = max(PureComp.Tmax_C);
    elseif isnan(Tmax2)
    Tmax2=max(PureComp.Tmax_C);
    end
    if T4 < Tmin1
    disp (['Re-run but enter T in oC greater than or Equal to: ', num2str(Tmin1),' oC'])
    return
    elseif T4 < Tmin2
    disp (['Re-run but enter T in oC greater than or Equal to: ', num2str(Tmin2),' oC'])
    return
    elseif T4>Tmax1
    disp (['Re-run but enter T in oC less than or Equal to: ', num2str(Tmax1),' oC'])
    return
    elseif T4>Tmax2
    disp (['Re-run but enter T in oC less than or Equal to: ', num2str(Tmax2),' oC'])
    end
    Psat1=10.0^(Aterm1-Bterm1/(T4+Cterm1));
    Psat2=10.0^(Aterm2-Bterm2/(T4+Cterm2));
    Patm1=Psat1/760.0;
    Patm2=Psat2/760.0;
    PDew=1.0/((Y1/Patm1)+((1-Y1)/Patm2));
```

```
    X1 = Y1 * PDew / Patm1;
    X2 = (1.0-Y1)*PDew/Patm2;
    SumX=X1+X2;
    PBubble=X1*Patm1+X2*Patm2;
    disp(['The material #1 is:',PureComp.Material(CompNum1)]);
    disp(['The material #2 is:',PureComp.Material(CompNum2)]);
    disp(['The Molecular Formula for #1 = ', PureComp.Formula(CompNum1)]);
    disp(['The Molecular Formula for #2 = ', PureComp.Formula(CompNum2)]);
    disp(['The given Tsat in oC = ',num2str(T4)]);
    disp(['The estimated PBubble in atm = ',num2str(PBubble)]);
    disp(['The estimated PDew in atm = ',num2str(PDew)]);
    disp(['The estimated X1 = ',num2str(X1)]);
    disp(['The sum of X1+X2 = ',num2str(SumX)]);
    disp(['The given Y1 = ',num2str(Y1)]);
case 5
    descrip=descrip5;
    disp(['You have chosen Case: ', descrip]);
    X1= str2double(input('Input the liquid mole-fraction of Component #1: ', 's'));
    T5= str2double(input('Input the temperature in oC for the system: ', 's'));
    Tmin1=PureComp.Tmin_C(CompNum1);
    Tmax1=PureComp.Tmax_C(CompNum1);
    Tmin2=PureComp.Tmin_C(CompNum2);
    Tmax2=PureComp.Tmax_C(CompNum2);
    if isnan(Tmax1)
    Tmax1 = max(PureComp.Tmax_C);
    elseif isnan(Tmax2)
    Tmax2=max(PureComp.Tmax_C);
    end
    if T5 < Tmin1
    disp (['Re-run but enter T in oC greater than or Equal to: ', num2str(Tmin1),' oC'])
    return
    elseif T5 < Tmin2
    disp (['Re-run but enter T in oC greater than or Equal to: ', num2str(Tmin2),' oC'])
    return
    elseif T5>Tmax1
    disp (['Re-run but enter T in oC less than or Equal to: ', num2str(Tmax1),' oC'])
    return
    elseif T5>Tmax2
    disp (['Re-run but enter T in oC less than or Equal to: ', num2str(Tmax2),' oC'])
    end
    Psat1=10.0^(Aterm1-Bterm1/(T5+Cterm1));
    Psat2=10.0^(Aterm2-Bterm2/(T5+Cterm2));
    Patm1=Psat1/760.0;
    Patm2=Psat2/760.0;
    PBubble=X1*Patm1+(1.0-X1)*Patm2;
    Y1 = X1 * Patm1 / PBubble;
    Y2 = (1.0-X1)*Patm2/PBubble;
    SumY=Y1+Y2;
    PDew=1.0/((Y1/Patm1)+(Y2/Patm2));
    disp(['The material #1 is:',PureComp.Material(CompNum1)]);
    disp(['The material #2 is:',PureComp.Material(CompNum2)]);
    disp(['The Molecular Formula for #1 = ', PureComp.Formula(CompNum1)]);
    disp(['The Molecular Formula for #2 = ', PureComp.Formula(CompNum2)]);
    disp(['The given Tsat in oC = ',num2str(T5)]);
    disp(['The estimated PBubble in atm = ',num2str(PBubble)]);
    disp(['The estimated PDew in atm = ',num2str(PDew)]);
    disp(['The estimated Y1 = ',num2str(Y1)]);
    disp(['The Sum of Y1+Y2 = ',num2str(SumY)]);
    disp(['The given X1 = ',num2str(X1)]);
```

```
  case 6
    descrip=descrip6;
    disp(['You have chosen Case: ', descrip]);
    X1= str2double(input('Input the liquid mole-fraction of Component #1: ', 's'));
    Y1= str2double(input('Input the vapor mole-fraction of Component #1: ', 's'));
    Tmin1=PureComp.Tmin_C(CompNum1);
    Tmax1=PureComp.Tmax_C(CompNum1);
    Tmin2=PureComp.Tmin_C(CompNum2);
    Tmax2=PureComp.Tmax_C(CompNum2);
    if isnan(Tmax1)
    Tmax1 = max(PureComp.Tmax_C);
    elseif isnan(Tmax2)
    Tmax2=max(PureComp.Tmax_C);
    end
    % Initial guess for temperature
    Tmin=max(Tmin1,Tmin2);
    Tmax=min(Tmax1,Tmax2);
    T6=(Tmin+Tmax)/2.0;
    if T6 < Tmin
    T6 = Tmin;
    elseif T6 > Tmax
    T6 = Tmax;
    end
    % Tolerance for sum of mole fractions
    tol = 1.0e-5;
    % Iteration counter and flag
    iteration = 1;
    found_solution = false;
while ~found_solution && iteration < 20000
   % Interpolate vapor pressures for current temperature
   Psat1=10.0^(Aterm1-Bterm1/(T6+Cterm1));
   Patm1=Psat1/760.0;
   Psat2=10.0^(Aterm2-Bterm2/(T6+Cterm2));
   Patm2=Psat2/760.0;
   PBubble=X1*Patm1+(1.0-X1)*Patm2;
   PDew=1.0/((Y1/Patm1)+((1-Y1)/Patm2));
   % Update temperature based on sum of mole fractions
   if abs(PBubble-PDew)<=tol
   found_solution = true;
   elseif (PBubble-PDew) < 0
   T6 = T6 - 0.05; % Slightly decrease T
   elseif (PBubble-PDew) > 0
   T6 = T6 + 0.05; % Slightly increase T
   end
   iteration = iteration + 1;
end
if found_solution
   disp(['**Solution found after ', num2str(iteration), ' iterations.']);
   disp(['Estimated Temperature: ', num2str(T6), '°C']);
   disp('The Liquid composition is below: ');
   disp([PureComp.Material(CompNum1),': Given X1 is: ', num2str(X1)]);
   disp('The Vapor composition is below: ');
   disp([PureComp.Material(CompNum1),': Given Y1 is: ', num2str(Y1)]);
   disp(['Estimated PBubble = ',num2str(PBubble), ' atm']);
   disp(['Estimated Pdew = ',num2str(PDew), ' atm ']);
else
   disp('Solution not found after 20000 iterations. Please adjust search parameters.');
end
end
```

Appendix F

Plot of Bubble-point and Dew-point P/T for Ideal Binary Mixtures

F.1 Human Language Description for P_BP_DP_Plot.m File

1) **Load Data**:
   - Attempt to read data from an Excel file named 'AntoineEq.xlsx'.
   - If that fails, try loading a MAT file named 'PurComp.mat'.
   - If both fail, display an error message.
2) **Get Information for First Component**:
   - Ask the user to search for a component by either name or formula.
   - Display a list of matching compounds and their formulas.
   - Prompt the user to select the desired compound by its number.
   - Extract the Antoine equation coefficients (A, B, C) for the chosen component.
3) **Get Information for Second Component**:
   - Repeat the search and selection process for the second component.
4) **Validate Temperature Input**:
   - Get the system temperature in Celsius from the user.
   - Check if the temperature falls within the valid range for both components.
   - If not, display an error message and prompt the user to re-enter the temperature.
5) **Calculate Vapor Pressures**:
   - Calculate the vapor pressures (Psat1, Psat2) in mmHg for both components at the given temperature using their Antoine equations.
   - Convert vapor pressures to atmospheres (Patm1, Patm2).

To my dear readers,
When copying MATLAB code or text, please right-click and attempt both the "Copy with Formatting" and "Copy" options from the context menu. This approach is effective within Adobe Acrobat. If neither option works, kindly carefully review any MATLAB error messages and compare the pasted text to the original to identify and correct any discrepancies. Alternatively, a soft copy of the MATLAB code is available in the Wiley book companion site of this book for your convenience.

MATLAB-based Computations of Chemical Engineering Principles, First Edition. Kamal I. M. Al-Malah.

Companion website: www.wiley.com/go/al-malah/matlabbasedcomputations

6) **Get System Pressure**:
   - Get the system pressure in atmospheres from the user.

7) **Define Functions**:
   - **Define Two Functions**:
     - **PBubble(X)**: Calculates the bubble point pressure for a given liquid composition X.
     - **PDew(Y)**: Calculates the dew point pressure for a given vapor composition Y.

8) **Create Data for Plotting**:
   - Generate 100 evenly spaced points between 0 and 1 for both X and Y.
   - Calculate the corresponding values of PBubble and PDew for those points.
   - Create a horizontal line representing the system pressure.

9) **Generate Plot**:
   - Create a new figure window.
   - Plot PBubble(X) in red, PDew(Y) in blue dashed line, and the system pressure line in black dashed line.
   - Add labels, title, legend, and grid lines.
   - Customize the plot appearance.

F.2 P_BP_DP_Plot.m MATLAB Code

```
clear variables;
try
   % Attempt to read the Excel file first
   PureComp = readtable('AntoineEq.xlsx','Format','auto');
catch ME
   % If reading the Excel file fails, try loading the MAT file
   disp('Failed to read Excel file; it will continue with the saved MATLAB file.');
   try
      load('PurComp.mat', 'PureComp');
      disp('Loaded PureComp from MAT file.');
   catch ME2
      % If loading the MAT file also fails, display an error message
      disp('Failed to read MAT file.');
   end
end
% Get information for the first component
disp('Search for the first component');
myvar = PureComp.Compound;
myvar2 = PureComp.Formula;
SearchKey= str2double(input('Input "1" to search by Compound Name "2" to search
by Formula: ', 's'));
if SearchKey==1
CompName= input('Input the name of the Compound, say acetone: ', 's');
matchStart = regexpi(myvar, CompName);
elseif SearchKey==2
FormName=input('Input the chemical formula, say C3H6O: ', 's');
matchStart = regexpi(myvar2, FormName);
else
```

```
disp('You did not enter 1 or 2');
return
end
tempindex=1;
flag=0;
myindx(1,1)=1;
for idx=1:size(matchStart,1)
   if cell2mat(matchStart(idx,1))>=1
      myindx(tempindex,1)=idx;
      tempindex=tempindex+1;
      flag=1;
   end
end
if flag==0
   disp ('No Entry was Found!; Good Luck and Try Again.')
   return
end
for idex2=1:size(myindx,1)
   value=myindx(idex2,1);
   display([value,PureComp.Compound(value),PureComp.Formula(value)])
end
CompNum1= str2double(input('Input the number for the chosen compound, say
"3384" for acetone: ', 's'));
Aterm1= PureComp.A(CompNum1);
Bterm1= PureComp.B(CompNum1);
Cterm1= PureComp.C(CompNum1);
% Get information for the second component
disp('Search for the second component');
SearchKey2= str2double(input('Input "1" to search by Compound Name "2" to search
by Formula: ', 's'));
if SearchKey2==1
CompName2= input('Input the name of the Compound, say benzene: ', 's');
matchStart = regexpi(myvar, CompName2);
elseif SearchKey2==2
FormName2=input('Input the chemical formula, say C6H6: ', 's');
matchStart = regexpi(myvar2, FormName2);
else
disp('You did not enter 1 or 2');
return
end
tempindex=1;
flag=0;
myindx(1,1)=1;
for idx=1:size(matchStart,1)
   if cell2mat(matchStart(idx,1))>=1
      myindx(tempindex,1)=idx;
      tempindex=tempindex+1;
      flag=1;
   end
end
if flag==0
   disp ('No Entry was Found!; Good Luck and Try Again.')
   return
end
```

```
for idex2=1:size(myindx,1)
   value=myindx(idex2,1);
   display([value,PureComp.Compound(value),PureComp.Formula(value)])
end
CompNum2= str2double(input('Input the number for the chosen compound, say
"3415" for benzene: ', 's'));
Aterm2= PureComp.A(CompNum2);
Bterm2= PureComp.B(CompNum2);
Cterm2= PureComp.C(CompNum2);
Tmin1=PureComp.Tmin_C(CompNum1);
Tmax1=PureComp.Tmax_C(CompNum1);
Tmin2=PureComp.Tmin_C(CompNum2);
Tmax2=PureComp.Tmax_C(CompNum2);
if isnan(Tmax1)
      Tmax1 = max(PureComp.Tmax_C);
elseif isnan(Tmax2)
      Tmax2=max(PureComp.Tmax_C);
end
Tmin=max(Tmin1,Tmin2);
Tmax=min(Tmax1,Tmax2);
if isnan(Tmax)
   Tmax = max(PureComp.Tmax_C);
end
T= str2double(input('Input system temperature in oC to create 2-in-1 plot: ', 's'));
if T < Tmin
 disp (['Re-run but enter T in oC greater than or Equal to: ', num2str(Tmin),' oC'])
return
elseif T>Tmax
 disp (['Re-run but enter T in oC less than or Equal to: ', num2str(Tmax),'  oC'])
return
end
Psat1 = 10.0.^(Aterm1 - Bterm1./(T+Cterm1));
Psat2 = 10.0.^(Aterm2 - Bterm2./(T+Cterm2));
Patm1 = Psat1/760.0;
Patm2 = Psat2/760.0;
P= str2double(input('Input system pressure in atm to create a horizontal tie
line: ', 's'));
PBubble = @(X) X*Patm1+(1-X)*Patm2;  % Define PBubble
PDew = @(Y) 1./(Y/Patm1+(1.0-Y)/Patm2);  % Define PDew
% Define X and Y range
X = linspace(0, 1, 100);  % Create 100 evenly spaced points between 0 and 1
Y = linspace(0, 1, 100);  % Create 100 evenly spaced points between 0 and 1
% Calculate Y values for both functions
y1 = PBubble(X);
y2 = PDew(Y);
% Create a horizontal line for P=1.0 atm
y3=linspace(P, P, 100);  % Create 100 evenly spaced points between 1 and 1
% Create the combined plot
figure;  % Create a new figure window
hold on;  % Allow multiple plots on the same axes
plot(X, y1, 'r', 'LineWidth', 3);  % Plot Pbubble(X) in red
plot(Y, y2, 'b--', 'LineWidth', 3);  % Plot Pdew(Y) in blue dashed line
plot(X, y3, 'k--', 'LineWidth', 3);  % Plot P=1.0 atm in black dashed line
% Add labels and title
xlabel('X,Y','FontSize',15);
```

```
ylabel('P','FontSize',15);
name1=string(PureComp.Material(CompNum1));
name2=string(PureComp.Material(CompNum2));
ttl="P_{BP} & P_{DP}: " + name1 + "+" + name2 +" at " + num2str(T) +"^{o}C";
title(ttl,FontSize=15);
% Add legend
legend('P_{Bubble}(X)', 'P_{Dew}(Y)');
% Customize grid and appearance
grid on;
box on;
hold off; % Release hold on the current axes
```

F.3 Human Language Description for T_BP_DP_Plot.m File

1) **Load Data**:
   - Attempt to read data from an Excel file named 'AntoineEq.xlsx'.
   - If that fails, try loading a MAT file named 'PurComp.mat'.
   - If both fail, display an error message.

2) **Get Information for Components**:
   - **For Both the First and Second Components**:
     - Prompt the user to search by either name or formula.
     - Display a list of matching compounds and their formulas.
     - Ask the user to select the desired compound by its number.
     - Extract the Antoine equation coefficients (A, B, C) for the chosen compound.

3) **Get System Pressure**:
   - Ask the user to input the system pressure in atmospheres.

4) **Define Functions**:
   - Define Psat1(T) and Psat2(T) to calculate the vapor pressures of the components in mmHg using their Antoine equations.
   - Define Patm1(T) and Patm2(T) to convert vapor pressures to atmospheres.

5) **Calculate Bubble Point Temperatures (T vs. X)**:
   - Define a function f(T, X) to represent the equation for Pbubble.
   - Generate a range of X values (0–1).
   - For each X value, use fzero to find the corresponding T value that satisfies the Pbubble equation.

6) **Plot Bubble Point Curve**:
   - Create a plot of T (temperature) vs. X (liquid composition).
   - Hold the plot for additional data.

7) **Calculate Dew Point Temperatures (T vs. Y)**:
   - Define a function f(T, Y) to represent the equation for PDew.
   - Generate a range of Y values (0–1).
   - For each Y value, use fzero to find the corresponding T value that satisfies the PDew equation.

8) **Plot Dew Point Curve**:
   - Add the plot of T vs. Y (dew point curve) to the existing plot.

9) **Customize Plot**:
   - Add labels for the axes and a title indicating the system pressure.
   - Include a legend to distinguish between the bubble point and dew point curves.
   - Release the hold on the plot.

F.4 T_BP_DP_Plot.m MATLAB Code

```
clear variables;
try
   % Attempt to read the Excel file first
   PureComp = readtable('AntoineEq.xlsx','Format','auto');
catch ME
   % If reading the Excel file fails, try loading the MAT file
   disp('Failed to read Excel file; it will continue with the saved MATLAB
file.');
   try
      load('PurComp.mat', 'PureComp');
      disp('Loaded PureComp from MAT file.');
   catch ME2
      % If loading the MAT file also fails, display an error message
      disp('Failed to read MAT file.');
   end
end
% Get information for the first component
disp('Search for the first component');
myvar = PureComp.Compound;
myvar2 = PureComp.Formula;
SearchKey= str2double(input('Input "1" to search by Compound Name "2" to search
by Formula: ', 's'));
if SearchKey==1
CompName= input('Input the name of the Compound, say acetone: ', 's');
matchStart = regexpi(myvar, CompName);
elseifSearchKey==2
FormName=input('Input the chemical formula, say C3H6O: ', 's');
matchStart = regexpi(myvar2, FormName);
else
disp('You did not enter 1 or 2');
return
end
tempindex=1;
flag=0;
myindx(1,1)=1;
for idx=1:size(matchStart,1)
   if cell2mat(matchStart(idx,1))>=1
      myindx(tempindex,1)=idx;
      tempindex=tempindex+1;
      flag=1;
   end
end
```

```
if flag==0
   disp ('No Entry was Found!; Good Luck and Try Again.')
   return
end
for idex2=1:size(myindx,1)
   value=myindx(idex2,1);
   display([value,PureComp.Compound(value),PureComp.Formula(value)])
end
CompNum1= str2double(input('Input the number for the chosen compound, say
"3384" for acetone: ', 's'));
Aterm1= PureComp.A(CompNum1);
Bterm1= PureComp.B(CompNum1);
Cterm1= PureComp.C(CompNum1);
% Get information for the second component
disp('Search for the second component');
SearchKey2= str2double(input('Input "1" to search by Compound Name "2" to search
by Formula: ', 's'));
if SearchKey2==1
CompName2= input('Input the name of the Compound, say benzene: ', 's');
matchStart = regexpi(myvar, CompName2);
elseifSearchKey2==2
FormName2=input('Input the chemical formula, say C6H6: ', 's');
matchStart = regexpi(myvar2, FormName2);
else
disp('You did not enter 1 or 2');
return
end
tempindex=1;
flag=0;
myindx(1,1)=1;
for idx=1:size(matchStart,1)
   if cell2mat(matchStart(idx,1))>=1
      myindx(tempindex,1)=idx;
      tempindex=tempindex+1;
      flag=1;
   end
end
if flag==0
   disp ('No Entry was Found!; Good Luck and Try Again.')
   return
end
for idex2=1:size(myindx,1)
   value=myindx(idex2,1);
   display([value,PureComp.Compound(value),PureComp.Formula(value)])
end
CompNum2= str2double(input('Input the number for the chosen compound, say
"3415" for benzene: ', 's'));
Aterm2= PureComp.A(CompNum2);
Bterm2= PureComp.B(CompNum2);
Cterm2= PureComp.C(CompNum2);

Pbubble_target= str2double(input('Input system pressure in atm to create plots: ',
's'));
%Pbubble_target = 1.0;  % Target Pbubble

% Define Psat1, Psat2, PMOH, and PEOH as functions of T
```

```
Psat1 = @(T) 10.0.^(Aterm1 - Bterm1./(T+Cterm1));
Psat2 = @(T) 10.0.^(Aterm2 - Bterm2./(T+Cterm2));
Patm1 = @(T) Psat1(T)/760.0;
Patm2 = @(T) Psat2(T)/760.0;
Tmin1=PureComp.Tmin_C(CompNum1);
Tmax1=PureComp.Tmax_C(CompNum1);
Tmin2=PureComp.Tmin_C(CompNum2);
Tmax2=PureComp.Tmax_C(CompNum2);
if isnan(Tmax1)
      Tmax1 = max(PureComp.Tmax_C);
elseifisnan(Tmax2)
      Tmax2=max(PureComp.Tmax_C);
end
Tmin=max(Tmin1,Tmin2);
Tmax=min(Tmax1,Tmax2);
Tguess=(Tmin+Tmax)/2.0;
% Define the function to find the root of the equation
f = @(T, X) Pbubble_target - (X.*Patm1(T) + (1-X).*Patm2(T));
% Generate vectors for X and T
X_values = 0:0.01:1;
T_values = zeros(size(X_values));
% Find T for each X value using fzero (corrected usage)
for i = 1:length(X_values)
   options = optimset('Display', 'off');  % Prevent excessive output
   T_values(i) = fzero(f, Tguess, options, X_values(i));  % Pass options
structure as the 3rd argument
end
% Plot T vs. X
plot(X_values, T_values,'r', 'LineWidth', 3);
hold on;
% Define constants
PDew_target = Pbubble_target;  % Target PDew
% Define the new function to find the root
f = @(T, Y) PDew_target - 1./((Y./Patm1(T)) + ((1-Y)./Patm2(T)));
% Generate vectors for Y and T
Y_values = 0:0.01:1;
T_values = zeros(size(Y_values));
% Find T for each Y value using fzero
for i = 1:length(Y_values)
   options = optimset('Display', 'off');
   T_values(i) = fzero(f, Tguess, options, Y_values(i));
end
plot(Y_values, T_values,'b', 'LineWidth', 3);
xlabel('Y','FontSize',15);
ylabel('T (^{o}C)','FontSize',15);
name1=string(PureComp.Material(CompNum1));
name2=string(PureComp.Material(CompNum2));
ttl="T_{BP} & T_{DP}: " + name1 + "+" + name2 +" at " + num2str(Pbubble_target)
+" atm";
title(ttl,FontSize=15);
% Add legend
legend('T_{Bubble}(X)', 'T_{Dew}(Y)');
grid on;
box on;
hold off; % Release hold on the current axes
```

Appendix G

Henry's Law and Magnus Law

G.1 Temperature-dependent of Gas Solubility in Water

Table G.1 shows temperature-dependent of gas solubility in water. The constants can be used to calculate solubility by the equation $\ln X = A + B/T + C \ln T + DT$, where T is in K and X is the mole fraction of the solute dissolved in water when the solute partial pressure is 1 atm. $K_{P/X}$ is the inverse of X, which represents Henry's constant based on liquid mole-fraction of gas in water.

Table G.1 Temperature-dependent gas solubility in water.

Name	Formula	A	B	C	D	T_{min} (K)	T_{max} (K)
Acetylene	C_2H_2	−156.51	8,160.2	21.403	0	274	343
CO_2	CO_2	−159.854	8,741.68	21.6694	−0.001103	273	353
CO	CO	−171.764	8,296.9	23.3376	0	273	353
Ethane	C_2H_6	−250.812	12,695.6	34.7413	0	275	323
Ethylene	C_2H_4	−153.027	7,965.2	20.5248	0	287	346
Helium	He	−105.977	4,259.62	14.0094	0	273	348
Hydrogen	H_2	−125.939	5,528.45	16.8893	0	273	345
Methane	CH_4	−338.217	13,282.1	51.9144	−0.042583	273	523
Nitrogen	N_2	−181.587	8,632.13	24.7981	0	273	350
Oxygen	O_2	−171.254	8,391.24	23.24323	0	273	333

Source: Green and Perry [1].

To my dear readers,
When copying MATLAB code or text, please right-click and attempt both the "Copy with Formatting" and "Copy" options from the context menu. This approach is effective within Adobe Acrobat. If neither option works, kindly carefully review any MATLAB error messages and compare the pasted text to the original to identify and correct any discrepancies. Alternatively, a soft copy of the MATLAB code is available in the Wiley book companion site of this book for your convenience.

MATLAB-based Computations of Chemical Engineering Principles, First Edition. Kamal I. M. Al-Malah.

Companion website: www.wiley.com/go/al-malah/matlabbasedcomputations

G.2 MATLAB Code for Calculating Henry's Constants as *f*(*T*)

```
clear variables;
gas_data = struct(...
    'name', {'Acetylene', 'CO2', 'CO', 'C2H6', ...
             'Ethene', 'He', 'H2', 'CH4', 'N2', 'O2'}, ...
    'A', [-156.51, -159.854, -171.764, -250.812, -153.027, -105.9768, ...
          -125.939, -338.217, -181.587, -171.2542], ...
    'B', [8160.2, 8741.68, 8296.9, 12695.6, 7965.2, 4259.62, 5528.45, ...
          13282.1, 8632.13, 8391.24], ...
    'C', [21.403, 21.6694, 23.3376, 34.7413, 20.5248, 14.0094, 16.8893, ...
          51.9144, 24.7981, 23.24323], ...
    'D', [0, -1.1026e-3, 0, 0, 0, 0, 0, -0.0425831, 0, 0], ...
    'Tmin', [274, 273, 275, 273, 287, 273, 273, 273, 273, 273], ...
    'Tmax', [343, 353, 323, 348, 346, 348, 345, 523, 350, 333]);
disp('Available gases:');
for i = 1:length(gas_data)
    disp(gas_data(i).name);  % Access the 'name' field using dot notation
end
selected_gas = input('Enter gas name: ', 's');
% Find selected gas data in the table
gas_index = find(strcmpi(selected_gas, {gas_data.name}));
if isempty(gas_index)
    error('Invalid gas name!');
end
% Extract data for the selected gas
gas_name = gas_data(gas_index).name;
A = gas_data(gas_index).A(gas_index);
B = gas_data(gas_index).B(gas_index);
C = gas_data(gas_index).C(gas_index);
D = gas_data(gas_index).D(gas_index);
Tmin = gas_data(gas_index).Tmin(gas_index);  % Choose Tmin for the given gas
Tmax = gas_data(gas_index).Tmax(gas_index);  % Choose Tmax for the given gas
% Get temperature input from user
T = str2double(input('Input the temperature in K for the system: ', 's'));
% Check if temperature is within valid range for the selected gas
if T < Tmin || T > Tmax
    error('Temperature is outside valid range!');
end
% Calculate mole fraction using the equation
ln_x = A + B/T + C*log(T) + D*T;
x = exp(ln_x);
% Calculate Henry's constants at the given T for the selected gas
KP_X=1/x;
KP_C = (KP_X / 55.5);
KC_P = (1/ KP_C);
KC_C=(1.357E3/KP_X);
% Display results
fprintf('Gas name: %s\n', gas_name);
fprintf('Temperature: %.2f K\n', T);
fprintf('Mole fraction of solute: %.4e\n', x);
fprintf('K_{P/C} Henry Constant: %.4e\n', KP_C);
fprintf('K_{C/P} Henry Constant: %.4e\n', KC_P);
fprintf('K_{P/X} Henry Constant: %.4e\n', KP_X);
fprintf('K_{C/C} Henry Constant: %.4e\n', KC_C);
```

G.3 Human Language Algorithm for Calculating Air/Water Dew-point Temperature Using Magnus Equation

1) **Clear the Workspace**:
   - Remove any existing variables or data from the MATLAB workspace to start with a clean slate.
2) **Prompt for Input**:
   - Ask the user to enter two values:
     - o Dry bulb temperature (TDB) in degrees Celsius
     - o Relative humidity (RH) as a percentage
3) **Error Handling**:
   - Use a try-catch block to handle potential input errors:
   - If the user enters invalid data (e.g., non-numeric characters), display an error message and ask them to try again.
4) **Check for Numeric Values**:
   - If the input values are not numeric (NaN), display an error message and stop the code.
5) **Calculate Intermediate Terms**:
   - Calculate two intermediate terms using the provided formulas:
     - o $\text{NumTerm} = 243.04 * (\log(\text{RH}/100) + (17.625 * \text{TDB})/(243.04 + \text{TDB}))$
     - o $\text{DenomTerm} = 17.625 - \log(\text{RH}/100) - (17.625 * \text{TDB})/(243.04 + \text{TDB})$
6) **Calculate Dew Point Temperature**:
   - **Calculate the Dew Point Temperature (TDP) Using the Formula**:
     - o $\text{TDP} = \text{NumTerm}/\text{DenomTerm}$
7) **Display Results**:
   - Show the entered values of dry bulb temperature (TDB) and relative humidity (RH).
   - Display the calculated dew point temperature (TDP) in degrees Celsius.

G.4 MATLAB Code for Magnus Equation: $T_{DP} = f(T_{DB}, \text{RH})$

```
clear variables;
try
    TDB= str2double(input('Input the dry bulb temperature oC, TDB: ', 's'));
    RH= str2double(input('Input the percent relative humidity, on 100%
scale, RH%: ', 's'));
catch ME
    % If reading the Excel file fails, try loading the MAT file
    disp('Failed to accept correct format of TDB and/or RH. Please, try
another time!');
end
if isnan(TDB) || isnan(RH)
    disp('Give me a break; the input is not a numeric number. Try again')
    % Proceed with your code here, using input_arg as needed
    return
end
```

```
NumTerm=243.04*(log((RH)/(100.0))+(17.625*TDB)/(243.04+TDB));
DenomTerm= 17.625-log((RH)/(100.0))-(17.625*TDB)/(243.04+TDB);
TDP= NumTerm./DenomTerm;
disp(['The entered dry bulb T in oC, TDB = ',num2str(TDB),' oC']);
disp(['The entered percent relative humidity, RH% = ', num2str(RH),' %']);
disp(['The estimated dew point T, TDP in oC = ',num2str(TDP), ' oC']);
```

Reference

1 Green, D.W. and Perry, R.H. (2008). *Perry's Chemical Engineers' Handbook*, 8e Table 2–123. New York, USA: McGraw-Hill, Inc. https://doi.org/10.1036/0071422943.

Appendix H

Sensible Heat and Latent Heat Calculations

The database contains 301compounds which can be liquids or solids. For each compound, the specific heat and the acceptable temperature range are stated. You can search by entering the keyword and the program will look up for all permutations containing the keyword. Your duty then is to select the one of your choice. The heat duty is calculated based on the assumption that $T_2 > T_1$. If interested in the cooling process (i.e., $T_2 < T_1$), then do the heating step and take the negative value for the calculated heat duty. The reason for restricting $T_2 > T_1$ is to assure that the user will not enter temperature values below or above the acceptable/applicable temperature range which is dictated by the melting and boiling point values of the given substance, in general

H.1 Human Language Algorithm for Calculating the Sensible Heat

1) **Data Loading (with Error Handling)**:
   - The code first attempts to read the data from an Excel file named "**SpecHeat.xlsx**" using readtable.
   - If reading the Excel file fails, it displays an error message and tries to load the data from a MAT file named "**SpHData.mat**" using load.
   - If both attempts fail, it displays another error message and terminates.
2) **User Interaction and Data Selection**:
   - If data loading is successful, the code prompts the user for a search keyword (converted to lowercase).
   - It then searches for the keyword within the "Compound" column (also converted to lowercase) and stores the matching row indices in matchRows.

To my dear readers,
When copying MATLAB code or text, please right-click and attempt both the "Copy with Formatting" and "Copy" options from the context menu. This approach is effective within Adobe Acrobat. If neither option works, kindly carefully review any MATLAB error messages and compare the pasted text to the original to identify and correct any discrepancies. Alternatively, a soft copy of the MATLAB code is available in the Wiley book companion site of this book for your convenience.

MATLAB-based Computations of Chemical Engineering Principles, First Edition. Kamal I. M. Al-Malah.

Companion website: www.wiley.com/go/al-malah/matlabbasedcomputations

- If no matches are found, it displays a message and exits the program.
- Otherwise, it iterates through the matching rows and displays the compound names and corresponding row numbers for the user's selection.

3) **User Input and Data Extraction**:
   - The code prompts the user to enter the row number for the desired compound.
   - It checks if the input is a valid number (not NaN). If not, it displays an error message and exits.
   - If the input is valid, it extracts the relevant data (specific heat, phase, minimum and maximum temperature) for the chosen compound from the selected row.

4) **User Input for Temperature Range and Mass**:
   - The code prompts the user to enter two temperatures (T_1 and T_2) for the calculation.
   - It validates the user input to ensure T_1 is greater than or equal to T_{min} (minimum temperature) and T_2 is lower than or equal to Tmax (maximum temperature).
   - Finally, it prompts the user to enter the mass of the compound in kilograms.

5) **Sensible Heat Calculation based on Phase (with Unit Conversions)**:
   - The code checks the phase of the compound ("S" for solid, "L" for liquid).
   - Based on the phase, it calculates the sensible heat (Q12) using the formula Q12 = mass * specific heat * $(T_2 - T_1)$.
   - The code then converts the calculated sensible heat from Joules (kJ) to kilocalories (kcal), British thermal units (BTU), and pound-force feet (lbf-ft) using the following conversion factors:
     - o Qkcal = Q12/4.184
     - o Qbtu = Qkcal * 3.965666
     - o Qlb_ft = Qkcal * 3085.96
   - Finally, it displays the calculated sensible heat values in all four units (kJ, kcal, BTU, lbf-ft) along with the compound name.

6) **Error Handling**:
   - The entire code block is wrapped in a try ... catch block to handle potential errors during execution (e.g., invalid user input, file access issues).
   - If an error occurs, it displays an error message with the specific details (ME.message) for debugging purposes.

H.2 MATLAB Code for Sensible Heat Calculation of Solids and Liquids

```
clear variables;
try
    % Attempt to read the Excel file first
    SpHdata = readtable('SpecHeat.xlsx','Format','auto');
    catch ME
    % If reading the Excel file fails, try loading the MAT file
    disp('Failed to read Excel file; it will continue with the saved MATLAB file.');
    try
        load('SpHData.mat', 'SpHdata');
        disp('Loaded SpHdata from MAT file.');
    catch ME2
```

```
      % If loading the MAT file also fails, display an error message
      disp('Failed to read MAT file.');
   end
end
try

   % Get search keyword from the user
   searchKeyword = lower(input('Enter a search keyword: ', 's'));
   SpHdata.Compound = lower(SpHdata.Compound);

   % Find rows containing the search keyword in the 'Compound' column
   matchRows = find(contains(SpHdata.Compound, searchKeyword));

   if isempty(matchRows)
      disp('No entries found matching the search keyword.');
      return;
   else
      % Display matching compounds and row numbers (looping approach)
      disp('Matching entries:');
      for i = 1:length(matchRows)
         disp([' - ', SpHdata.Compound{matchRows(i)}, ' (Row ', num2str(matchRows(i)),
')']);
      end

   end
rowNumber = input('Enter the row number for the compound: ');

   % Check for NAN input
   if isnan(rowNumber)
      disp('Terminating program');
      return;
   end

   % Get relevant data for the selected compound
   selectedCompound = SpHdata(rowNumber,:);
   spHeat = selectedCompound.SpHeat;
   phase = selectedCompound.Phase;
   Tmin = selectedCompound.Tmin;
   Tmax = selectedCompound.Tmax;
   disp(['Tmin: ', num2str(Tmin), ' oC']);
   disp(['Tmax: ', num2str(Tmax), ' oC']);
   % User input for T1 and T2
   T1 = input('Enter T1 (must be higher than or equal to Tmin): ');
   while T1 < Tmin
     T1 = input('Invalid T1. Enter a value higher than or equal to Tmin: ');
   end

   T2 = input('Enter T2 (must be lower than or equal to Tmax): ');
   while T2 > Tmax
     T2 = input('Invalid T2. Enter a value lower than or equal to Tmax: ');
   end
   mass = input('Enter the mass of the compound in kg: ');

   % Calculate sensible heat based on phase
   switch string(phase)
     case 'S'
       % Code for solid phase
       % ...
       disp('Sensible heat calculation for solid phase: ...');
       cmpname=selectedCompound.Compound;
       disp(['The compound names is: ',cmpname]);
       Q12=mass*spHeat*(T2-T1);
```

```
            Qkcal=Q12/4.184;
            Qbtu=Qkcal*3.965666;
            Qlb_ft=Qkcal*3085.96;
            disp(['QkJ: ', num2str(Q12),' kJ']);
            disp(['Qkcal: ', num2str(Qkcal),' kcal']);
            disp(['Qbtu: ', num2str(Qbtu),' BTU']);
            disp(['Qlb_ft: ', num2str(Qlb_ft),' lbf-ft'])
        case 'L'
            % Code for liquid phase
            % ...
            disp('Sensible heat calculation for liquid phase: ...');
            cmpname=selectedCompound.Compound;
            disp(['The compound names is: ',cmpname]);
            Q12=mass*spHeat*(T2-T1);
            Qkcal=Q12/4.184;
            Qbtu=Qkcal*3.965666;
            Qlb_ft=Qkcal*3085.96;
            disp(['QkJ: ', num2str(Q12),' kJ']);
            disp(['Qkcal: ', num2str(Qkcal),' kcal']);
            disp(['Qbtu: ', num2str(Qbtu),' BTU']);
            disp(['Qlb_ft: ', num2str(Qlb_ft),' lbf-ft']);
        otherwise
            error('Invalid phase value');
    end
catch ME
    disp(['Error: ', ME.message]);
end
```

H.3 Human-Language Algorithm of the MATLAB Code for Evaluating Water/Steam Thermodynamic Properties

1) **Data Acquisition (Try-catch Blocks)**:
   - Attempt to read data from an Excel file named "**SteamTable.xlsx**" using the **readtable** function.
   - **If Reading the Excel File Fails**:
     - o Display a message indicating the error.
     - o Try to load data from a MAT file named "**SteamTable.mat**" using the load function.
     - o **If Loading the MAT File also Fails**:
       - ▪ Display an error message indicating both file reading failures.
   - **If Data is Successfully Loaded (From Either Excel or MAT File)**:
     - o **Extract Specific Columns from the Data Table**:
       - ▪ **Pressure (MPa)**: P_MPa
       - ▪ **Temperature (°C)**: T_C (assuming conversion from a string)
       - ▪ **Specific Volume (m^3/kg)**: V_m3_kgInv (assuming conversion from a string)
       - ▪ **Internal Energy (kJ/kg)**: U_kJ_kgInv (assuming conversion from a string)
       - ▪ **Enthalpy (kJ/kg)**: H_kJ_kgInv (assuming conversion from a string)
       - ▪ **Entropy (kJ/kg∗K)**: S_kJ_kgInv_Kinv (assuming conversion from a string)
       - ▪ **Phase**: Phase

2) **User Input and Validation**:
   - Prompt the user to enter a temperature value (°C) within the range 0–2000.
   - If the entered temperature is outside the valid range, display an error message and exit the program.
   - Prompt the user to enter a pressure value (MPa) within the range 0.01–1000.
   - If the entered pressure is outside the valid range, display an error message and exit the program.

3) **Data Interpolation (If Temperature Doesn't Match an Exact Value)**:
   - Find the indices of pressure values in the data that match the user-entered pressure.
   - Based on these pressure indices, identify the two closest temperature values in the data that bracket the user-entered temperature.
   - If either the lower or upper temperature is not found (i.e., the user-entered temperature falls outside the data range), display an error message and exit the program.
   - **Use Linear Interpolation to Estimate the Following Properties at the User-entered Temperature**:
     - Specific volume (m^3/kg)
     - Internal energy (kJ/kg)
     - Enthalpy (kJ/kg)
     - Entropy (kJ/kg∗K)

4) **Property Lookup (If Temperature Matches an Exact Value)**:
   - If the user-entered temperature exactly matches one of the temperature values in the data, directly retrieve the corresponding property values (specific volume, internal energy, enthalpy, and entropy) from the data table.

5) **Results and Phase Check**:
   - Display the user-entered temperature and pressure.
   - **If the Data Indicates a Two-phase Region (Saturated Vapor and Saturated Liquid) at the Given Pressure and Temperature**:
     - Prompt the user to enter a steam quality value (x) between 0 and 1.
     - If the entered steam quality is outside the valid range, display an error message and exit the program.
     - Calculate the properties (specific volume, internal energy, enthalpy, and entropy) based on the steam quality and the interpolated or retrieved values from step 3 or 4.
     - Display a message indicating the program is operating in the two-phase region with the entered steam quality.

   - **If the Data Indicates a Single-phase Region (Or the Temperature Matches an Exact Value)**:
     - **If the Exact Temperature Case (Step 4)**:
       - Display a message indicating the temperature matches an exact value in the data.
     - Display the phase information from the data.
     - Display the interpolated or retrieved property values (specific volume, internal energy, enthalpy, and entropy).

6) **Program Termination**:
   - After displaying the results, terminate the program.

H.4 MATLAB Code for Evaluating Liquid Water/Steam Properties at different *T* and *P*

```
clear variables;
try
   % Attempt to read the Excel file first
   data = readtable('SteamTable.xlsx','Format','auto');
   catch ME
   % If reading the Excel file fails, try loading the MAT file
   disp('Failed to read Excel file; it will continue with the saved MATLAB file.');
   try
      load('SteamTable.mat', 'data');
      disp('Loaded SpHdata from MAT file.');
   catch ME2
      % If loading the MAT file also fails, display an error message
      disp('Failed to read MAT file.');
   end
end
% Provide data
P_MPa=data.P_MPa;
T_C=data.T_C;
V_m3_kgInv=str2double(data.V_m3_kgInv);
U_kJ_kgInv=str2double(data.U_KJ_kgInv);
H_kJ_kgInv=str2double(data.H_kJ_kgInv);
S_kJ_kgInv_Kinv=str2double(data.S_kJ_kgInv_Kinv);
Phase=data.Phase;
% Interpolation function
interpolate = @(x, y, xi) interp1(x, y, xi, 'linear', 'extrap');
temperature = input('Enter T between [0 2,000] oC): ');
if temperature < 0 || temperature>2000
       disp('Wrong T value');
       return
end
pressure = input('Enter P between [0.01 1,000] MPa): ');
if pressure < 0.01 || pressure>1000
       disp('Wrong P value');
       return
end
indices = find(P_MPa == pressure); %and find (T_C==temperature);
lower_index_temp = find(T_C(indices) <= temperature, 1, 'last');
upper_index_temp = find(T_C(indices) >= temperature, 1, 'first');
if isempty(lower_index_temp) || isempty(upper_index_temp)
   disp('Temperature or pressure is outside the provided range.');
   return;
end
lower_index = indices(lower_index_temp);
upper_index = indices(upper_index_temp);
 if strcmp(Phase(lower_index), 'saturated vapor')  strcmp(Phase(upper_index),
'saturated liquid')
      % Ask for steam quality
      x = input('Enter steam quality (x): ');
      if x < 0 || x>1
         disp('Wrong x value');
         return
      end
disp('Performing calculations based on steam quality...');
V_m3_kgInv=V_m3_kgInv(upper_index)+x*abs((V_m3_kgInv(upper_index)-
V_m3_kgInv(lower_index)));
U_kJ_kgInv=U_kJ_kgInv(upper_index)+x*abs((U_kJ_kgInv(upper_index)-
```

```
U_kJ_kgInv(lower_index)));
H_kJ_kgInv=H_kJ_kgInv(upper_index)+x*abs((H_kJ_kgInv(upper_index)-
H_kJ_kgInv(lower_index)));
S_kJ_kgInv_Kinv=S_kJ_kgInv_Kinv(upper_index)+x*abs((S_kJ_kgInv_Kinv(upper_index)-
S_kJ_kgInv_Kinv(lower_index)));
disp(['Thermodynamic properties at T:', num2str(temperature), ' °C:']);
disp(['Thermodynamic properties at P: ', num2str(pressure), ' MPa:']);
disp(['You are trapped in the 2-phase region at steam quality: ', num2str(x), '']);
disp(['Specific volume: ', num2str(V_m3_kgInv),' m3/kg']);
disp(['Internal energy: ', num2str(U_kJ_kgInv), ' kJ/kg']);
disp(['Enthalpy: ', num2str(H_kJ_kgInv), ' kJ/kg']);
disp(['Entropy: ', num2str(S_kJ_kgInv_Kinv), ' kJ/(kg.K)']);
return
end
if lower_index == upper_index
   % Temperature matches one of the provided values exactly
   disp('Temperature matches one of the provided values exactly.');
   disp('Performing calculations for exact temperature...');
   disp(['Entered Temperature: ', num2str(temperature), ' °C']);
   disp(['Entered Pressure: ', num2str(pressure), ' MPa']);
   disp(['Phase @(P,T): ', Phase(lower_index)]);
   disp(['Specific volume: ', num2str(V_m3_kgInv(lower_index)),' m3/kg']);
   disp(['Internal energy: ', num2str(U_kJ_kgInv(lower_index)),' kJ/kg']);
   disp(['Enthalpy: ', num2str(H_kJ_kgInv(lower_index)),' kJ/kg']);
   disp(['Entropy: ', num2str(S_kJ_kgInv_Kinv(lower_index)),' kJ/(kg.K)']);
else
   disp('Interpolating properties between two temperatures...');
   % Interpolate all properties
   V_m3_kgInv_interp = interpolate(T_C([lower_index, upper_index]),
V_m3_kgInv([lower_index, upper_index]), temperature);
   U_kJ_kgInv_interp = interpolate(T_C([lower_index, upper_index]),
U_kJ_kgInv([lower_index, upper_index]), temperature);
   H_kJ_kgInv_interp = interpolate(T_C([lower_index, upper_index]),
H_kJ_kgInv([lower_index, upper_index]), temperature);
   S_kJ_kgInv_Kinv_interp = interpolate(T_C([lower_index, upper_index]),
S_kJ_kgInv_Kinv([lower_index, upper_index]), temperature);
   % Display interpolated properties
   disp(['Interpolated properties at ', num2str(temperature), ' °C']);
   disp(['Interpolated properties at ', num2str(pressure), ' MPa']);
   if strcmp(Phase(lower_index), 'saturated vapor') && strcmp(Phase(upper_index),
'vapor')
   %disp(['Phase @lower index: ', Phase(lower_index)]);
   disp(['Phase @(P,T): ', Phase(upper_index)]);
   else
   disp(['Phase @(T,P): ', Phase(lower_index)]);
   end
   disp(['Specific volume: ', num2str(V_m3_kgInv_interp), ' m3/kg']);
   disp(['Internal energy: ', num2str(U_kJ_kgInv_interp), ' kJ/kg']);
   disp(['Enthalpy: ', num2str(H_kJ_kgInv_interp),' kJ/kg']);
   disp(['Entropy: ', num2str(S_kJ_kgInv_Kinv_interp),' kJ/(kg.K)']);
end
```

Index

MATLAB-based Computations of Chemical Engineering Principles, First Edition. Kamal I. M. Al-Malah.

Companion website: www.wiley.com/go/al-malah/matlabbasedcomputations

d

p

r

s

Printed and bound by CPI Group (UK) Ltd, Croydon, CR0 4YY

21/09/2026

14965823-0001